Codes and Turbo Codes

Springer

Paris
Berlin
Heidelberg
New York
Hong Kong
Londres
Milan
Tokyo

Claude Berrou (Ed.)

Codes and Turbo Codes

Claude Berrou
Télécom Bretagne
CS 83818
29238 Brest Cedex 3
France

ISBN : 978-2-8178-0038-7 Springer Paris Berlin Heidelberg New York

Printed in France
Springer-Verlag France is a member of the group Springer Science + Business Media

First edition in French © Springer-Verlag France 2007
ISBN: 978-2-287-32739-1

Cover design: Jean-François MONTMARCHÉ
Cover illustration: Jean-Noël JAFFRY

Codes and Turbo Codes

under the direction of Claude Berrou (Télécom Bretagne)

The following have contributed to this work:
- Karine Amis,
- Matthieu Arzel,
- Catherine Douillard,
- Alain Glavieux †,
- Alexandre Graell i Amat,
- Frédéric Guilloud,
- Michel Jézéquel,
- Sylvie Kerouédan,
- Charlotte Langlais,
- Christophe Laot,
- Raphaël Le Bidan,
- Émeric Maury,
- Youssouf Ould-Cheikh-Mouhamedou,
- Samir Saoudi,
- Yannick Saouter,
all at Télécom Bretagne,
- Gérard Battail,
at Télécom ParisTech,
- Emmanuel Boutillon,
at the Université de Bretagne Sud,

with the invaluable assistance of Josette Jouas, Mohamed Koubàa and Nicolas Puech.

Translation: Janet Ormrod (Télécom Bretagne).

Cover illustration: Jean-Noël Jaffry (Télécom Bretagne).

Any comments on the contents of this book can be sent to this e-mail address:
`turbocode@mlistes.telecom-bretagne.eu`

"The oldest, shortest words — yes and no —
are those which require the most thought"

Pythagoras, fifth century BC

To our late lamented colleagues and friends,
Alain Glavieux and Gérard Graton.

Foreword

What is commonly called the information age began with a double big bang. It was 1948 and the United States of America was continuing to invest heavily in high-tech research, the first advantages of which had been reaped during the Second World War. In the *Bell Telephone Laboratories*, set up in New Jersey, to the south of New York, several teams were set up around brilliant researchers, many of whom had been trained at MIT (*Massachusetts Institute of Technology*). That year two exceptional discoveries were made, one technological and the other theoretical, which were to mark the 20th century. For, a few months apart, and in the same institution John Bardeen, Walter Brattain and William Shockley invented the transistor while Claude Elwood Shannon established information and digital communications theory. This phenomenal coincidence saw the birth of near-twins: the semi-conductor component which, according to its conduction state (on or off), is able to materially represent binary information ("0" or "1") and the *Shannon* or *bit* (short for *b*inary un*it*), a unit that measures information capacity.

Today we can recognize the full importance of these two inventions that enabled the tremendous expansion of computing and telecommunications, to name but these two. Since 1948, the meteoric progress of electronics, then of micro-electronics, has provided engineers and researchers in the world of telecommunications with a support for their innovations, in order to continually increase the performance of their systems. Who could have imagined, only a short while ago, that a television programme could be transmitted via a pair of telephone wires? In short, Shockley and his colleagues, following Gordon Moore's law (which states that the number of transistors on a silicon chip doubles every 18 months), gradually provided the means to solve the challenge issued by Shannon, thanks to algorithms that could only be more and more complex. A typical example of this is the somewhat late invention of turbo codes and iterative processing in receivers, which could only be imagined because the dozens or hundreds of thousands of transistors required were available.

Experts in micro-electronics foresee the ultimate limits of CMOS technology at around 10 billion transistors per square centimetre, in around 2015. This is about the same as the number of neurons in the human brain (which will, however, remain incomparably more powerful, due to its extraordinary network of connections - several thousand synapses per neuron). Billions of transistors on the same chip means that there will be easily enough room for algorithms that require the greatest calculating resources, at least among those algorithms that are known today. To repeat the slogan of one integrated circuit manufacturer, "the limit lies not in the silicon but in your imagination". Even so, and to be honest, let us point out that designing and testing these complex functions will not be easy.

However, we are already a long way from the era when Andrew Viterbi, concluding the presentation of his famous algorithm in 1967, showed scepticism that matched his modesty: "Although this algorithm is rendered impractical by the excessive storage requirements, it contributes to a general understanding of convolutional codes and sequential decoding through its simplicity of mechanization and analysis" [1]. Today, a Viterbi decoder takes up a tenth of a square millimetre in a cellphone.

Among the results presented by Shannon in his founding paper [2], the following is particularly astonishing: *in a digital transmission in the presence of perturbation, if the average level of the latter does not exceed a certain power threshold, by using appropriate coding, the receiver can identify the original message without any errors.* By coding, here and throughout this book, we mean error-correcting coding, that is, the redundant writing of binary information. Source coding (digital compression), cryptographic coding, and any other meaning that the term coding might have, are not treated in *Codes and Turbo codes.*

For thousands of researchers and engineers, the theoretical result established by Shannon represented a major scientific challenge since the economic stakes are considerable. Improving the error correction capability of a code means, for the same quality of received information (for example, no more than one erroneous bit out of 10,000 received in digital telephony), enabling the transmission system to operate in more severe conditions. It is then possible to reduce the size of antennas or of solar panels and the weight of power batteries. In space systems (satellites, probes, etc.), the savings can be measured in hundreds of thousands of dollars since the weight of the equipment and the power of the launcher are thus notably reduced. In mobile telephone (cellphone) systems, improving the code also enables operators to increase the potential number of users in each cell. Today, rare are those telecommunications systems that do not integrate an error-correcting code in their specifications.

Another field of application for error-correcting codes is that of mass memories: computer hard drives, CD-ROMs, DVDs and so on. The progress made in the last few years in miniaturizing the elementary magnetic or optical memorization patterns has been accompanied by the normal degradation of energy available when the data is being read and therefore a greater vulnerability to perturbations. Added to this are the increased effects of interference between neighbours. Today, it is essential to use tried and tested techniques in telecommunications systems, especially coding and equalization, in order to counter the effects induced by the miniaturization of these storage devices. Although *Codes and Turbo codes* does not explicitly tackle these applications, the concepts developed and the algorithms presented herein are also a topical issue for mass memory providers.

This book therefore deals mainly with error-correction coding, also called channel coding, and with its applications to digital communications, in association with modulation. The general principles of writing redundant information and most of the techniques imagined up until 1990 to protect digital transmissions, are presented in the first half of the book (chapters 1 to 6). In this first part, one chapter is also dedicated to the different modulation techniques without which the coded signals could not be transported in real transmission environments. The second part (chapters 7 to 11) deals with turbo codes, invented more recently (1990-93), whose correction capability, neighbouring on the theoretical limits predicted by Shannon, have made them a coding standard in more and more applications. Different versions of turbo codes, as well as the important family of LDPC codes, are presented. Finally, certain techniques using the principles of turbo-decoding, like turbo-equalization and multi-user turbo-detection, are introduced at the end of the book.

A particular characteristic of this book, in comparison with the way in which the problem of coding may be tackled elsewhere, is its concern with applications. Mathematical aspects are dealt with only for the sake of necessity, and certain results, which depend on complex developments, will have to be taken as given. On the other hand, practical considerations, particularly concerning the processing algorithms and circuits, are fully detailed and commented upon. Many examples of performance are given, for different coding and coded modulation schemes.

The book's authors are lecturers and researchers well-known for their expertise in the domain of algorithms and the associated circuits for communications. They are, in particular, the inventors of turbo codes and responsible for generalizing the "turbo principle" to different functions of data processing in receivers. Special care has been taken in writing this collective work vis-à-vis the unity of point of view and the coherence of notations. Certain identical or similar concepts may, however, be introduced several times and in different ways, which – we hope – does not detract from the pedagogy of the work, for pedagogy is the art of repetition. The aim of *Codes and turbo codes* is for it to be a book not only for learning about error-correction coding and decoding, a precious source of information about the many techniques imagined since the middle of the twentieth century, but also for addressing problems that have not yet been completely resolved.

[1] A. J. Viterbi, "Error Bounds for Convolutional Codes and an Asymptotically Optimum Decoding algorithm", *IEEE Trans. Inform. Theory*, vol. IT-13, pp. 260-269, Apr. 1967.

[2] C. E. Shannon, "A Mathematical Theory of Communication", *Bell System Technical Journal*, Vol. 27, July and October 1948.

Contents

Chapter 1

Introduction

Redundancy, diversity and parsimony are the keywords of error correction coding. To these, on the decoding side, can be added efficiency, that is, making the most of all the information available. To illustrate these concepts, consider a simple situation in everyday life.

Two people are talking near a road where there is quite heavy traffic. The noise of the engines more or less disrupts their conversation, with peaks of perturbation noise corresponding to the vehicles going past. First assume that one of the people regularly transmits one letter chosen randomly: "a", "b" ... or any of the 26 letters of the alphabet, with the same probability (that is, 1/26). The message does not contain any other information and there is no connection between the sounds transmitted. If the listener doesn't read the speaker's lips, he will certainly often be at a loss to recognize certain letters. So there will be transmission errors.

Now, in another scenario, one of the two people speaks in full sentences, on a very particular topic, for example, the weather. In spite of the noise, the listener understands what the speaker says better that when he says individual letters, because the message includes *redundancy*. The words are not independent and the syllables themselves are not concatenated randomly. For example, we know that after a subject we generally have a verb, and we guess that after "clou", there will be "dy" even if we cannot hear properly, etc. This redundancy in the construction of the message enables the listener to understand it better in spite of the difficult transmission conditions.

Suppose that we want to improve the quality of the transmission even further, in this conversation that is about to take an unexpected turn. To be sure of being understood, the speaker repeats some of the words, for example "dark dark". However, after the double transmission, the receiver has understood "dark lark". There is obviously a mistake somewhere, but is it "dark" or "lark" that the receiver is supposed to understand? No error correction is possible using this repetition technique, except maybe to transmit the word more than twice. "dark lark dark" can, without any great risk of error, be translated as "dark".

More elaborate coding involves transmitting the original message no longer accompanied by the same word but by a synonym or an equivalent: "dark dusk", for example. If we receive "dark dust" or "park dusk", correction is possible by referring to a dictionary of equivalences. The decoding rule would then be as follows: in the case of an error (if the two words received are not directly equivalent) and if we can find two equivalent words by changing at most one letter in the received message, then the correction is adopted, ("dust" becomes "dusk") and the first of the two words ("dark") is accepted as the original message. The same would be true if "park dusk" was received, where the first word would now be corrected. Of course, if a large number of errors alter the transmission and we receive "park dust", we will probably no longer understand anything at all. So there is a limit to this error correction capability. This is the famous *Shannon limit*, that in theory no correcting code can exceed.

Compared to simple repetition, coding by equivalence, which is more efficient, uses the *diversity* effect. In this analogy with conversation, this diversity is expressed by a lexicographical property: two distinct words with a close spelling (dark and lark) are unlikely to have two equivalents ("dusk" and "bird" for example) that also have a close spelling. Diversity, as presented above, thus involves constructing a redundant message in a way that minimizes any ambiguity on reception. This is also called temporal diversity as equivalent words in the message are transmitted at different instants and undergo perturbations of unequal intensities. For example, "dark" and "dusk" could be transmitted when a motorbike or a bicycle, respectively, passed by. In telecommunications systems, we can search for complementary diversity effects. Frequential diversity involves cutting up and sending the message in frequency bands that are not perturbed at the same instant in the same way. As for using several emission and/or reception antennas, this offers spatial diversity as the paths between antennas do not have the same behaviour. Jointly exploiting these three types of diversity: temporal, frequential and spatial, leads to highly efficient communications systems.

Finally, the desire for *parsimony*, or economy, is imposed by the limitation of resources, either temporal or frequential, of the transmission. The choice of "dark dusk" is thus certainly more judicious, from the concision point of view, than "dark night-fall". However, we sense that the latter message might be more resistant to multiple errors because it is more redundant (the reception and resolution of "dirk might-fall" is not problematic if we use the decoding law mentioned above and extend the correction capability to two errors). Searching for performance via the redundancy rate, and the parsimony constraint are therefore in total opposition.

Redundant coding is generally simple to implement and the corresponding software or hardware has low complexity. Decoding, however, requires computation techniques that can be costly, even if, in fact, the number of instructions in the program (typically several hundred, in high level computing language) or the silicon surface occupied (typically several square millimetres) remains low.

1.1 Digital messages

A digital message is a sequence of characters or symbols taken from an alphabet of finite size. Genetic information (DNA), to take a natural example, uses an alphabet of four characters, denoted A, T, G and C, that stand for the initials of their nitrogen bases (adenine, thymine, guanine, and cytosine). The first digital transmission technique was Morse code (1832), with its two-character audio alphabet: *TIT* or dot, a short tone lasting four hundredths of a second and *TAT* or dash, a long tone lasting twelve hundredths of a second. Samuel F. B. Morse could well have called these characters 0 and 1, which today are the universal names used in any alphabet with two elements or binary alphabet. The binary elements, 0 and 1, were first called *bits* by J. W. Tukey (1943), as a contraction of *binary digit*, after rejecting *bigit* and *binit*. Shannon borrowed the term when he wished to introduce the concept of unit of information. Today, it is preferable to refer to this unit of information as the *Shannon*, to distinguish it from the bit, which has acquired a more electronic meaning.

An alphabet having M symbols is called an M-ary alphabet. It can be transcribed into a binary alphabet by representing each of the M symbols by a word of m bits, with:

$$\begin{aligned} & m = \left\lfloor \log_2(M) \right\rfloor + 1 && \text{if } M \text{ is not a power of 2} \\ \text{or:} \quad & \\ & m - \log_2(M) && \text{if } M \text{ is a power of 2} \end{aligned} \tag{1.1}$$

where $\lfloor x \rfloor$ denotes the whole part of x. Multimedia messages (voice, music, fixed and moving images, text etc.) transiting through communication systems or stocked in mass memories, are exclusively binary. However, in this book we shall sometimes have to consider alphabets with more than two elements. This will be the case in Chapter 4, to introduce certain algebraic codes. In Chapters 2 and 10, which deal with modulations, the alphabets, which we then more concretely call constellations, contain a number of symbols that are a power of 2, that is, we have precisely: $m = \log_2(M)$.

Correction coding techniques are only implemented on digital messages. However, there is nothing against constructing a redundant analogue message. For example, an analogue signal, in its temporal dimension, accompanied or followed by its frequential representation obtained thanks to the Fourier transform, performs a judiciously redundant coding. However, this technique is not very simple and the decoder remains to be invented.

Furthermore, the digital messages that we shall be considering in what follows, before the coding operation has been performed, will be assumed to be made up of binary elements that are mutually independent and taking the values 0 and 1 with the same probability, that is, 1/2. The signals that are produced by a sensor like a microphone or a camera, and then digitized to become binary sequences, do not generally satisfy these properties of independence and equiprobability. It is the same with text (for example, the recurrence of "e" in an English text is on average 5 times higher than that of "f"). The effects of dependency or disparity in the original message, whether they be of physical, orthographical or semantic origin or whatever, cannot be exploited by the digital communication system, which transmits 0s and 1s independently of their context. To transform the original message into a message fulfilling the conditions of independence and equiprobability, an operation called source coding, or digital compression, can be performed. Today, compression norms like JPEG, MPEG, ZIP, MUSICAM, etc. are well integrated into the world of telecommunications, the Internet in particular. At the output of the source encoder, the statistical properties of independence and equiprobability are generally respected and the compressed message can then be processed by the channel encoder, which will add redundancy mathematically exploitable by the receiver.

1.2 A first code

Figure 1.1 shows an electronic circuit that performs a very simple, easily decodable correction encoding. The code implemented is an extended Hamming code, which was used in teletext (one of the first digital communication systems), which will also be presented in Chapter 4.

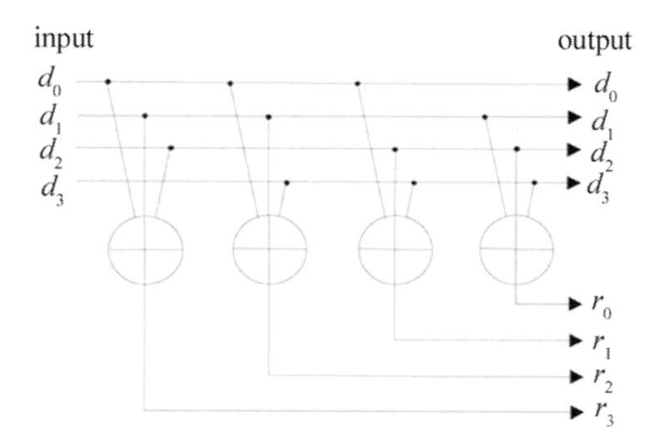

Figure 1.1 – Extended Hamming encoder: a simple, easily decodable code.

The encoder contains four identical operators performing the exclusive-or function. The exclusive-or (XOR) of k binary values b_0, b_1, ..., b_{k-1} is calculated as:

$$\text{XOR}(b_0, b_1, ..., b_{k-1}) = b_0 \oplus b_1 \oplus ... \oplus b_{k-1} = \sum_{p=0}^{k-1} b_p \text{ modulo } 2 \tag{1.2}$$

It is therefore quite simply 0 if the number of logical 1s appearing in the input sequence is even and 1 in the opposite case. In the sequel, when modulo 2 additions are performed and if there is no ambiguity in the notation, the term modulo 2 may be omitted.

The encoder transforms the message containing four data bits: $\mathbf{d} = (d_0, d_1, d_2, d_3)$ into a word of eight bits: $\mathbf{c} = (d_0, d_1, d_2, d_3, r_0, r_1, r_2, r_3)$, called *codeword.* The codeword is therefore *separable* into one part that is the information coming from the source, called the *systematic*[1] part and a part added by the encoder, called the *redundant* part. Any code producing codewords of this form is called a systematic code. Most codes, in practice, are systematic but there is one important exception in the family of convolutional codes (Chapter 5).

The law for the construction of the redundant part by the particular encoder of Figure 1.1 can be simply written as:

$$r_j = d_j + \sum_{p=0}^{3} d_p \qquad (j = 0, ..., 3) \tag{1.3}$$

Table 1.1 shows the sixteen possible values of $\mathbf{c}$, that is, the set $\{\mathbf{c}\}$ of codewords.

d_0	d_1	d_2	d_3	r_0	r_1	r_2	r_3	d_0	d_1	d_2	d_3	r_0	r_1	r_2	r_3
0	0	0	0	0	0	0	0	1	0	0	0	0	1	1	1
0	0	0	1	1	1	1	0	1	0	0	1	1	0	0	1
0	0	1	0	1	1	0	1	1	0	1	0	1	0	1	0
0	0	1	1	0	0	1	1	1	0	1	1	0	1	0	0
0	1	0	0	1	0	1	1	1	1	0	0	1	1	0	0
0	1	0	1	0	1	0	1	1	1	0	1	0	0	1	0
0	1	1	0	0	1	1	0	1	1	1	0	0	0	0	1
0	1	1	1	1	0	0	0	1	1	1	1	1	1	1	1

Table 1.1 – The sixteen codewords that the encoder in Figure 1.1 can produce.

We can first note that the coding law is *linear*: the sum of two codewords is also a codeword. It is the linearity of relation (1.3) that guarantees the linearity of the encoding. All the codes that we shall consider in what follows are linear as they are all based on two linear operations: addition and permutation (including shifting). Since the code is linear and the transmission of a codeword might be affected by a process that is also linear (the addition of a perturbation: noise, interference, etc.), the choice of a codeword, to explain or justify the properties of the code, is completely indifferent. It is the "all zero" codeword that will play this "representative" or reference role for all the codewords vis-à-vis the general properties of the encoder/decoder pair. At reception, the presence of 1 will therefore be representative of transmission errors.

[1] We can also say information part because it is made up of bits of information coming from the source.

The number of 1s contained in a codeword that is not "all zero" is called the Hamming weight and is denoted w_H. We can distinguish the weight relating to the systematic part ($w_{H,s}$) and the weight relating to the redundant part ($w_{H,r}$). We note, in Table 1.1, that w_H is at least equal to 4. Because of linearity, this also means that the number of bits that differ in two codewords is also at least equal to 4. The number of bits that are different, when we compare two binary words, is called the Hamming distance. The smallest of all the distances between all the codewords, considered two by two, is called the minimum Hamming distance (MHD) and denoted d_{min}. Linearity means that it is also the smallest of the Hamming weights in the list of codewords excluding the "all zero". d_{min} is an essential parameter for characterizing a particular code, since the correction capability of the corresponding decoder is directly linked to it.

We now write a decoding law for the code of Figure 1.1:

"After receiving the word $\mathbf{c'} = (d_0', d_1', d_2', d_3', r_0', r_1', r_2', r_3')$ transmitted by the encoder and possibly altered during transmission, the decoder chooses the closest codeword $\hat{\mathbf{c}}$ in the sense of the Hamming distance".

The job of the decoder is therefore to run through Table 1.1 and, for each of the sixteen possible codewords, to count the number of bits that differ from $\mathbf{c}'$. The $\hat{\mathbf{c}}$ codeword selected is the one that differs least from $\mathbf{c}'$. When several solutions are possible, the decoder selects one at random. Mathematically, this is written:

$$\hat{\mathbf{c}} = \mathbf{c} \in \{\mathbf{c}\} \text{ such that } \sum_{j=0}^{3} d_j \oplus d_j^{'} + \sum_{j=0}^{3} r_j \oplus r_j^{'} \text{ is minimum} \tag{1.4}$$

A decoder capable of implementing such a process is called a *maximum likelihood* (ML) decoder as all the cases, here the sixteen possible codewords, are run through in the search for a solution and no other more efficient decoding process exists. With codes other than the very simple Hamming code, the message to be encoded contains far more than four bits (in practice, it goes from a few tens to a few tens of thousands of bits) and ML decoding is impossible to execute as the number of codewords is too high.

Assume that the Hamming encoder transmits the "all zero" word that we have chosen as the reference and that some of the 0s have been inverted before reception, on the *transmission channel*. How many errors can the decoder correct, based on law (1.4)? If $\mathbf{c}'$ contains a single 1 (a single error), the "all zero" word is closer to $\mathbf{c}'$ than any other codeword that possesses at least four 1s. Correction is thus possible. If $\mathbf{c}'$ contains two 1s, for example in the first two places ($d_0 = d_1 = 1$), the decoder is faced with a dilemma: four codewords are at a distance of 2 from the received word. It must therefore choose $\hat{\mathbf{c}}$ at random, from among the four possible solutions, with three risks out of four of being erroneous. In this situation, we can also ensure that the decoder does not give a solution but merely indicates its dilemma: it then plays the role of non-correctable *error detector*. Finally, if **c**' contains three 1s, the decoder will find a single codeword at a distance of 1 and it will propose this codeword as the most probable solution, but it will be erroneous.

The error correction capability of the extended Hamming code is therefore $t = 1$ errors. More generally, the error correction capability of a code with a minimum distance d_{min} is:

$$t = \left\lfloor \frac{d_{min} - 1}{2} \right\rfloor \tag{1.5}$$

Note that the correction capability of the code given as the example in this introduction is not decreased if we remove one (any) of the redundancy symbols. The MHD passes from 4 to 3 but the correction capability is still of one error. This shortened version is in fact the original Hamming code, the first correcting code in the history of information theory (1948).

In a given family of codes, we describe a particular version of it by the shortcut (n, k, d_{min}) where n and k are the lengths of the codewords and of the source messages, respectively. Up to now, we have thus just defined two Hamming codes denoted (8, 4, 4) and (7, 4, 3). The second seems more interesting as it offers the same error correction capability ($t = 1$) with a $\tau = \frac{n-k}{k}$ redundancy rate of 0.75, instead of 1 for the first one. However, the code (7, 4, 3) cannot play the role of an error detector: if the received word contains two errors, the decoder will decide in favour of the single, erroneous codeword that is to be found at a Hamming distance of 1.

Rather than redundancy rate, we usually prefer to use the notion of coding rate, denoted R, and defined by:

$$R = \frac{k}{n} = \frac{1}{1 + \tau} \tag{1.6}$$

The product Rd_{min} will appear in the sequel as an essential figure of merit vis-à-vis a perturbation caused by additive noise with a Gaussian distribution.

1.3 Hard input decoding and soft input decoding

We continue this presentation of the basic principles of coding and decoding by using the example of the extended Hamming code.

The decoding principle defined by (1.4) assumes that the received codeword **c**' is composed of binary values, that is, the transmission of data is carried out according to the law of perturbation given by the diagram of Figure 1.2. A 0 becomes a 1 and vice-versa, with a probability p and the binary values are correctly received with the complementary probability 1 - p. Such a transmission channel is said to be a *binary symmetric channel* and the decoder performs what we call *hard input decoding*. In certain communications systems (optical fibres, switched networks, etc.) and in most storage hardware, the decoders can effectively exploit only binary information.

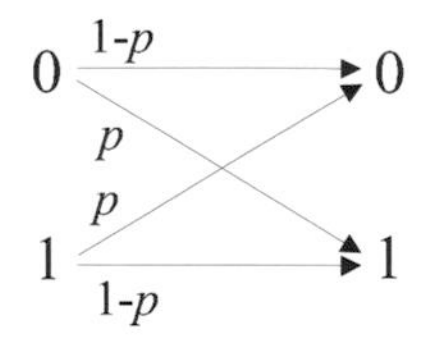

Figure 1.2 – Binary symmetric channel with error probability p.

When the signal received by the decoder comes from a device capable of producing estimations of an analogue nature on the binary data transmitted, the error correction capability of the decoder can be greatly improved. To show this using the example of the extended Hamming code, we must first change alphabet and adopt an antipodal (or symmetric) binary alphabet. We will make the transmitted values x = -1 and $x = +1$ correspond to the systematic binary data $d = 0$ and $d = 1$, respectively. Similarly, we will make the transmitted values y = -1 and $y = +1$ correspond to the redundant binary data $r = 0$ and $r = 1$, respectively. We then have:

$$\begin{aligned} x &= 2d - 1 = -(-1)^d \\ y &= 2r - 1 = -(-1)^r \end{aligned} \quad (1.7)$$

Figure 1.3 gives an example of a transmission during which the transmitted values -1 and +1 are altered by additive noise of an analogue type. The values at the output of the transmission channel are then real variables, which must in practice be clipped then quantified when the decoder is a digital processor. The number of quantization bits, denoted N_q, does not need to be high: 3, 4 or 5 bits are sufficient to closely represent analogue samples. One time out of two on average, the noise is favourable as it has the same sign as the transmitted value. In the other case, the amplitude of the signal is attenuated and, when this unfavourable noise is large, the sign can be inverted. An immediate decision taken per threshold (that is, is it larger or smaller than the analogue zero?) would then lead to an erroneous binary value being given.

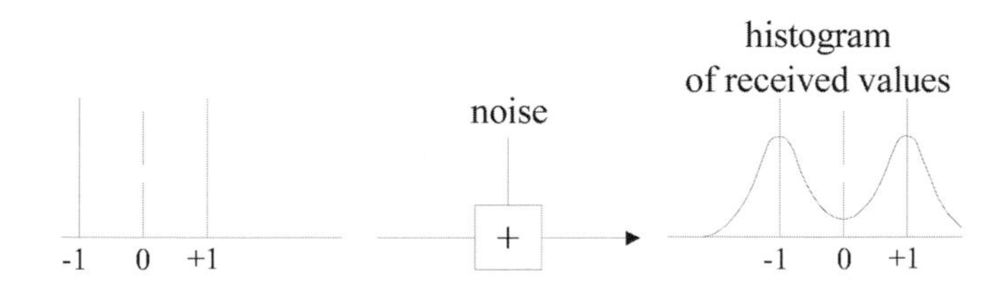

Figure 1.3 – Transmission channel with additive noise of an analogue type.

Since the decoder has information about the degree of reliability of the values received, called *soft* or *weighted* values in what follows, the decoding of the extended Hamming code according to law (1.4) is no longer optimal. The law of maximum likelihood decoding to implement in order to exploit these weighted values depends on the type of noise. An important case in practice is additive white Gaussian noise (AWGN).

u is a random Gaussian variable with mean μ and variance σ^2 when its probability density $p(u)$ can be expressed in the form:

$$p(u) = \frac{1}{\sigma\sqrt{2\pi}} \exp\left(-\frac{(u-\mu)^2}{2\sigma^2}\right) \tag{1.8}$$

The AWGN is a perturbation which, after adapted filtering and periodic sampling (see Chapter 2), produces independent samples whose amplitude follows probability density law (1.8), with zero mean and variance:

$$\sigma^2 = \frac{N_0}{2} \tag{1.9}$$

where N_0 is the noise power spectral density.

A transmission channel on which the only alteration of the signal comes from an AWGN is called a *Gaussian channel.* At the output of such a channel, the ML decoding is based on the exhaustive search for the codeword that is at *the smallest Euclidean distance* from the received word. Denoting X and Y the received values corresponding to the transmitted symbols x and y respectively, the soft input decoder of the extended Hamming code therefore chooses:

$$\hat{\mathbf{c}} = \mathbf{c} \in \{\mathbf{c}\} \text{ such that } \sum_{j=0}^{3} (x_j - X_j)^2 + \sum_{j=0}^{3} (y_j - Y_j)^2 \text{ is minimum} \tag{1.10}$$

Since the values transmitted are all such that $x_j^2 = 1$ or $y_j^2 = 1$ and all the Euclidean distances contain X_j^2 and Y_j^2, the previous law can be simplified as:

$$\hat{\mathbf{c}} = \mathbf{c} \in \{\mathbf{c}\} \text{ such that } -\sum_{j=0}^{3} 2x_j X_j - \sum_{j=0}^{3} 2y_j Y_j \text{ is minimum}$$

or as:

$$\hat{\mathbf{c}} = \mathbf{c} \in \{\mathbf{c}\} \text{ such that } \sum_{j=0}^{3} x_j X_j + \sum_{j=0}^{3} y_j Y_j \text{ is maximum} \tag{1.11}$$

Minimizing the Euclidean distance between two codewords $\mathbf{c}$ and $\mathbf{c}$' therefore means maximizing the *scalar product* $\langle \mathbf{x}, \mathbf{X} \rangle + \langle \mathbf{y}, \mathbf{Y} \rangle = \sum_{j=0}^{3} x_j X_j + \sum_{j=0}^{3} y_j Y_j$ where $\mathbf{x}$, $\mathbf{X}$, $\mathbf{y}$ and $\mathbf{Y}$ represent the transmitted and received sequences of the systematic and redundant parts.

In a *Digital Signal Processor* (DSP) or in an *Application Specific Integrated Circuit* (ASIC), it may be useful to have only to deal with positive numbers. Law (1.11) can then be implemented as:

$$\hat{\mathbf{c}} = \mathbf{c} \in \{\mathbf{c}\} \text{ such that } \sum_{j=0}^{3} (V_{max} + x_j X_j) + (V_{max} + y_j Y_j) \text{ is maximum} \qquad (1.12)$$

where $[-V_{max}, V_{max}]$ is the interval of the values that the input samples X_j and Y_j of the decoder can take after the clipping operation.

In Figure 1.4, the "all zero" codeword has been transmitted and received with three alterations in the first three positions. These three alterations have inverted the signs of the symbols but their amplitudes are at a fairly low level: 0.2, 0.4 and 0.1. Hard input decoding produces an erroneous result as the closest codeword in terms of the Hamming distance is $(1,1,1,0,0,0,0,1)$. However, soft input decoding according to (1.11) does produce the "all zero" word, whose maximum scalar product is:

$$\begin{aligned} &(-1)(0.2) + (-1)(0.4) + (-1)(0.1) + (-1)(-1) \\ &+ (-1)(-1) + (-1)(-1) + (-1)(-1) + (-1)(-1) \quad = \quad 4.3 \end{aligned}$$

in comparison with:

$$\begin{aligned} &(+1)(0.2) + (+1)(0.4) + (+1)(0.1) + (-1)(-1) \\ &+ (-1)(-1) + (-1)(-1) + (-1)(-1) + (+1)(-1) \quad = \quad 3.7 \end{aligned}$$

for the competitor word $(1,1,1,0,0,0,0,1)$.

This simple example shows the interest of not excluding the reliability information in decision taking, whenever possible.

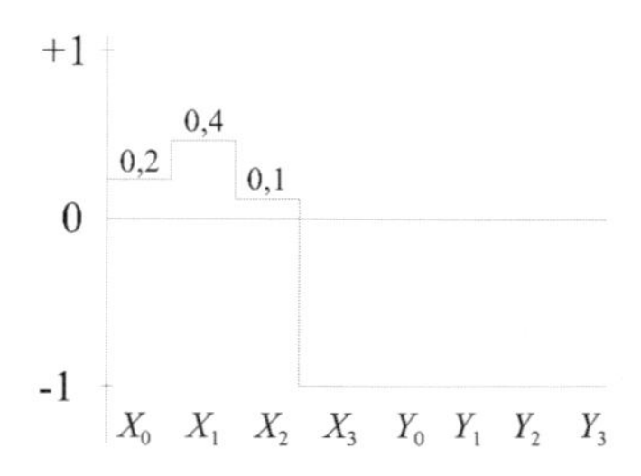

Figure 1.4 – The "all zero" word (emission of symbols with the value -1) has been altered during the transmission on the first three positions. The hard input decoding is erroneous, but not the soft input decoding.

The ML decoding rules that we have just used in a specific example are easily generalizable. However, we do realize that beyond a certain length of message, such a decoding principle is unrealistic. Applying ML decoding to codewords containing 240 bits in the systematic part, for example, would mean considering as many codewords as atoms in the visible universe (10^{80}). In spite of this, for most of the codes known, non-exhaustive decoding methods have been imagined, enabling us to get very close to the optimal result of the ML method.

1.4 Hard output decoding and soft output decoding

When the output of the decoder is not directly transmitted to a recipient but must be used by another processor whose performance is improved thanks to weighted inputs, this upstream decoder can be required to elaborate such weighted values. We thus distinguish the *hard output* when the decoder provides logical 0s and 1s from the *soft output*. In the latter case, the decoder accompanies its binary decisions with reliability measures or weights. The output scale of weighted values is generally the same as the input scale $[-V_{max}, V_{max}]$.

For the extended Hamming code decoder, it is relatively easy to build weighted decisions. When the decoder has calculated the sixteen scalar products, it lists them in decreasing order. In the first position, we find the scalar product of the most likely codeword, that is, the one that decides the signs of the weighted decisions at the output. Then, for each of the four bits of the systematic part, the decoder looks for the highest scalar product that corresponds to a competitor codeword in which the information bit in question is opposite that of the binary decision. The weight associated with this binary decision is then the difference between the maximum scalar product and the scalar product corresponding to the competitor word. A supplementary division by 2 puts the weighted output on the input scale. This process is optimal for an AWGN perturbation. Taking the example of Figure 1.4 (which is not typical of an AWGN) again, the weight associated with the decisions on the first three bits would be identical and equal to $(4.3 - 3.7)/2 = 0.3$.

From a historical point of view, the first decoding methods were of the hard input and output type. It was the Viterbi algorithm, detailed in Chapter 5, that popularized the idea of soft input decoding. Then turbo codes, which are decoded by repeated processing and require weighted values at every level of this processing, made soft input and output decoders popular. The generic abbreviation used to qualify these decoders is SISO for *Soft-Input/Soft-Output*.

1.5 The performance measure

The performance of an encoder/decoder pair is first judged in terms of residual errors at the output of the decoder, when we have fixed a specific evaluation framework: type of perturbation, length of message, rate of redundancy or coding rate, etc. Other aspects, like the complexity of the decoding, the latencies introduced by the encoder and the decoder, the degree of flexibility of the code (in particular its ability to conform to different lengths of message and/or to different coding rates) are also to be considered more or less closely, depending on the specific constraints of the communication system.

The residual errors that the decoder has not managed to correct are measured by using two parameters. The binary error rate (BER) is the ratio between the number of residual binary errors and the total number of bits of information transmitted. The word, block or packet error rate (PER) is the number of codewords badly decoded (at least one of the bits of information is wrong) out of the total number of codewords transmitted. The ratio between BER and PER is the average density of errors δ_e in the systematic part of a badly decoded word:

$$\delta_e = \frac{\bar{w}}{k} = \frac{\text{BER}}{\text{PER}} \tag{1.13}$$

where $\bar{w} = k\delta_e$ is the average number of erroneous information bits in the systematic part of a badly decoded block.

Figure 1.5 gives a typical example of the graphic representation of the performance of error correction coding and decoding. The ordinate gives the BER on a logarithmic scale and the abscissa carries the signal to noise ratio $\frac{E_b}{N_0}$, expressed in decibels (dB). N_0 is defined by (1.9) and E_b is the energy received per bit of information. If E_s is the energy received for each of the symbols of the codeword, E_s and E_b are linked by:

$$E_s = RE_b \tag{1.14}$$

The comparison of different coding and decoding processes or the variation in performance of a particular process with the coding rate are always defined with the same global reception energy. When there is no coding, the energy per received codeword is kE_b. With coding, which increases the number of bits transmitted, the energy kE_b is to be distributed between the n bits of the codeword, which justifies relation (1.14). The reference of energy to be considered, independent of the code and of the rate, is therefore E_b.

In Figure 1.5 are plotted the curves for the error correction of the (8, 4, 4) and (7, 4, 3) Hamming codes that have been dealt with throughout this introduction on a Gaussian channel. Hard-input decoding according to (1.4) and soft-input decoding according to (1.11) are considered. Also shown in the diagram is the curve for the binary error probability P_e that is obtained on this channel without using coding[2]. This curve is linked to the complementary error function $\text{erfc}(x)$ given by the relation (2.74) of Chapter 2: $P_e = \frac{1}{2}\text{erfc}\sqrt{\frac{E_b}{N_0}}$. With a low error

[2] The distinction between P_e and BER is only traditional: the value of P_e is given by an equation whereas the BER is obtained by measuring or simulation.

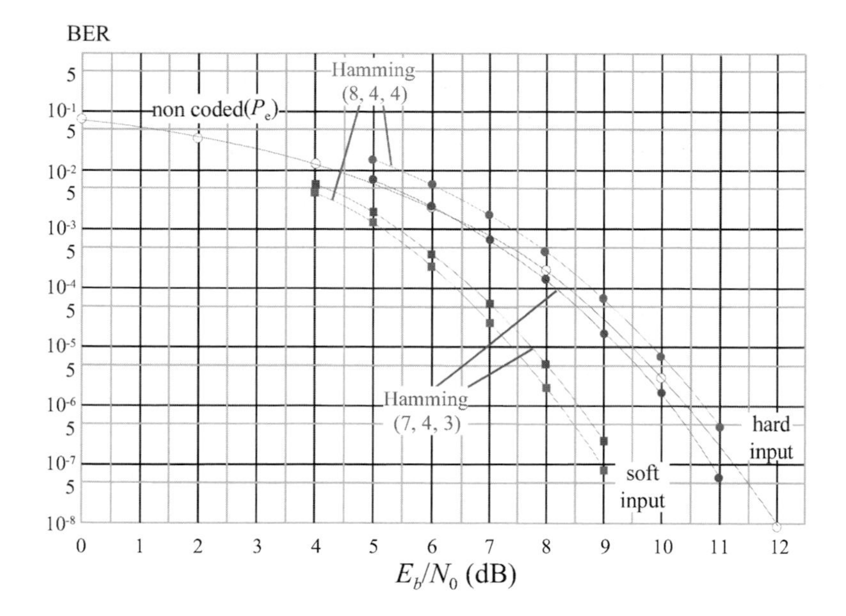

Figure 1.5 – Error correction capability of (8, 4, 4) and (7, 4, 3) Hamming codes on a Gaussian channel, with hard input and soft input decoding.

rate, the *asymptotic performance* of P_e is approximated by:

$$P_e \approx \frac{1}{2} \frac{\exp\left(-\frac{E_b}{N_0}\right)}{\sqrt{\pi \frac{E_b}{N_0}}} \tag{1.15}$$

To evaluate the probability $P_{e,word}$ that the soft-input decoder of a code with rate R with minimum distance d_{min} produces an erroneous codeword, in the previous equation we replace $\frac{E_b}{N_0}$ by $Rd_{\min}\frac{E_b}{N_0}$ and we introduce a multiplicative coefficient denoted $N(d_{min})$:

$$P_{e,word} = \frac{1}{2} N(d_{\min}) \operatorname{erfc} \sqrt{Rd_{\min} \frac{E_b}{N_0}} \approx \frac{1}{2} N(d_{\min}) \frac{\exp\left(-Rd_{\min}\frac{E_b}{N_0}\right)}{\sqrt{\pi Rd_{\min}\frac{E_b}{N_0}}} \tag{1.16}$$

The replacement of E_b by RE_b comes from (1.14) for the energy received by symbol is E_s. The multiplication by d_{min} is explained by the ML decoding rule (relation (1.11)), through which the decoder can discriminate the correct codeword and its closest competitor codewords thanks to d_{min} distinct values. Finally, the coefficient $N(d_{min})$, called *multiplicity*, takes into account the number of competitor codewords that are the minimum distance away. For example, in the case of the extended Hamming code, we have $N(d_{min} = 4) = 14$ (see Table 1.1).

To obtain the error binary probability $P_{e,bit}$, it suffices to multiply $P_{e,word}$ by the mean error density δ_e defined by (1.13):

$$P_{e,bit} \approx \frac{1}{2}\delta_e N(d_{\min}) \frac{\exp\left(-Rd_{\min}\frac{E_b}{N_0}\right)}{\sqrt{\pi Rd_{\min}\frac{E_b}{N_0}}} \tag{1.17}$$

Reading Table 1.1, we note that average number of errors in the 14 competitor words of weight 4, at the minimum distance from the "all zero" word, is 2. Equation (1.19) applied to the extended Hamming code is therefore:

$$P_{e,bit} \approx \frac{1}{2} \times \frac{2}{4} \times 14 \frac{\exp\left(-\frac{1}{2} \times 4 \times \frac{E_b}{N_0}\right)}{\sqrt{\pi \frac{1}{2} \times 4 \times \frac{E_b}{N_0}}} = 3.5 \frac{\exp\left(-2\frac{E_b}{N_0}\right)}{\sqrt{2\pi \frac{E_b}{N_0}}}$$

This expression gives $P_{e,bit} = 2.8\ 10^{-5}$, $1.8\ 10^{-6}$ et $6.2\ 10^{-8}$ for $\frac{E_b}{N_0} = 7$, 8 and 9 dB respectively, which corresponds to the results of the simulation of Figure 1.5. Such agreement between equations and experimentation cannot be found so clearly for more complex codes. In particular, finding the competitor codewords at distance d_{min} may not be sufficient and we then have to consider words at distance $d_{min} + 1$, $d_{min} + 2$ etc.

For a same value of P_e and $P_{e,bit}$ provided by the relations (1.15) and (1.17) respectively, the signal to noise ratios $\left.\frac{E_b}{N_0}\right|_{NC}$ and $\left.\frac{E_b}{N_0}\right|_{C}$ without coding (NC) and with coding (C) are such that:

$$Rd_{\min} \left.\frac{E_b}{N_0}\right|_C - \left.\frac{E_b}{N_0}\right|_{NC} = \log\left(\delta_e N(d_{\min}) \sqrt{\frac{\left.\frac{E_b}{N_0}\right|_{NC}}{Rd_{\min} \left.\frac{E_b}{N_0}\right|_C}}\right) \tag{1.18}$$

If $\delta_e N(d_{\min})$ is not too far from unity, this relation can be simplified as:

$$Rd_{\min} \left.\frac{E_b}{N_0}\right|_C - \left.\frac{E_b}{N_0}\right|_{NC} \approx 0$$

The *asymptotic gain*, expressed in dB, provides the gap between $\left.\frac{E_b}{N_0}\right|_{NC}$ and $\left.\frac{E_b}{N_0}\right|_{C}$:

$$G_a = 10\log\left(\frac{\left.\frac{E_b}{N_0}\right|_{NC}}{\left.\frac{E_b}{N_0}\right|_C}\right) \approx 10\log\left(Rd_{\min}\right) \tag{1.19}$$

As mentioned above, Rd_{min} appears as a figure of merit which, in a link budget with a low error rate, fixes the gain that a coding process can provide on a

Gaussian channel when the decoder is soft input. This is a major parameter for communication system designers. For types of channel other than Gaussian channels (Rayleigh, Rice, etc.), the asymptotic gain is always higher than what is approximately given by (1.19).

In Figure 1.5, the soft input decoding of the (8, 4, 4) Hamming code gives the best result, with an observed asymptotic gain of the order of 2.4 dB, in accordance with relation (1.18) that is more precise than (1.19). The (7, 4, 3) code is slightly less efficient since the product Rd_{min} is 12/7 instead of 2 for the (8, 4, 4) code. On the other hand, hard input decoding is unfavourable to the extended code as it does not offer greater correction capability in spite of a higher redundancy rate. This example is atypical: in the very large majority of practical cases, the hierarchy of codes that can be established based on their performance on a Gaussian channel, with soft input decoding, is respected for other types of channels.

1.6 What is a good code?

Figure 1.6 represents three possible behaviours for an error correcting code and its associated decoder, on a Gaussian channel. To be concrete, the information block is assumed to be length $k = 1504$ bits (188 bytes, a typical length for MPEG 1 compression) and the coding rate 1/2.

Curve 1 corresponds to the ideal system. There are in fact limits to the correction capacity of any code. These limits, whose first values were established by Shannon (1947-48) and which have been refined since then for practical situations, are said to be impassable. They depend on the type of noise, on the size of codeword and on the rate. Their main values are given in Chapter 3.

Curve 2 describes a behaviour having what we call good *convergence* but with mediocre MHD. Good convergence means that the error rate greatly decreases close to the theoretical limit (this region of great decrease is called the *waterfall*) but the MHD is not sufficient to maintain a steep slope down to very low error rates. The asymptotic gain, approximated by (1.19), is reached at a binary error rate of the order of 10^{-5} in this example. Beyond that, the curve remains parallel to the curve of error rates without coding: $P_e = \frac{1}{2}\text{erfc}\sqrt{\frac{E_b}{N_0}}$. The asymptotic gain is here of the order of 7.5 dB. This kind of behaviour, which was not encountered before the 1990s, is typical of coding systems implementing an iterative decoding technique (turbo codes, LDPC, etc.), when the MHD is not very high.

Curve 3 shows performance with mediocre convergence and high asymptotic gain. A typical example of this is the concatenation of a Reed-Solomon code and of a convolutional code. Whereas the MHD can be very large (around 100, for example), the decoder can benefit only relatively far from the theoretical limit. It is therefore not the quality of the code that is in question but the decoder, which cannot exploit all the information available at reception.

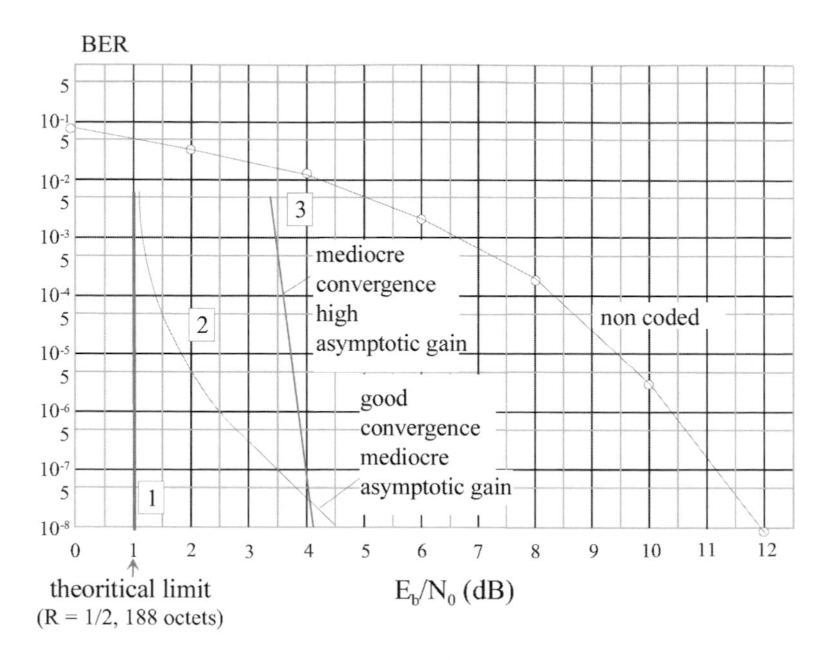

Figure 1.6 – Possible behaviours for a coding/decoding scheme on a Gaussian channel ($k = 1504$ bits, $R = 1/2$).

The search for the ideal encoder/decoder pair, since Shannon's work, has always had to face this dilemma: good convergence *versus* high MHD. Excellent algebraic codes like BCH or Reed-Solomon codes were fairly rapidly elaborated in the history of the correction coding (see Chapter 4). MHDs are high (and even sometimes optimal) but it is not always easy to implement soft input decoding. In addition, algebraic codes are generally "sized" for a specific length of codeword and coding rate, which limits their fields of application. In spite of this, algebraic codes are of great use in applications that require very low error rates, especially mass memories and/or when soft information is not available.

It is only recently, with the introduction of iterative probabilistic decoding (turbo decoding), that we have been able to obtain efficient error correction close to the theoretical limit. And it is even more recently that we have been able to obtain sufficient MHDs to avoid a change in slope that is penalizing for the performance curve.

It is not easy to find a simple answer to the question posed at the beginning of this section. Performance is, of course, the main criterion: for a given error rate, counted either in BER or in PER and for a fixed coding rate, a good code is first the one whose decoder offers a good error correction capability close to the corresponding theoretical limit. One preliminary condition of this is obviously the existence of a decoding algorithm (random codes do not have a decoder, for example) and that the software and/or hardware of this algorithm should not be too complex. Furthermore, using soft inputs could be an imperative that might not be simple to satisfy.

Other criteria like decoding speed (that fixes the throughput of the information decoded), latency (the delay introduced by the decoding process) or flexibility (the ability of the code to be defined for various word lengths and coding rates) are also to be taken into account in the context of the application targeted.

Finally, non-technical factors may also be very important. Technological maturity (do applications and standards already exist?), the cost of components, possible intellectual property rights, strategic preferences or force of habit are elements that carry weight when choosing a coding solution.

1.7 Families of codes

Up until the last few years, codes were traditionally classified in two families that were considered to be quite distinct due to their principles and their applications: *algebraic codes* (also called *block codes*) and *convolutional codes.* This distinction was mainly based on three observations:

- algebraic codes are appropriate for protecting blocks of data independent one from the other whereas convolutional codes are suitable for protecting continuous flows of data,
- the coding rates of algebraic codes are rather close to unity, whilst convolutional codes have lower rates,
- block code decoding is rather of the hard input decoding type, and that of convolutional codes is almost always soft input decoding.

Today, these distinctions are tending to blur. Convolutional codes can easily be adapted to encode blocks and most decoders of algebraic codes accept soft inputs. Via concatenation (Chapter 6), algebraic code rates can be lowered to values comparable with those of convolutional codes. One difference remains, however, between the two sub-families: the number of possible logical states of algebraic encoders is generally very high, which prevents decoding by exhaustive state methods. Decoding algorithms are based on techniques specific to each code. Convolutional encoders have a limited number of states, 2 to 256 in practice, and their decoding uses a complete representation of states, called a trellis (Chapter 5). It is for this reason that the book is structured in a traditional manner, which, for the time being, makes the distinction between algebraic codes and convolutional codes.

Modern coding requires concatenated or composite structures, which use several elementary encoders and whose decoding is performed by repeated passages in the associated decoders and by an exchange of information of a probabilistic type. Turbo codes (1993) opened the way to this type of iterative processing and, since then, many composite structures based on iterative decoding have been imagined (or rediscovered). Three of these are detailed in this book: turbo codes, both their original version and their recent evolution, iteratively decoded product codes and LDPC (*Low Density Parity Check*) codes. All these codes have been adopted in international norms, and understanding their coding processes and their decoding algorithms is a basis wide enough to tackle any other principle of distributed coding and of associated iterative decoding.

Chapter 2

Digital communications

2.1 Digital Modulations

2.1.1 Introduction

The function of modulation has its origin in radio-electric communications. An emitter can radiate an electromagnetic wave only in a limited, and generally narrow, portion of the spectrum, which can roughly be described as a frequency "window" with Δf centred on a frequency f_0, with $\Delta f << f_0$. The messages to be transmitted, that can be either analogue (for example speech) or digital (for example Morse code), are represented by signals that occupy only the bottom of the frequency spectrum. The spectrum of the signal, coming from a microphone in the case of speech, does not extend beyond a few kilohertz. The same thing applies for a signal that represents the two short (Tit) or long (Tat) elements of Morse "code", since the speed of handling several dozen signs per second is very small compared to the frequency f_0 that is measured in hundreds of kilohertz or in megahertz. Another use for modulation is frequency multiplexing which enables several simultaneous communications on the same wideband (cable or optical fibre) support, that are easily separated due to the fact that they each occupy a specific bandwidth, not connected to that of any other.

A sinusoidal wave f_0 can be represented by the function

$$s(t) = a\cos(2\pi f_0 t + \varphi) \tag{2.1}$$

where t denotes the time and f_0 is constant. Modulation involves making one or other of parameters a the amplitude, and φ the phase, depend on the signal to be transmitted. The modulated signal $s(t)$ then has a narrow spectrum centred on f_0, which is what we want.

The signal to be transmitted will in the sequel be called the *modulating signal.* Modulation makes one of the parameters a and φ vary as a function of the modulating signal if the latter is analogue. In the case of a digital signal, the modulating signal is a series of elements of a finite set, or symbols, applied to the modulator at discrete instants that are called *significant instants.* This series is called the *digital message* and we assume that the symbols are binary data applied periodically at the input of the modulator, every T_b seconds, therefore with a binary rate $D = 1/T_b$ bits per second. The binary data of this series are assumed to be independent and identically distributed (iid). A given digital, for example binary, message can be replaced by its "mth extension" obtained by grouping the initial symbols into packets of m. Then the symbols are numbers with m binary digits (or m-tuples), the total number of which is $M = 2^m$, applied to the modulator at significant instants with period mT_b. If the mth extension is completely equivalent to the message (of which it is only a different description), then the signal modulated by the original message and the signal modulated by its mth extension do not have the same properties, in particular concerning their bandwidth, since the larger m is, the narrower the bandwidth. The choice of integer m thus allows the characteristics of the modulated message to be varied.

Consider the complex signal

$$\sigma(t) = a \exp[j(2\pi f_0 t + \varphi)] \tag{2.2}$$

whose $s(t)$ is the real part, where j is the solution to the equation $x^2 + 1 = 0$. We can represent $\sigma(t)$ as the product

$$\sigma(t) = \alpha \exp(2\pi j f_0 t), \tag{2.3}$$

where only the first factor

$$\alpha = a \exp(j\varphi) \tag{2.4}$$

depends on the parameters a and φ that represent the data to be transmitted. The values taken by this first factor for all possible values of the parameters can be represented by points in the complex plane. The set of these points is then called a *constellation* and the complex plane is called a *Fresnel plane.* The modulated signal (2.1) is the real part of the complex signal $\sigma(t)$ defined by (2.3).

If the correspondence established between the modulating signal and the variable parameters is instantaneous, the modulation is said to be *memoryless.* It can be useful for this correspondence to be established between the variable parameters and a function of the values taken later by the modulating signal. For example, if the latter is analogue, a conventional modulation process (called frequency) involves making φ vary proportionally to the integral of the modulating signal in relation to time. In the same way, in the case of a digital modulation, the constellation point can be chosen to represent the symbol present at the instant considered, and the modulation is then said to be *memoryless*, or a symbol obtained by combining it with other later symbols. A modulation can therefore be either analogue or digital, with memory or memoryless. In all that follows below, we restrict ourselves to digital modulations that differ from each other according to their form or number of points in their constellation, and perhaps by a memory effect. When the latter is obtained by combining the symbol applied with later symbols, this can be interpreted as a preliminary transformation of the digital message. Moreover, it is often necessary to ensure the continuity of the phase to improve the shape of the spectrum of the modulated signals, which implies a memory effect.

The choice of a modulation system depends on many factors. Modulated signals are emitted on an imperfect channel, perturbed by the addition of parasitic signals collectively called *noise* and often, in radio-electricity, affected by variations in the amplitude of the received signal, due for example to a rapid change in the propagation conditions, a phenomenon called fading. In spite of these channel faults, we want to receive modulating messages with a small error probability, which implies that the signals associated with them should be as different as possible. On the other hand, whether it be radio-electricity or multiplexing, the radio-electric spectrum is common to several users, each of which perturbs the others. Therefore we wish to concentrate the power emitted in a frequency interval that is limited as far as possible. This implies choosing the modulation parameters in order to give the spectrum the most appropriate shape. The signal spectrum of the general (2.3) shape is made up of a *main lobe* centred on f_0 that concentrates most of the power emitted, and *tails* or *sidelobes* where the spectral density decreases more or less rapidly in relation to the central frequency f_0. Whatever the modulation system, the width of the main lobe is proportional to the *symbol rate* $R = D/m = 1/mT_b$, expressed in symb/s. The decrease in the spectral density far enough away from the central frequency depends only on the discontinuities of the modulating signal and its derivations. It varies in $1/(f - f_0)^{2(d+1)}$ where d is the smallest order of a derivation of the discontinuous signal ($d = 0$ if the modulated signal itself is discontinuous). We note that it is only possible to increase the value of d by introducing an increasing delay between the instant when a modulating symbol is applied to the modulator and the characteristic instant corresponding to it.

The main parameters associated with modulation are therefore the size and the shape of the constellation used (on which the error probability depends, on a given channel), the width of the main lobe of the spectrum of the modulated spectrum and the decrease in its spectral density away from the central frequency. They are largely in conflict: for example, we can only reduce the width of the central lobe by increasing the size of the constellation, to the detriment of the error probability for the same power. The choice of a modulation system can therefore only result from a compromise adapted to a particular application. Apart from the parameters indicated, the complexity of implementation must be taken into account. For example, shaping that improves the decrease in the secondary lobes of the spectrum by the increase of order d in the first discontinuous derivation, or increasing the size of the constellation to lower the central lobe of the spectrum, mean an increase in the complexity of the modulator.

2.1.2 Linear Memoryless Modulations

Amplitude-shift keying with M states: M-ASK

Let us first note a particular case of amplitude modulation: On Off Keying (OOK), for which a in the expression (2.1) takes the value 0 or A. One of the binary states therefore corresponds to an extinction of the carrier. The presence or absence of the carrier can be recognized independently of the knowledge of the phase φ, by measuring the energy of the received signal during a short interval of time on the scale of period $1/f_0$ of the carrier wave, which is called *incoherent detection*. Figure 2.1 shows the time interval of a signal modulated by On Off Keying.

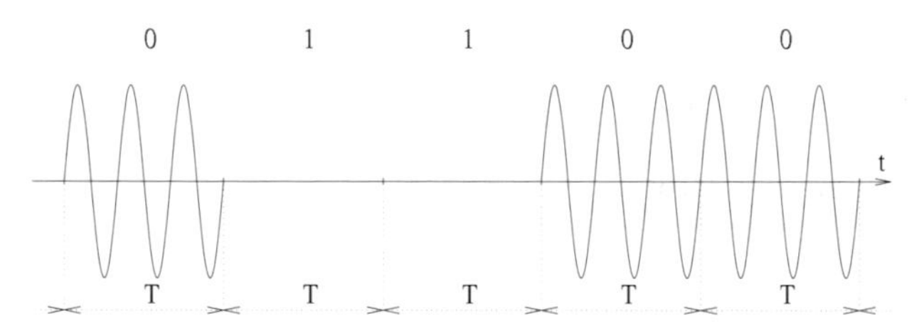

Figure 2.1 – On Off Keying (OOK).

In the general case of amplitude-shift keying (ASK) with M states, the amplitude of the carrier is the modulated value $a_j = A_j h(t)$ for $j = 1, 2, ..., M$ where A_j takes a value among $M = 2^m$ values according to the group of data presented at the input of the modulator and $h(t)$ is a rectangular pulse with unit amplitude and width T. The modulator thus provides signals of the form:

$$s_j(t) = A_j h(t) \cos(2\pi f_0 t + \varphi_0) \tag{2.5}$$

$$\text{with } A_j = (2j - 1 - M)A \quad j = 1,\, 2,\, \cdots,\, M \ \text{ where } A \text{ is constant} \tag{2.6}$$

$$\begin{aligned} \text{and } h(t) &= 1 \quad \text{for} \quad t \in [0, T[\\ &= 0 \quad \text{elsewhere} \end{aligned} \tag{2.7}$$

Amplitude A_j of the modulated signal is constant for a width T then changes value; the modulated signal thus transmits $\log_2(M)$ binary data every T seconds. We can note that half the nominal amplitudes A_j are negative. The identity $a \cos \varphi = -a \cos(\varphi + \pi)$ thus implies *coherent* demodulation where the phase φ is known. The phase often only being known to a multiple of π, a binary message where the symbols are ± 1 will only be demodulated to one sign.

For an M-ASK signal, the different states of the modulated signal are situated on a straight line and its constellation is therefore one-dimensional in the Fresnel plane. There are many ways to make the association between the value of the amplitude of the modulated signal and the particular realization of a group of data of $m = \log_2(M)$ data. In general, we associate with two adjacent values taken by the amplitude, two groups of data that differ by only one binary value. This particular association is called *Gray coding*. It enables the errors made by the receiver to be minimized. Indeed, when the receiver selects an amplitude adjacent to the emitted amplitude because of noise, which corresponds to the most frequent situation, we make only one error for $m = \log_2(M)$ data transmitted. We show in Figure 2.2 two examples of signal constellations modulated in amplitude by Gray coding.

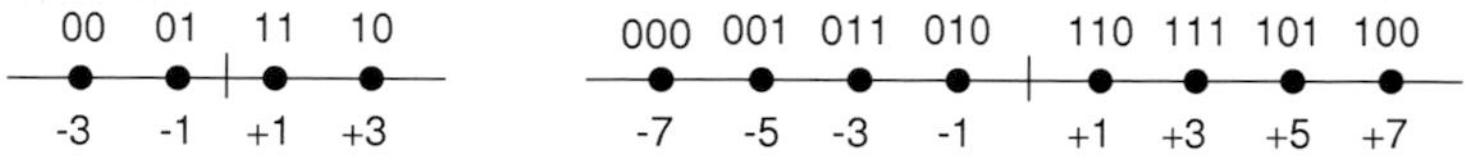

00 01 11 10
-3 -1 +1 +3

000 001 011 010 110 111 101 100
-7 -5 -3 -1 +1 +3 +5 +7

Figure 2.2 – Example of 4-ASK and 8-ASK signal constellations with Gray coding

The mean energy E_s used to transmit an M-ary symbol is equal to:

$$E_s = \int_0^T \mathrm{E}\left\{A_j^2\right\} \cos^2(2\pi f_0 t + \varphi_0)dt$$

where $\mathrm{E}\left\{A_j^2\right\}$, the expectation of A_j^2 has the expression $A^2(M^2 - 1)/3$.
Making the hypothesis that $f_0 >> \frac{1}{T}$, the previous relation gives the mean energy E_s:

$$E_s = \frac{A^2 T}{2} \frac{(M^2 - 1)}{3} \tag{2.8}$$

The mean energy E_b used to transmit a bit is:

$$E_b = \frac{E_s}{\log_2(M)} \tag{2.9}$$

For a transmission with a continuous data stream, the amplitude modulated signal can be written in the form:

$$S(t) = A \sum_i a_i h(t - iT) \cos(2\pi f_0 t + \varphi_0) \tag{2.10}$$

where the $\{a_i\}$ are a sequence of M-ary symbols, called modulation symbols, which take the values $(2j - 1 - M)$, $j = 1, 2, \cdots, M$. In the expression of the modulated signal, i is the time index.

The signal $S(t)$ can again be written in the form:

$$S(t) = \Re_e \left\{s_e(t) \exp\left(j(2\pi f_0 t + \varphi_0)\right)\right\} \tag{2.11}$$

where $s_e(t)$ is the complex envelope of the signal $S(t)$ with:

$$s_e(t) = A \sum_i a_i h(t - iT) \tag{2.12}$$

Taking into account the fact that the data d_i provided by the source of information are *iid*, the modulation symbols a_i are independent, with zero mean and variance equal to $(M^2 - 1)/3$.

It can be shown that the power spectral density (psd) of the signal $S(t)$ is equal to:

$$\gamma_S(f) = \frac{1}{4}\gamma_{s_e}(f - f_0) + \frac{1}{4}\gamma_{s_e}(f + f_0) \tag{2.13}$$

with:

$$\gamma_{s_e}(f) = \frac{M^2 - 1}{3} A^2 T \left(\frac{\sin \pi f T}{\pi f T}\right)^2 \tag{2.14}$$

The psd of $s_e(t)$ expressed in dB is shown in Figure 2.3 as a function of the normalized frequency fT, for $M = 4$ and $A^2T = 1$.

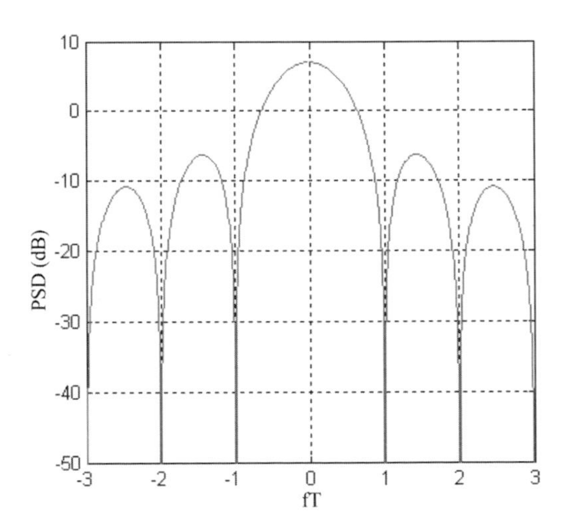

Figure 2.3 – Power spectral density (psd) of the complex envelope of a signal ASK-4, with $A^2T = 1$.

The psd of $S(t)$ is centred on the frequency carrier f_0 and its envelope decreases in f^2. It is made up of a main lobe of width $2/T$ and of sidelobes that periodically have zero crossing at $f_0 \pm k/T$.

Note

The bandwidth is, strictly speaking, infinite. In practice, we can decide only to transmit a percentage of the power of the signal $S(t)$ and in this case, the bandwidth is finite. If, for example, we decide to transmit 99% of the power of the modulated signal, which results in only a low distortion of the signal $S(t)$, then the bandwidth is about $8/T$ where $1/T$ is the symbol rate. We shall see in Section 2.3 that it is possible to greatly reduce this band without degrading the performance of the modulation. This remark is valid for all linear modulations.

Phase Shift Keying with M states (M-PSK)

For this modulation, also called *Phase Shift Keying* (PSK), it is the phase of the carrier that is the modulated value. The modulator provides signals of the form:

$$s_j(t) = Ah(t)\cos(2\pi f_0 t + \varphi_0 + \phi_j) \tag{2.15}$$

where f_0 is the carrier frequency, φ_0 its phase and $h(t)$ a rectangular pulse of unit amplitude and width T. The modulated phase ϕ_j takes a value among $M = 2^m$ with:

$$\phi_j = (2j+1)\frac{\pi}{M} + \theta_0 \quad 0 \leq j \leq (M-1) \tag{2.16}$$

The different states of the phase are equidistributed on a circle of radius A. The phase θ_0 is fixed at $-\pi/2$ for a 2-PSK modulation and at 0 for an M-PSK modulation with $M > 2$ states.

The modulated signal can again be written in the form:

$$s_j(t) = Ah(t)\left[\cos\phi_j \cos(2\pi f_0 t + \varphi_0) - \sin\phi_j \sin(2\pi f_0 t + \varphi_0)\right] \tag{2.17}$$

In this form, the M-PSK signal can be expressed as the sum of two quadrature carriers, $\cos(2\pi f_0 t + \varphi_0)$ and $-\sin(2\pi f_0 t + \varphi_0)$, the amplitude modulated by $\cos\phi_j$ and $\sin\phi_j$ with $\cos^2\phi_j + \sin^2\phi_j = 1$. We can check that when M is a multiple of 4, the possible values of the amplitude of the two carriers are identical.

In Figure 2.4 we show two constellations of a phase modulated signal with Gray coding. The constellations have two dimensions and the different states of the modulated signal are on a circle of radius A. We say that the constellation is circular.

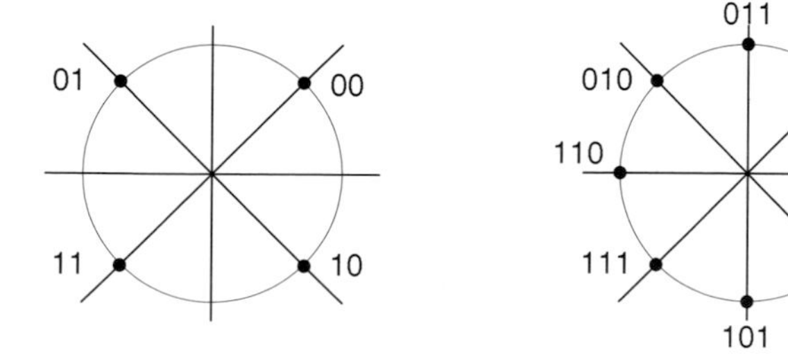

Figure 2.4 – Examples of constellations of a phase modulated signal with Gray coding.

The energy E_s for transmitting a phase state, that is, a group of $\log_2(M)$ binary data, is equal to:

$$E_s = \int_0^T A^2 \cos^2(2\pi f_0 t + \varphi_0 + \phi_j)dt = \frac{A^2 T}{2} \quad \text{if } f_0 >> 1/T \tag{2.18}$$

Energy E_s is always the same whatever the phase state transmitted. The energy used to transmit a bit is $E_b = E_s/\log_2(M)$.

For the transmission of a continuous data stream, the modulated signal can be written in the form:

$$S(t) = A\left[\sum_i a_i h(t-iT)\cos(2\pi f_0 t + \varphi_0) - \sum_i b_i h(t-iT)\sin(2\pi f_0 t + \varphi_0)\right] \tag{2.19}$$

where the modulation symbols a_i and b_i take their values in the following sets:

$$\begin{array}{ll} a_i \in \left\{\cos\left[(2j+1)\frac{\pi}{M} + \theta_0\right]\right\} & 0 \leq j \leq (M-1) \\ b_i \in \left\{\sin\left[(2j+1)\frac{\pi}{M} + \theta_0\right]\right\} & 0 \leq j \leq (M-1) \end{array} \tag{2.20}$$

The signal $S(t)$ can again be written in the form given by (2.11) with:

$$s_e(t) = A\sum_i c_i h(t-iT), \quad c_i = a_i + jb_i \tag{2.21}$$

Taking into account the fact that the data d_i provided by the source of information are *iid*, the modulation symbols c_i are independent, with zero mean and unit variance.

The psd of the signal $S(t)$ is again equal to:

$$\gamma_S(f) = \frac{1}{4}\gamma_{s_e}(f - f_0) + \frac{1}{4}\gamma_{s_e}(f + f_0)$$

with this time:

$$\gamma_{s_e}(f) = A^2 T\left(\frac{\sin \pi f T}{\pi f T}\right)^2 \tag{2.22}$$

the psd looking like that of Figure 2.3.

Quadrature Amplitude Modulation using two quadrature carriers (M-QAM)

For this modulation, also called *Quadrature Amplitude Modulation* (M-QAM), it is two quadrature carriers $\cos(2\pi f_0 t + \varphi_0)$ and $-\sin(2\pi f_0 t + \varphi_0)$ that are amplitude modulated. The modulator provides signals of the form:

$$s_j(t) = A_j^c h(t)\cos(2\pi f_0 t + \varphi_0) - A_j^s h(t)\sin(2\pi f_0 t + \varphi_0) \tag{2.23}$$

where f_0 is the frequency of the carrier, φ_0 its phase and $h(t)$ a rectangular pulse of unit amplitude and width T.

Two situations can arise depending on whether the length m of the groups of data at the input of the modulator is even or not. If m is even, then $M = 2^m$ is a perfect square (4, 16, 64, 256, ...); in the opposite case, M is simply a power of two (8, 32, 128, ...).

When m is even, the group of data can be separated into two sub-groups of length $m/2$, each being associated respectively with amplitudes A_j^c and A_j^s that take their values in the set $(2j - 1 - \sqrt{M})A, \quad j = 1, 2, \cdots, \sqrt{M}$. In Figure 2.5 are represented the constellations of the 16-QAM and 64-QAM modulations. These constellations are said to be square.

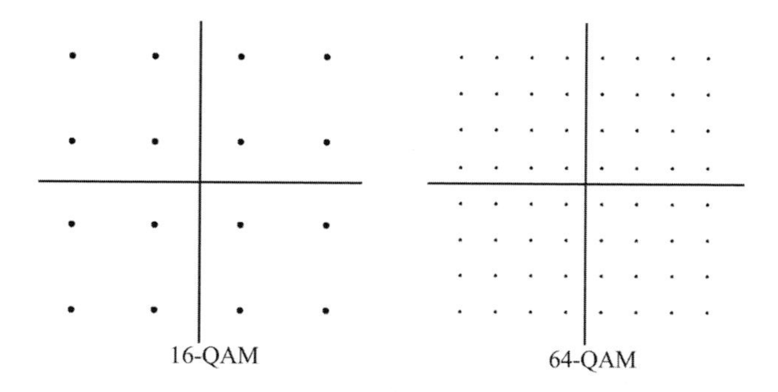

Figure 2.5 – Constellations of two QAM-type modulations.

When m is odd, the M-QAM signal can no longer be obtained as a combination of two quadrature amplitude-modulated carriers. However, we can build the M-QAM signal from an N-QAM signal modulated classically on two quadrature carriers, where N is a square immediately higher than M by preventing $(N - M)$ states. For example, 32-QAM modulation can be obtained from 36-QAM modulation where A_j^c and A_j^s take the values $(\pm A, \pm 3A, \pm 5A)$ by preventing the four states of amplitude $(\pm 5A, \pm 5A)$ for the pairs (A_j^c and A_j^s). The constellation of the 32-QAM modulation is shown in Figure 2.6.
The M-QAM signal can again be written in the form:

$$s_j(t) = V_j h(t) \cos(2\pi f_0 t + \varphi_0 + \phi_j) \tag{2.24}$$

with:

$$V_j = \sqrt{(A_j^c)^2 + (A_j^s)^2} \qquad \phi_j = \tan^{-1} \frac{A_j^s}{A_j^c}$$

In this form, the M-QAM modulation can be considered as a modulation combining phase and amplitude. Assuming that the phase takes $M_1 = 2^{m_1}$ states and the amplitude $M_2 = 2^{m_2}$ states, the modulated signal transmits $\log_2(M_1 M_2) = m_1 + m_2$ data every T seconds. Figure 2.7 shows the constellation

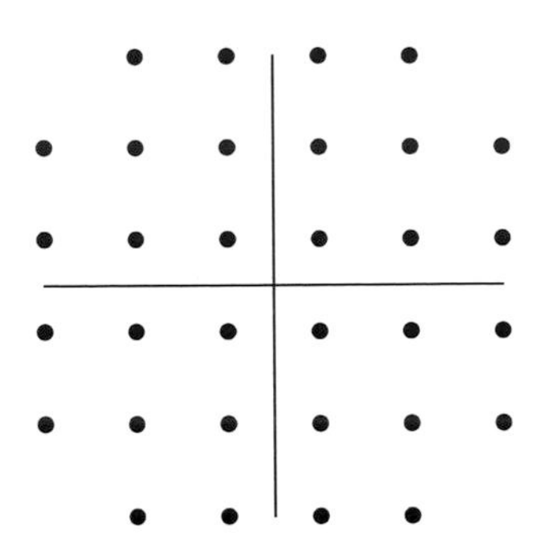

Figure 2.6 – Constellation of the 32-QAM modulation.

of a modulation combining phase and amplitude for $M = 16$ ($M_1 = 4$, $M_2 = 4$).

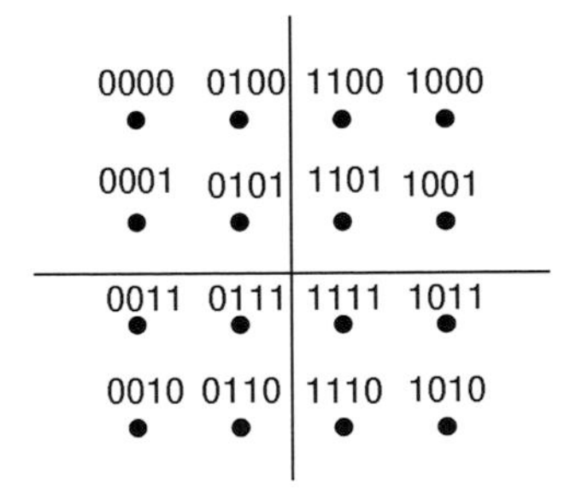

Figure 2.7 – Constellation of a modulation combining phase and amplitude for $M = 16$.

The average energy E_s to transmit the pair (A_j^c, A_j^s), that is, a group of $\log_2(M)$ binary data, is equal to:

$$E_s = \int_0^T \mathrm{E}\left\{V_j^2\right\} \cos^2(2\pi f_0 t + \varphi_0 + \phi_j)dt \tag{2.25}$$

For a group of data of even length m, $\mathrm{E}\left\{V_j^2\right\} = 2A^2(M-1)/3$ and thus, for $f_0 >> 1/T$, the average energy E_s is equal to:

$$E_s = A^2T\frac{M-1}{3} \tag{2.26}$$

The average energy used to transmit a bit is $E_b = E_s/\log_2(M)$.

For a continuous data stream, the signal can be written in the form:

$$S(t) = A \left[\sum_i a_i h(t-iT)\cos(2\pi f_0 t + \varphi_0) - \sum_i b_i h(t-iT)\sin(2\pi f_0 t + \varphi_0) \right] \tag{2.27}$$

where the modulation symbols a_i and b_i take the values $(2j-1-\sqrt{M})$, for $j = 1, 2, \cdots, \sqrt{M}$ and for $M = 2^m$ with even m. The signal $S(t)$ can be expressed by the relations (2.11) and (2.21):

$$S(t) = \Re_e \{ s_e(t) \exp j(2\pi f_0 t + \varphi_0) \}$$

with $s_e(t) = A \sum_i c_i h(t-iT)$, $\quad c_i = a_i + jb_i$

The binary data d_i provided by the information source being *iid*, the modulation symbols c_i are independent, with zero mean and variance equal to $2(M-1)/3$. The psd of the signal $S(t)$ is again given by (2.13) with:

$$\gamma_{s_e}(f) = \frac{2(M-1)}{3} A^2 T \left(\frac{\sin \pi f T}{\pi f T} \right)^2 \tag{2.28}$$

The spectral width of a modulated M-QAM signal is therefore, to within an amplitude, the same as that of M-ASK and M-PSK signals.

2.1.3 Memoryless modulation with M states (M-FSK)

For this modulation, also called *Frequency Shift Keying* (M-FSK), it is the frequency that is the modulated value. The modulator generates signals of the form:

$$s_j(t) = Ah(t)\cos(2\pi(f_0 + f_j)t + \varphi_j) \tag{2.29}$$

where $f_j = j\Delta f$, $j = 1, 2, \cdots, M$ and $h(t)$ is a rectangular pulse with unit amplitude and width T. The φ_j are random independent phases with constant realization on the interval $[0, T[$. The $s_j(t)$ signals can therefore be generated by independent oscillators since there is no relation between the phases φ_j.

Let us compute the correlation coefficient between two modulated signals taking different frequency states.

$$\rho_{j,n} = \int_0^T A^2 \cos(2\pi(f_0 + j\Delta f)t + \varphi_j)\cos(2\pi(f_0 + n\Delta f)t + \varphi_n)dt \tag{2.30}$$

After integration, and assuming $f_0 >> 1/T$, we obtain:

$$\rho_{j,n} = \frac{A^2 T}{2} \left(\frac{\sin(2\pi(j-n)\Delta f T + \varphi_j - \varphi_n)}{2\pi(j-n)\Delta f T} - \frac{\sin(\varphi_j - \varphi_n)}{2\pi(j-n)\Delta f T} \right) \tag{2.31}$$

Choosing $\Delta f = 1/T$, $\rho_{j,n} = 0\ \forall j \neq n$ and the modulated M signals are orthogonal. Orthogonal signals are generally chosen since signals of different frequencies can easily be separated at reception. At instants iT where the M-FSK signal changes frequency, the modulated signal presents a discontinuity since the phases φ_j are independent. We then speak of discontinuous-phase frequency modulation. Figure 2.8 gives an example of a 2-FSK signal.

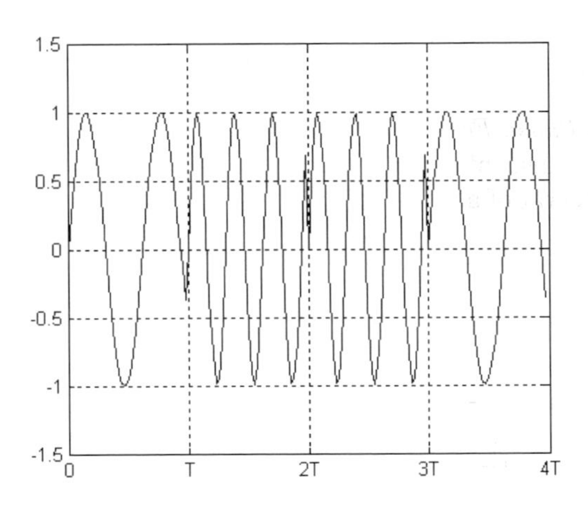

Figure 2.8 – Frequency modulated signal with discontinuous phase.

The energy E_s used to transmit a group of data is equal to:

$$E_s = \int_0^T A^2 \cos^2(2\pi(f_0 + \frac{j}{T})t + \varphi_j)dt = \frac{A^2T}{2} \quad \text{if } f_0 >> 1/T \tag{2.32}$$

and the energy E_b used to transmit a bit is equal to $E_s/\log_2(M)$.

For a continuous flow transmission of data, the modulated signal can be written in the form:

$$S(t) = A\sum_i h(t - iT)\cos(2\pi(f_0 + \frac{a_i}{T})t + \varphi_i) \tag{2.33}$$

where the modulation symbol a_i is equal to $1, 2, \cdots, M$, $M = 2^m$. Every T seconds, the modulated signal $S(t)$ transmits a group of $\log_2(M)$ binary data.

For 2-FSK modulation, the power spectral density of the signal $S(f)$ is equal to:

$$\gamma_S(f) = \frac{1}{4}(\gamma(f - f_1) + \gamma(f + f_1) + \gamma(f - f_2) + \gamma(f + f_2)) \tag{2.34}$$

where $f_1 = f_0 + 1/T$ and $f_2 = f_0 + 2/T$ and:

$$\gamma(f) = \frac{A^2T}{4}\left(\frac{\sin \pi fT}{\pi fT}\right)^2 + \frac{A^2}{4}\delta(f) \tag{2.35}$$

where $\delta(f)$ is the Dirac distribution. The psd of a 2-FSK signal has a continuous part and a discrete part. Limiting ourselves to the two main lobes of this power spectral density, the band of frequencies occupied by a 2-FSK signal is $3/T$, that is, three times the symbol rate. Let us recall that at a same symbol rate, an M-PSK or an M-QAM signal occupies a bandwidth of only $2/T$. The discrete part corresponds to two spectral lines situated in f_1 and f_2.

2.1.4 Modulations with memory by continuous phase frequency shift keying (CPFSK)

For *Continuous Phase Frequency Shift Keying* (CPFSK), the modulated signal does not show any discontinuities at the instants of frequency changes. Figure 2.9 shows a time interval of a CPFSK signal for $M = 2$. The CPFSK signal has the expression:

$$S(t) = A\cos(2\pi f_0 t + \phi_0 + \phi(t)) \tag{2.36}$$

where f_0 is the frequency of the carrier and ϕ_0 its phase.

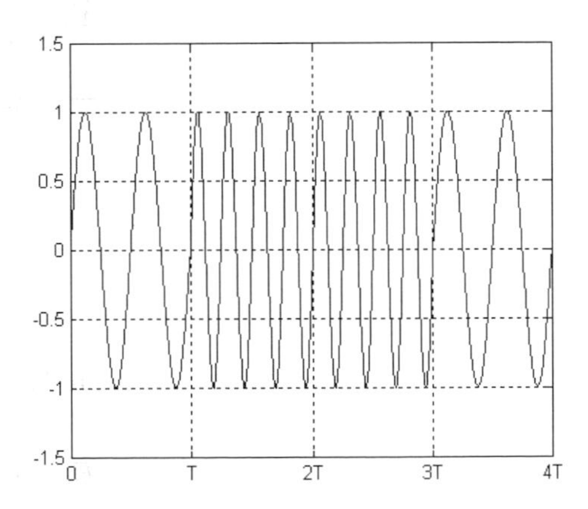

Figure 2.9 – Time interval of a 2-CPFSK signal.

The instantaneous frequency deviation is:

$$f(t) = \frac{1}{2\pi}\frac{d\phi}{dt} = h\sum_i a_i g(t - iT) \tag{2.37}$$

where h is called the modulation index and the M-ary symbols a_i take their values in the alphabet $\{\pm 1, \pm 3, \cdots, \pm(2p+1), \cdots, \pm(M-1)\}$; $M = 2^n$. The function $g(t)$ is causal and has a finite pulse widtht.

$$\begin{aligned} g(t) &\neq 0 \quad t \in [0, LT[\quad L \text{ integer} \\ &= 0 \quad \text{elsewhere} \end{aligned} \tag{2.38}$$

Putting:

$$q(t) = \int_0^t g(\tau)d\tau$$

and for considerations of normalization, by imposing that:

$$q(t) = \frac{1}{2} \text{ if } \quad t \geq LT$$

the phase $\phi(t)$ on the interval $[iT, (i+1)T[$ is equal to:

$$\phi(t) = 2\pi h \sum_{n=i-L+1}^{i} a_n q(t - nT) + \pi h \sum_{n=-\infty}^{i-L} a_n \tag{2.39}$$

When $L = 1$, the continuous phase frequency modulations are said to be *full response* whereas for $L > 1$, they are said to be *partial response.*

To illustrate continuous phase-frequency-shift keying, we are going to consider three examples, *Minimum Shift Keying* (MSK) modulation, *L-ary Raised Cosine* (LRC) modulation and *Gaussian Minimum Shift Keying* (GMSK) modulation.

Continuous phase-frequency-shift keying with modulation index $h = 1/2$: Minimum Shift Keying (MSK)

For this modulation, the index h is equal to $1/2$ and the symbols a_i are binary (± 1). The function $g(t)$ is a rectangular pulse of amplitude $1/2T$ and width T. Thus the function $q(t)$ is equal to:

$$\begin{array}{lcll} q(t) & = & 0 & t \leq 0 \\ q(t) & = & \frac{t}{2T} & 0 \leq t \leq T \\ q(t) & = & \frac{1}{2} & t \geq T \end{array} \tag{2.40}$$

MSK modulation is full response continuous phase frequency shift keying modulation $(L = 1)$.

On the interval $[iT, (i+1)T[$, the phase $\phi(t)$ of the MSK signal has the expression:

$$\phi(t) = \frac{\pi}{2} a_i \frac{(t - iT)}{T} + \frac{\pi}{2} \sum_{n=-\infty}^{i-1} a_n \tag{2.41}$$

The evolution of the phase $\phi(t)$ as a function of time is shown in Figure 2.10. We can note that the phase $\phi(t)$ varies linearly over a time interval T and that there is no discontinuity at instants iT.

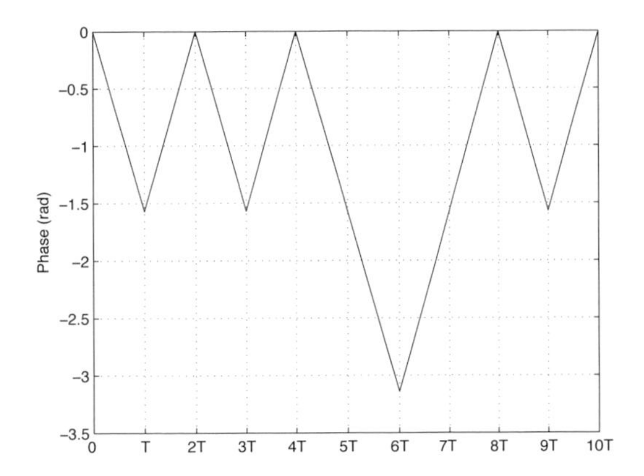

Figure 2.10 – Evolution of the phase of an MSK signal as a function of time.

Using the expressions (2.36) and (2.41), the MSK signal can be written in the form:

$$S(t) = A\cos\left[2\pi(f_0 + \frac{a_i}{4T})t - i\frac{\pi}{2}a_i + \theta_i + \varphi_0\right] \quad iT \leq t < (i+1)T \qquad (2.42)$$

with:

$$\theta_i = \frac{\pi}{2}\sum_{n=-\infty}^{i-1} a_i \qquad (2.43)$$

The MSK signal uses two frequencies to transmit the binary symbols $a_i = \pm 1$.

$$\begin{aligned} f_1 &= f_0 + \tfrac{1}{4T} \quad \text{if} \quad a_i = +1 \\ f_2 &= f_0 - \tfrac{1}{4T} \quad \text{if} \quad a_i = -1 \end{aligned} \qquad (2.44)$$

We can verify that the two frequency signals f_1 and f_2 are orthogonal and that they present a frequency deviation $\Delta f = f_1 - f_2 = 1/2T$ minimal. This minimum deviation is at the origin of the name of MSK modulation.

The modulated MSK signal can also be written in the form:

$$\begin{aligned} S(t) = A\Big[&\sum_i c_{2i-1}h(t-2iT)\cos\tfrac{\pi t}{2T}\cos(2\pi f_0 t + \varphi_0) \\ &- \sum_i c_{2i}h(t-(2i+1)T)\sin\tfrac{\pi t}{2T}\sin(2\pi f_0 t + \varphi_0)\Big] \end{aligned} \qquad (2.45)$$

where the symbols c_i are deduced from the symbols a_i by transition coding.

$$c_{2i} = a_{2i}c_{2i-1} \quad \text{and} \quad c_{2i-1} = a_{2i-1}c_{2i-2} \qquad (2.46)$$

and $h(t)$ is a rectangular pulse of unit amplitude, but of width $2T$.

$$\begin{aligned} h(t) &= 1 \quad \text{if } t \in [-T, T[\\ &= 0 \quad \text{elsewhere} \end{aligned} \tag{2.47}$$

MSK modulation can be seen as an amplitude modulation of the terms

$$\cos\frac{\pi t}{2T}\cos(2\pi f_0 t + \varphi_0) \text{ and } -\sin\frac{\pi t}{2T}\sin(2\pi f_0 t + \varphi_0)$$

by two bit streams $u_c(t) = \sum_i c_{2i-1}h(t - 2iT)$ and $u_s(t) = \sum_i c_{2i}h(t - (2t + 1)T)$ whose transitions are shifted by T. Each bit stream enables a bit to be transmitted every $2T$ seconds and thus, the binary rate of an MSK modulation is $D = 1/T_b$ with $T = T_b$.

MSK modulation is a particular case of CP M-FSK modulation since it is linear. The psd is given by:

$$\gamma_S(f) = \frac{1}{4}\gamma(f - f_0) + \frac{1}{4}\gamma(-f - f_0) \tag{2.48}$$

with:

$$\gamma(f) = \frac{16A^2T}{\pi^2}\left(\frac{\cos 2\pi fT}{1 - 16f^2T^2}\right)^2 \tag{2.49}$$

Figure 2.11 shows the psd of the complex envelope of an MSK signal expressed in dB as a function of the normalized frequency fT_b. We have also plotted the psd of the complex envelope of a 4-PSK signal. In order for the comparison of these two power spectral densities to make sense, we have assumed that the rate transmitted was identical for these two modulations (that is, $T = 2T_b$ for the 4-PSK modulation).

The width of the main lobe of the power spectral density of an MSK modulation is $3/2T_b$ whereas it is only $1/T_b$ for 4-PSK modulation. Thus, for a same rate transmitted, the main lobe of MSK modulation occupies 50% more bandwidth than that of 4-PSK modulation. However, the envelope of the psd of an MSK signal decreases in f^{-4} whereas it only decreases in f^{-2} for 4-PSK modulation. One of the consequences of this is that the bandwidth B that contains 99% of the power of the modulated signal for MSK is $1.2/T_b$ whereas it is around $8/T_b$ for 4-PSK.

L-ary Raised Cosine modulation (L-RC)

For this modulation, the function $g(t)$ has the expression:

$$\begin{aligned} g(t) &= \frac{1}{2LT}(1 - \cos\frac{2\pi t}{LT}) \quad \text{for } 0 \leq t \leq LT \\ &= 0 \quad \text{elsewhere} \end{aligned} \tag{2.50}$$

The larger parameter L is, the faster the power spectral density of this modulation decreases. For example, the psd is -40 dB for a $2RC$ ($h = 1/2$, $a_i = \pm 1$)

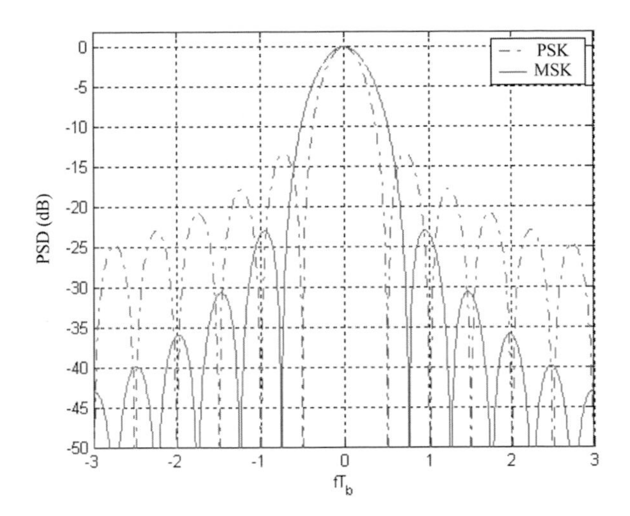

Figure 2.11 – Power spectral density of the complex envelope of MSK and 4-PSK signals.

modulation for $fT = 1.2$ whereas for a $4RC$ ($h = 1/2$, $a_i = \pm 1$) modulation, this level is reached for $fT = 0.7$.
The function $q(t)$ is equal to:

$$\begin{aligned} q(t) &= \frac{t}{2LT} - \frac{1}{4\pi}\sin\frac{2\pi t}{LT} \quad & 0 \leq t \leq LT \\ &= \frac{1}{2} & t > LT \end{aligned} \tag{2.51}$$

Gaussian minimum shift keying modulation (GMSK)

The index h of this modulation is equal to 1/2 and the symbols a_i are binary (± 1). The function $g(t)$ is defined as follows:

$$g(t) = h(t) * \chi(t) \tag{2.52}$$

where $h(t)$ is a rectangular pulse of amplitude $1/T$ ($T = T_b$) on the interval $[-T/2, T/2]$ and $\chi(t)$ is the impulse response of a Gaussian filter with passband at -3 dB equal to B_g:

$$\chi(t) = \sqrt{\frac{2\pi}{\ln 2}} B_g \exp(-2\pi^2 B_g^2 t^2)/\ln 2 \tag{2.53}$$

After resolution of the convolution product, the function $g(t)$ can be written in the form:

$$g(t) = \frac{1}{2T}\left[\operatorname{erf}\left(\pi B_g \sqrt{\frac{2}{\ln 2}}(t + \frac{T}{2})\right) - \operatorname{erf}\left(\pi B_g \sqrt{\frac{2}{\ln 2}}(t - \frac{T}{2})\right)\right] \tag{2.54}$$

where $\mathrm{erf}(x)$ represents the error function defined by:

$$\mathrm{erf}(x) = \frac{2}{\sqrt{\pi}} \int_0^x \exp(-u^2)du$$

In Figure 2.12, we have plotted $g(t)$ as a function of the normalized variable t/T, for different values of the normalized passband $B_N = B_g T$. We note that the graph of the function $g(t)$ has been shifted by $2T$ for $B_N = 0.2$ and by $1.5T$ for $B_N = 0.3$.

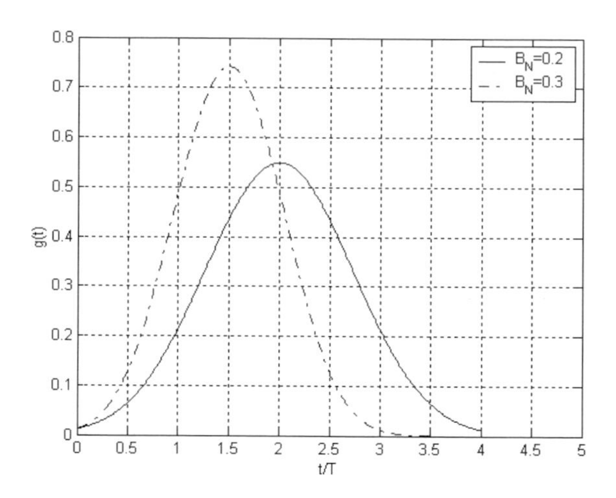

Figure 2.12 – Variation of the function $g(t)$ for two values of B_N.

The term B_N allows the time spreading of function $g(t)$ to be fixed. Thus for $B_N = 0.2$, this function is approximately of width $4T$ whereas its width is only $3T$ for $B_N = 0.3$. When B_N tends towards infinity, it becomes a recatngular pulse with width T (the case of MSK modulation). GMSK modulation is therefore a partial response continuous phase modulation ($L > 1$).

On the interval $[iT, (i+1)T[$, the phase $\phi(t)$ of the GMSK signal is equal to:

$$\phi(t) = \pi \sum_{n=i-L+1}^{i} a_n q(t - nT) + \frac{\pi}{2} \sum_{n=-\infty}^{i-L} a_n \qquad (2.55)$$

where $L = 3$ if $B_N = 0.3$ and $L = 2$ if $B_N = 0.2$.
Thus on an interval $[iT, (i+1)T[$, the phase $\phi(t)$ of the GMSK signal depends on symbol a_i but also on the symbols prior to symbol a_i $(a_{i-1}, a_{i-2}, \cdots, a_{i-L+1})$. This non-linear modulation presents an memory effect that gives it good spectral properties. GMSK modulation, with a normalized passband $B_N = 0.3$ was

chosen for the GSM (Groupe Spécial Mobile and later *Global System for Mobile communications*) system. We note that there is no simple expression of the power spectral density of a GMSK signal. For values of the normalized passband of 0.3 or of 0.2, the power spectral density of the GMSK signal does not show sidelobes and its decrease as a function of frequency is very rapid. Thus at -10 dB the band occupied by the GMSK signal is approximately 200 kHz, and at -40 dB 400 kHz for a rate $D = 271$ kbit/s.

2.2 Structure and performance of the optimal receiver on a Gaussian channel

The object of this chapter is to determine the structure and the performance of the optimal receiver for memory and memoryless modulations on an additive white Gaussian noise (AWGN) channel. The type of receiver considered is the *coherent receiver* where the frequency and the phase of the signals transmitted by the modulator are assumed to be known by the receiver. Indeed, a coherent receiver is capable of locally generating signals having the same frequency and the same phase as those provided by the modulator, unlike the so-called *non-coherent* or *differential receiver.*

Generally, the receiver is made up of first a demodulator, the aim of which is to convert the modulated signal into a baseband signal, and second, a decision circuit in charge of estimating the blocks of data transmitted. The receiver is optimal in the sense that it guarantees a minimal error probability on the estimated blocks of data.

2.2.1 Structure of the coherent receiver

Let $s_j(t)$, $j = 1, 2, \cdots, M$ be the signals transmitted on the transmission channel perturbed by a AWGN $b(t)$, with zero mean and of power spectral density equal to $N_0/2$. On the time interval $[0, T[$, the signal received by the receiver is equal to:

$$r(t) = s_j(t) + b(t)$$

The M $s_j(t)$ signals define a space of dimension $N \leq M$, and can be represented in the form of a series of normed and orthogonal weighted functions $\nu_p(t)$.

$$s_j(t) = \sum_{p=1}^{N} s_{jp}\nu_p(t)$$

where s_{jp} is a scalar equal to the projection of the signal $s_j(t)$ on the function $\nu_p(t)$.

$$s_{jp} = \int_0^T s_j(t)\nu_p(t)dt$$

The noise can also be represented in the form of a series of normed and orthogonal functions but of infinite length (Karhunen Loeve expansion). When the noise is white, we show that the normed and orthogonal functions can be chosen arbitrarily. We are therefore going to take the same orthonormed functions as those used to represent the $s_j(t)$ signals, but after extension to infinity of this base of functions:

$$b(t) = \sum_{p=1}^{\infty} b_p \nu_p(t) = \sum_{p=1}^{N} b_p \nu_p(t) + b'(t)$$

where b_p is a scalar equal to the projection of $b(t)$ on the function $\nu_p(t)$.

$$b_p = \int_0^T b(t)\nu_p(t)dt$$

The quantities b_p are random non-correlated Gaussian variables, with zero mean and variance $\sigma^2 = N_0/2$.

$$\mathrm{E}\left\{b_p b_n\right\} = \int_0^T\!\!\int_0^T \mathrm{E}\left\{b(t)b(t')\right\}\nu_p(t)\nu_n(t')dtdt'$$

The noise being white, $\mathrm{E}\left\{b(t)b(t')\right\} = \frac{N_0}{2}\delta(t-t')$ and thus:

$$\mathrm{E}\left\{b_p b_n\right\} = \frac{N_0}{2}\int_0^T \nu_p(t)\nu_n(t)dt = \frac{N_0}{2}\delta_{n,p} \qquad (2.56)$$

where $\delta_{n,p}$ is the Kronecker symbol, equal to 1 if $n = p$ and to 0 if $n \neq p$.

Using the representations of the $s_j(t)$ signals and of the $b(t)$ noise by their respective series, we can write:

$$r(t) = \sum_{p=1}^{N}(s_{jp} + b_p)\nu_p(t) + \sum_{p=N+1}^{\infty} b_p\nu_p(t) = \sum_{p=1}^{N} r_p\nu_p(t) + b'(t)$$

Conditionally to the emission of the signal $s_j(t)$, the quantities r_p are random Gaussian variables, with mean and variance $N_0/2$. They are non-correlated to the noise $b'(t)$. Indeed, we have:

$$\mathrm{E}\left\{r_p b'(t)\right\} = \mathrm{E}\left\{(s_{jp} + b_p)\sum_{n=N+1}^{\infty} b_n\nu_n(t)\right\} \quad \forall p = 1, 2, \cdots, N$$

Taking into account the fact that the variables b_n, whatever n is, are zero mean and non-correlated, we obtain:

$$\mathrm{E}\left\{r_p b'(t)\right\} = \sum_{n=N+1}^{\infty} \mathrm{E}\left\{b_p b_n\right\} \nu_n(t) = 0 \quad \forall p = 1,\, 2,\, \cdots,\, N \tag{2.57}$$

The quantities r_p and the noise $b'(t)$ are therefore independent since Gaussian.

In conclusion, the optimal receiver can base its decision only on the quantities r_p, $p = 1,\, 2,\, \cdots,\, N$ with:

$$r_p = \int_0^T r(t)\nu_p(t)dt \tag{2.58}$$

Passing the signal $r(t)$ provided by the transmission channel to the N quantities r_p is called demodulation.

Example

Let us consider an M-PSK modulation for which the $s_j(t)$ signals are of the form:

$$s_j(t) = Ah(t)\cos(2\pi f_0 t + \varphi_0 + \phi_j)$$

The signals $s_j(t)$ define a space with $N = 2$ dimensions if $M > 2$. The normed and orthogonal functions $\nu_p(t)$, $p = 1,\, 2$ can be expressed respectively as:

$$\begin{aligned} \nu_1(t) &= \sqrt{\tfrac{2}{T}}\cos(2\pi f_0 t + \varphi_0) \\ \nu_2(t) &= \sqrt{\tfrac{2}{T}}\sin(2\pi f_0 t + \varphi_0) \end{aligned}$$

and the signals $s_j(t)$ can be written:

$$s_j(t) = A\sqrt{\frac{T}{2}}\cos\phi_j h(t)\nu_1(t) - A\sqrt{\frac{T}{2}}\sin\phi_j h(t)\nu_2(t)$$

After demodulation, the observation $R = (r_1, r_2)$ is equal to:

$$r_1 = A\sqrt{\frac{T}{2}}\cos\phi_j + b_1 \quad r_2 = A\sqrt{\frac{T}{2}}\sin\phi_j + b_2$$

The observation $R = (r_1, r_2)$ depends only on the states of phase ϕ_j and on the noise. We say that observation $R = (r_1, r_2)$ is in *baseband* since independent of the carrier frequency f_0.

The demodulation operation requires knowledge of the frequency f_0 and the phase φ_0 of the carrier, the signals $\nu_p(t)$ having to be synchronous with the carrier generated by the modulator. That is the reason why we speak of synchronous demodulation or coherent demodulation.

The N integrators of the demodulator can be replaced by N $h(T-t)$ impulse response filters, each followed by a sampler at time $t = T$.

$$s_j(t)\nu_j(t) * h(T-t) = \int_{-\infty}^{+\infty} s_j(\tau)\nu_j(\tau)h(T-t+\tau)d\tau$$

where $*$ represents the convolution product.
Sampling at $t = T$, we obtain:

$$s_j(t)\nu_j(t) * h(T-t)\,|_{t=T} = \int_{0}^{T} s_j(\tau)\nu_j(\tau)d\tau$$

which is equal to the output of the integrator.
The filter $h(T-t)$ is called the *filter matched* to waveform $h(t)$ of width T. We can show that this filter maximizes the signal to noise ratio at its output at time $t = T$.

For a continuous data stream, the integration is performed on each interval $[iT,\ (i+1)T[\ i = 1,\ 2,\ \cdots$ and, if we use matched filters, the sampling is realized at time $(i+1)T$.

After demodulation, the receiver must take a decision about the group of data transmitted on each time interval $[iT,\ (i+1)T[$. To do this, it searches for the most probable signal $s_j(t)$ by using the maximum *a posteriori* (MAP) probability criterion:

$$\hat{s}_j(t) \quad \text{if} \quad \Pr\{s_j(t)\,|\,R\} > \Pr\{s_p(t)\,|\,R\} \quad \forall p \neq j \quad p = 1,\ 2,\ \cdots,\ M$$

where $\hat{s}_j(t)$ is the signal that was transmitted and $R = (r_1 \cdots r_p \cdots r_N)$ the output of the demodulator. To simplify the notations, the time reference has been omitted for the components of observation R. $\Pr\{s_j(t)/R\}$ denotes the probability of $s_j(t)$ conditionally to the knowledge of observation R.

Using Bayes' rule, the MAP criterion can again be written:

$$\hat{s}_j(t) \quad \text{if} \quad \pi_j p(R|s_j(t)) > \pi_p p(R|s_p(t)) \quad \forall p \neq j \quad p = 1,\ 2,\ \cdots,\ M$$

where $\pi_j = \Pr\{s_j(t)\}$ represents the *a priori* probability of transmitting the signal $s_j(t)$ and $p(R|s_j(t))$ is the probability density of observation R conditionally to the emission of the signal $s_j(t)$ by the modulator.

Taking into account the fact that the components $r_p = s_{jp} + b_p$ of observation R conditionally to the emission of the signal $s_j(t)$ are non-correlated Gaussian, with mean s_{jp} and variance $N_0/2$, we can write:

$$\hat{s}_j(t) \quad \text{if} \quad \pi_j \prod_{p=1}^{N} p(r_p|s_j(t)) > \pi_n \prod_{p=1}^{N} p(r_p|s_n(t))$$
$$\forall n \neq j \quad p = 1,\ 2,\ \cdots,\ M$$

Replacing the probability densities by their respective expression we obtain:

$$\hat{s}_j(t) \quad \text{if} \quad \pi_j \left(\frac{1}{\sqrt{\pi N_0}}\right)^N \exp\left(-\frac{1}{N_0}\sum_{p=1}^{N}(r_p - s_{jp})^2\right)$$
$$> \pi_n \left(\frac{1}{\sqrt{\pi N_0}}\right)^N \exp\left(-\frac{1}{N_0}\sum_{p=1}^{N}(r_p - s_{np})^2\right)$$

After simplification:

$$\hat{s}_j(t) \quad \text{if} \quad \sum_{p=1}^{N} r_p s_{jp} + C_j > \sum_{p=1}^{N} r_p s_{np} + C_n \ \forall n \neq j \ n = 1, 2, \cdots, M \tag{2.59}$$

where $C_j = \frac{N_0}{2}\ln(\pi_j) - \frac{E_j}{2}$ with $E_j = \sum_{p=1}^{N}(s_{jp})^2$.

Noting that:

$$\int_0^T r(t)s_j(t)dt = \int_0^T \sum_{p=1}^{N} r_p \nu_p(t) \sum_{m=1}^{N} s_{jm}\nu_m(t)dt$$

and recalling that the functions $\nu_p(t)$ are normed and orthogonal, we obtain:

$$\int_0^T r(t)s_j(t)dt = \sum_{p=1}^{N} r_p s_{jp}$$

In the same way:

$$\int_0^T s_j^2(t)dt = \int_0^T \sum_{p=1}^{N} s_{jp}\nu_p(t) \sum_{m=1}^{N} s_{jm}\nu_m(t)dt$$

and finally:

$$\int_0^T s_j^2(t)dt = \sum_{p=1}^{N} s_{jp}^2$$

Taking into account the above, the MAP criterion can again be written in the form:

$$\hat{s}_j(t) \text{ if } \int_0^T r(t)s_j(t)dt + C_j > \int_0^T r(t)s_n(t)dt + C_n \quad \forall n \neq j \quad n = 1, 2, \cdots, M \tag{2.60}$$

where $C_j = \frac{N_0}{2}\ln(\pi_j) - \frac{E_j}{2}$ with $E_j = \int_0^T s_j^2(t)dt$.

If all the $s_j(t)$ signals are transmitted with the same probability ($\pi_j = 1/M$), the term C_j reduces to $-E_j/2$. In addition, if all the $s_j(t)$ signals have in addition the same energy $E_j = E$ (the case of phase or frequency shift keying modulation), then the MAP criterion is simplified and becomes:

$$\hat{s}_j(t) \text{ if } \int_0^T r(t)s_j(t)dt > \int_0^T r(t)s_n(t)dt \quad \forall n \neq j \quad n = 1, 2, \cdots, M \tag{2.61}$$

2.2.2 Performance of the coherent receiver

Amplitude shift keying with M states

For an M-ASK modulation, the $s_j(t)$ signals are of the form:

$$s_j(t) = A_j h(t)\cos(2\pi f_0 t + \varphi_0)$$

with:

$$A_j = (2j - 1 - M)A \quad j = 1, 2, \cdots, M$$

They define a space of dimension $N = 1$ and thus observation R at the output of the demodulator reduces to a component r.

$$r = \int_0^T r(t)\nu(t)dt$$

with $\nu(t) = \sqrt{\frac{2}{T}}\cos(2\pi f_0 t + \varphi_0)$.

For a continuous data stream, the estimation of the symbols a_i is done by integrating the product $r(t)\nu(t)$ on each time interval $[iT, (i+1)T[$. If a matched filter is used rather than an integrator, the sampling at the output of the filter is realized at time $(i+1)T$.

On the interval $[0, T[$, assuming *iid* information data d_i, all the amplitude states have the same probability and decision rule (2.59) leads to:

$$\hat{A}_j \text{ if } rs_j - \frac{1}{2}s_j^2 > rs_n - \frac{1}{2}s_n^2 \quad \forall n \neq j \tag{2.62}$$

where $\hat{A}_j$ represents the amplitude estimated with:

$$s_j = \int_0^T A_j \cos(2\pi f_0 t + \varphi_0)\nu(t)dt = A_j\sqrt{\frac{T}{2}} \text{ if } f_0 >> \frac{1}{T} \tag{2.63}$$

The coherent receiver, shown in Figure 2.13, takes its decision by comparing observation r to a set of $(M-1)$ thresholds of the form:

$$\begin{aligned} -(M-2)A\sqrt{\tfrac{T}{2}}, \cdots, -2pA\sqrt{\tfrac{T}{2}}, \cdots, -2A\sqrt{\tfrac{T}{2}}, 0, \\ 2A\sqrt{\tfrac{T}{2}}, \cdots, 2pA\sqrt{\tfrac{T}{2}}, \cdots, (M-2)A\sqrt{\tfrac{T}{2}} \end{aligned} \tag{2.64}$$

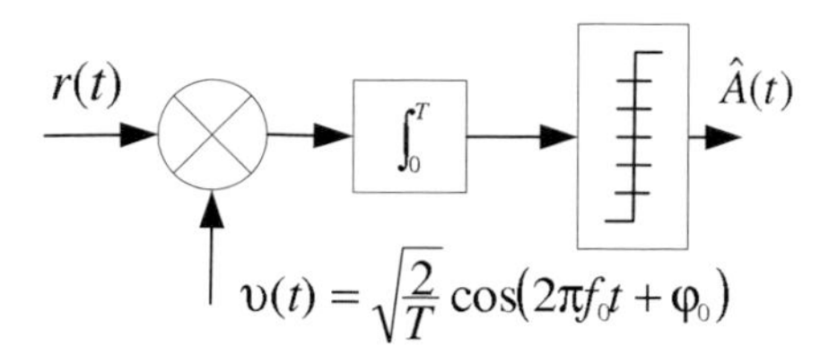

Figure 2.13 – Coherent receiver for M-ASK modulation.

Example

Consider a 4-ASK modulation, the three thresholds being $-2A\sqrt{\frac{T}{2}}$, 0, $2A\sqrt{\frac{T}{2}}$. The decisions are the following:

$$\begin{aligned} \hat{A}_j &= -3A \quad \text{if} \quad r < -2A\sqrt{\tfrac{T}{2}} \\ \hat{A}_j &= -A \quad \text{if} \quad -2A\sqrt{\tfrac{T}{2}} < r < 0 \\ \hat{A}_j &= A \quad \text{if} \quad 0 < r < 2A\sqrt{\tfrac{T}{2}} \\ \hat{A}_j &= 3A \quad \text{if} \quad r > 2A\sqrt{\tfrac{T}{2}} \end{aligned}$$

The emission of a state of amplitude A_j corresponds to the transmission of a group of $\log_2(M)$ binary data d_i. The error probability on a group of data can be calculated as the average value of the conditional error probabilities Pe_{2j-1-M} given by:

$$Pe_{2j-1-M} = \Pr\left\{\hat{A}_j \neq (2j-1-M)A \,|\, A_j = (2j-1-M)A\right\}$$

The mean error probability on the symbols, denoted Pes, is equal to:

$$Pes = \frac{1}{M}\sum_{j=1}^{M} Pe_{2j-1-M}$$

The conditional error probabilities can be classified into two types. The first type corresponds to the probabilities that the observation is higher or is lower than a certain threshold and the second type, to the probabilities that the observation does not fall between two thresholds.

TYPE 1: Probabilities that the observation is higher or is lower than a threshold

$$\begin{aligned} Pe_{(M-1)} &= \Pr\left\{\hat{A}_j \neq (M-1)A \,|\, A_j = (M-1)A\right\} \\ &= \Pr\left\{r < (M-2)A\sqrt{\tfrac{T}{2}} \,|\, A_j = (M-1)A\right\} \end{aligned}$$

$$\begin{aligned} Pe_{-(M-1)} &= \Pr\left\{\hat{A}_j \neq -(M-1)A \,|\, A_j = -(M-1)A\right\} \\ &= \Pr\left\{r > -(M-2)A\sqrt{\tfrac{T}{2}} \,|\, A_j = -(M-1)A\right\} \end{aligned}$$

TYPE 2: Probabilities that the observation does not fall between two thresholds

$$Pe_{2j-1-M} = \Pr\left\{\hat{A}_j \neq (2j-1-M)A \,|\, A_j = (2j-1-M)A\right\}$$

$$Pe_{2j-1-M} = 1- \Pr\left\{(2j-2-M)A\sqrt{\tfrac{T}{2}} < r < (2j-M)A\sqrt{\tfrac{T}{2}} \,| \right. \\ \left. A_j = (2j-1-M)A\right\}$$

Observation r is Gaussian conditionally to a realization of the amplitude A_j, with mean $\pm A_j\sqrt{T/2}$ and variance $N_0/2$. The conditional probabilities have the expressions:

$$Pe_{(M-1)} = Pe_{-(M-1)} = \frac{1}{2}\text{erfc}\sqrt{\frac{A^2T}{2N_0}}$$

$$Pe_{(2j-1-M)} = \text{erfc}\sqrt{\frac{A^2T}{2N_0}}$$

where the complementary error function is always defined by:

$$\text{erfc}(x) = 1 - \text{erf}(x) = \frac{2}{\sqrt{\pi}}\int_x^{+\infty} \exp(-u^2)du$$

To calculate the mean error probability on the groups of data, we have two conditional probabilities of type 1, and $(M-2)$ conditional probabilities of type 2.

$$Pes = \frac{M-1}{M}\text{erfc}\sqrt{\frac{A^2T}{2N_0}}$$

Introducing the average energy $E_s = \frac{A^2T}{2}\frac{(M^2-1)}{3}$ received by group of data, the mean error probability is again equal to:

$$Pes = \frac{M-1}{M}\text{erfc}\sqrt{\frac{3}{M^2-1}\frac{E_s}{N_0}}$$

The symbol error probability Pes can also be expressed as a function of the average energy $E_b = E_s/\log_2(M)$ received by transmitted bit d_i.

$$Pes = \frac{M-1}{M}\operatorname{erfc}\sqrt{\frac{3\log_2(M)}{M^2-1}\frac{E_b}{N_0}}$$

or again as a function of the received mean power P and of the transmitted bit rate $D = 1/T_b$:

$$Pes = \frac{M-1}{M}\operatorname{erfc}\sqrt{\frac{3\log_2(M)}{M^2-1}\frac{P}{N_0 D}} \tag{2.65}$$

Figure 2.14 provides the mean error probability Pes as a function of the signal to noise ratio E_b/N_0 for different values of the parameter M.

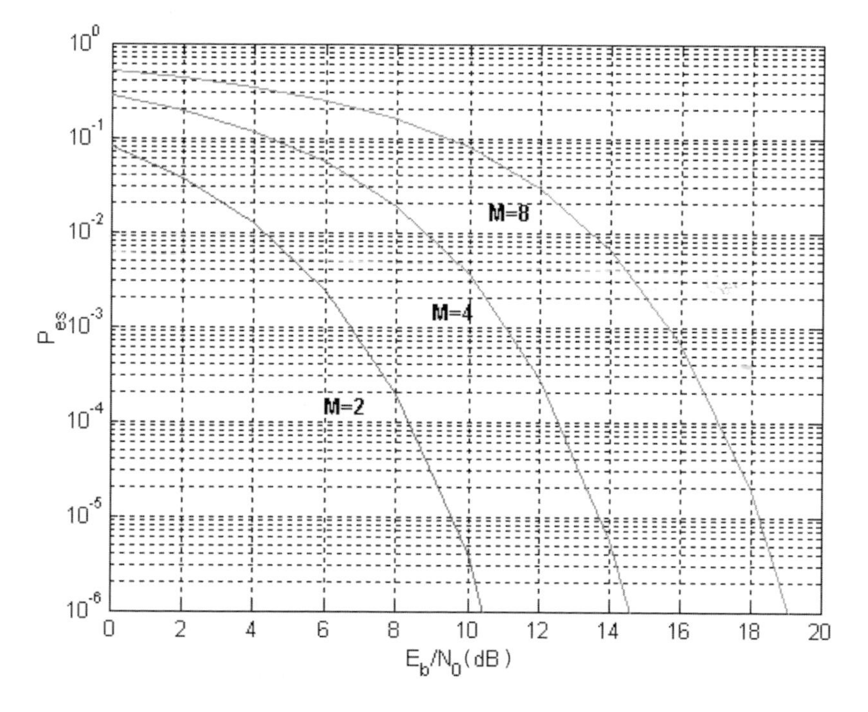

Figure 2.14 – Mean error probability Pes as a function of the signal to noise ratio E_b/N_0 for different values of parameter M of an M-ASK modulation.

The bit error probability Peb can be deduced from the mean error probability Pes in the case where Gray coding is used and under the hypothesis of a sufficiently high signal to noise ratio. Indeed, in this case we generally have an erroneous bit among the $\log_2(M)$ data transmitted. (We assume that the amplitude of the received symbol has a value immediately lower or higher than the value of the transmitted amplitude).

$$Peb \cong \frac{Pes}{\log_2(M)} \quad \text{if} \quad \frac{E_b}{N_0} >> 1 \tag{2.66}$$

Phase shift keying with M states

For M-PSK modulation , the $s_j(t)$ signals are of the form:

$$s_j(t) = A\cos(2\pi f_0 t + \varphi_0 + \phi_j) \tag{2.67}$$

with:

$$\phi_j = (2j+1)\frac{\pi}{M} + \theta_0 \quad j = 0,\, 1,\, \cdots,\, (M-1)$$

The $s_j(t)$ signals, for $M > 2$, define a two-dimensional space. Observation R at the output of the demodulator is therefore made up of two components (r_1, r_2) with:

$$r_1 = \int_0^T r(t)\nu_1(t)dt \quad r_2 = \int_0^T r(t)\nu_2(t)dt$$

where $\nu_1(t) = \sqrt{\frac{2}{T}}\cos(2\pi f_0 t + \varphi_0)$ et $\nu_2(t) = -\sqrt{\frac{2}{T}}\sin(2\pi f_0 t + \varphi_0)$.
Using decision rule (2.50) and assuming the information data *iid*, all the states of phase have the same probability and the decision is the following:

$$\hat{\phi}_j \quad \text{if} \quad \sum_{p=1}^{2} r_p s_{jp} > \sum_{p=1}^{2} r_p s_{np} \quad \forall n \neq j \tag{2.68}$$

with:

$$s_{j1} = A\sqrt{\frac{T}{2}}\cos\phi_j \quad \text{and} \quad s_{j2} = A\sqrt{\frac{T}{2}}\sin\phi_j \quad \text{if} \quad f_0 >> \frac{1}{T} \tag{2.69}$$

Taking into account the expressions of s_{j1} and of s_{j2}, the decision rule can again be written:

$$\hat{\phi}_j \quad \text{if} \quad r_1\cos\phi_j + r_2\sin\phi_j > r_1\cos\phi_n + r_2\sin\phi_n \quad \forall n \neq j \tag{2.70}$$

The coherent receiver for an M-PSK modulation is represented in Figure 2.15. It is made up of two components called the phase component (projection of the received signal on $\nu_1(t) = \sqrt{2/T}\cos(2\pi f_0 t + \varphi_0)$) and the quadrature component (projection of the received signal on $\nu_2(t) = \sqrt{2/T}\sin(2\pi f_0 t + \varphi_0)$) and a decision circuit.

The emission by the modulator of a phase state corresponds to the transmission of a group of $\log_2(M)$ information bit. The error probability on a group of

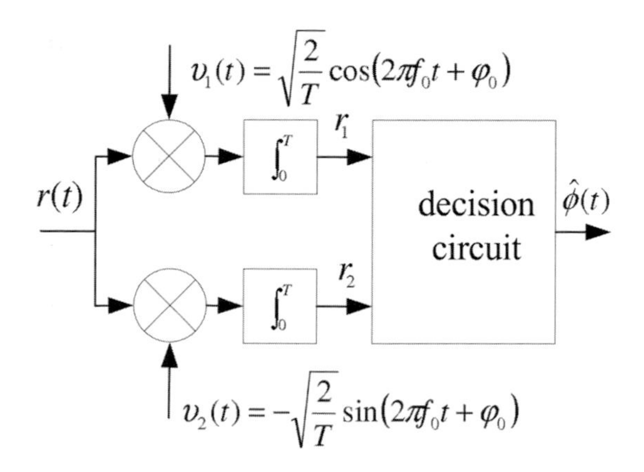

Figure 2.15 – Coherent receiver for M-PSK modulation.

binary data, whatever the value of M, does not have an analytical expression. However, at high signal to noise ratios, this probability is well approximated by the following expression:

$$Pes \cong \text{erfc}\left[\sqrt{\log_2(M)\frac{E_b}{N_0}}\sin\frac{\pi}{M}\right] \quad \text{if} \quad \frac{E_b}{N_0} >> 1 \tag{2.71}$$

Noting that $E_b = PT_b$ and $D = 1/T_b$, the relation E_b/N_0 is again equal to P/N_0D where P is the received power of the modulated signal.

For Gray coding, the bit error probability with a high signal to noise ratio is equal to:

$$Peb \cong \frac{Pes}{\log_2(M)} \quad \text{if} \quad \frac{E_b}{N_0} >> 1 \tag{2.72}$$

Case of 2-PSK modulation

For this modulation, phase ϕ_j takes the values 0 or π. Each phase state is therefore associated with a bit. Adopting the following coding:

$$\phi_j = 0 \rightarrow d_i = 1 \quad \phi_j = \pi \rightarrow d_i = 0$$

the decision rule for 2-PSK modulation is simple:

$$\hat{d}_i = 1 \quad \text{if} \quad r_1 > 0 \qquad \hat{d}_i = 0 \quad \text{if} \quad r_1 < 0 \tag{2.73}$$

Observation r_2 is not used for decoding the data d_i since the space defined by the signals modulated with two phase states has dimension $N = 1$.

For 2-PSK modulation, there is an exact expression of the bit error probability Peb. Assuming the binary data *iid*, this error probability is equal to:

$$Peb = \frac{1}{2}\Pr\{r_1 > 0|\phi_j = \pi\} + \frac{1}{2}\Pr\{r_1 < 0|\phi_j = 0\}$$

Output r_1 of the demodulator is:

$$r_1 = \pm\sqrt{E_b} + b$$

where $E_b = A^2T/2$ is the energy received per information bit transmitted and b is an AWGN, with zero mean and variance equal to $N_0/2$.

$$Peb = \tfrac{1}{2}\tfrac{1}{\sqrt{\pi N_0}}\int_0^{\infty}\exp(-\tfrac{1}{N_0}(r_1 + \sqrt{E_b})^2)dr_1$$
$$+\tfrac{1}{2}\tfrac{1}{\sqrt{\pi N_0}}\int_{-\infty}^{0}\exp\left(-\tfrac{1}{N_0}(r_1 - \sqrt{E_b})^2\right)dr_1$$

Introducing the complementary error function, the bit error probability Peb is equal to:

$$Peb = \frac{1}{2}\mathrm{erfc}\sqrt{\frac{E_b}{N_0}} \tag{2.74}$$

Case of 4-PSK modulation

For this modulation, phase ϕ_j takes four values $\pi/4$, $3\pi/4$, $5\pi/4$, $7\pi/4$. With each state of the phase are associated two binary data. For equiprobable phase states, the MAP criterion leads to the following decision rules:

$$\hat{\phi}_j = \tfrac{\pi}{4} \quad \text{if} \quad r_1 > 0;\, r_2 > 0$$

$$\hat{\phi}_j = \tfrac{3\pi}{4} \quad \text{if} \quad r_1 < 0;\, r_2 > 0$$

$$\hat{\phi}_j = \tfrac{5\pi}{4} \quad \text{if} \quad r_1 < 0;\, r_2 < 0$$

$$\hat{\phi}_j = \tfrac{7\pi}{4} \quad \text{if} \quad r_1 > 0;\, r_2 < 0$$

Considering the following Gray coding:

$$\frac{\pi}{4} \to 11 \quad \frac{3\pi}{4} \to 01 \quad \frac{5\pi}{4} \to 00 \quad \frac{7\pi}{4} \to 10$$

The estimation of the binary data can be performed by separately comparing outputs r_1 and r_2 of the demodulator to a threshold fixed to zero. The coherent receiver for a 4-PSK modulation is represented in Figure 2.16.

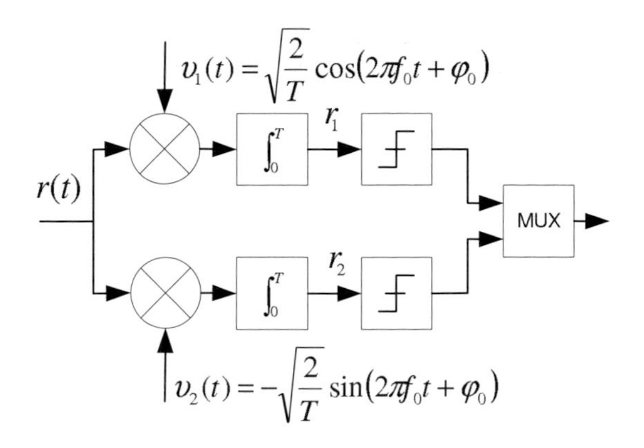

Figure 2.16 – Coherent receiver for 4-PSK modulation.

An exact expression of *Peb* can however be given by observing simply that the coherent receiver for a 4-PSK modulation is made up of two components identical to that of a 2-PSK receiver. The bit error probability *Peb* for a 4-PSK modulation is thus the same as for a modulation 2-PSK, that is:

$$Peb = \frac{1}{2}\text{erfc}\sqrt{\frac{E_b}{N_0}} \tag{2.75}$$

The symbol error probability *Pes* is equal to:

$$Pes = 1 - (1 - Peb)^2$$

For high signal to noise ratios, the error probability *Peb* is much lower than unity and thus, for 4-PSK modulation:

$$Pes = 2Peb \quad \text{if} \quad \frac{E_b}{N_0} >> 1 \tag{2.76}$$

Figure 2.17 shows the error probability *Pes* as a function of the relation E_b/N_0 for different values of the parameter M[1].

Amplitude modulation on two quadrature carriers – M-QAM

For M-QAM modulation, the $s_j(t)$ signals provided by the modulator are of the form:

$$s_j(t) = A_{jc}h(t)\cos(2\pi f_0 t + \varphi_0) - A_{js}h(t)\sin(2\pi f_0 t + \varphi_0) \tag{2.77}$$

Two situations can arise depending whether the length m of the groups of data at the input of the modulator is even or not. When $M = 2^m$ with even m,

[1] For $M = 2$, it is the exact relation 2.74 that is used since $Pes = Peb$, for $M > 2$, Pes is provided by Equation (2.71).

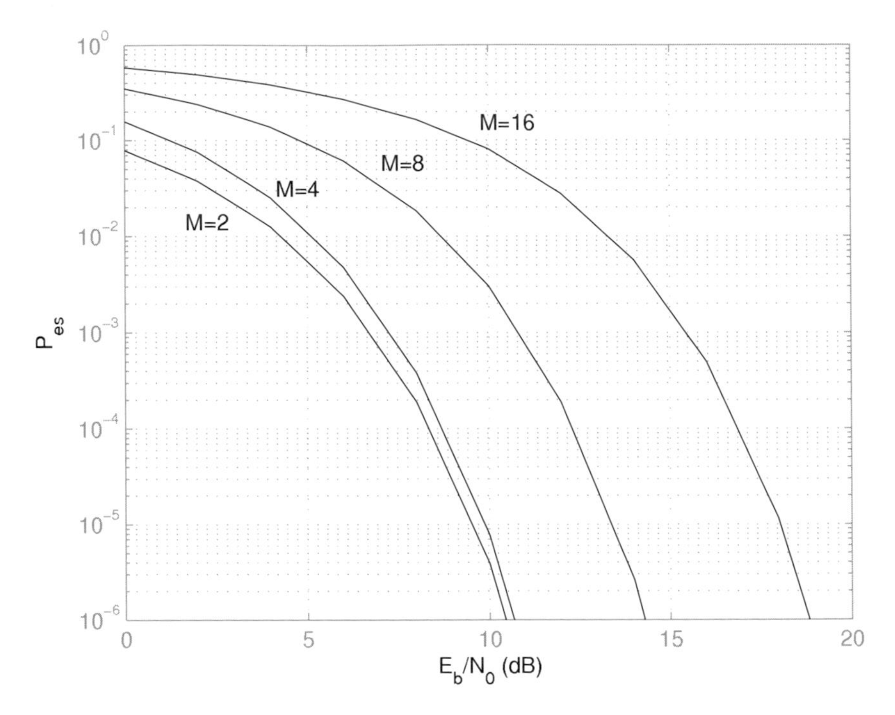

Figure 2.17 – Error probability Pes as a function of the relation E_b/N_0 for different values of M of an M-PSK modulation.

the group of data can be separated into two sub-groups of length $m/2$, each sub-group being associated respectively with amplitudes A_{jc} and A_{js}, with:

$$\begin{aligned} A_{jc} &= (2j - 1 - \sqrt{M})A \quad j = 1, 2, \cdots, \sqrt{M} \\ A_{js} &= (2j - 1 - \sqrt{M})A \quad j = 1, 2, \cdots, \sqrt{M} \end{aligned} \tag{2.78}$$

In this case, the M-QAM modulation is equivalent to two $\sqrt{M}$-ASK modulations having quadrature carriers. The coherent receiver for an M-QAM modulation is made up of two components called phase and quadrature, and each component, similar to a receiver for modulation $\sqrt{M}$-ASK, performs the estimation of a group of $m/2$ binary data. The receiver is shown in Figure 2.18.

The error probability on a group of $m/2$ binary data is equal to the error probability of an ASK$-\sqrt{M}$ modulation, that is:

$$Pe_{m/2} = \frac{\sqrt{M}-1}{\sqrt{M}} \text{erfc} \sqrt{\frac{3\log_2(\sqrt{M})}{M-1} \frac{E_b}{N_0}} \tag{2.79}$$

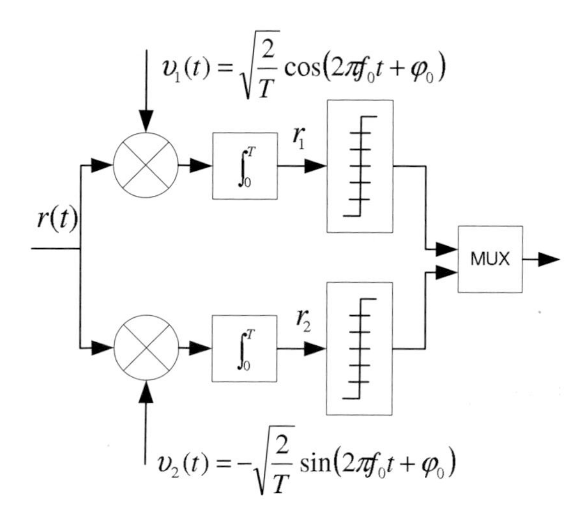

Figure 2.18 – Coherent receiver for M-QAM modulation with $M = 2^m$ and even m.

The symbol error probability Pes on the group of m binary data is therefore equal to:

$$Pes = 1 - (1 - Pe_{m/2})^2$$

When m is odd, the probability of error Pes on a group of m binary data can be upper bound by the following expression:

$$Pes \leq 2\text{erfc}\sqrt{\frac{3\log_2(\sqrt{M})}{M-1}\frac{E_b}{N_0}} \tag{2.80}$$

The bit error probability Peb can be deduced from Pes if Gray coding is used and for a sufficiently high signal to noise ratio:

$$Peb = \frac{Pes}{\log_2(M)} \tag{2.81}$$

For the values of $M \geq 8$, the performance of M-PSK and M-QAM modulations can easily be compared. Indeed, by making the following approximation:

$$\sin\left(\frac{\pi}{M}\right) \cong \frac{\pi}{M} \quad \text{if} \quad M \geq 8$$

the bit error probability for an M-PSK modulation can be written:

$$Peb = \frac{1}{\log_2(M)}\text{erfc}\sqrt{\log_2(M)\frac{\pi^2}{M^2}\frac{E_b}{N_0}} \quad \text{if} \quad \frac{E_b}{N_0} >> 1 \tag{2.82}$$

Neglecting the coefficients that weight the complementary error functions, M-PSK modulation requires the relation E_b/N_0 to be increased by:

$$10\log\left(\frac{3M^2}{2(M-1)\pi^2}\right)\text{ dB}$$

to obtain the same error probability as M-QAM modulation. If, for example, we compare the performance of 16-PSK modulation with that of 16-QAM modulation, we note that the former requires about 4 dB more for the relation E_b/N_0 to obtain the same error probabilities.

Frequency shift keying – M-FSK

For an M-FSK modulation, the modulator provides signals of the form:

$$s_j(t) = Ah(t)\cos(2\pi f_j t + \varphi_j) \tag{2.83}$$

where the frequencies f_j are chosen in such a way that the M $s_j(t)$ signals are orthogonal. The space defined by these signals therefore has dimension $N = M$ and the vectors $\nu_j(t)$ are of the form:

$$\nu_j(t) = \sqrt{\frac{2}{T}}\cos(2\pi f_j t + \varphi_j) \quad j = 1,\, 2,\, \cdots M \tag{2.84}$$

Assuming the information data d_i *iid*, the $s_j(t)$ signals have the same probability. In addition, they have the same energy E and thus, the MAP criterion leads to the following decision rule:

$$\hat{s}_j(t) \quad \text{if} \quad \sum_{p=1}^{M} r_p s_{jp} > \sum_{p=1}^{M} r_p s_{np} \quad \forall n \neq j \tag{2.85}$$

where s_{jp} is equal to:

$$s_{jp} = \int_0^T s_j(t)\nu_p(t)dt = A\sqrt{\frac{T}{2}}\delta_{jp} \tag{2.86}$$

Taking into account the expression of s_{jp}, decision rule (2.85) is simplified and becomes:

$$\hat{s}_j(t) \quad \text{if} \quad r_j > r_n \;\forall n \neq j \tag{2.87}$$

The optimal coherent receiver for an M-FSK modulation is shown in Figure 2.19.

Conditionally to the emission of the signal $s_j(t)$, the M outputs of the demodulator are of the form:

$$r_j = \sqrt{E_s} + b_j \quad r_p = b_p \quad \forall p \neq j$$

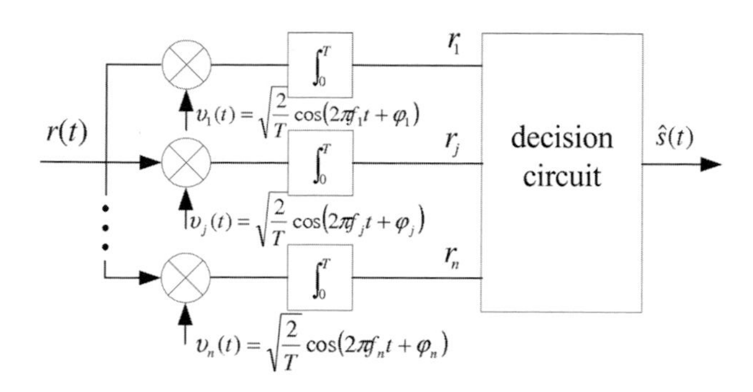

Figure 2.19 – Coherent receiver for M-FSK modulation.

where b_j and b_p are AWGN, with zero mean and variance equal to $N_0/2$. The probability of a correct decision on a group of binary data, conditionally to the emission of the signal $s_j(t)$ is equal to:

$$Pc_j = \int_{-\infty}^{+\infty} \Pr\{b_1 < r_j, \cdots, b_p < r_j, \cdots, b_M < r_j\}\, p(r_j)dr_j$$

The noises being non-correlated and therefore independent, since they are Gaussian, we have:

$$\Pr\{b_1 < r_j, \cdots, b_p < r_j, \cdots, b_M < r_j\} = \left(\int_{-\infty}^{r_j} \frac{1}{\sqrt{\pi N_0}} \exp\left(-\frac{b^2}{N_0}\right) db\right)^{M-1}$$

and thus the probability of a correct decision is equal to:

$$Pc_j = \int_{-\infty}^{+\infty} \left(\int_{-\infty}^{r_j} \frac{1}{\sqrt{\pi N_0}} \exp\left(-\frac{1}{N_0}b^2\right) db\right)^{M-1} \frac{1}{\sqrt{\pi N_0}} \exp\left(-\frac{1}{N_0}(r_j - \sqrt{E_s})^2\right) dr_j$$

After changing variables, the probability of a correct decision can be expressed as a function of the relation E_s/N_0.

$$Pc_j = \frac{1}{\sqrt{2\pi}} \int_{-\infty}^{+\infty} \left(\int_{-\infty}^{y} \frac{1}{\sqrt{2\pi}} \exp\left(-\frac{x^2}{2}\right) dx\right)^{M-1} \exp\left(-\frac{1}{2}(y - \sqrt{\frac{E_s}{N_0}})^2\right) dy \quad (2.88)$$

The probability of a correct decision is the same whatever the transmitted signal. The signals $s_j(t)$ being equiprobable, the mean probability of a correct decision Pc is therefore equal to the conditional probability Pc_j. The symbol error probability is then equal to:

$$Pes = 1 - Pc$$

The error probability can also be expressed as a function of the relation E_b/N_0 where E_b is the energy used to transmit a bit with $E_b = E_s/\log_2(M)$.

We can also try to determine the bit error probability Peb. All the $M-1$ groups of erroneous data appear with the same probability:

$$\frac{Pes}{M-1} \tag{2.89}$$

In a group of erroneous data, we can have k erroneous data among m and this can occur in $\begin{pmatrix} m \\ k \end{pmatrix}$ possible ways. Thus, the average number of erroneous data in a group is:

$$\sum_{k=1}^{m} k \binom{m}{k} \frac{Pes}{M-1} = m\frac{2^{m-1}}{2^m-1}Pes$$

and finally the bit error probability is equal to:

$$Peb = \frac{2^{m-1}}{2^m-1}Pes \tag{2.90}$$

where $m = \log_2(M)$.

The error probability for an M-FSK modulation does not have a simple expression and we have to resort to digital computation to determine this probability as a function of the relation E_b/N_0. We show that for a given error probability Peb, the relation E_b/N_0 necessary decreases when M increases. We also show that probability Pes tends towards a value arbitrarily small when M tends towards infinity, and for $E_b/N_0 = 4\ln 2$ dB, that is, -1.6 dB.
For a binary transmission ($M = 2$), there is an expression of the error probability Peb.
Let us assume that the signal transmitted is $s_1(t)$, we then have:

$$r_1 = \sqrt{E_b} + b_1 \quad r_2 = b_2$$

The decision can be taken by comparing $z = r_1 - r_2$ to a threshold fixed to zero. The error probability Peb_1 conditionally to the emission of $s_1(t)$, is equal to:

$$Peb_1 = \Pr\{z < 0 \,|\, s_1(t)\}$$

Assuming the two signals identically distributed, error probability Peb has the expression:

$$Peb = \frac{1}{2}(Peb_1 + Peb_2)$$

The noises b_1 and b_2 are non-correlated Gaussian, with zero mean and same variance equal to $N_0/2$. The variable z, conditionally to the emission of the signal $s_1(t)$, is therefore Gaussian, with mean $\sqrt{E_b}$ and variance N_0. Thus probability Peb_1 is equal to:

$$Peb_1 = \int_{-\infty}^{0} \frac{1}{\sqrt{2\pi N_0}} \exp\left(-\frac{(z-\sqrt{E_b})^2}{2N_0}\right) dz$$

Introducing the complementary error function, Peb_1 is written:

$$Peb_1 = \frac{1}{2}\operatorname{erfc}\sqrt{\frac{E_b}{2N_0}}$$

It is easy to check that the error probability conditionally to the emission of signal $s_2(t)$ is identical to the error probability conditionally to the emission of signal $s_1(t)$ and thus, we obtain:

$$Peb = \frac{1}{2}\operatorname{erfc}\sqrt{\frac{E_b}{2N_0}} \tag{2.91}$$

If we compare the performance of 2-FSK modulation to that of 2-PSK we note that the former requires 3 dB more signal to noise ratio to obtain the same performance as the latter.

Continuous phase frequency shift keying – CPFSK

For continuous phase frequency shift keying, the modulated signal has the expression:

$$S(t) = A\cos(2\pi f_0 t + \phi(t)) \tag{2.92}$$

where the phase $\phi(t)$, on the interval $[iT, (i+1)T[$, is equal to :

$$\phi(t) = 2\pi h \sum_{n=i-L+1}^{i} a_n q(t-nT) + \theta_{i-L} \tag{2.93}$$

with:

$$\theta_{i-L} = \pi h \sum_{n=-\infty}^{i-L} a_n$$

h is the modulation index and the symbols a_i are M-ary in the general case. They take their values in the alphabet $\{\ \pm 1, \pm 3, \cdots, \pm(2p+1), \cdots, \pm(\ M\ -1\ )\ \}$; $M = 2^m$.
If the modulation index $h = m/p$ where m and p are relatively prime integers, phase θ_{i-L} takes its values in the following sets:

$$\begin{aligned} &\theta_{i-L} \in \left\{0, \tfrac{\pi m}{p}, \tfrac{2\pi m}{p}, \cdots, \tfrac{(p-1)\pi m}{p}\right\} \quad \text{if} \quad m \text{ even} \\ &\theta_{i-L} \in \left\{0, \tfrac{\pi m}{p}, \tfrac{2\pi m}{p}, \cdots, \tfrac{(2p-1)\pi m}{p}\right\} \quad \text{if} \quad m \text{ odd} \end{aligned} \tag{2.94}$$

The evolution of phase $\phi(t)$ can be represented by a trellis whose states are defined by $(a_{i-L+1}, a_{i-L+2}, \cdots, a_{i-1}; \theta_{i-L})$, that is:

$$\begin{aligned} &(M^{L-1}p) \text{ states} \quad \text{if } m \text{ even} \\ &(M^{L-1}2p) \text{ states} \quad \text{if } m \text{ odd} \end{aligned} \tag{2.95}$$

Note that the complexity of the trellis increases very rapidly with the parameters M and L. For example, for a modulation with quaternary symbols ($M = 4$) with a partial response with modulation index $h = 1/2$ and with parameter $L = 4$, the trellis has 256 states. For MSK and GMSK, the symbols a_i are

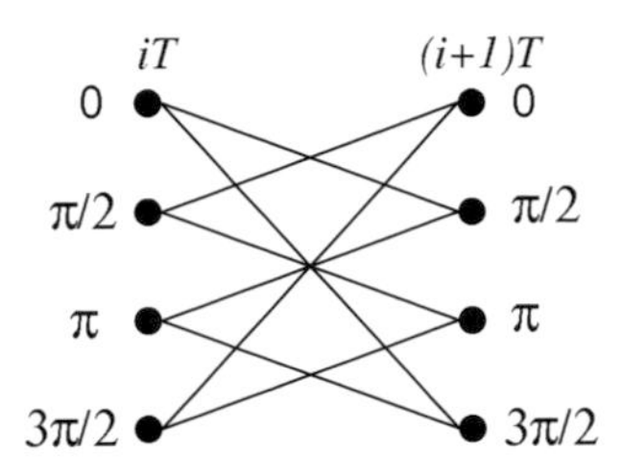

Figure 2.20 – Trellis associated with phase $\phi(t)$ for MSK modulation.

binary ($M = 2$) and the modulation index h is 1/2, that is, $m = 1$ and $p = 2$. Phase θ_{i-L} therefore takes its values in the set $\{0, \pi/2, \pi, 3\pi/2\}$ and the trellis associated with phase $\phi(t)$ has $2^{L-1} \times 4$ states. Figure 2.20 shows the trellis associated with phase $\phi(t)$ for MSK modulation.
To decode symbols a_i we use the Viterbi algorithm whose principle is recalled below. For each time interval $[iT, (i+1)T[$, proceed in the following way:

- for each branch l leaving a state of the trellis at instant iT calculate metric z_i^l as defined later, that is, for MSK and GMSK, $2^L \times 4$ metrics have to be calculated;
- for each path converging to instant $(i+1)T$ towards a state of the trellis, calculate the cumulated metric, then select the path with the largest cumulated metric, called the survivor path;
- among the survivor paths, trace back along s branches of the path having the largest cumulated metric and decode symbol a_{i-s};
- continue the algorithm on the following time interval.

Branch metric z_i^l has the expression:

$$z_i^l = \int_{iT}^{(i+1)T} r(t)\cos(2\pi f_0 t + \phi_i^l(t) + \varphi_0)dt$$

where $r(t) = s(t) + b(t)$ is the signal received by the receiver and $b(t)$ is a AWGN, with zero mean and power spectral density equal to $N_0/2$. Quantity $\phi_i^l(t)$ represents a realization of phase $\phi(t)$ associated with branch l of the trellis on time interval $[iT, (i+1)T[$.
Taking into account the fact that the noise can be put in the form $b(t) = b_c(t)\cos(2\pi f_0 t + \varphi_0) - b_s(t)\sin(2\pi f_0 t + \varphi_0)$ and that $f_0 >> 1/T$, the branch metric can again be written:

$$z_i^l = \int_{1T}^{(i+1)T} r_c(t)\cos\phi_i^l(t)dt + \int_{iT}^{(i+1)T} r_s(t)\sin\phi_i^l(t)dt \tag{2.96}$$

where the signals $r_c(t)$ and $r_s(t)$ are obtained after transposition into baseband of $r(t)$(multiplying $r(t)$ by $\cos(2\pi f_0 t + \varphi_0)$ and $-\sin(2\pi f_0 t + \varphi_0)$ respectively, then lowpass filtering).

$$\begin{aligned} \cos\phi_i^l(t) &= \cos\left(2\pi h \sum_{n=i-L+1}^{i} a_n^l q(t-nT) + \theta_{i-L}^l\right) \\ \sin\phi_i^l(t) &= \sin\left(2\pi h \sum_{n=i-L+1}^{i} a_n^l q(t-nT) + \theta_{i-L}^l\right) \end{aligned} \tag{2.97}$$

Putting:

$$\psi_i^l(t) = 2\pi h \sum_{n=i-L+1}^{i} a_n^l q(t-nT)$$

branch metric z_i^l can again be written in the form:

$$z_i^l = \cos\theta_{i-L}^l A_l + \sin\theta_{i-L}^l B_l \tag{2.98}$$

with:

$$A_i^l = \int_{iT}^{(i+1)T} (r_c(t)\cos\psi_i^l(t) + r_s(t)\sin\psi_i^l(t))dt$$

$$B_i^l = \int_{iT}^{(i+1)T} (r_s(t)\cos\psi_i^l(t) - r_c(t)\sin\psi_i^l(t))dt$$

For MSK modulation, it is possible to decode symbols a_i by using a receiver similar to that of 4-PSK modulation. Indeed, the *MSK* signal can be written in the following form:

$$\begin{aligned} S(t) = A\Big[&\sum_i c_{2i-1}h(t-2iT)\cos\tfrac{\pi t}{2T}\cos(2\pi f_0 t + \varphi_0) \\ &- \sum_i c_{2i}h(t-(2i+1)T)\sin\tfrac{\pi t}{2T}\sin(2\pi f_0 t + \varphi_0)\Big] \end{aligned} \tag{2.99}$$

where symbols c_i are deduced from symbols a_i by a coding by transition.

$$c_{2i} = a_{2i}c_{2i-1} \quad \text{and} \quad c_{2i-1} = a_{2i-1}c_{2i-2} \tag{2.100}$$

and $h(t)$ is a unit amplitude rectangular pulse shape of width $2T$:

$$\begin{aligned} h(t) &= 1 \text{ if } t \in [-T, T[\\ &= 0 \text{ elsewhere} \end{aligned}$$

The coherent receiver for MSK is shown in Figure 2.21. It comprises two matched filters at $h(t)$ with waveform $h(2T-t)$. Symbols c_{2i-1} and c_{2i} are decoded by comparing samples taken at the output of the matched filters, at time $2iT$ and $(2i+1)T$ respectively.

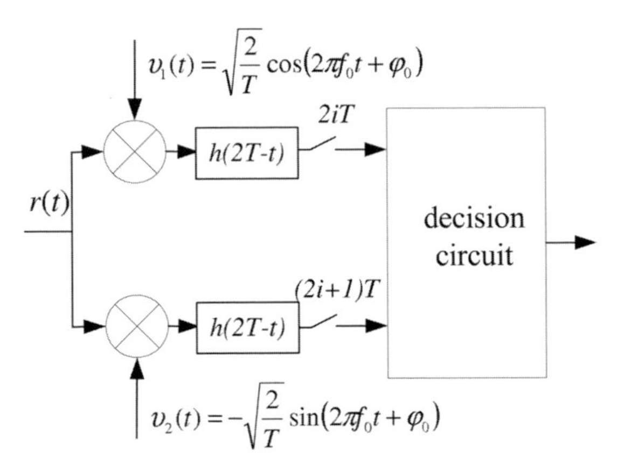

Figure 2.21 – Coherent receiver for MSK modulation.

It is easy to show that the error probabilities on binary symbols c_{2i-1} and c_{2i} are identical and equal to:

$$Pe_{c_i} = \frac{1}{2}\text{erfc}\sqrt{\frac{E_b}{N_0}} \tag{2.101}$$

where E_b is the energy used to transmit a binary symbol c_i.
To obtain binary data a_i from the symbols c_i, at the output of the coherent receiver we have to use a differential decoder given by the following equations:

$$a_{2i} = c_{2i}c_{2i-1} \quad \text{and} \quad a_{2i-1} = c_{2i-1}c_{2I-2}$$

The bit error probability Peb on a_i is:

$$Peb = 1 - (1 - Pe_{c_i})^2$$

thus for $Pe_{c_i} << 1$, a good approximation of the bit error probability Peb is:

$$Peb \approx 2Pe_{c_i} \tag{2.102}$$

As a first approximation, the performance of the MSK modulation is identical to that of the 4-PSK modulation.

2.3 Transmission on a band-limited channel

2.3.1 Introduction

In this chapter, so far we have assumed that the bandwidth allocated to the transmission is infinite. We will now envisage a more realistic situation where a bandwidth W is available to transmit the modulated signal. In this band W the channel is assumed to be frequency non-selective. We shall restrict ourselves to the case of linear modulations of the M-ASK, M-PSK and M-QAM types that have a power spectral density made up of a main lobe of width $2/T$, where $1/T$ is the modulation speed, and sidelobes with zero crossing at n/T. The bandwidth of a linearly modulated signal is therefore, strictly speaking, infinite. The modulated signal must consequently be filtered by an emission filter before being placed at the input of the transmission channel. We are now going to determine what the minimum band W is necessary to transmit the modulated signal without degradation of the performance compared to a transmission with an infinite bandwidth. The frequency response of the emission and reception filters will also be established.

In what follows, we will consider the complex envelope of the modulated signal and the equivalent baseband response of the emission filter. Without compromising the generality of our remarks, this avoids introducing the carrier frequency, which complicates the notations.

The complex envelope of an M-ASK, M-PSK and M-QAM signal has the expression:

$$s_e(t) = A\sum_i c_i h(t - iT) \tag{2.103}$$

where $h(t)$ is a unit amplitude rectangular pulse shape of width T, and $c_i = a_i + jb_i$ is a modulation symbol with:

M-ASK	$c_i = a_i$	$b_i = 0$
M-PSK	$a_i = \cos\phi_i$	$b_i = \sin\phi_i$
M-QAM	a_i and b_i	symbols$\sqrt{M} -$ ary

Let $g(t)$ be the impulse response of the emission passband filter centred on the carrier frequency. This waveform can be written:

$$g(t) = g_c(t)\cos(2\pi f_0 t + \theta_0) - g_s(t)\sin(2\pi f_0 t + \theta_0) \tag{2.104}$$

or equivalently:

$$g(t) = \Re_e\left\{g_e(t)\exp\left[j(2\pi f_0 t + \theta_0)\right]\right\} \tag{2.105}$$

where $g_e(t) = g_c(t) + jg_s(t)$ is the baseband-equivalent waveform of the emission filter. The output $e(t)$ of the emission filter is equal to:

$$e(t) = A \sum_i c_i z(t - iT) \tag{2.106}$$

where $z(t) = h(t) \otimes g_e(t)$ is, in the general case, a complex waveform while $h(t)$ is real.

2.3.2 Intersymbol interference

After passing through the emission filter, the modulated signal has a bandwidth W and, thus, the signal at the output of the transmission channel has the expression:

$$r(t) = e(t) + b(t) \tag{2.107}$$

where $b(t)$ is a complex AWGN, with zero mean and power spectral density equal to $2N_0$.

The coherent receiver uses a reception filter followed by a sampler at time $t_0 + nT$, where t_0 can be chosen arbitrarily. The output of the reception filter with impulse response $g_r(t)$ has the expression:

$$y(t) = A \sum_i c_i x(t - iT) + b'(t) \tag{2.108}$$

where:

$$\begin{aligned} x(t) &= z(t) \otimes g_r(t) \\ b'(t) &= b(t) \otimes g_r(t) \end{aligned}$$

Sampling signal $y(t)$ at time $t_0 + nT$, we obtain:

$$y(t_0 + nT) = A \sum_i c_i x(t_0 + (n - i)T) + b'(t_0 + nT) \tag{2.109}$$

Considering that in the general case $x(t) = p(t) + jq(t)$ is a complex waveform, the sample $y(t_0 + nT)$ can again be written in the form:

$$\begin{aligned} y(t_0 + nT) = Ac_n p(t_0) &+ A \sum_{i \neq 0} c_{n-i} p(t_0 + iT) \\ &+ jA \sum_i c_{n-i} q(t_0 + iT) + b'(t_0 + nT) \end{aligned} \tag{2.110}$$

The first term $Ac_n p(t_0)$ represents the desired information for the decoding of the symbol c_n, the following two terms being Intersymbol Interference (ISI) terms. Let us examine the outputs of the two components of the receiver, called the in-phase and quadrature components, corresponding to the real part and to the imaginary part of $y(t_0 + nT)$ respectively. We can notice that the in-phase component (respectively the quadrature component) depends on symbols

a_i (respectively on symbols b_i) but also on symbols b_i (respectively on symbols a_i). We sometimes say that there is crosstalk between the two components of the receiver. Of course, ISI is a phenomenon that can only degrade the quality of the transmission. This is the reason why it is important to define the condition to be satisfied in order to cancel the ISI. But before that, we are going to indicate a simple way to characterize the ISI by tracing the eye diagram, thus called by analogy with the shape of the human eye, at the output of the reception filter of the receiver's in-phase and quadrature components.
The eye diagram is the figure obtained by superposing all the plots or realizations of the non-noised signal $y_c(t)$ where $y_c(t)$ is the real part of $y(t)$. We also obtain the eye diagram from non-noised $y_s(t)$ where $y_s(t)$ is the imaginary part of $y(t)$.

$$\begin{aligned} y_c(t) &= A\sum_i a_i p(t-iT) - A\sum_i b_i q(t-iT) \\ y_s(t) &= A\sum_i b_i p(t-iT) + A\sum_i a_i q(t-iT) \end{aligned} \tag{2.111}$$

Let us analyse, for example, output $y_c(t)$ of the reception filter of the in-phase component on time interval $[t_1+nT,\ t_1+(n+1)T[$ where t_1 represents an arbitrary time. Replacing t by $t+t_1+nT$, signal $y_c(t)$ can be written:

$$\begin{aligned} y_c(t+t_1+nT) &= A\sum_i a_{n-i} p(t+t_1+iT) \\ &\quad -A\sum_i b_{n-i} q(t+t_1+iT) \quad \text{for} \quad 0 \le t \le T \end{aligned} \tag{2.112}$$

Assuming $x(t+t_1+iT)$ to be negligible outside the interval $[t_1 - L_1T,\ t_1+L_2T]$, each sum of the previous expression includes (L_1+L_2+1) terms, that is, $4^{(L_1+L_2+1)}$ possible plots for the eye diagram. If we shift the observation interval by a quantity T, the number of plots that make up the eye diagram is always $4^{(L_1+L_2+1)}$, the eye diagram is therefore a figure that is repeated every T seconds. Its analysis can, thus, be limited to an interval of time T.

The eye diagram can be visualized on an oscilloscope. Indeed, the different plots of $p(t)$ and $q(t)$ can remain on the screen if the scanning speed of the oscilloscope is fast enough compared to the remanence time of the cathode ray tube or, better, if it is a deep memory oscilloscope. We show the eye diagram of an ISI and ISI-free 2-PSK modulation in Figure 2.22. For this modulation, the symbols c_i are real and the receiver has a single component.

$$y_c(t+t_1+nT) = A\sum_i a_{n-i} p(t+t_1+iT) \text{ for } 0 \le t \le T \tag{2.113}$$

In the absence of ISI, at the sampling time, all the plots of $p(t)$ pass through a single point. The more open the eye diagram at the sampling time, the greater the immunity of the transmission to noise. In the same way, the greater the horizontal aperture of the eye diagram, the less sensitive the transmission is to

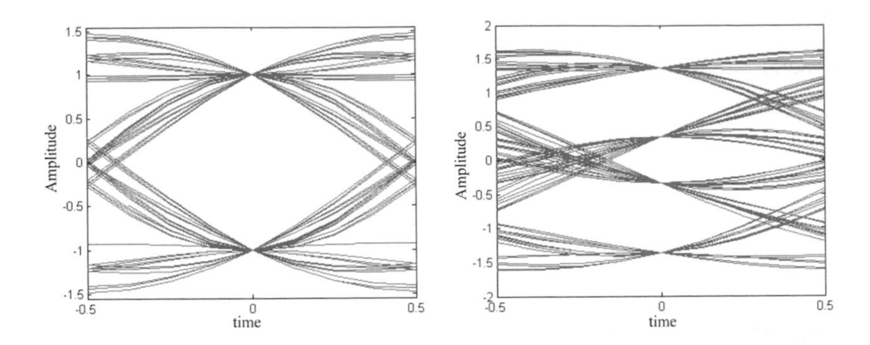

Figure 2.22 – Eye diagram of an (a) ISI and (b) ISI-free 2-PSK modulation.

positioning errors of the sampling time. In the presence of ISI, the different plots of $p(t)$ no longer pass through a single point at the sampling time and the ISI contributes to closing the eye diagram.

The output of the reception filter at time $t_0 + nT$ of the in-phase component of the receiver is equal to:

$$y_c(t_0 + nT) = Aa_n p(t_0) + A \sum_{i \neq 0} a_{n-i} p(t_0 + iT) \\ - A \sum_i b_{n-i} q(t_0 + iT) \tag{2.114}$$

For a M-QAM modulation, the useful signal is $A(2j - 1 - \sqrt{M})p(t_0)$, for $j = 1, \cdots, \sqrt{M}$ and the decision is taken by comparing signal $y_c(t_0 + nT)$ to a set of thresholds separated by $2p(t_0)$. There will be errors in the absence of noise if the ISI, for certain configurations of symbols c_i, is such that the received signal is situated outside the correct decision zone. This occurs if the ISI is higher in absolute value to $p(t_0)$. This situation is translated by the following condition:

$$\text{Max}_{a_{n-i}, b_{n-i}} \left| \sum_{i \neq 0} a_{n-i} p(t_0 + iT) - \sum_i b_{n-i} q(t_0 + iT) \right| > |p(t_0)|$$

Taking into account the fact that the largest value taken by symbols a_i and b_i is $\sqrt{M} - 1$, the previous condition becomes:

$$D_{\text{max}} = (\sqrt{M} - 1) \frac{\left(\sum_{i \neq 0} |p(t_0 + iT)| + \sum_i |q(t_0 + iT)| \right)}{|p(t_0)|} \geq 1 \tag{2.115}$$

Quantity D_{max} is called the maximum distortion. When the maximum distortion is greater than unity, the eye diagram is closed at the sampling time and errors are possible, even in the absence of noise.

2.3.3 Condition of absence of ISI: Nyquist criterion

The absence of ISI is translated by the following conditions:

$$p(t_0 + iT) = 0 \quad \forall i \neq 0 \tag{2.116}$$

$$q(t_0 + iT) = 0 \quad \forall i \tag{2.117}$$

which can again be written by using the complex signal $x(t) = p(t) + jq(t)$

$$x(t_0 + iT) = p(t_0)\delta_{0,i} \quad \forall i \tag{2.118}$$

where $\delta_{0,i}$ is the Kronecker symbol.

Let us introduce the sampled signal $x_E(t)$, defined by:

$$x_E(t) = x(t) \sum_i \delta(t - t_0 - iT) \tag{2.119}$$

We can notice that the impulse train $u(t) = \sum_i \delta(t - t_0 - iT)$ is periodic, of period T. It can therefore be decomposed into the Fourier series:

$$u(t) = \frac{1}{T} \sum_i \exp\left(-j2\pi \frac{i}{T} t_0\right) \exp\left(-j2\pi \frac{i}{T} t\right) \tag{2.120}$$

Since we are seeking to determine the minimal bandwidth W necessary to transmit the ISI-free modulated signal, it is wise to work in the frequency domain. Taking the Fourier transform denoted $X_E(f)$ of relation (2.119) and taking into account the previous expression of $u(t)$, we obtain:

$$X_E(f) = \frac{1}{T} \sum_i \exp\left(-j2\pi \frac{i}{T} t_0\right) X(f - \frac{i}{T}) \tag{2.121}$$

The sampled signal, according to relation (2.119), can also be written:

$$x_E(t) = \sum_i x(t_0 + iT)\delta(t - t_0 - iT) \tag{2.122}$$

which, after Fourier transform and taking into account the condition of absence of ISI, becomes:

$$X_E(f) = p(t_0) \exp(-j2\pi f t_0) \tag{2.123}$$

The equality of the relations (2.121) and (2.123) gives:

$$\sum_i \exp\left(j2\pi(f - \frac{i}{T})t_0\right) X(f - \frac{i}{T}) = Tp(t_0)$$

Putting:

$$X^{t_0}(f) = \frac{X(f)}{p(t_0)} \exp(j2\pi f t_0)$$

the condition for the absence of ISI can be expressed from $X^{t_0}(f)$ by the following relation:

$$\sum_{i=-\infty}^{+\infty} X^{t_0}(f - \frac{i}{T}) = T \tag{2.124}$$

This condition of absence of ISI is called the *Nyquist criterion.*

Let us recall that the transmission channel with spectrum $C(f)$ has a passband W.

$$C(f) = 0 \quad \text{for} \quad |f| > W \tag{2.125}$$

Let us consider relation (2.124) for three situations.

1. $X^{t_0}(f)$ has a bandwidth $W < 1/2T$. Relation (2.124) being a sum of functions shifted by $1/T$, there are no functions $X^{t_0}(f)$ that enable the Nyquist criterion to be satisfied. The bandwidth W necessary for ISI-free transmission is therefore higher than or equal to $1/2T$.

2. $X^{t_0}(f)$ has a bandwidth $W = 1/2T$. There is a single solution that satisfies the Nyquist criterion:

$$\begin{aligned} X^{t_0}(f) &= T \quad |f| \leq W \\ &= 0 \quad \text{elsewhere} \end{aligned} \tag{2.126}$$

or again:

$$\begin{aligned} X(f) &= Tp(t_0)\exp(-j2\pi t_0) \quad |f| \leq W \\ &= 0 \quad \text{elsewhere} \end{aligned} \tag{2.127}$$

which, in the time domain, gives:

$$x(t) = p(t_0)\frac{\sin\left[\pi(t - t_0)/T\right]}{\pi(t - t_0)/T} \tag{2.128}$$

This solution corresponds to a strictly speaking non-causal waveform $x(t)$. However, since the function $\sin y/y$ decreases fairly rapidly as a function of its argument y, it is possible, by choosing t_0 large enough, to consider $x(t)$ as practically causal. With this solution, the eye diagram has a horizontal aperture that tends towards zero and thus, any imprecision about the sampling time can lead to errors even in the absence of noise. In conclusion, this solution is purely theoretical and therefore has no practical applications.

3. $X^{t_0}(f)$ has a bandwidth $W > 1/2T$. In this case, there are many solutions that enable the Nyquist criterion to be satisfied. Among these solutions, the most commonly used is the raised-cosine function defined by:

$$
\begin{aligned}
X^{t_0}(f) &= T && \text{if} \quad 0 \leq |f| \leq \tfrac{1-\alpha}{2T} \\
&\tfrac{T}{2}\left[1+\sin\left(\tfrac{\pi T}{\alpha}\left(\tfrac{1}{2T}-|f|\right)\right)\right] && \text{if} \quad \tfrac{1-\alpha}{2T} \leq |f| \leq \tfrac{1+\alpha}{2T} \\
&0 && \text{if} \quad |f| > \tfrac{1+\alpha}{2T}
\end{aligned}
\tag{2.129}
$$

or again:

$$X(f) = p(t_0)X^{t_0}(f)\exp(-j2\pi f t_0) \tag{2.130}$$

whose waveform is:

$$x(t) = p(t_0)\frac{\sin\frac{\pi(t-t_0)}{T}}{\frac{\pi(t-t_0)}{T}}\frac{\cos\frac{\pi\alpha(t-t_0)}{T}}{1-4\alpha^2\frac{(t-t_0)^2}{T^2}} \tag{2.131}$$

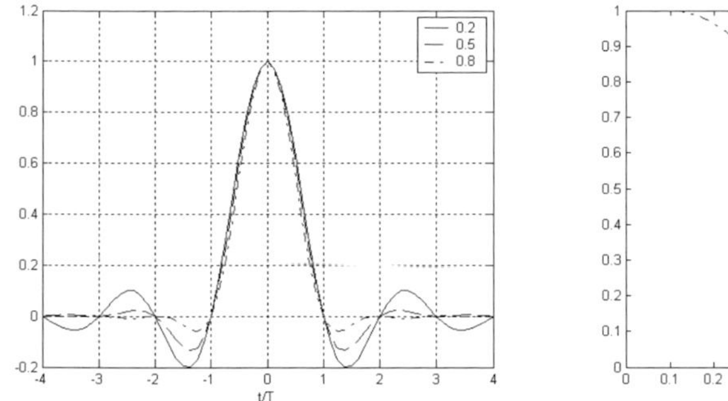

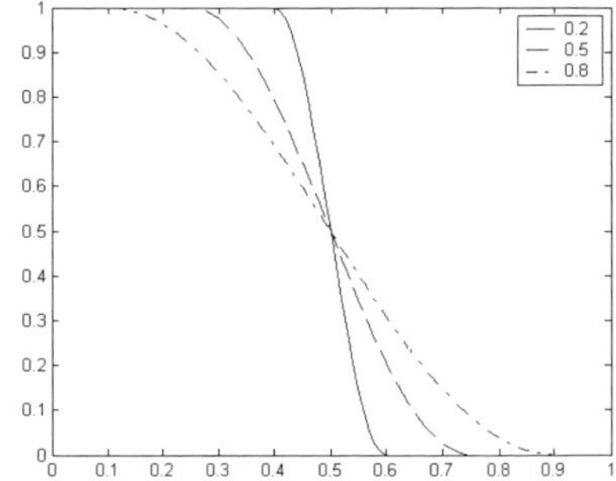

Figure 2.23 – Frequency and time domain characteristics of a raised-cosine function for different values of the roll-off factor α.

Figure 2.23 shows the frequency domain $X^{t_0}(f)$ and time domain $x(t)$ characteristics of a raised-cosine function for different values of α, called the *roll-off* factor.

The bandwidth of the raised-cosine function is $W = (1+\alpha)/2T$; $0 \leq \alpha \leq 1$. Function $x(t)$ is again non-causal in the strict sense but the more the roll-off factor increases, the greater this function decreases. Thus, by choosing t_0 large enough, implementing a raised cosine becomes possible. Figure 2.24 plots the eye diagrams obtained with raised-cosine functions for different values of roll-off factor.

All the plots of $x(t)$ pass through a single point at the sampling time $t_0 + nT$, whatever the value of the roll-off factor. Note that the larger the roll-off factor, the greater the horizontal aperture of the eye diagram. For $\alpha = 1$, the aperture of the eye is maximum and equal to T; the sensitivity to any imprecision about the sampling time is thus minimum.

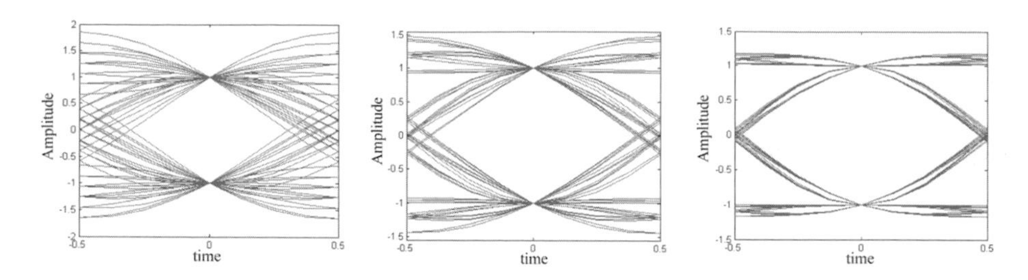

Figure 2.24 – Eye diagrams for modulations with (2-PSK or 4-PSK) binary symbols for different values of roll-off factor α $(0.2, 0.5, 0.8)$.

Having determined the global spectrum $X(f)$ that the transmission chain must satisfy in order to guarantee the absence of ISI, we will now establish the frequency characteristics of the emission and reception filters.

Optimal distribution of filtering between transmission and reception

We have seen that the reception filter must be matched to the waveform placed at its input, that is, in our case:

$$g_r(t) = z(t_0 - t) \tag{2.132}$$

where $z(t)$ results from filtering $h(t)$ by the transmission filter:

$$z(t) = h(t) \otimes g_e(t)$$

the frequency characteristic $G_r(f)$ of the reception filter being equal to:

$$G_r(f) = Z^*(f) \exp(-j2\pi f t_0) \tag{2.133}$$

where $*$ represents the conjugate operator.

Of course, the global characteristic of the transmission chain must satisfy the Nyquist criterion, which is translated by:

$$Z(f)G_r(f) = p(t_0)CS_\alpha(f) \exp(-j2\pi f t_0) \tag{2.134}$$

where $CS_\alpha(f) = X^{t_0}(f)$ is the raised-cosine spectrum of roll-off factor α. In the previous relation we considered that the channel transmits the signal placed at its input in its entirety.

Expressing function $Z^*(f)$ from the previous relation,

$$Z^*(f) = \frac{p(t_0)CS_\alpha(f)}{G_r^*(f)} \exp(j2\pi f t_0) \tag{2.135}$$

then replacing $Z^*(f)$ by its expression in relation (2.133), we obtain the magnitude spectrum of the reception filter:

$$|G_r(f)| = \sqrt{p(t_0)}\sqrt{CS_\alpha(f)} \tag{2.136}$$

The magnitude spectrum of the emission filter is obtained without difficulty from:

$$|G_e(f)| = \frac{|Z(f)|}{|H(f)|} \tag{2.137}$$

Using relation (2.135), we can determine the module of $|Z(f)|$.

$$|Z(f)| = \sqrt{p(t_0)}\sqrt{CS_\alpha(f)} \tag{2.138}$$

Replacing $|Z(f)|$ by its value in relation (2.137), we obtain the magnitude spectrum of the reception filter:

$$|G_e(f)| = \frac{\sqrt{p(t_0)}\sqrt{CS_\alpha(f)}}{|H(f)|} \tag{2.139}$$

We have obtained the magnitude spectrum of the emission and reception filters. These filters are therefore defined to within one arbitrary phase. Distributing time t_0 between the emission and reception filters, we obtain:

$$\begin{aligned} G_r(f) &= \sqrt{p(t_0)}\sqrt{CS_\alpha(f)}\exp(-j2\pi f t_1) \\ G_e(f) &= \frac{\sqrt{p(t_0)}\sqrt{CS_\alpha(f)}}{H(f)}\exp(-j2\pi f(t_0 - t_1)) \end{aligned} \tag{2.140}$$

where t_1 is a time lower than t_0 that can be chosen arbitrarily.

The characteristics of filters $G_e(f)$ and $G_r(f)$ show that Nyquist filtering must be equidistributed between emission and reception. The term $H(f)$ that appears at the denominator of the emission filter means that the output of this filter no longer depends on the waveform used by the modulator. Spectrum $H(f)$, a Fourier transform of a rectangular pulse shape of width T, has spectral zeros crossing at frequencies n/T which, strictly speaking, makes it impossible to implement filter $G_e(f)$. However, filter $G_e(f)$ must be implemented only in the $[-(1+\alpha)/2T;\ (1+\alpha)/2T]$ band where $H(f)$ has no spectral zero-crossing. To finish, let us determine the PSD of the modulated signal at the output of the emission filter and check that it has bandwidth $W = (1+\alpha)/2T$.

The PSD of the baseband modulated signal at the input of the emission filter is equal to:

$$\gamma_{s_e}(f) = 2\frac{(M-1)}{3}A^2T\left(\frac{\sin \pi f T}{\pi f T}\right)^2 \tag{2.141}$$

The PSD of the baseband modulated signal at the output of the emission filter has the expression:

$$\gamma_e(f) = \gamma_{s_e}(f)\,|G_e(f)|^2$$

Replacing $|G_e(f)|$ by its expression, we obtain:

$$\gamma_e(f) = 2\frac{(M-1)}{3}\frac{A^2}{T}p(t_0)CS_\alpha(f) \tag{2.142}$$

Considering the modulated signal on a frequency carrier, its PSD is given by expression (2.13):

$$\gamma(f) = \frac{1}{4}\gamma_e(f - f_0) + \frac{1}{4}\gamma_e(f + f_0)$$

In conclusion, the modulated signal on a frequency carrier uses the $f_0-(1+\alpha)/2T$ to $f_0+(1+\alpha)/2T$ band, that is, a bandwidth $2W = (1+\alpha)/T$ or again $(1+\alpha)R_m$ where R_m is the symbol rate. The spectral efficiency of the M-QAM modulation expressed in bit/s/Hz is then equal to:

$$\eta = \frac{D}{2W} = \frac{R_m \log_2(M)}{(1+\alpha)R_m} = \frac{\log_2(M)}{(1+\alpha)} \tag{2.143}$$

where D is the bit rate of the transmission expressed in bit/s.

The spectral efficiency increases as a function of the number of states M of the modulation but the performance of the modulation, in terms of error probability, decreases as a function of this parameter M.

2.3.4 Expression of the error probability in presence of Nyquist filtering

Let us determine the bit error probability provided by the source by considering, for example, a 4-PSK modulation. In this case, each symbol a_i (respectively b_i) transmits a bit d_i every T seconds. The error probability on the data d_i is therefore identical to the error probability on the symbols a_i or b_i. Let us calculate, for example, the error probability on symbol a_i.

The output of the reception filter of the in-phase component at sampling time $t_0 + nT$ is equal to:

$$y_c(t_0 + nT) = Aa_n p(t_0) + b'_c(t_0 + nT) \tag{2.144}$$

where $b'_c(t_0 + nT)$ is the real part of the noise $b'(t_0 + nT)$.

Assuming the data d_n *iid*, the error probability on the symbols a_n has the expression:

$$\begin{aligned} Pe_{a_n} = \tfrac{1}{2}\Pr\{y_c(t_0+nT) > 0|a_n = -1\} \\ \tfrac{1}{2}\Pr\{y_c(t_0+nT) < 0|a_n = +1\} \end{aligned} \tag{2.145}$$

Since $y_c(t_0 + nT)$ is Gaussian, with mean $Aa_n p(t_0)$ and variance $\sigma^2_{b'_c}$, the error probability Pe_{a_n} is equal to:

$$Pe_{a_n} = \frac{1}{2}\text{erfc}\frac{Ap(t_0)}{\sigma_{b'_c}\sqrt{2}} \tag{2.146}$$

The variance $\sigma^2_{b'_c}$ of the noise $b'_c(t_0 + nT)$ is equal to:

$$\sigma^2_{b'_c} = N_0 \int_{-\infty}^{+\infty} |G_r(f)|^2 df = N_0 \int_{-\infty}^{+\infty} p(t_0) CS_\alpha(f) df = N_0 p(t_0) \tag{2.147}$$

Let us introduce the power transmitted at the output of the emission filter:

$$P = \frac{1}{4}\int_{-\infty}^{+\infty} \gamma_e(f - f_0)df + \frac{1}{4}\int_{-\infty}^{+\infty} \gamma_e(f + f_0)df = \frac{1}{2}\int_{-\infty}^{+\infty} \gamma_e(f)df \qquad (2.148)$$

Replacing $\gamma_e(f)$ by its expression, we obtain:

$$P = \int_{-\infty}^{+\infty} \frac{A^2}{T} p(t_0) CS_\alpha(f)df = \frac{A^2}{T} p(t_0) \qquad (2.149)$$

Using expressions (2.147) and (2.149), the error probability is equal to:

$$Pe_{a_n} = \frac{1}{2}\mathrm{erfc}\sqrt{\frac{PT}{2N_0}} \qquad (2.150)$$

The energy E_b used to transmit an information bit d_n is:

$$E_b = PT_b \qquad (2.151)$$

where T_b is the inverse of the bit rate of the transmission.

For a 4-PSK modulation, $T = 2T_b$ and the bit error probability d_n is finally:

$$Pe_{d_n} = \frac{1}{2}\mathrm{erfc}\sqrt{\frac{E_b}{N_0}} \qquad (2.152)$$

The error probability in the presence of Nyquist filtering for a 4-PSK modulation is identical to that obtained for a transmission on an infinite-bandwidth channel. This result is also true for the other linear modulations of the M-ASK, M-PSK and M-QAM type.

To conclude this section, we can say that filtering according to the Nyquist criterion of a linear modulation makes it possible to reduce the bandwidth necessary for its transmission to $(1 + \alpha)R_m$, where R_m is the symbol rate. This filtering does not degrade the performance of the modulation, that is, it leads to the same bit error probability as that of a transmission on an infinite bandwidth channel.

2.4 Transmission on fading channels

2.4.1 Characterization of a fading channel

Let us consider a transmission over a multipath channel where the transmitter, which is mobile compared to the receiver, provides a non-modulated signal $s(t) = A\exp(j2\pi ft)$ with amplitude A and frequency f. Signal $s(t)$ propagates by being reflected on different obstacles and, thus, the receiver receives M copies of the signal transmitted, each copy being affected by an attenuation $\rho_n(t)$, with delay $\tau_n(t)$ and a Doppler frequency shift $f_n^d(t)$. The attenuations, delays and Doppler frequencies are functions of time in order to take into account the time-varying channel. To simplify the notations, in the following we will omit variable t for the attenuations, the delays and the Doppler frequencies.

Let $r(t)$ be the response of the transmission channel to the signal $s(t)$, which is generally written in the form:

$$r(t) = \sum_{n=1}^{M} \rho_n \exp\left[j2\pi(f_n^d + f)(t - \tau_n)\right] \tag{2.153}$$

Making $s(t)$ appear, the received signal can again be written:

$$r(t) = \sum_{n=1}^{M} \rho_n \exp\left[j2\pi(f_n^d t - (f_n^d + f)\tau_n)\right] s(t) \tag{2.154}$$

and thus the frequency response of the transmission channel is defined by:

$$c(f,t) = \sum_{n=1}^{M} \rho_n \exp\left[-j2\pi(f\tau_n - f_n^d t + f_n^d \tau_n)\right] \tag{2.155}$$

The multipath channel is generally frequency selective, that is, it does not transmit all the frequency components of the signal placed at its input in the same way, certain components being more attenuated than others. The channel will therefore create distortions of the transmitted signal. In addition, their evolution over time can be more or less rapid.
To illustrate the frequency selectivity of a multipath channel, we have plotted in Figure 2.25 the power spectrum of the frequency response of this channel for $M = 2$, in the absence of a Doppler frequency shift ($f_n^d = 0$) and fixing τ_1 to zero.

$$|c(f)|^2 = \rho_1^2\left[(1 + \alpha\cos 2\pi f\tau_2)^2 + \alpha^2 \sin^2 2\pi f\tau_2\right] \tag{2.156}$$

with $\alpha = \rho_2/\rho_1$.

Two parameters are now introduced: coherence bandwidth B_c and coherence time t_c that allow the transmission channel to be characterized in relation to the frequency selectivity and its evolution speed.

Coherence bandwidth

There are several definitions of the coherence bandwidth but the most common definition is:

$$B_c \approx \frac{1}{T_m} \tag{2.157}$$

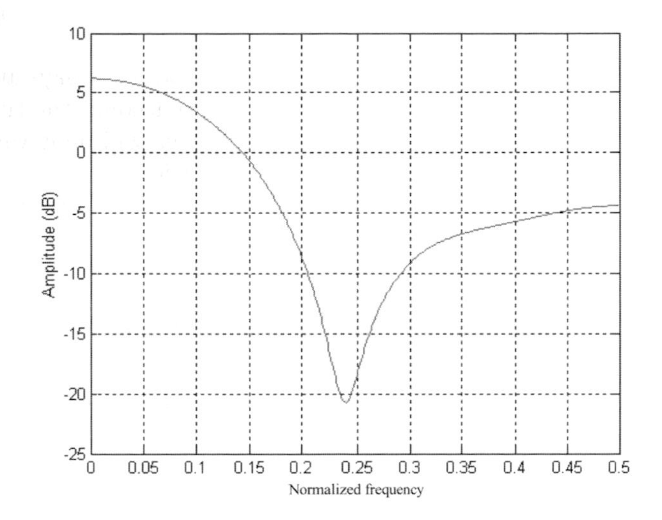

Figure 2.25 – Frequency response of a multipath channel.

where T_m is the multipath spread of the channel as a function of the delays τ_n of the different paths.

Two modulated signals whose carrier frequencies are separated by a quantity higher than the coherence band of the channel are attenuated by the latter in a non-correlated way. Thus, at a given instant, if one of the two signals is strongly attenuated by the channel, it is highly probable that the other will be little affected by the channel.

When band B occupied by a modulated signal is lower than the coherence band, the channel is frequency non-selective. It shows a flat frequency response and a linear phase on the bandwidth of the modulated signal. The modulated signal just undergoes attenuation and dephasing on passing through the channel. We will illustrate this point by considering the transmission of an M-QAM signal by a frequency non-selective channel.

Let $s(t)$ be an M-QAM signal provided at the input of the frequency non-selective multipath channel.

$$s(t) = A\left[\sum_i a_i h(t-iT)\cos 2\pi f_0 t - \sum_i b_i h(t-iT)\sin 2\pi f_0 t\right]$$

and let $r(t)$ be the response of the channel at $s(t)$:

$$\begin{aligned} r(t) = A\sum_{n=1}^{M}\rho_n \Big[&\sum_i a_i h(t-\tau_n-T)\cos\left(2\pi(f_0+f_n^d)(t-\tau_n)\right) \\ &- \sum_i b_i h(t-\tau_n-T)\sin\left(2\pi(f_0+f_n^d)(t-\tau_n)\right)\Big] \end{aligned}$$

For a frequency non-selective channel, we have $B < B_c$. Noting that band B is proportional to $1/T$, this leads to $T > T_m$, or again $T >> \tau_n \quad \forall n$. In the expression of $r(t)$, we can therefore neglect τ_n in front of T which gives:

$$r(t) = A \sum_{n=1}^{M} \rho_n \left[\sum_i a_i h(t-T) \cos(2\pi f_0 t + \varphi_n(t)) \right. \\ \left. - \sum_i b_i h(t-T) \sin(2\pi f_0 t + \varphi_n(t)) \right]$$

with:

$$\varphi_n(t) = f_n^d t - (f_0 + f_n^d)\tau_n$$

Putting:

$$a_c(t) = \sum_{n=1}^{M} \rho_n \cos \varphi_n(t) \text{ and } a_s(t) = \sum_{n=1}^{M} \rho_n \sin \varphi_n(t)$$

and:

$$\cos \phi(t) = \frac{a_c(t)}{\sqrt{a_c^2(t) + a_s^2(t)}} \text{ and } \sin \phi(t) = \frac{a_s(t)}{\sqrt{a_c^2(t) + a_s^2(t)}}$$

signal $r(t)$ can again be written:

$$r(t) = A\alpha(t) \left[\sum_i a_i h(t-iT) \cos(2\pi f_0 t + \phi(t)) \right. \\ \left. - \sum_i b_i h(t-iT) \sin(2\pi f_0 t + \phi(t)) \right] \tag{2.158}$$

with $\alpha(t) = \sqrt{a_c^2(t) + a_s^2(t)}$.

For a frequency non-selective multipath channel, the modulated M-QAM signal only undergoes an attenuation $\alpha(t)$ and dephasing $\phi(t)$.

Modelling the attenuations ρ_n, the delays τ_n, and the Doppler frequencies f_n^d by mutually independent random variables then, for large enough M and for a given t, $a_c(t)$ and $a_s(t)$ tend towards non-correlated random Gaussian variables (central limit theorem). The attenuation $\alpha(t)$, for a given t, follows a Rayleigh law and the phase $\phi(t)$ is equidistributed on $[0, 2\pi[$.

$$p(\alpha) = \frac{2\alpha}{\sigma_\alpha^2} \exp\left(-\frac{\alpha^2}{\sigma_\alpha^2}\right) \quad \alpha \geq 0 \tag{2.159}$$

with $\sigma_\alpha^2 = E\left\{\alpha^2\right\}$.

The attenuation $\alpha(t)$ can take values much lower than unity and, in this case, the information signal received by the receiver is very attenuated. Its level is then comparable to, if not lower than, that of the noise. We say that the transmission channel shows deep Rayleigh fading.

If band B occupied by the modulated signal is higher than the coherence band, the channel is frequency selective. Its frequency response, on band B, is no longer flat and some spectral components of the modulated signal can be very attenuated. The channel introduces a distortion of the modulated signal which results in the phenomenon of Intersymbol Interference (ISI). In the presence of ISI, the signal at a sampling time is a function of the symbol of the modulated signal at this time but also of the symbols prior to and after this time. ISI appears as noise that is added to the additive white Gaussian noise and, of course, degrades the performance of the transmission.

Coherence time

The coherence time t_c of a fading channel is defined by:

$$t_c \approx \frac{1}{B_d} \tag{2.160}$$

where B_d is the Doppler band of the channel that is well approximated by $f^d_{\max}$ with:

$$f^d_{\max} = \mathrm{Max} f^d_n \tag{2.161}$$

The coherence time of the channel is a measure of its evolution speed over time. If t_c is much higher than the symbol period T of the modulated signal, the channel is said to be slow-fading. For a frequency non-selective slow-fading channel, attenuation $\alpha(t)$ and phase $\phi(t)$ are practically constant over one or more symbol periods T.

A channel is frequency non-selective slow-fading if it satisfies the following condition:

$$T_M B_d < 1 \tag{2.162}$$

2.4.2 Transmission on non-frequency-selective slow-fading channels

Performance on a Rayleigh channel

For this channel, the modulated signal undergoes an attenuation $\alpha(t)$ and a random dephasing $\phi(t)$ of constant realizations over a duration higher than or equal to T. Considering a coherent receiver, the error probability per binary data, conditionally to a realization α of the attenuation $\alpha(t)$, is equal to:

$$\text{2-PSK or 4-PSK modulation} \quad Peb(\alpha) = \frac{1}{2}\mathrm{erfc}\sqrt{\frac{\alpha^2 E_b}{N_0}} \tag{2.163}$$

$$\text{2-FSK modulation} \quad Peb(\alpha) = \frac{1}{2}\mathrm{erfc}\sqrt{\frac{\alpha^2 E_b}{2N_0}} \tag{2.164}$$

We obtain the error probability Peb by averaging $Peb(\alpha)$ over the different realizations of $\alpha(t)$.

$$Peb = \int_0^\infty Peb(\alpha)p(\alpha)d\alpha \tag{2.165}$$

where $p(\alpha)$ is the probability density of α.

Taking into account the fact that, for a given t, $\alpha(t)$ is a Rayleigh random variable with probability density

$$p(\alpha) = \frac{\alpha}{\sigma_\alpha^2} \exp\left(-\frac{\alpha^2}{2\sigma_\alpha^2}\right) \quad \alpha \geq 0$$

the probabilities Peb have the expressions:

$$\text{2-PSK or 4-PSK modulation} \quad Peb = \frac{1}{2}\left(1 - \sqrt{\frac{\bar{E}_b/N_0}{1+\bar{E}_b/N_0}}\right) \tag{2.166}$$

$$\text{2-FSK modulation} \quad Peb = \frac{1}{2}\left(1 - \sqrt{\frac{\bar{E}_b/N_0}{2+\bar{E}_b/N_0}}\right) \tag{2.167}$$

where $\bar{E}_b$ is the average energy per transmitted bit :

$$\bar{E}_b = \mathrm{E}\left\{\alpha^2 \frac{A^2 T_b}{2}\right\} = A^2 T_b \sigma_\alpha^2 \tag{2.168}$$

For high $\bar{E}_b/N_0$, the error probabilities can be approximated by:

$$\text{2-PSK or 4-PSK} \quad Peb \approx \frac{1}{4\bar{E}/N_0} \tag{2.169}$$

$$\text{2-FSK} \quad Peb \approx \frac{1}{2\bar{E}/N_0} \tag{2.170}$$

On a Rayleigh fading channel, the performance of the different receivers is severely degraded compared to those obtained on a Gaussian channel (with identical $E_b/N0$ at the input). Indeed, on a Gaussian channel, the error probabilities Peb decrease exponentially as a function of the signal to noise ratio E_b/N_0 whereas on a Rayleigh fading channel, the decrease in the probability Peb is proportional to the inverse of the average signal to noise ratio, $\bar{E}_b/N_0$. To improve the performance on a Rayleigh fading channel, we use two techniques, which we can combine, diversity and, of course, channel coding (which is, in fact, diversity of information).

Performance on a Rayleigh channel with diversity

Diversity involves repeating the same message (or copies coming from channel coding) several times by using the carrier frequencies separated by a quantity higher than or equal to the coherence band B_c of the channel. In this case, we speak of frequency diversity. An alternative to this approach involves transmitting a same message several times on a same carrier but on time intervals separated by a quantity that is higher than or equal to the coherence time t_c of the channel. This is time diversity. Finally, we can transmit a message a single time and use several sensors spaced typically by a few wavelengths from the carrier of the modulated signal. In this case, we have space diversity.

Let us assume that we use a 2-PSK modulation to transmit the information message and a diversity of order L. On the time interval $[iT, (i+1)T[$ and considering a coherent reception, after demodulation we have L observations of the form:

$$r_i^n = \alpha_i^n \sqrt{E_b} \cos \varphi_i + b_i^n \quad n = 1, 2, \cdots, L \tag{2.171}$$

where α_i^n is a Rayleigh attenuation, φ_i the phase (0 or π) carrying the information to transmit and b_i^n a white Gaussian noise, with zero mean and variance equal to $N_0/2$. The L attenuations α_i^n are mutually independent as well as the L terms of noise b_i^n. These L attenuations can be seen as L independent subchannels, also called diversity branches. E_b is thus the energy used to transmit one bit per diversity branch.

To take a decision in the presence of diversity, we construct the variable Z_i in the following way:

$$Z_i = \sum_{n=1}^{L} r_i^n \cdot \alpha_i^n$$

The bit error probability Peb in presence of diversity is then equal to:

$$Peb = \frac{1}{2} \Pr(Z_i > 0 | \varphi_i = \pi) + \frac{1}{2} \Pr(Z_i < 0 | \varphi_i = 0) \tag{2.172}$$

Conditionally to *one* realization of the L attenuations α_i^n, the decision variable Z_i is Gaussian with mean:

$$\begin{aligned} E\{Z_i\} &= \sqrt{E_b} \sum_{n=1}^{L} (\alpha_i^n)^2 \quad \text{if} \quad \varphi_i = 0 \\ E\{Z_i\} &= -\sqrt{E_b} \sum_{n=1}^{L} (\alpha_i^n)^2 \quad \text{if} \quad \varphi_i = \pi \end{aligned} \tag{2.173}$$

and variance:

$$\sigma_Z^2 = \frac{N_0}{2} \sum_{n=1}^{L} (\alpha_i^n)^2 \tag{2.174}$$

Under this hypothesis, the error probability $Peb(\rho)$ is equal to :

$$Peb(\rho) = \frac{1}{2}\text{erfc}\sqrt{\rho} \tag{2.175}$$

with:

$$\rho = \frac{E_b}{N_0}\sum_{n=1}^{L}(\alpha_i^n)^2$$

By averaging the probability $Peb(\rho)$ on the different realizations of the random variable ρ, we obtain the bit error probability in presence of diversity of order L:

$$Peb = \int_0^{\infty} Peb(\rho)p(\rho)d\rho$$

The random variable ρ follows a χ^2 law with probability density:

$$p(\rho) = \frac{1}{(L-1)!\bar{m}^L}\rho^{L-1}\exp -\frac{\rho}{\bar{m}} \tag{2.176}$$

where $\bar{m}$ is equal to:

$$\bar{m} = \frac{E_b}{N_0}\text{E}\left\{(\alpha_j^n)^2\right\}$$

After integration, the bit error probability is equal to:

$$Peb = \left(\frac{1-\eta}{2}\right)^L \sum_{n=0}^{L-1}\left(\begin{array}{c} L-1+n \\ n \end{array}\right)\left(\frac{1+\eta}{2}\right)^n \tag{2.177}$$

with:

$$\eta = \sqrt{\frac{\bar{E}_b/LN_0}{1+\bar{E}_b/LN_0}} \quad \text{and} \quad \left(\begin{array}{c} L-1+n \\ n \end{array}\right) = \frac{(L-1+n)!}{n!(L-1)!}$$

where $\bar{E}_b$ is the average total energy used to transmit an information bit ($\bar{E}_b = LE_b$).

For a high signal to noise ratio, an approximation of the bit error probability Peb is given by:

$$Peb \approx \left(\begin{array}{c} 2L-1 \\ L \end{array}\right)\left(\frac{1}{4\bar{E}_b/LN_0}\right)^L \quad \text{pour} \quad \frac{\bar{E}_b}{LN_0} >> 1 \tag{2.178}$$

In the presence of diversity, the bit error probability Peb decreases following the inverse of the signal to noise ratio to the power of L.

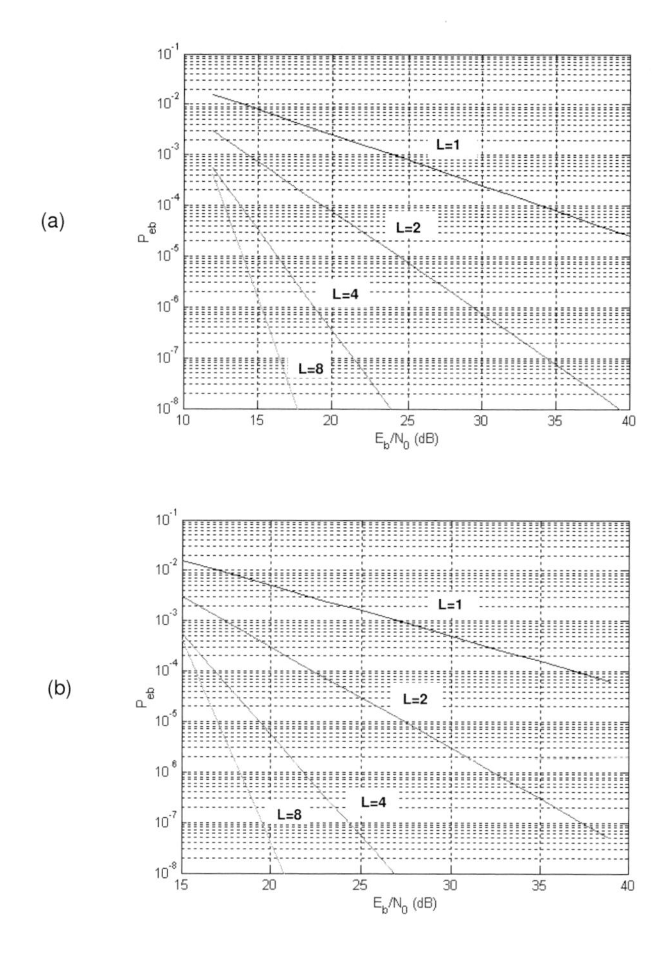

Figure 2.26 – Performance of a) 2-PSK and b) 2-FSK modulations in the presence of diversity.

For an 2-FSK modulation, calculating the bit error probability in the presence of coherent reception is similar to that of 2-PSK modulation. We obtain the following result:

$$Peb = \left(\frac{1-\eta}{2}\right)^L \sum_{n=0}^{L-1} \left(\begin{array}{c} L-1+n \\ n \end{array} \right) \left(\frac{1+\eta}{2}\right)^n \qquad (2.179)$$

with this time:

$$\eta = \sqrt{\frac{\bar{E}_b/LN_0}{2+\bar{E}_b/LN_0}}$$

With a high signal to noise ratio, a good approximation of the error probability Peb is given by:

$$Peb \approx \binom{2L-1}{L}\left(\frac{1}{2\bar{E}_b/LN_0}\right)^L \quad \text{for} \quad \frac{\bar{E}_b}{LN_0} >> 1 \tag{2.180}$$

Note that diversity like the type presented here is a form of coding that uses a repetition code and weighted decoding at reception. Figure 2.26 shows the performance of 2-PSK and 2-FSK modulations in the presence of diversity.

Transmission on a slow-fading frequency-selective channel

Different transmission strategies are possible. We can use a waveform at emission that is only slightly, or not at all, sensitive to the selectivity of the channel, or we can correct the effects of the selectivity of the channel at reception.
Multicarrier transmission

Multicarrier transmission uses a multiplex of orthogonal carriers that are digitally phase modulated (M-PSK), frequency modulated (M-FSK) or amplitude and phase modulated (M-QAM). This waveform, called a "parallel" waveform and known as *Orthogonal Frequency Division Multiplexing* (OFDM), enables us to avoid the frequency selectivity of transmission channels.

We have seen that a channel is frequency selective for a modulated signal with bandwidth B if its coherence bandwidth $B_c \approx 1/T_M$ is lower than B. Let us recall, on the other hand, that the bandwidth B of a digitally modulated signal is proportional to its symbol rate $1/T$. To build a type of waveform that is not sensitive to the frequency selectivity of the channel, we can proceed in the following way.

Let us divide the bit rate $D = 1/T_b$ to be transmitted into N sub-rates $D' = 1/NT_b$, each elementary rate feeding a modulator with M states, with carrier frequency f_i. The symbol rate of the modulated carriers is then equal to $R = 1/T$ with $T = NT_b \log_2(M)$. Choosing N large enough, the symbol rate of the modulated carriers can become as low as required, and the bandwidth B of a modulated carrier then becomes far lower than the coherence bandwidth B_c of the channel. Proceeding thus, the channel is frequency non-selective for the modulated carriers of the multiplex. For a multipath channel each modulated carrier is weighted by an attenuation that follows a Rayleigh law and with a dephasing equidistributed on $[0, 2\pi[$. Of course, not all the modulated carriers are affected in the same way at the same time by the channel; some are strongly attenuated whereas others are less so. The performance in terms of error probability per modulated carrier is that of a non-frequency-selective slow-fading Rayleigh channel.

To avoid large packets of errors at reception, we can make sure that information data that follow each other are transmitted by carriers affected differently by the channel, that is, spaced by a quantity at least equal to the coherence band

B_c of the channel. We can also transmit these data by using the same carrier but at a time separated by a quantity at least equal to the coherence time t_c of the channel. This way of proceeding amounts to performing frequency interleaving combined with time interleaving.

Implementing multicarrier transmissions

Considering an M-QAM modulation on each carrier, the OFDM signal has the expression:

$$s(t) = A \sum_{i=-\infty}^{+\infty} \sum_{n=0}^{N-1} \Re\left\{c_{n,i} h(t-iT) \exp(j2\pi f_n t)\right\} \tag{2.181}$$

where $c_{n,i} = a_{n,i} + jb_{n,i}$ is a complex modulation symbol, $h(t)$ a unit amplitude rectangular pulse shape of width T, and N is the number of carriers with frequency f_n.

Considering time interval $[iT, (i+1)T[$, the signal $s(t)$ is equal to:

$$s(t) = A \sum_{n=0}^{N-1} \Re\left\{c_{n,i} \exp(j2\pi f_n t)\right\} \quad \forall t \in [iT, (i+1)T[\tag{2.182}$$

The implementation of an OFDM signal requires N M-QAM modulators with carrier frequency f_n to be realized. We can show that these N modulators can be realized from an inverse discrete Fourier transform, which allows a reasonable complexity of implementation.

Considering N orthogonal carriers, the frequencies f_n must be separated by at least $1/T$.

$$f_n = \frac{n}{T} \qquad n = 0, 1, \cdots, (N-1) \tag{2.183}$$

The power spectral density $\gamma_{OFDM}(f)$ of an OFDM signal in baseband is proportional to:

$$\gamma_{OFDM}(f) \propto \sum_{n=0}^{N-1} \left[\frac{\sin \pi(f - n/T)T}{\pi(f - n/T)T}\right]^2 \tag{2.184}$$

which gives a flat spectrum in the frequency band $B = (N-1)/T$.

The signal $s(t)$ can be sampled at frequency f_e on condition that f_e satisfies the Nyquist criterion, that is:

$$f_e \geq \frac{2(N-1)}{T} \tag{2.185}$$

We can choose $f_e = 2N/T$, and thus signal $s(t)$ sampled at time lT_e with $T_e = 1/f_e$ is equal to:

$$s_l = s(lT_e) = A \sum_{n=0}^{N-1} \Re\left\{c_{n,i} \exp\left(j2\pi \frac{nl}{2N}\right)\right\} \quad l = 0, 1, \cdots, (2N-1) \tag{2.186}$$

In this form, s_l is not an inverse discrete Fourier transform. Introducing virtual symbols in the following way:

$$c_{2N-n,\,i} = c^*_{n,\,i} \quad n = 1, \cdots, (N-1) \quad c_{N,i} = \Im\left\{c_{0,\,i}\right\} \tag{2.187}$$

and replacing $c_{0,i}$ by $\Re\left\{c_{0,i}\right\}$, we can show that the sampled signal s_l can effectively be written in the form of an inverse discrete Fourier transform:

$$s_l = A \sum_{n=0}^{2N-1} c_{n,\,i} \exp\left(j2\pi \frac{nl}{2N}\right) \tag{2.188}$$

On each interval of duration T we perform an inverse discrete Fourier transform on $2N$ points that enable the N M-QAM baseband signals to be obtained. A digital-analogue converter followed by a frequency transposition enables the modulated signals to be obtained on a carrier frequency.

At reception, after amplification and transposition of the OFDM signal into baseband, the decoding of the modulation symbols $c_{n,i} = a_{n,i} + jb_{n,i}$ can also be performed by a discrete Fourier transform.

We have seen that the duration T of the modulation symbols increases with N and, for large values of N, can become comparable to the coherence time t_c of the transmission channel. In this case, the hypothesis of a slow-fading channel is no longer satisfied. There are therefore limits to the choice of parameter N. For a multipath channel, if the choice of N does not enable $B < B_c$ to be obtained, the channel remains frequency-selective in relation to the modulated carriers, and intersymbol interference appears.

To avoid residual intersymbol interference without increasing N, we can use the guard interval principle. For a multipath channel, the propagation paths are received at the receiver with delays τ_n. Calling τ_{Max} the largest of these delays, guard interval Δ must satisfy the following inequality:

$$\Delta \geq \tau_{\text{Max}} \tag{2.189}$$

We put:

$$T = t_s + \Delta$$

In the presence of a guard interval, the modulation symbols are still of duration T but the discrete Fourier transform at reception is realized on the time intervals $[iT + \Delta,\, (i+1)T[$. Proceeding thus, we can check that on this time interval only the modulation symbol transmitted between iT and $(i+1)T$ is taken into account for the decoding: there is therefore no intersymbol interference.

The introduction of a guard interval has two consequences. The first is that only a part of the energy transmitted on emission is exploited on reception. Indeed, we transmit each modulation symbol over a duration T and we recover this same symbol from an observation of duration $t_s = T - \Delta$. The loss, expressed in dB, is therefore equal to:

$$10 \log \left(\frac{1}{1 + \Delta/t_s} \right) \tag{2.190}$$

The second consequence is that the orthogonality of the carriers must be ensured so as to be able to separate these carriers at reception, that is, separating them by a quantity $1/t_s$. The band of frequencies occupied by the signal OFDM with a guard interval is therefore:

$$B = \frac{N-1}{t_s} \tag{2.191}$$

that is, an expansion of the bandwidth compared to a system without a guard interval of $1 + \Delta/t_s$.

In the presence of a guard interval, we should therefore choose Δ so as to minimize the degradations of the signal to noise ratio and of the spectral efficiency, that is, choose the smallest possible Δ compared to duration t_s.

Transmission with equalization at reception

In the presence of a frequency selective channel, we can use a linear single-carrier (M-PSK, M-QAM) modulation at emission and correct the ISI created by the channel by an equalizer. Some equalizer architectures are presented in Chapter 11.

If we compare the OFDM approach and single-carrier transmission with equalization, in terms of complexity of implementation, the difficulty of realizing an OFDM transmission lies in the modulator whereas with equalization, it lies in the receiver.

Chapter 3

Theoretical limits

The recent invention of turbo codes and the rediscovery of LDPC codes have brought back into favour the theoretical limits of transmission which were reputed to be inaccessible until now. This chapter provides the conceptual bases necessary to understand and compute these limits, in particular those that correspond to real transmission situations with messages of finite length and binary modulations.

3.1 Information theory

3.1.1 Transmission channel

A *channel* is any environment where symbols can be propagated (telecommunications) or recorded (mass memories). For example, the symbols 0 and 1 of the binary alphabet can be represented by the polarity of a voltage applied to one end of a pair of conducting wires, stipulating for example that $+V$ corresponds to 1 and $-V$ to 0. Then, the polarity measure at the other end will show which binary symbol was emitted. At the emitter side, the polarity is changed at regularly spaced intervals to represent the bits of a message and will enable this message to be reconstituted at the receiver side. This scheme is far too simple to illustrate modern telecommunications systems but, generally, it is the sign of a real physical value that represents a binary symbol at the output of the channel. Usually, a binary symbol is represented by a certain waveform and the operation that associates a sequence of waveforms with the sequence of symbols of the message is the modulation. Modulation was the subject of the previous chapter.

We consider a situation where the channel is not very reliable, that is, where the observation at the receiving end does not enable the bit really emitted to be identified with certitude because an interference value, noise, independent of the emitted message and random, is added to the useful value (spurious effects of attenuation can also be added, like on the Rayleigh channel). Thermal noise is well represented by a Gaussian random process. When demodulation is performed in an optimal way, it results in a random Gaussian variable whose sign represents the best hypothesis concerning the binary symbol emitted. The channel is then characterized by its *signal to noise ratio*, defined as the ratio of the power of the useful signal to that of the perturbing noise. For a given signal to noise ratio, the decisions taken on the binary symbols emitted are assigned a constant error probability, which leads to the simple model of the binary symmetric channel.

3.1.2 An example: the binary symmetric channel

This is the simplest channel model, and it has already been mentioned in Section 1.3. A channel can generally be described by the probabilities of the transition of the symbols that are input, towards the symbols that are output. A binary symmetric channel is thus represented in Figure 3.1. This channel is *memoryless*, in the sense that it operates separately on the successive input bits. Its input symbol X and its output symbol Y are both binary. If $X = 0$ (respectively $X = 1$), there exists a probability p that $Y = 1$ (resp. $Y = 0$). p is called the error probability of the channel.

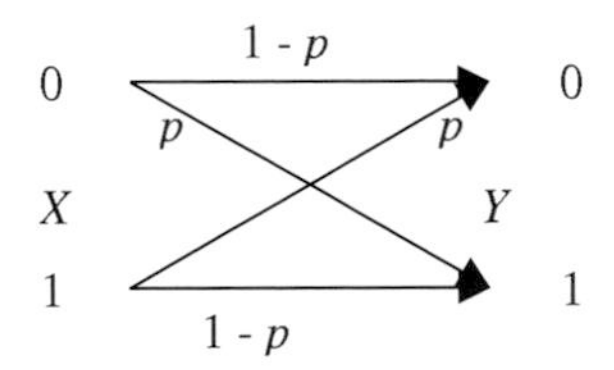

Figure 3.1 – Binary symmetric channel with error probability p. The transition probabilities of an input symbol towards an output symbol are equal two by two.

Another description of the same channel can be given in the following way: let E be a binary random variable taking value 1 with a probability $p < 1/2$ and value 0 with the probability $1-p$. The hypothesis that $p < 1/2$ does not restrict the generality of the model because changing the arbitrary signs 0 and 1 leads to replacing an initial error probability $p > 1/2$ by $1 - p < 1/2$. The behaviour of the channel can be described by the algebraic expression $Y = X \oplus E$, where X and Y are the binary variables at the input and at the output of the channel respectively, E a binary error variable, and $\oplus$ represents the modulo 2 addition.

Configurations of errors on the binary symmetric channel

Let us now suppose that we no longer consider a particular single symbol, but a set of n symbols (consecutive or not) making up a *word*, denoted

$\mathbf{x} = (x_1 x_2 \ldots x_n)$. The operation of the channel is described by the vector addition modulo 2 of $\mathbf{x}$ and of an error vector $\mathbf{e} = (e_1 e_2 \ldots e_n)$:

$$\mathbf{y} = \mathbf{x} \oplus \mathbf{e} \tag{3.1}$$

with $\mathbf{y} = (y_1 y_2 \ldots y_n)$, the notation $\oplus$ now designating the modulo 2 addition of two words, symbol to symbol. The hypothesis that the binary symmetric channel is memoryless means that the random variables e_i, $i = 1...n$, are mutually independent. The number of configurations of possible errors is 2^n, and their probability, for an error probability p of the given channel, depends only on the weight $w(e)$ of the configuration of errors e realized, defined as the number of 1 symbols that it contains. Thus, the probability of the appearance of a particular configuration of errors of weight $w(e)$ affecting a word of length n equals:

$$P_{\mathbf{e}} = p^{w(\mathbf{e})}(1-p)^{n-w(\mathbf{e})} \tag{3.2}$$

As p was assumed to be lower than 1/2, probability P_e is a decreasing function of the weight $w(\mathbf{e})$, whatever n.

The probability of the appearance of any configuration of errors of weight w equals:

$$P_w = \binom{n}{w} p^w (1-p)^{n-w} \tag{3.3}$$

The weight of the error configurations thus follows a Bernoulli distribution whose mathematical expectation (or mean) is np and the variance (the expectation of the square of the difference between its effective value and its mean) is $np(1-p)$.

Mutual information and capacity of the binary symmetric channel

To characterize a channel, we first have to measure the quantity of information that a symbol Y leaving a channel provides, on average, about the corresponding symbol that enters, X. This value called *mutual information* and whose unit is the Shannon (Sh), is defined for a discrete input and output channel by:

$$I(X;Y) = \sum_X \sum_Y \Pr(X,Y) \log_2 \frac{\Pr(X\,|Y\,)}{\Pr(X)} = \sum_X \sum_Y \Pr(X,Y) \log_2 \frac{\Pr(X,Y)}{\Pr(X)\Pr(Y)} \tag{3.4}$$

In this expression, the sums are extended to all the discrete values that X and Y can take in a given alphabet. $\Pr(X,Y)$ denote the joint probability of X and Y, $\Pr(X|Y)$ the probability of X conditionally to Y (that is, when Y is given), $\Pr(X)$ and $\Pr(Y)$ are the marginal probabilities of X and Y (that is, of each of the variables X and Y whatever the value taken by the other: $\Pr(X) = \sum_Y \Pr(X,Y)$ and $\Pr(Y) = \sum_X \Pr(X,Y)$). These different probabilities are linked according to Bayes' law:

$$\Pr(X,Y) = \Pr(X\,|Y\,)\Pr(Y) = \Pr(Y\,|X\,)\Pr(X) \tag{3.5}$$

The first equality in (3.4) defined $I(X;Y)$ as the logarithmic increase of the probability of X that results on average from the data Y, that is, the average quantity of information that the knowledge of Y provides about that of X. The second equality in (3.4), deduced from the first using (3.5), shows that this value is symmetric in X and in Y. The quantity of information that Y provides about X is therefore equal to what X provides about Y, which justifies the name of mutual information.

Mutual information is not sufficient to characterize the channel because the former also depends on the *entropy* of the source, that is, the quantity of information that it produces on average per emitted symbol. Entropy, that is, in practice the average number of bits necessary to represent each symbol, is defined by:

$$H(X) = \sum_X \Pr(X) \log_2 (\Pr(X))$$

The *capacity* of a channel is defined as the maximum of the mutual information of its input and output random variables with respect to all the possible probability distributions of the input variables, and it could be demonstrated that this maximum is reached for a symmetric memoryless channel when the input variable of the channel, X, has equiprobable values (which also causes the entropy of the source to be maximum). For example, for the binary symmetric channel, the capacity is given by:

$$C = 1 + p\log_2 (p) + (1-p)\log_2 (1-p) \text{ (Sh)} \tag{3.6}$$

This capacity is maximum for $p = 0$ (then it equals 1 Sh, like the entropy of the source: the channel is then "transparent") and null for $p = 1/2$, which is what we could expect since then there is total incertitude.

3.1.3 Overview of the fundamental coding theorem

The simplest code that we can imagine is the repetition code that involves emitting information bits in the form of several identical symbols. Hard decoding is performed according to the principle of a majority vote, and soft decoding by the simple addition of the samples received. If the channel is Gaussian, repetition coding provides no gain in the case of soft decoding. For example, transmitting the same symbol twice, allocating each of them half of the available energy and then reconstituting the emitted symbol by addition does not give a better result than transmitting a single symbol with all the energy available. As for the majority vote, it can only be envisaged from a triple emission and in all cases, on a Gaussian channel this procedure degrades the budget link in relation to the non-coded solution. It should however be noted that repeating messages is a widespread technique, not as a procedure for error correction coding, but as a technique for recovering packets of erroneous messages or messages lost during transmission. This technique called ARQ (Automatic Repeat reQuest) cannot be implemented in all systems, in particular in point to multipoint links (e.g. television broadcasting).

The codes are ideally efficient only if their codewords are long, in the sense that the error probability can be made arbitrarily small only if the length of these codewords tends towards infinity. In addition, a good code must keep an emission or coding rate $R = k/n$ non-null when the number k of information bits tends towards infinity. That an error-free communication is effectively possible asymptotically for a non-null emission rate is a major result of information theory, called the *fundamental theorem of channel coding*, which preceded attempts to construct practical codes, thus of finite length. This theorem was a powerful incentive in the search for ever more efficient new codes. Moreover, it presented engineers with a challenge, insofar as the proof of the theorem was based on *random coding*, whose decoding is far too complex to be envisaged in practice.

Although the mathematical proof of the fundamental theorem in its most general form contains fairly difficult mathematics, we believe that it can be easily understood with the help of the *law of large numbers*. This law simply says that experimental realizations have frequencies, defined as the ratio of the number of occurrences noted to the total number of attempts, which tend towards the probabilities of the corresponding events when the number of attempts tend towards infinity. Let us consider, for example, the game of heads and tails. With an "honest" coin, after 10000 throws we could theoretically arrive at the sequence consisting exclusively of all heads (or all tails), but with a probability that is only $2^{-10000} \approx 10^{-3010}$ (in comparison, one second represents about 10^{-18} of the time that has elapsed since the creation of the universe). In stark contrast, the probability that the frequency of the heads (or tails) is close to the mean 1/2 (belonging for example to the interval 0,47-0,53) is in the neighbourhood of 1. In a similar way, an error configuration with a weight close to np when n symbols are emitted on a binary symmetric channel of error probability p is very likely, on condition that the message sent is sufficiently long.

3.1.4 Geometrical interpretation

Consider the finite space S_n of the codewords of n bits having the minimum Hamming distance d. It contains 2^n elements that are said to be its points. In geometrical terms, saying that the number of errors is close to np with high probability means that the received word is represented by a point that, with high propability it is very close to the surface of a hypersphere with n dimensions in S_n, centred on the emitted word and whose radius is equal to the expected mean number of errors np. If the minimum distance d of the code is higher than twice this number, the point on the surface of this hypersphere is closer to the word effectively emitted than to any other codeword and therefore identifies it without ambiguity. The optimal decoding rule, which was presented in Chapter 1, can therefore be stated thus:

"*Choose the codeword closest to the received word*"

The larger n is, the smaller the probability that this rule has an erroneous result is, and this probability tends towards 0 (assuming that p is kept constant) when n tends towards infinity, provided that $d > 2np$. So d has also to tend towards infinity.

Still in geometrical terms, the construction of the best possible code can therefore be interpreted as involving choosing $M < 2^n$ points belonging to S_n in such a way that they are as far away as possible from each other (note that the inequality $M < 2^n$ implies that the code is necessarily redundant). For a given value of the error probability p of the channel (still assumed to be binary symmetric) it is clear that there is a limit to the number M of points that can be placed in S_n while maintaining the distance between these points higher than $2np$. Let $M_{\max}$ be this number. The value

$$C = \lim_{n \to \infty} \frac{\log_2 (M_{\max})}{n}$$

measures in shannons the greatest quantity of information per symbol that can be communicated without any errors through the channel, and it happens to coincide with the capacity of the channel defined in Section 3.1. No explicit procedure making it possible to determine $M_{\max}$ points in S_n while maintaining the distance between these points higher than $2np$ is generally known, except in a few simple, not very useful, cases.

3.1.5 Random coding

Random coding, that is, the construction of a code by randomly choosing its elements, is a way of choosing M scattered points in the space S_n. This method is optimal for the distribution of distances, when n tends towards infinity. Random coding enables the points to be, on average, equally distributed in all the n dimensions of S_n and it reaches a mean emission rate equal to the capacity of the channel. For a code containing M codewords of length n, it means randomly drawing each bit of a codeword independently of the others with the probability 1/2 that it is 0 or 1, the M codewords that make up the code being drawn in the same way independently from each other. The probability of a particular codeword $\mathbf{c}$ is $P_c = 2^{-n}$. We thus obtain codewords whose weights follow a Bernoulli distribution and the probability of obtaining any codeword of weight w is given by (3.3) for $p = 1/2$, that is:

$$P_w = \binom{n}{w} 2^{-n} \tag{3.7}$$

The mathematical expectation, or mean, of this weight is $n/2$ and its variance equals $n/4$. For very large n, a good approximation of the weight distribution of the codewords obtained by random coding is a Gaussian distribution. If we replace w/n by the continuous random variable X, the probability that $X \in (x, x + \mathrm{d}x)$ is $p_X(x)\mathrm{d}x$, where:

$$p_X(x) = \sqrt{\frac{2n}{\pi}} \exp\left[2n(x - 1/2)^2\right] \tag{3.8}$$

This function has a maximum at $x = 1/2$, therefore for $w = n/2$, and takes symmetric decreasing values when x diverges from $1/2$. It is centred around its maximum $x = 1/2$ and the width of the region where it takes non-negligible values decreases as $1/\sqrt{n}$, and therefore tends towards 0 when n tends towards infinity.

Unfortunately, decoding a code obtained by random coding is impossible in practice since decoding a *single* received word would imply comparing it to *all* the codewords. Since long words are necessary for good performance, the number of codewords (2^{Rn}), and therefore the number of necessary combinations, is considerable if Rn is large, which is the case in practice. This is why research on error correcting codes has been directed towards non-random coding rules offering the path to decoding with reasonable complexity.

No general way is known for constructing a code having $M_{\max}$ codewords, for an arbitrary value of n and a given error probability p. We know with certitude, or we conjecture, that a small number of schemes are optimal for given values of M and n, for a few simple channels. In the absence of a general rule for building optimal codes, research has focused on codes satisfying a simpler criterion: that of minimum distance, that is, the greater a code's minimum distance, the better it is. The pertinence of this criterion was not questioned until the end of the 1980's. This criterion does not take into account the number of codewords at the minimum distance from a given word (or multiplicity), whereas a large value for this number leads to a degradation in performance. Turbo codes, which will be examined in the following chapters, were not initially built to satisfy this criterion. Their minimum distance can be small (at least if we compare it to the known bounds on the largest minimum distance possible and in particular the Gilbert-Varshamov bound which we shall see later) but their multiplicity is also very small. These properties mean that there can be an *error floor*, that is, a far less rapid decrease in the error probability of their decoding as a function of the signal to noise ratio when the latter is large, than when it is small. This error floor phenomenon can also be visible with LDPC codes, although the latter can be designed on the criterion of minimum distance. Be that as it may, in the case of turbo codes like in that of LDPC, since the finality of correction coding is to improve communications when the channel is bad, we could say that these codes are good when they are useful and mediocre when they are less useful.

Codes imitating random coding

A simple idea is to try to build codes "imitating" random coding, in a certain sense. Since the performance of a code depends essentially on the distribution of its distances, and that of a linear code on the distribution of its weights, we can undertake to build a linear code having a weight distribution close to that of random coding. This idea has not been much exploited directly, but we can interpret turbo codes as being a first implementation. Before returning to the design of coding procedures, we will make an interesting remark concerning codes that imitate random coding.

The probability of obtaining a codeword of length n and weight w by randomly drawing the bits 0 and 1 each with a probability of $1/2$, independently of each other, is given by (3.7). Drawing a codeword 2^k times, we obtain an average number of words of weight w equal to:

$$N_{w,k} = \binom{n}{w} 2^{-(n-k)}$$

Assuming that n, k and w are large, we can express $\binom{n}{w}$ approximately, using the Stirling formula:

$$\binom{n}{w} \approx \frac{1}{\sqrt{2\pi}} \frac{n^{n+1/2}}{w^{w+1/2}\,(n-w)^{n-w+1/2}}$$

The minimal weight obtained on average, that is $w_{\min}$, is the largest number such that $N_{w_{\min},k}$ has value 1 for the best integer approximation. The number $N_{w_{\min},k}$ is therefore small. It will be sufficient for us to put it equal to a constant λ close to 1, which it will not be necessary to detail further because it will be eliminated from the calculation. We must therefore have:

$$2^{-(n-k)} \frac{1}{\sqrt{2\pi}} \frac{n^{n+1/2}}{w_{\min}^{w_{\min}+1/2}\,(n-w_{\min})^{n-w_{\min}+1/2}} = \lambda$$

Taking the base 2 logarithms and ignoring the constant in relation to n, k and $w_{\min}$ that tend towards infinity, we obtain:

$$1 - \frac{k}{n} \approx H_2\left(w_{\min}/n\right)$$

where $H_2(\cdot)$ is the *binary entropy* function:

$$\begin{aligned} H_2(x) &= -x\log_2(x) - (1-x)\log_2(1-x) & \text{for } 0 < x < 1 \\ &= 0 & \text{for } x = 0 \text{ or } x = 1 \end{aligned}$$

The weight $w_{\min}$ is the average minimal weight of a code obtained by drawing at random. Among the set of all the linear codes thus obtained (weights and distances therefore being merged), there is at least one whose minimum distance d is higher than or equal to the average weight $w_{\min}$, so that we have:

$$1 - \frac{k}{n} \leq H_2\left(d/n\right) \tag{3.9}$$

This is the asymptotic form of the Gilbert-Varshamov bound that links the minimum distance d of the code having the greatest minimum distance possible, given the parameters k and n. It is a lower bound but, in its asymptotic form, it is very close to equality. A code whose minimum distance verifies this bound with equality is considered to be good for the minimum distance criterion. This shows that a code built with a weight distribution close to that of random coding is also good for this criterion.

3.2 Theoretical limits to performance

3.2.1 Binary input and real output channel

Only the case of the binary symmetric channel, with constant error probability p, has been considered so far. Instead of admitting a constant error probability, we can consider that the error probability in fact varies from one symbol to another because the noise sample that affects the received value varies randomly. Thus, in the presence of Gaussian noise, the value leaving the optimal demodulator is a Gaussian random variable whose sign represents the optimal decision. We will consider the channel that has this real random variable as its output value, that we denote a. It can be shown that this value is linked to the optimal decision $\hat{x}$, that is, to the best hypothesis concerning the emitted bit x, and to the "instantaneous" error probability p_a, according to the relation:

$$a = -(-1)^{\hat{x}} \ln\left(\frac{1-p_a}{p_a}\right) \tag{3.10}$$

which means, assuming p_a lower than 1/2:

$$p_a = \frac{1}{\exp\left(-(-1)^{\hat{x}}a\right)+1} = \frac{1}{\exp\left(|a|\right)+1} \tag{3.11}$$

We mean by instantaneous error probability the error probability p_a that affects the received symbol when the real value measured at the output of the channel is a. The inequality $p_a < 1/2$ makes $\ln\left(\frac{1-p_a}{p_a}\right)$ positive and then the best decision is $\hat{x} = 1$ when a is positive and $\hat{x} = 0$ when a is negative. In addition, the absolute value $|a| = \ln\left(\frac{1-p_a}{p_a}\right)$ is a decreasing function of the error probability of the decision, and it therefore measures its reliability. It is null for $p_a = 1/2$ and tends towards infinity when error probability p_a tends towards 0 (the decision then becomes absolutely reliable). The real quantity that (3.10) defines is called *relative value* or more often *log likelihood ratio* (LLR) of the corresponding binary symbol.

The capacity of the channel thus defined can be calculated as the maximum with respect to X of the mutual information $I(X;Y)$, defined by generalizing (3.4) to real $Y = a$. This generalization is possible but the expression of the capacity thus obtained will not be given here. We merely note that this capacity is higher than that of the binary symmetric channel that is deduced from it by taking a hard decision, that is, restricted to the binary symbol $Y = \hat{x}$, by a factor that increases when the signal to noise ratio of the channel decreases. It reaches $\pi/2$ when we make this ratio tend towards 0, if the noise is Gaussian. For a given signal to noise ratio, the binary input continuous output channel is therefore better than the binary symmetric channel that can be deduced from it by taking hard decisions. This channel is also simpler than the hard decision channel, since it does not have any means to take a binary decision according to the received real value. Taking a hard decision means losing the information carried by the individual variations of this value, which explains that the capacity of the soft output channel is higher.

3.2.2 Capacity of a transmission channel

Here we will consider the most general case where the input and the output of the channel are no longer only scalar values but can be vectors whose dimension N is a function of the modulation system. For example, we will have $N = 1$ for an amplitude modulation and $N = 2$ for a phase modulation with a 4-point constellation. X and Y are therefore replaced by $\mathbf{X}$ and $\mathbf{Y}$.

Capacity was introduced in Section 3.1 for a discrete input and output channel, and is defined as the maximum of the mutual information of its input and output variables, with respect to all the possible probability distributions of the variables. For any dimension of the signal space, the law remains:

$$C = \max_{p(\mathbf{X})} I\left(\mathbf{X};\mathbf{Y}\right) \tag{3.12}$$

where $I(\mathbf{X};\mathbf{Y})$ is the mutual information between $\mathbf{X}$ and $\mathbf{Y}$. When the input and the output of the channel are real values, and no longer discrete values, the probabilities are replaced by probability densities and the sums in relation (3.4) become integrals. For realizations $\mathbf{x}$ and $\mathbf{y}$ of the random variables $\mathbf{X}$ and $\mathbf{Y}$, we can write the mutual information as a function of the probabilities of $\mathbf{x}$ and $\mathbf{y}$:

$$I\left(\mathbf{X};\mathbf{Y}\right) = \underbrace{\int_{-\infty}^{+\infty} \cdots \int_{-\infty}^{+\infty}}_{2N \text{ times}} p\left(\mathbf{x}\right) p\left(\mathbf{y}\left|\mathbf{x}\right.\right) \log_2 \frac{p\left(\mathbf{y}\left|\mathbf{x}\right.\right)}{p\left(\mathbf{y}\right)} \mathrm{d}\mathbf{x}\mathrm{d}\mathbf{y} \tag{3.13}$$

To determine C, we therefore have to maximize (3.13) which is valid for all types of inputs (continuous, discrete) of any dimension N. In addition, the maximum is reached for equiprobable inputs (see Section 3.1), for which we have:

$$p\left(\mathbf{y}\right) = \frac{1}{M} \sum_{i=1}^{M} p\left(\mathbf{y}\left|\mathbf{x}_i\right.\right)$$

where M is the number of symbols or modulation order. (3.13) can then be written in the form:

$$C = \log_2(M) - \frac{1}{M}\sum_{i=1}^{M} \underbrace{\int_{-\infty}^{+\infty} \cdots \int_{-\infty}^{+\infty}}_{N \text{ times}} p(\mathbf{y}\,|\mathbf{x}_i\,) \log_2 \left(\frac{\sum_{j=1}^{M} p(\mathbf{y}\,|\mathbf{x}_j\,)}{p(\mathbf{y}\,|\mathbf{x}_i\,)} \right) \mathbf{dy} \quad (3.14)$$

According to the additional information available about the transmission, such as the type of noise on the channel, possible fading, the type of input and output (continuous, discrete) and the modulation used, (3.14) can be particularized.

Shannon limit of a band-limited continuous input and output Gaussian channel

Consider the case of a Gaussian channel, with continuous input and output. The Shannon bound [3.3] giving the maximum capacity C of such a channel is reached taking at its input a white Gaussian noise of null mean and variance σ^2, described by independent probabilities on each dimension, that is, such that:

$$p(\mathbf{x}) = \prod_{n=1}^{N} p(x_n)$$

where $\mathbf{x} = [x_1 x_2 \ldots x_N]$ is the input vector and $p(x_n) = N(0, \sigma^2)$. The mutual information is reached for equiprobable inputs, and denoting $N_0/2$ the variance of the noise, (3.14) after development gives:

$$C = \frac{N}{2} \log_2 \left(1 + \frac{2\sigma^2}{N_0}\right).$$

This relation is modified to make the mean energy E_b of each of the bits and consequently the signal to noise ratio $\frac{E_b}{N_0}$. For N=2, we have:

$$C_b = \log_2 \left(1 + R\frac{E_b}{N_0}\right) \quad (3.15)$$

the capacity being expressed in bit per second per Hertz and per couple dimension. Taking $R = 1$, this leads to the ratio E_b/N_0 being limited by the normalized Shannon limit, as shown in Figure 3.2.

Capacity of a discrete input Gaussian channel

The discrete input, denoted $\mathbf{x} = \mathbf{x}_i\,, i = 1, \cdots, M$, is typically the result of a modulation performed before transmission. The inputs $\mathbf{x}_i$ belong to a set of M

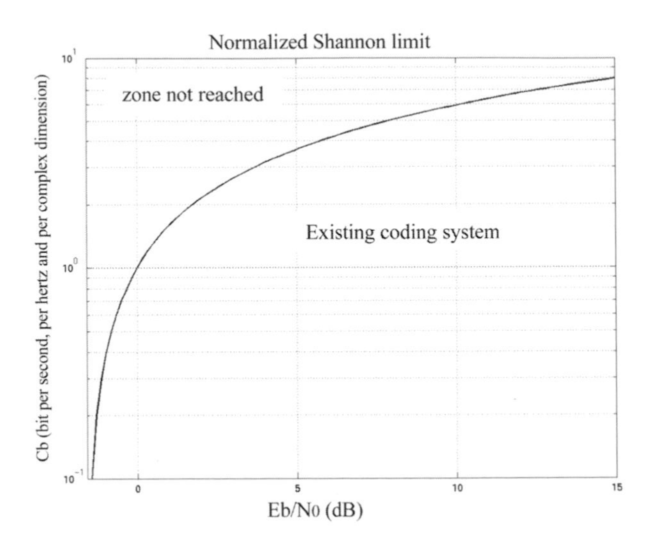

Figure 3.2 – Normalized Shannon limit.

discrete values, M being the modulation order, and have dimension N, that is $\mathbf{x}_i = [x_{i1}, x_{i2}, \cdots, x_{iN}]$. The transition probability of the Gaussian channel is:

$$p(\mathbf{y}\,|\mathbf{x}_i) = \prod_{n=1}^{N} p(y_n\,|x_{in}) = \prod_{n=1}^{N} \frac{1}{\sqrt{\pi N_0}} \exp\left(\frac{-(y_n - x_{in})^2}{N_0}\right)$$

and we assume inputs taking the M different possible values equiprobably. Denoting $\mathbf{d}_{ij} = (\mathbf{x}_i - \mathbf{x}_j)/\sqrt{N_0}$ the vector of dimension N relative to the distance between the symbols $\mathbf{x}_i$ and $\mathbf{x}_j$, and $\mathbf{t}$ an integration vector of dimension N, we obtain a simplified expression of (3.14), representing the capacity of a discrete input Gaussian channel, for any type of modulation:

$$C = \log_2(M) - \frac{(\sqrt{\pi})^{-N}}{M} \sum_{i=1}^{M} \underbrace{\int_{-\infty}^{+\infty} \cdots \int_{-\infty}^{+\infty}}_{N \text{ times}} \exp\left(-|\mathbf{t}|^2\right) \log_2\left[\sum_{j=1}^{M} \exp\left(-2\mathbf{t}\mathbf{d}_{ij} - |\mathbf{d}_{ij}|^2\right)\right] \mathbf{dt} \quad (3.16)$$

C being expressed in bit/symbol. We note that $\mathbf{d}_{ij}$ increases when the signal to noise ratio increases (N_0 decreases) and the capacity tends towards $\log_2(M)$. The different possible modulations only appear in the expression of $\mathbf{d}_{ij}$. The discrete sums from 1 to M represent the possible discrete inputs. For the final calculation, we express $\mathbf{d}_{ij}$ as a function of E_s/N_0 according to the modulation, E_s being the energy per symbol, and the capacity of the channel can be determined using a computer. Figure 3.3 gives the result of the calculation for some PSK and QAM modulations.

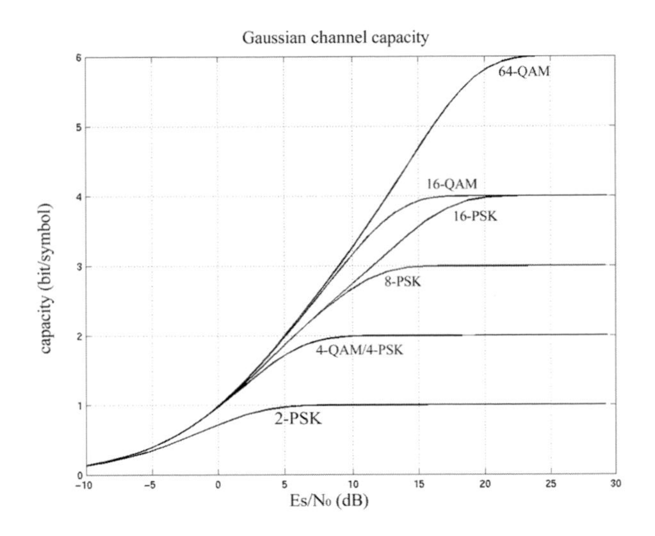

Figure 3.3 – Capacity of some modulations.

Capacity of the Rayleigh channel

Let there be a Rayleigh channel whose attenuation is denoted α. For discrete equiprobable inputs (a case similar to the Gaussian channel treated above), (3.14) is always applicable. There are two cases, conditioned by the knowledge of the attenuation α of the channel, or not.

In the case where α is not known *a priori*, we write the conditional probability density of the Rayleigh channel in the form:

$$\begin{aligned} p(\mathbf{y}\,|\mathbf{x}_i\,) &= \int\limits_{-\infty}^{+\infty} p(\alpha)p(\mathbf{y}\,|\mathbf{x}_i, \alpha\,)\mathrm{d}\alpha \\ &= \int\limits_{0}^{+\infty} \frac{1}{\sqrt{2\pi\sigma^2}} 2\alpha \exp\left(-\alpha^2\right) \exp\left(-\frac{|\mathbf{y}-\alpha\mathbf{x}_i|^2}{2\sigma^2}\right) \mathrm{d}\alpha \end{aligned}$$

One development of this expression means that we can explicitly write this conditional probability density that turns out to be independent of α:

$$\begin{aligned} p(\mathbf{y}\,|\mathbf{x}_i\,) &= \sqrt{\frac{2}{\pi}} \frac{\sigma e^{-\frac{|\mathbf{y}|^2}{2\sigma^2}}}{|\mathbf{x}_i|^2+2\sigma^2} + 2\frac{\mathbf{x}_i \mathbf{y} e^{-\frac{|\mathbf{y}|^2}{|\mathbf{x}_i|^2+2\sigma^2}}}{\left(|\mathbf{x}_i|^2+2\sigma^2\right)^{3/2}} \\ &\times \left[1 - \frac{1}{2}\mathrm{erfc}\left(\frac{\mathbf{x}_i \mathbf{y}}{\sigma\sqrt{2\left(|\mathbf{x}_i|^2+2\sigma^2\right)}}\right)\right] \end{aligned}$$

which is sufficient to enable the capacity to be evaluated by using (3.14).
In the case where the attenuation is known, the probability density for a particular realization of α can be written:

$$p(\mathbf{y}|\mathbf{x}_i, \alpha) = \frac{1}{\sqrt{2\pi\sigma^2}} \exp\left(-\frac{|\mathbf{y} - \alpha\mathbf{x}_i|^2}{2\sigma^2}\right)$$

The instantaneous capacity C_α for this particular realization of α is first calculated and then we have to average C_α over the set of realizations of α in order to obtain the capacity of the channel:

$$C_\alpha = \frac{1}{M}\sum_{i=1}^{M}\int_{-\infty}^{+\infty} p(\mathbf{y}|\mathbf{x}_i, \alpha)\log_2\left(\frac{p(\mathbf{y}|\mathbf{x}_i, \alpha)}{p(\mathbf{y}|\alpha)}\right)\mathrm{d}\mathbf{y}$$

$$C = \int_0^{+\infty} C_\alpha p(\alpha)\,\mathrm{d}\alpha = E[C_\alpha]$$

3.3 Practical limits to performance

In the sections above, we obtained the theoretical limits for performance which are subject to certain hypotheses that are not realistic in practice, in particular the transmission of infinite length data blocks. In the great majority of communication systems today, it is a sequence of data blocks that is transmitted, these blocks being of very variable size depending on the system implemented. Logically, limited size block transmission leads to a loss of performance compared to infinite size block transmission, because the quantity of redundant information contained in the codewords is lower.

Another parameter used to specify the performance of real transmission systems is the packet error rate (PER), which corresponds to the proportion of blocks of wrong data (containing at least one binary error after decoding).

What follows contains some results on the Gaussian channel, for two cases: the binary input and continuous output Gaussian channel, and the continuous input and output Gaussian channel. The case of the continuous input can be assimilated to that of a modulation with an infinite number of states M. The fewer states we have to describe the input, the less efficient the communication. Consequently, a binary input channel gives a lower bound on the practical performance of the set of modulations, whereas a continuous input channel gives its higher limit.

3.3.1 Gaussian binary input channel

Initial work on this channel was done by Gallager [3.2]. We denote again $p(\mathbf{y}|\mathbf{x})$ the probability of transition on the channel, and we consider information messages of size k. Assuming that a message, chosen arbitrarily and equiprobably among 2^k, is encoded then transmitted through the channel, and assuming that we use maximum likelihood decoding, then the coding theorem provides a bound on the mean error probability of decoding the correct codeword. In [3.2], it is shown that it is possible to limit the PER in the following way, for whatever value of variable ρ, $0 \leq \rho \leq 1$:

$$\mathrm{PER} \leq (2^k - 1)^\rho \sum_{\mathbf{y}} \left[\sum_{\mathbf{x}} \Pr(\mathbf{x}) p(\mathbf{y}|\mathbf{x})^{1/1+\rho} \right]^{1+\rho} \tag{3.17}$$

In the case of a channel with equiprobable binary inputs, the probability of each of the inputs is 1/2 and the vectors $\mathbf{x}$ and $\mathbf{y}$ can be treated independently in x and y scalar values. Considering that (3.17) is valid for any ρ, in order to obtain the closest upper bound to the PER, we must minimize the right-hand side of (3.17) as a function of ρ. Introducing the rate $R = k/n$, it therefore means minimizing for $0 \leq \rho \leq 1$, the expression :

$$\left\{ 2^{\rho R} \int_{-\infty}^{+\infty} \frac{1}{2} \left(\frac{1}{\sigma\sqrt{2\pi}} \right)^{\frac{1}{1+\rho}} \times \left[\exp\left(-\frac{(y-1)^2}{2\sigma^2(1+\rho)} \right) + \exp\left(-\frac{(y+1)^2}{2\sigma^2(1+\rho)} \right) \right] dy \right\}^k$$

The explicit value of σ is known for binary inputs (2-PSK and 4-PSK modulations): $\sigma = (2RE_b/N_0)^{-1/2}$. An exploitable expression of Gallager's upper bound on the PER of a binary input channel is then:

$$e^{-k\frac{E_b}{N_0}} \min_{0 \leqslant \rho \leqslant 1} \left\{ \int_0^{+\infty} 2^{\rho R+1} \frac{\exp^{-y^2}}{\sqrt{\pi}} \left(\cosh\left(\frac{y\sqrt{4RE_b/N_0}}{1+\rho} \right) \right)^{1+\rho} dy \right\}^k \tag{3.18}$$

This expression links the PER, the rate, the size k of the messages and the signal to noise ratio E_b/N_0, for a Gaussian binary input channel. It gives an upper bound of the PER and not an equality. This equation is not very well adapted to all cases. In particular, simulations show that for a rate close to 1, the bound is far too lax and does not give really useful results.

If we want to determine the penalty associated with a given packet size, we can compare the result obtained by evaluating (3.18) with the result obtained by computing the capacity that considers infinite size packets

3.3.2 Gaussian continuous input channel

In the case of a continuous input channel, we consider the case contrasting with that of the binary input channel, that is, we will obtain an upper bound on the practical limits of performance (all the modulations show performance lower than a continuous input channel). Any modulation used will give performance lower bounded by the limit obtained by a binary input and upper bounded by a continuous input.

The first results were given by Shannon [3.4] and by the so-called *sphere-packing bound* method which provides a lower bound on the error probability of random codes on a Gaussian channel. We again assume maximum likelihood decoding. A codeword of length n is a sequence of n whole numbers. Geometrically, this codeword can be assimilated to a point in an n-dimensional Euclidean space and the noise can be seen as a displacement of this point towards a neighbouring point following a Gaussian distribution (see Section 3.1.4). Denoting P the power of the emitted signal, all the codewords are situated on the surface of a sphere of radius $\sqrt{nP}$.

Observing that we have a code with 2^k points (codewords), each at a distance $\sqrt{nP}$ from the origin in n-dimensional space, any two points are equidistant from the origin, and consequently, the bisector of these two points (a hyperplane of dimension $n-1$) passes through the origin. Considering the set of 2^k points making up the code, all the hyperplanes pass through the origin and form pyramids with the origin as the summit. The error probability, after decoding, is $\Pr(e) = \frac{1}{2^k}\sum\limits_{i=1}^{2^k} \Pr(e_i)$, where $\Pr(e_i)$ is the probability that the point associated with the codeword i is moved by the noise outside the corresponding pyramid.

The principle of Shannon's sphere-packing bound involves this geometrical vision of coding. However, it is very complex to keep the 'pyramid' approach and the solid angle pyramid Ω_i, around the codeword i, is replaced by a cone with the same summit and the same solid angle Ω_i (Figure 3.4).

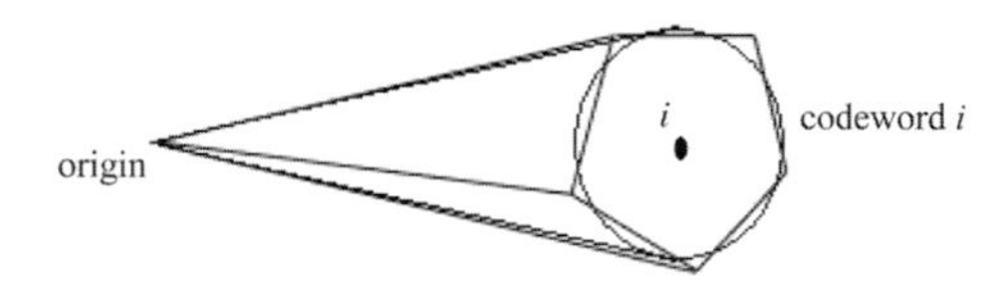

Figure 3.4 – Assimilation of a pyramid with one cone in Shannon's so-called sphere-packing approach.

It can be shown that the probability that the signal remains in the cone is higher than the probability that it remains in the same solid angle pyramid. Consequently, the error probability can be lower-bounded in the following way:

$$\Pr(e) \geq \frac{1}{2^k}\sum_{i=1}^{2^k} Q^*(\Omega_i) \tag{3.19}$$

denoting $Q^*(\Omega_i)$ the probability that the noise moves point i out of the solid angle cone Ω_i (therefore a decoding error is made on this point). We also observe that, if we consider the set of codewords equally distributed on the surface of the sphere of radius $\sqrt{nP}$, the decoding pyramids form a partition of this same sphere, and therefore the solid angle of this sphere Ω_0 is the sum of all the solid angles of the Ω_i pyramids. We can thus replace the solid angles Ω_i by the mean solid angle $\Omega_0/2^k$.

This progression, which leads to a lower bound on the error probability for an optimal decoding of random codes on the Gaussian channel, is called the sphere-packing bound because it involves restricting the coding to an n-dimensional sphere and the effects of the noise to movements on this sphere.

Mathematical simplifications give an exploitable form of the lower bound on the packet error rate (PER):

$$\begin{aligned}
&\ln(\mathrm{PER}) \geq \tfrac{k}{R}\left[\ln\left(G\left(\theta_i, A\right)\sin\theta_i\right) - \tfrac{1}{2}\left(A^2 - AG\left(\theta_i, A\right)\cos\theta_i\right)\right] \\
&\theta_i \approx \arcsin\left(2^{-R}\right) \\
&G\left(\theta_i, A\right) \approx \left(A\cos\theta_i + \sqrt{A^2\cos^2\theta_i + 4}\right)/2 \\
&A = \sqrt{2RE_b/N_0}
\end{aligned} \tag{3.20}$$

These expressions link the size k of the messages, the signal to noise ratio E_b/N_0 and the coding rate R. For high values of R and for block sizes k lower than a few tens of bits, the lower bound is very far from the real PER.

Asymptotically, for block sizes tending towards infinity, the bound obtained by (3.20) tends towards the Shannon limit for a continuous input and output channel such as presented in Section 3.2. In the same way as for the binary input channel, if we wish to quantify the loss caused by the transmission of finite length packets, we must normalize the values obtained by evaluating (3.20) by removing the Shannon limit (3.15) from them, the penalty having to be null when the packet sizes tend towards infinity. The losses due to the transmission of finite length packets in comparison with the transmission of a continuous flow of data are less in the case of a continuous input channel than in the case of a binary input channel.

3.3.3 Some examples of limits

Figure 3.5 below gives an example of penalties caused by the transmission of blocks of size k lower than 10000 bits, in the case of continuous input and in the case of binary input. These penalty values should be combined with the values of capacities presented in Figure 3.3, in order to obtain the absolute limits. As we have already mentioned, this figure is to be considered with caution for small values of k and high PER.

The results obtained concern the Gaussian channel. It is theoretically possible to consider the case of fading channels (Rayleigh, for example) but the computations become complicated and the results very approximate.

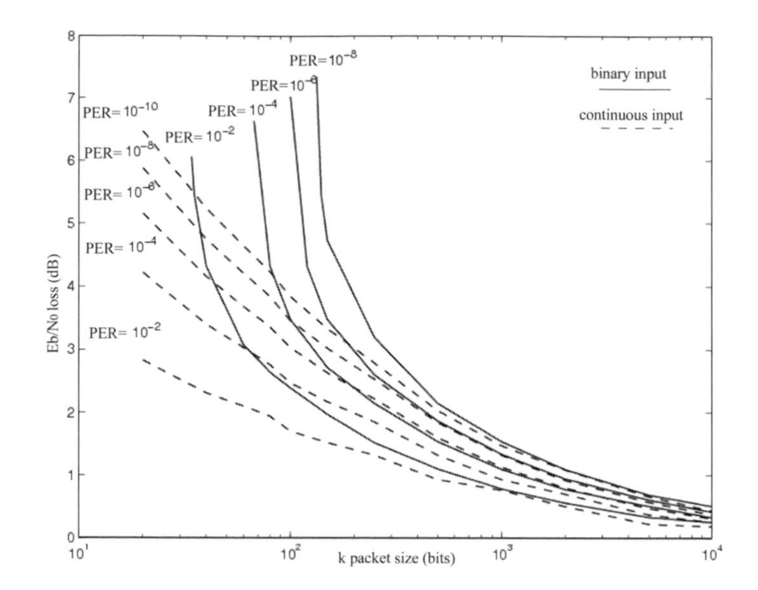

Figure 3.5 – Penalties in $Eb/N0$ for the transmission of finite length packets for the continuous input channel and the binary input channel as a function of size k (information bits), for a coding rate 5/6 and different PER.

3.4 Minimum distances required

So far, we have highlighted the theoretical limits and they have been calculated for the Gaussian channel. These limits determine boundaries, expressed in signal to noise ratio, between transmission channels at the output of which it is possible to correct the errors and channels for which this correction cannot be envisaged. Assuming that codes exist whose decoding can be performed close to these limits, the question now arises about how we can know which minimum Hamming distances (MHD) these codes should have in order to satisfy a given objective of error rates.

Here we present some results for the Gaussian channel and modulations currently used: 4-PSK, 8-PSK and 16-QAM.

3.4.1 MHD required with 4-PSK modulation

With maximum likelihood decoding after transmission on a Gaussian channel, the PER has a known upper bound, called the union bound:

$$\text{PER} \leq \frac{1}{2} \sum_{d \geq d_{\min}} N(d) \operatorname{erfc}\left(\sqrt{dR\frac{E_b}{N_0}}\right) \tag{3.21}$$

where erfc(x) denotes the complementary error function defined by $\mathrm{erfc}\,(x) = \frac{2}{\sqrt{\pi}} \int_x^\infty \exp\left(-t^2\right) \mathrm{d}t$. $d_{\min}$ is the minimum Hamming distance of the code associated with the modulation considered, 2-PSK or 4-PSK in the present case. $N(d)$ represents the multiplicities of the code (see Section 1.5). In certain cases, these multiplicities can be determined precisely (like for example simple convolutional codes, Reed-Solomon codes, BCH codes, etc. ...), and (3.21) can easily be evaluated. For other codes, in particular turbo codes, it is not possible to determine these multiplicities easily and we have to consider some realistic hypotheses in order to get round the problem. The hypotheses that we adopt for turbo codes and for LDPC codes are the following [3.1]:

- *Hypothesis 1: Uniformity.* There exists at least one codeword of weight[1] $d_{\min}$ having an information bit d_i equal to "1", for any place i of the systematic part ($1 \leq i \leq k$).
- *Hypothesis 2: Unicity.* There is only one codeword of weight $d_{\min}$ such that $d_i =$"1".
- *Hypothesis 3: Non-overlapping.* The k codewords of weight $d_{\min}$ associated with the k bits of information are distinct.

Using these hypotheses and limiting ourselves to the first term of the sum in (3.21), the upper bound becomes an asymptotic approximation (low PERs):

$$\mathrm{PER} \approx \frac{k}{2} \mathrm{erfc}\left(\sqrt{d_{\min} R \frac{E_b}{N_0}}\right) \tag{3.22}$$

The three hypotheses, taken separately, are more or less realistic. Hypotheses 1 and 3 are somewhat pessimistic as to the quantity of codewords at the minimum distance. As for hypothesis 2, it is slightly optimistic. The three hypotheses together are suitable for an acceptable approximation of the multiplicity, especially since imprecision about the value of this multiplicity does not affect the quality of the final result. Indeed, the targeted minimum distance that we wish to determine from (3.22) appears in an exponential argument, whereas the multiplicity is a multiplying coefficient.

It is then possible to combine (3.22) with the results obtained in Section 3.3 which provide the signal to noise ratio limits. Giving E_b/N_0 the limit value beyond which using a code is not worthwhile, we can extract from (3.22) the MHD sufficient to reach a PER at that limit value. Given, on the one hand, that (3.22) assumes ideal (maximum likelihood) decoding and, on the other hand, that the theoretical limit is not reached in practice, the targeted MHD can be slightly lower than the result of this extraction.

Figure 3.6 presents some results obtained using this method.

[1] The codes being linear, distance and weight have the same meaning.

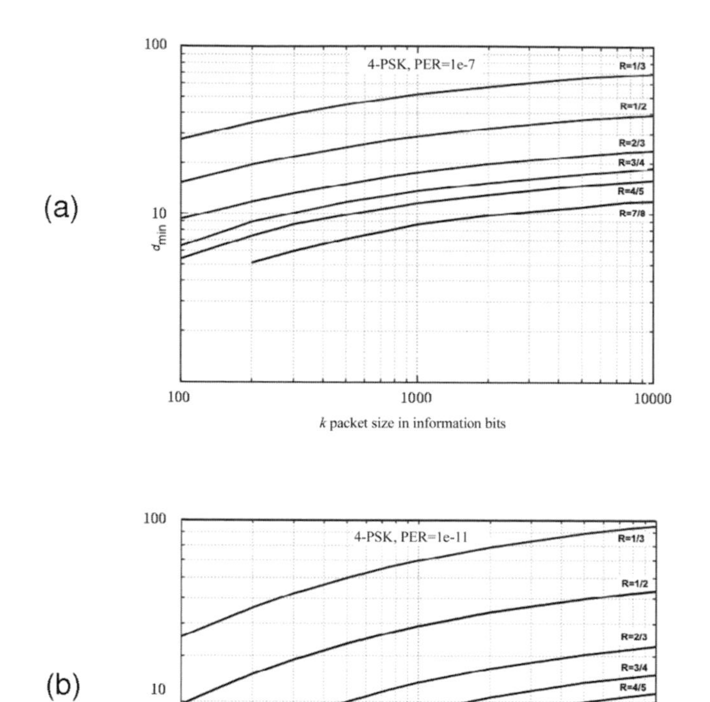

Figure 3.6 – Minimum distances required for 4-PSK modulation and a Gaussian channel as a function of packet size, for some coding rates and $PER = 10^{-7}$ and $PER = 10^{-11}$.

3.4.2 MHD required with 8-PSK modulation

Here we consider an 8-PSK modulation on a Gaussian channel implemented using the principle of the "pragmatic" approach, as presented in Figure 3.7. This approach first involves encoding the data flow in packets to produce codewords that are then randomly permuted by the interleaver Π, with a permutation law drawn randomly. The contents of the permuted codewords are then organized in groups of 3 bits using a Gray coding, before being modulated in 8-PSK. The demodulator provides the received symbols from which we extract the log likelihood ratios (LLRs) for all the bits of the packets. Finally, inverse interleaving and decoding complete the transmission chain.

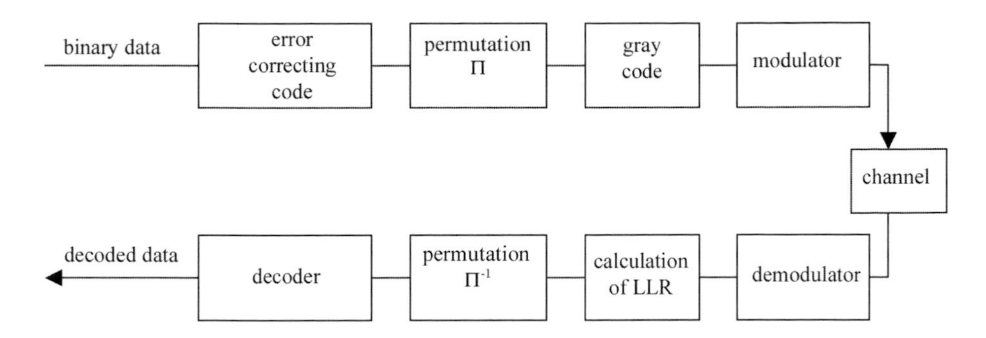

Figure 3.7 – Transmission scheme using the pragmatic approach.

The error probability P_e is the probability of deciding about an incorrect codeword instead of the codeword emitted. Let N_s be the number of modulated symbols that differ between the incorrectly decoded codeword and the codeword emitted. Also let $\{\phi_i\}$ and $\{\phi'_i\}$ $(1 \leq i \leq N_s)$ be the transmitted phase sequences for these symbols that differ. It is possible to express P_e as a function of these phases and of the signal to noise ratio:

$$P_e = \frac{1}{2}\text{erfc}\sqrt{\frac{E_s}{N_0}\left[\sum_{i=1,N_s} \sin^2\left(\frac{\varphi'_i - \varphi_i}{2}\right)\right]} \tag{3.23}$$

where E_s is the energy per symbol emitted and N_0 the monolateral noise power spectral density. It is however not possible to exploit (3.23) in the general case. We require an additional hypothesis, which is then added to the three hypotheses formulated in the previous section, and assume that N_S is much lower than the size of the interleaved codewords:

- *Hypothesis 4:* A symbol does not contain more than one opposite bit in the correct codeword and in the wrong codeword.

This hypothesis allows the following probabilities to be expressed:

$$\Pr\{\varphi_i - \varphi'_i = \pi/4\} = 2/3; \quad \Pr\{\varphi_i - \varphi'_i = 3\pi/4\} = 1/3$$

which means that two times out of three on average, the Euclidean distance between the concurrent symbols is $2\sqrt{\frac{E_s}{T}}\sin(\pi/8)$ and, one time out of three, is raised to $2\sqrt{\frac{E_s}{T}}\sin(3\pi/8)$ (Figure 3.8).

Considering the asymptotic case, that is, putting $N_s = d_{\min}$, yields:

$$\text{PER}_{\text{8-PSK},\Pi\,\text{random}} \approx$$
$$k\left(\tfrac{2}{3}\right)^{d_{\min}} \sum_{j=0}^{d_{\min}} \binom{d_{\min}}{j} \left(\tfrac{1}{2}\right)^{j+1} \text{erfc}\sqrt{\tfrac{E_s}{N_0}\left[j\sin^2\tfrac{3\pi}{8} + (d_{\min}-j)\sin^2\tfrac{\pi}{8}\right]} \tag{3.24}$$

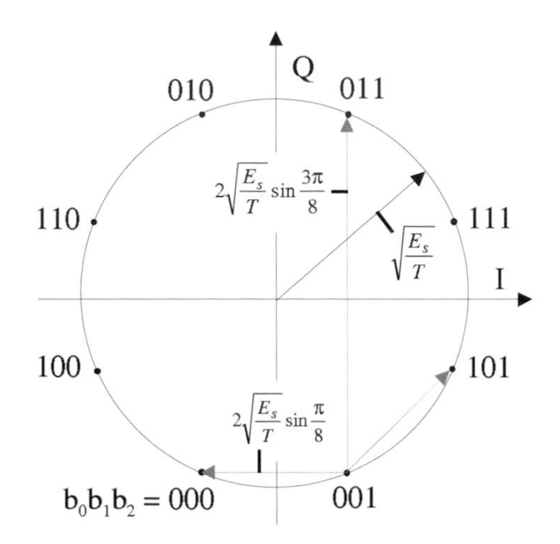

Figure 3.8 – 8-PSK constellation with Gray coding. E_s and T are the energy and the duration of a symbol, respectively.

This relation therefore makes it possible to establish a relation between the signal to noise ratio, the size of the information blocks and the PER. In the same way as in Section 3.4, we can combine this result with the limits on the signal to noise ratio to obtain the MHD targeted for a 8-PSK coded modulation using the pragmatic approach. Figure 3.9 presents some results obtained with this method.

3.4.3 MHD required with 16-QAM modulation

The same method as above, based on the same four hypotheses, can be applied to the case of 16-QAM modulation with pragmatic encoding. The constellation is a standard 16-state Gray constellation. For 75% of the bits making up the symbols, the minimum Euclidean distance is $\sqrt{2E_s/5}$ and for the remaining 25%, this distance is $3\sqrt{2E_s/5}$. Estimating the PER gives:

$$\text{PER} \approx k\left(\frac{3}{4}\right)^{d_{\min}} \sum_{j=0}^{d_{\min}} \binom{d_{\min}}{j} \left(\frac{1}{3}\right)^{j} \frac{1}{2}\text{erfc}\sqrt{(8j+d_{\min})\frac{E_s}{10N_0}} \tag{3.25}$$

Like for 4-PSK and 8-PSK modulations, this relation used jointly with signal to noise ratio limits makes it possible to obtain targeted MHD values for 16-QAM modulation (Figure 3.10).

Some observations can be made from the results obtained in Section 3.4. For example, in the particular case of 4-PSK modulation, for a rate $R = 1/2$, size $k = 4000$ bits and PER of 10^{-11}, Figure 3.6 provides a targeted MHD of 50. From the evaluation that we can make from the Gilbert-Varshamov bound

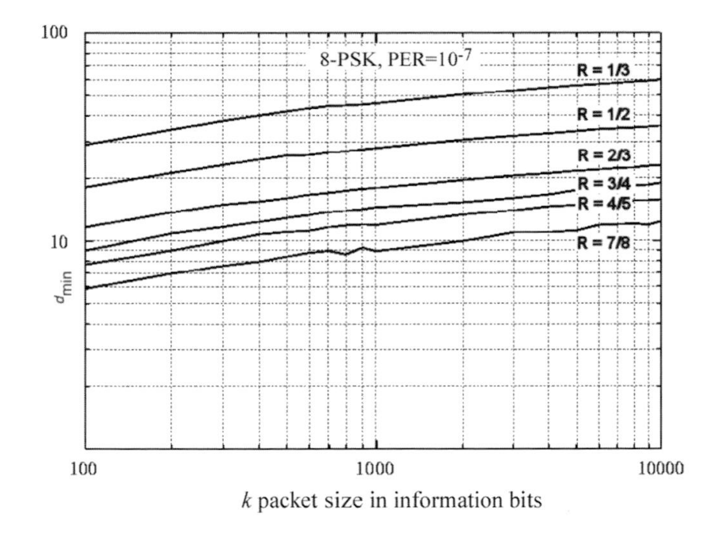

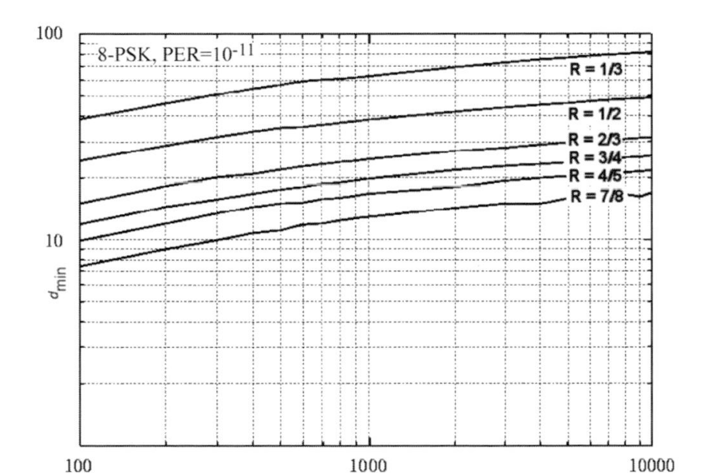

Figure 3.9 – Minimum distances required for 8-PSK modulation and a Gaussian channel as a function of packet size, for some coding rates and $PER = 10^{-7}$ and $PER = 10^{-11}$.

(relation (3.9)), random codes have a minimum distance of about 1000. There is therefore a great difference between what ideal (random) coding can offer and what we really need.

A second aspect concerns the dependency of the required MHD upon the modulation used, a dependency that turns out to be minimum. Thus, a code

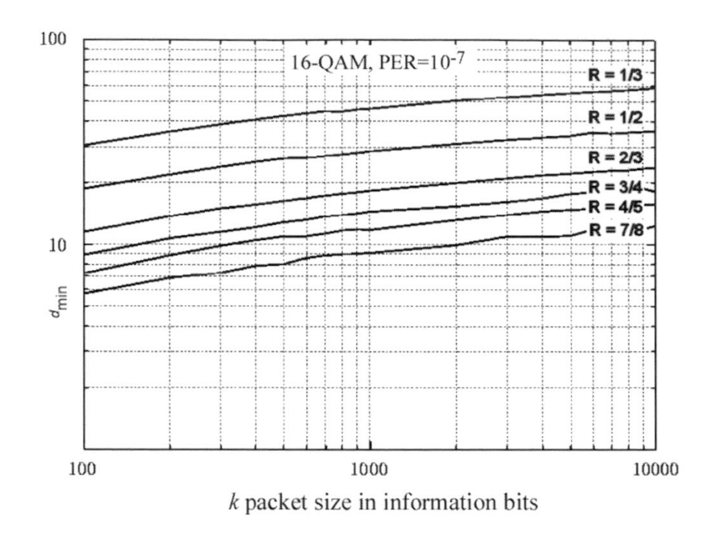

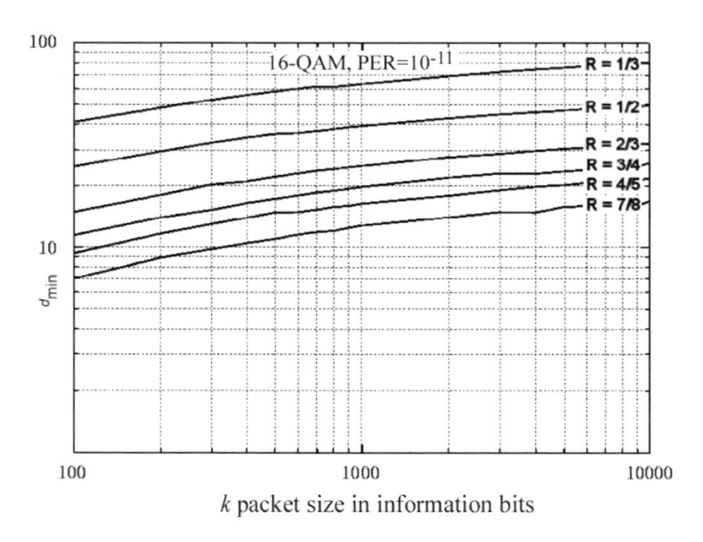

Figure 3.10 – Minimum distances required for 16-QAM modulation and a Gaussian channel as a function of the packet size, for some coding rates and $PER = 10^{-7}$ and $PER = 10^{-11}$.

having a minimum distance sufficient to reach the channel capacity with 4-PSK modulation will also satisfy specifications with the other modulations, for a certain size of message (larger than 1000 bits for $R = 1/2$) or for longer messages (over 5000 bits) if the rate is higher.

Bibliography

[3.1] C. Berrou, E. Maury, and H. Gonzalez. Which minimum hamming distance do we really need? In *Proceedings of the 3rd International Symposium on Turbo Codes & related topics (ISTC 2003)*, pages 141–148, Brest, France, Sept. 2003.

[3.2] R.E. Gallager. *Information Theory and Reliable Communications.* John Wiley & Sons, 1968.

[3.3] C.E. Shannon. A mathematical theory of communication. *Bell Systems Technical Journal*, 27, July-Oct. 1948.

[3.4] C.E. Shannon. Probability of error for optimal codes in a gaussian channel. *Bell Systems Technical Journal*, 38, 1959.

Chapter 4

Block codes

Block coding involves associating with a data block $\mathbf{d}$ of k symbols coming from the information source, a block $\mathbf{c}$, called the codeword, of n symbols with $n \geq k$. The $(n-k)$ is the amount of redundancy introduced by the code. Knowledge of the coding rule at reception enables errors to be detected and corrected, under certain conditions. The ratio k/n is called the coding rate of the code.

The message symbols of the information $\mathbf{d}$ and of the codeword $\mathbf{c}$ take their values in a finite field $\mathbf{F}_q$ with q elements, called a Galois field, whose main properties are given in the appendix to this chapter. We shall see that for most codes, the symbols are binary and take their value in the field $\mathbf{F}_2$ with two elements (0 and 1). This field is the smallest Galois field.

The elementary addition and multiplication operations in field F_2 are resumed in Table 4.1.

a	b	$a + b$	ab
0	0	0	0
0	1	1	0
1	0	1	0
1	1	0	1

Table 4.1 – Addition and multiplication in the Galois field $\mathbf{F}_2$

A block code of length n is an application g of the set $\mathbf{F}_q^k$ towards the set $\mathbf{F}_q^n$ that associates a codeword $\mathbf{c}$ with any block of data $\mathbf{d}$.

$$\begin{array}{rccc} g: & \mathbf{F}_q^k & \rightarrow & \mathbf{F}_q^n \\ & \mathbf{d} & \mapsto & \mathbf{c} = g(\mathbf{d}) \end{array}$$

The set of q^k codewords generally constitutes a very reduced subset of $\mathbf{F}_q^n$.

A block code with parameters (n, k), that we denote $C(n, k)$, is linear if the codewords are a vector subspace of $\mathbf{F}_q^n$, that is, if g is a linear application. A direct consequence of linearity is that the sum of two codewords is a codeword, and that the null word made up of n symbols at zero is always a codeword.

We will now consider linear block codes with binary symbols. Linear block codes with non binary symbols will be addressed later.

4.1 Block codes with binary symbols

In the case of a binary block code, the elements of $\mathbf{d}$ and $\mathbf{c}$ have values in $\mathbf{F}_2$. As g is a linear application, we will be able to describe the coding operation simply as the result of the multiplication of a vector of k symbols representing the data to be coded by a matrix representative of the code considered, called a code generator matrix.

4.1.1 Generator matrix of a binary block code

Let us denote $\mathbf{d} = [d_0 \cdots d_j \cdots d_{k-1}]$ and $\mathbf{c} = [c_0 \cdots c_j \cdots c_{n-1}]$ the dataword and the associated codeword. Expressing the vector $\mathbf{d}$ from a base $(\mathbf{e}_0, \ldots, \mathbf{e}_j, \ldots, \mathbf{e}_{k-1})$ of $\mathbf{F}_2^k$, we can write:

$$\mathbf{d} = \sum_{j=0}^{k-1} d_j \mathbf{e}_j \tag{4.1}$$

Taking into account the fact that application g is linear, the word $\mathbf{c}$ associated with $\mathbf{d}$ is equal to:

$$\mathbf{c} = g(\mathbf{d}) = \sum_{j=0}^{k-1} d_j g(\mathbf{e}_j) \tag{4.2}$$

Expressing the vector $g(\mathbf{e}_j)$ from a base $(\mathbf{e}'_0, \cdots, \mathbf{e}'_l, \cdots, \mathbf{e}'_{n-1})$ of $\mathbf{F}_2^n$, we obtain:

$$g(\mathbf{e}_j) = \sum_{l=0}^{n-1} g_{jl} \mathbf{e}'_l \tag{4.3}$$

The vectors $g(\mathbf{e}_j) = \mathbf{g}_j = (g_{j0} \cdots g_{jl} \cdots g_{j,n-1})$, $0 \leq j \leq k-1$ represent the k rows of matrix $\mathbf{G}$ associated with the linear application g.

$$\mathbf{G} = \begin{bmatrix} \mathbf{g_0} \\ \vdots \\ \mathbf{g_j} \\ \vdots \\ \mathbf{g_{k-1}} \end{bmatrix} = \begin{bmatrix} g_{0,0} & \cdots & g_{0,l} & \cdots & g_{0,n-1} \\ \vdots & \ddots & \vdots & \ddots & \vdots \\ g_{j,0} & \cdots & g_{j,l} & \cdots & g_{j,n-1} \\ \vdots & \ddots & \vdots & \ddots & \vdots \\ g_{k-1,0} & \cdots & g_{k-1,l} & \cdots & g_{k-1,n-1} \end{bmatrix} \tag{4.4}$$

Matrix $\mathbf{G}$ with k rows and n columns, having its elements $g_{jl} \in \mathbf{F}_2$ is called a generator matrix of the code $C(n,k)$. It associates the codeword $\mathbf{c}$ with the block of data $\mathbf{d}$ by the matrix relation:

$$\mathbf{c} = \mathbf{dG} \tag{4.5}$$

The generator matrix of a block code is not unique. Indeed, by permuting the vectors of the base $(\mathbf{e'}_0, \ldots, \mathbf{e'}_l, \ldots, \mathbf{e'}_{n-1})$ or of the base $(\mathbf{e}_0, \ldots, \mathbf{e}_j, \ldots, \mathbf{e}_{k-1})$, we obtain a new generator matrix $\mathbf{G}$ whose columns or rows have also been permuted. Of course, the permutation of the columns or the rows of the generator matrix always produces the same set of codewords; what changes is the association between the codewords and the k-uplets of data.

Note that the rows of the generator matrix of a linear block code are independent codewords, and that they make up a base of the vector subspace generated by the code. The generator matrix of a linear block code is therefore of rank k. A direct consequence is that any family made up of k independent codewords can be used to define a generator matrix of the code considered.

Example 4.1

Let us consider a linear block code called the parity check code denoted $C(n,k)$, with $k = 2$ and $n = k + 1 = 3$ (for a parity check code, the sum of the symbols of a codeword is equal to zero). We have four codewords:

Dataword	Codeword
00	000
01	011
10	101
11	110

To write a generator matrix of this code, let us consider, for example, the canonical base of $\mathbf{F}_2^2$:

$$\mathbf{e}_0 = \begin{bmatrix} 1 & 0 \end{bmatrix}, \mathbf{e}_1 = \begin{bmatrix} 0 & 1 \end{bmatrix}$$

and the canonical base of $\mathbf{F}_2^3$:

$$\mathbf{e'}_0 = \begin{bmatrix} 1 & 0 & 0 \end{bmatrix}, \mathbf{e'}_1 = \begin{bmatrix} 0 & 1 & 0 \end{bmatrix}, \mathbf{e'}_2 = \begin{bmatrix} 0 & 0 & 1 \end{bmatrix}$$

We can write:

$$g(\mathbf{e}_0) = [101] = 1.\mathbf{e'}_0 + 0.\mathbf{e'}_1 + 1.\mathbf{e'}_2$$
$$g(\mathbf{e}_1) = [011] = 0.\mathbf{e'}_0 + 1.\mathbf{e'}_1 + 1.\mathbf{e'}_2$$

A generator matrix of the parity check code is therefore equal to :

$$\mathbf{G} = \begin{bmatrix} 1 & 0 & 1 \\ 0 & 1 & 1 \end{bmatrix}$$

By permuting the first two vectors of the canonical base of $\mathbf{F}_2^3$, we obtain a new generator matrix of the same parity check code:

$$\mathbf{G}' = \begin{bmatrix} 0 & 1 & 1 \\ 1 & 0 & 1 \end{bmatrix}$$

In this example, we have just seen that the generator matrix of a block code is not unique. By permuting the rows or the columns of a generator matrix or by adding one or several other rows to a row, which means considering a new base in $\mathbf{F}_2^n$, it is always possible to write a generator matrix of a block code in the following form:

$$\mathbf{G} = \begin{bmatrix} \mathbf{I}_k & \mathbf{P} \end{bmatrix} = \begin{bmatrix} 1 & 0 & \cdots & 0 & p_{0,1} & \cdots & p_{0,l} & \cdots & p_{0,n-k} \\ 0 & 1 & \cdots & 0 & p_{1,1} & \cdots & p_{1,l} & \cdots & p_{1,n-k} \\ \vdots & \vdots & \cdots & \vdots & \vdots & \cdots & \vdots & \cdots & \vdots \\ 0 & 0 & \ldots & 1 & p_{k-1,1} & \cdots & p_{k-1,l} & \cdots & p_{k-1,n-k} \end{bmatrix} \quad (4.6)$$

where $\mathbf{I}_k$ is the identity matrix $k \times k$ and $\mathbf{P}$ a matrix $k \times (n-k)$ used to calculate the $(n-k)$ redundancy symbols.
Written thus, the generator matrix $\mathbf{G}$ is in a reduced form and produces codewords of the form:

$$\mathbf{c} = \begin{bmatrix} \mathbf{d} & \mathbf{d}P \end{bmatrix} \quad (4.7)$$

The code is therefore systematic. Following 4.7, the code is said to be systematic when there exist k indices $i_0, i_1, \ldots, i_{k-1}$, such that for any data word $\mathbf{d}$, the associated codeword $\mathbf{c}$ satisfies the relation:

$$c_{i_q} = d_q, \qquad q = 0, 1, \cdots, k-1.$$

4.1.2 Dual code and parity check matrix

Before tackling the notion of dual code, let us define the orthogonality between two vectors made up of n symbols. Two vectors $\mathbf{x} = [x_0 \cdots x_j \cdots x_{n-1}]$ and $\mathbf{y} = [y_0 \cdots y_j \cdots y_{n-1}]$ are orthogonal ($\mathbf{x} \perp \mathbf{y}$) if their scalar product denoted $\langle \mathbf{x}, \mathbf{y} \rangle$ is null.

$$\mathbf{x} \perp \mathbf{y} \Leftrightarrow \langle \mathbf{x}, \mathbf{y} \rangle = \sum_{j=0}^{n-1} x_j y_j = \mathbf{0}$$

With each linear block code $C(n,k)$, we can associate a dual linear block code that verifies that any word of the dual code is orthogonal to any word of the code $C(n,k)$. The dual of code $C(n,k)$ is therefore a vector subspace of $\mathbf{F}_2^n$ made up of 2^{n-k} codewords of n symbols. This vector subspace is the orthogonal of the vector subspace made up of 2^k words of the code $C(n,k)$. It results that any word $\mathbf{c}$ of code $C(n,k)$ is orthogonal to the rows of the generator matrix $\mathbf{H}$ of its dual code

$$\mathbf{c}\mathbf{H}^{\mathbf{T}} = \mathbf{0} \tag{4.8}$$

where T indicates the transposition.

A vector $\mathbf{y}$ belonging to $\mathbf{F}_2^n$ is therefore a codeword of code $C(n,k)$ if, and only if, it is orthogonal to the codewords of its dual code, that is, if:

$$\mathbf{y}\mathbf{H}^{\mathbf{T}} = \mathbf{0}$$

The decoder of a code $C(n,k)$ can use this property to verify that the word received is a codeword and thus to detect the presence of errors. That is why matrix $\mathbf{H}$ is called the parity check matrix of code $C(n,k)$.

It is easy to see that the matrices $\mathbf{G}$ and $\mathbf{H}$ are orthogonal ($\mathbf{G}\mathbf{H}^{\mathbf{T}} = \mathbf{0}$). Hence, when the code is systematic and its generator matrix is of the form $\mathbf{G} = [\mathbf{I}_k \;\; \mathbf{P}]$, we have:

$$\mathbf{H} = [\mathbf{P}^{\mathbf{T}} \; \mathbf{I}_{n-k}] \tag{4.9}$$

4.1.3 Minimum distance

Before recalling what the minimum distance of a linear block code is, let return to the notion of Hamming distance that measures the difference between two codewords. The Hamming distance, denoted d_H, is equal to the number of places where the two codewords have different symbols.

We can also define the Hamming weight, denoted w_H, of a codeword as the number of non-null symbols of this codeword. Thus, the Hamming distance between two codewords is also equal to the weight of their sum.

Example 4.2

Let there be two words $\mathbf{u} = [1101001]$ and $\mathbf{v} = [0101101]$. The Hamming distance between $\mathbf{u}$ and $\mathbf{v}$ is 2. Their sum $\mathbf{u} + \mathbf{v} = [1000100]$ has a Hamming weight 2.

The minimum distance $d_{\min}$ of a block code is equal to the smallest Hamming distance between its codewords.

$$d_{\min} = \min_{\mathbf{c} \neq \mathbf{c}'} d_H(\mathbf{c}, \mathbf{c}'), \qquad \forall \mathbf{c}, \mathbf{c}' \in C(n,k) \tag{4.10}$$

Taking into account the fact that the distance between two codewords is equal to the weight of their sum, the minimum distance of a block code is also equal to the minimum weight of its non-null codewords.

$$d_{\min} = \min_{\mathbf{c} \neq 0, \mathbf{c} \in C(n,k)} w_H(\mathbf{c}) \tag{4.11}$$

When the number of codewords is very high, searching for the minimum distance can be laborious. A first solution to get round this difficulty is to determine the minimum distance from the parity check matrix.

We have seen that $d_{\min}$ is equal to the minimum Hamming weight of the non-null codewords. Let us consider a codeword of weight $d_{\min}$. The orthogonality property $\mathbf{cH}^{\mathrm{T}} = \mathbf{0}$ implies that the sum of $d_{\min}$ columns of the parity check matrix is null. Thus $d_{\min}$ corresponds to the minimum number of linearly dependent columns of the parity check matrix.

A second solution to evaluate $d_{\min}$ is to use higher bounds of the minimum distance. A first bound can be expressed as a function of the k and n parameters of the code. For a linear block code whose generator matrix is written in the systematic form $\mathbf{G} = [\mathbf{I}_k \ \mathbf{P}]$, the $(n-k)$ columns of the matrix $\mathbf{I}_{n-k}$ of the parity check matrix ($\mathbf{H} = \left[\mathbf{P}^{\mathrm{T}} \ \mathbf{I}_{n-k}\right]$) being linearly independent, any column of $\mathbf{P}^{\mathrm{T}}$ can be expressed as at most a combination of these $(n-k)$ columns. The minimum distance is therefore upper bounded by:

$$d_{\min} \leq n - k + 1 \tag{4.12}$$

Another bound of the minimum distance, called the Plotkin bound, can be obtained by noting that the minimum distance is necessarily lower than the average weight of the non-null codewords. If we consider the set of codewords, it is easy to see that there are as many symbols at 0 as symbols at 1. Thus the sum of the weights of all the codewords is equal to $n2^{k-1}$. The number of non-null codewords being $2^k - 1$, the minimum distance can be upper bounded by:

$$d_{\min} \leq \frac{n2^{k-1}}{2^k - 1} \tag{4.13}$$

4.1.4 Extended codes and shortened codes

From a block code $C(n, k)$ with minimum distance $d_{\min}$ we can build a linear code $C(n+1, k)$ by adding to the end of each codeword a symbol equal to 1 (respectively to 0) if the codeword includes an odd (respectively even) number of 1s. This code is called an extended code and its minimum distance is equal to $d_{\min} + 1$ if $d_{\min}$ is an odd number.

The parity check matrix $\mathbf{H}_e$ of an extended code is of the form:

$$\mathbf{H}_e = \left[\begin{array}{ccc|c} & & & 0 \\ & \mathbf{H} & & \vdots \\ & & & 0 \\ 1 & \cdots & 1 & 1 \end{array} \right]$$

where $\mathbf{H}$ is the parity check matrix of code $C(n, k)$.

A systematic block code $C(n, k)$ with minimum distance $d_{\min}$ can be shortened by setting $s < k$ data symbols to zero. We thus obtain a systematic linear code $C(n - s, k - s)$. Of course the s symbols set to zero are not transmitted, but they are retained in order to calculate the $(n - k)$ redundancy symbols. The minimum distance of a shortened code is always higher than or equal to the distance of code $C(n, k)$.

4.1.5 Product codes

A product code is a code with several dimensions built from elementary codes. To illustrate these codes, let us consider a product code built from two systematic block codes $C_1(n_1, k_1)$ and $C_2(n_2, k_2)$.

Let there be a table with n_2 rows and n_1 columns. The k_2 first rows are filled with codewords of length n_1 generated by the code $C_1(n_1, k_1)$. The remaining $(n_2 - k_2)$ rows are filled by the redundancy symbols generated by the code $C_2(n_2, k_2)$; the k_2 symbols of each of the n_1 columns being the information bits of the code $C_2(n_2, k_2)$. We can show that the $(n_2 - k_2)$ rows of the table are codewords of code $C_1(n_1, k_1)$. It results that all the rows of the table are codewords of $C_1(n_1, k_1)$ and all the columns of the table are codewords of $C_2(n_2, k_2)$.

The parameters of the two-dimensional product code $C(n, k)$ with minimum distance $d_{\min}$ are equal to the product of the parameters of the elementary codes.

$$n = n_1 n_2 \qquad k = k_1 k_2 \qquad d_{\min} = d^1_{\min} d^2_{\min}$$

where $d^1_{\min}$ and $d^2_{\min}$ are the minimum distances of codes $C_1(n_1, k_1)$ and $C_2(n_2, k_2)$ respectively.

A two-dimensional product code can be seen as a double serial concatenation of two elementary codes (see Chapter 6). An encoder C_1 is fed with k_2 data blocks of length k_1 and it produces k_2 codewords of length n_1 that are written in rows in a matrix. The matrix is read column-wise and produces n_1 blocks of symbols of length k_2 that feed an encoder C_2. The latter in turn produces n_1 codewords of length n_2. Figure 4.1 illustrates the implementation of a two-dimensional product code built from two systematic block codes.

4.1.6 Examples of binary block codes

Parity check code

This code uses a redundancy binary symbol $(n = k + 1)$ determined in such a way as to ensure the nullity of the modulo 2 addition of the symbols of each codeword.

$$\mathbf{c} = \begin{bmatrix} d_0 & d_1 & \cdots & d_{k-2} & d_{k-1} & c_{n-1} \end{bmatrix} \quad \text{with} \quad c_{n-1} = \sum_{j=0}^{k-1} d_j$$

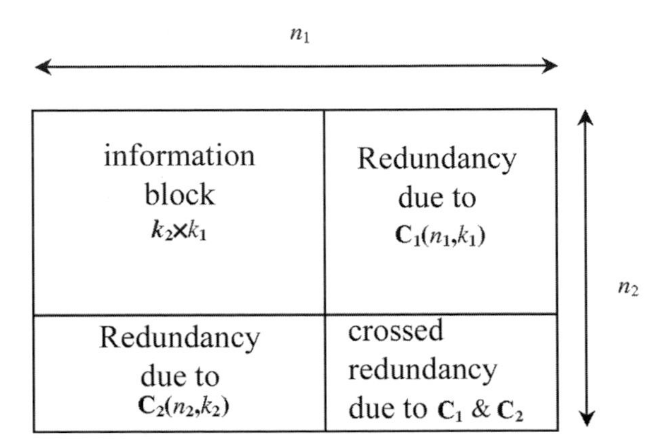

Figure 4.1 – Product code resulting from the serial concatenation of two systematic block codes.

where $\mathbf{d} = \begin{bmatrix} d_0 & d_1 & \cdots & d_{k-1} \end{bmatrix}$ represents the dataword. The minimum distance of this code is 2.

Example 4.3

A generator matrix **G** of this code for $n = 5$, $k = 4$ is equal to:

$$\mathbf{G} = \begin{bmatrix} 1 & 0 & 0 & 0 & 1 \\ 0 & 1 & 0 & 0 & 1 \\ 0 & 0 & 1 & 0 & 1 \\ 0 & 0 & 0 & 1 & 1 \end{bmatrix} = \begin{bmatrix} \mathbf{I}_4 & \mathbf{P} \end{bmatrix}$$

and the parity check matrix **H** is reduced to one vector.

$$\mathbf{H} = \begin{bmatrix} 1 & 1 & 1 & 1 & 1 \end{bmatrix} = \begin{bmatrix} \mathbf{P}^{\mathrm{T}} & \mathbf{I}_1 \end{bmatrix}$$

Repetition code

For this code with parameters $k = 1$ and $n = 2m + 1$, each bit coming from the information source is repeated an odd number of times. The minimum distance of this code is $2m + 1$. The repetition code $C(2m + 1, 1)$ is the dual code of the parity check code $C(2m + 1, 2m)$.

Example 4.4

The generator matrix and the parity check matrix of this code, for $k = 1$, $n = 5$, can be the following:

$$\mathbf{G} = \begin{bmatrix} 1 & 1 & 1 & 1 & 1 \end{bmatrix} = \begin{bmatrix} \mathbf{I}_1 & \mathbf{P} \end{bmatrix}$$

$$\mathbf{H} = \begin{bmatrix} 1 & 1 & 0 & 0 & 0 \\ 1 & 0 & 1 & 0 & 0 \\ 1 & 0 & 0 & 1 & 0 \\ 1 & 0 & 0 & 0 & 1 \end{bmatrix} = \begin{bmatrix} \mathbf{P}^{\mathrm{T}} & \mathbf{I}_4 \end{bmatrix}$$

Hamming code

For a Hamming code, the columns of the parity check matrix are the binary representations of the numbers from 1 to n. Each column being made up of $m = (n-k)$ binary symbols, the parameters of the Hamming code are therefore:

$$n = 2^m - 1 \qquad k = 2^m - m - 1$$

The columns of the parity check matrix being made up of all the possible combinations of $(n-k)$ binary symbols except $(00\cdots0)$, the sum of two columns is equal to one column. The minimum number of linearly dependent columns is 3. The minimum distance of a Hamming code is therefore equal to 3, whatever the value of parameters n and k.

Example 4.5

Let there be a Hamming code with parameter $m = 3$. The codewords and the datawords are then made up of $n = 7$ and $k = 4$ binary symbols respectively. The parity check matrix can be the following:

$$\mathbf{H} = \begin{bmatrix} 1 & 1 & 1 & 0 & 1 & 0 & 0 \\ 1 & 1 & 0 & 1 & 0 & 1 & 0 \\ 1 & 0 & 1 & 1 & 0 & 0 & 1 \end{bmatrix} = \begin{bmatrix} \mathbf{P}^{\mathrm{T}} & \mathbf{I}_3 \end{bmatrix}$$

and the corresponding generator matrix is equal to:

$$\mathbf{G} = \begin{bmatrix} 1 & 0 & 0 & 0 & 1 & 1 & 1 \\ 0 & 1 & 0 & 0 & 1 & 1 & 0 \\ 0 & 0 & 1 & 0 & 1 & 0 & 1 \\ 0 & 0 & 0 & 1 & 0 & 1 & 1 \end{bmatrix} = \begin{bmatrix} \mathbf{I}_4 & \mathbf{P} \end{bmatrix}$$

Maximum length code

The columns of the generator matrix of a maximum length code are the binary representations of the numbers from 1 to n. The parameters of this code are therefore $n = 2^m - 1, k = m$ and we can show that its minimum distance is 2^{k-1}. The maximum length code with parameters $n = 2^m - 1, k = m$ is the dual code of the Hamming code with parameters $n = 2^m - 1, k = 2^m - m - 1$, that is, the generator matrix of the one is the parity check matrix of the other.

Hadamard code

The codewords of a Hadamard code are made up of the rows of a Hadamard matrix and of its complementary matrix. A Hadamard matrix has n rows and n columns (n even) whose elements are 1s and 0s. Each row differs from the other rows at $n/2$ positions. The first row of the matrix is made up only of 0, the other rows having $n/2$ 0 and $n/2$ 1.
For $n = 2$, the Hadamard matrix is of the form:

$$\mathbf{M}_2 = \begin{bmatrix} 0 & 0 \\ 0 & 1 \end{bmatrix}$$

From a $\mathbf{M}_n$ matrix we can generate a $\mathbf{M}_{2n}$ matrix.

$$\mathbf{M}_{2n} = \begin{bmatrix} \mathbf{M}_n & \mathbf{M}_n \\ \mathbf{M}_n & \overline{\mathbf{M}}_n \end{bmatrix}$$

where $\overline{\mathbf{M}}_n$ is the complementary matrix of $\mathbf{M}_n$, that is, where each element at 1 (respectively at 0) of $\mathbf{M}_n$ becomes an element at 0 (respectively at 1) for $\overline{\mathbf{M}}_n$.

Example 4.6

If $n = 4$ $\mathbf{M}_4$ and $\overline{\mathbf{M}}_4$ have the form:

$$\mathbf{M}_4 = \begin{bmatrix} 0 & 0 & 0 & 0 \\ 0 & 1 & 0 & 1 \\ 0 & 0 & 1 & 1 \\ 0 & 1 & 1 & 0 \end{bmatrix} \quad \overline{\mathbf{M}}_4 = \begin{bmatrix} 1 & 1 & 1 & 1 \\ 1 & 0 & 1 & 0 \\ 1 & 1 & 0 & 0 \\ 1 & 0 & 0 & 1 \end{bmatrix}$$

The rows of $\mathbf{M}_4$ and $\overline{\mathbf{M}}_4$ are the codewords of a Hadamard code with parameters $n = 4$, $k = 3$ and with minimum distance equal to 2. In this particular case, the Hadamard code is a parity check code.

More generally, the rows of matrices $\mathbf{M}_n$ and $\overline{\mathbf{M}}_n$ are the codewords of a Hadamard code with parameters $n = 2^m, k = m+1$ and with minimum distance $d_{\min} = 2^{m-1}$.

Reed-Muller codes

A Reed-Muller code (RM) of order r and with parameter m, denoted $\mathrm{RM}_{r,m}$, has codewords of length $n = 2^m$ and the datawords are made up of k symbols with:

$$k = 1 + \binom{m}{1} + \cdots + \binom{m}{r}, \quad \text{with} \quad \binom{N}{q} = \frac{N!}{q!\,(N-q)!}$$

where $r < m$. The minimum distance of an RM code is $d_{\min} = 2^{m-r}$.
The generator matrix of an RM code of order r is built from the generator matrix of an RM code of order $r-1$ and if $\mathbf{G}^{(r,m)}$ represents the generator matrix of the Reed-Muller code of order r and with parameter m, it can be obtained from $\mathbf{G}^{(r-1,m)}$ by the relation:

$$\mathbf{G}^{(r,m)} = \begin{bmatrix} \mathbf{G}^{(r-1,m)} \\ \mathbf{Q}_r \end{bmatrix}$$

where $\mathbf{Q}_r$ is a matrix with dimensions $\binom{m}{r} \times n$.

By construction, $\mathbf{G}^{(0,m)}$ is a row vector of length n whose elements are equal to 1. The matrix $\mathbf{G}^{(1,m)}$ is obtained by writing on each column the binary representation of the index of the columns (from 0 to $n-1$). For example, for $m = 4$, the matrix $\mathbf{G}^{(1,m)}$ is given by:

$$G^{(1,4)} = \begin{bmatrix} 0&0&0&0&0&0&0&0&1&1&1&1&1&1&1&1 \\ 0&0&0&0&1&1&1&1&0&0&0&0&1&1&1&1 \\ 0&0&1&1&0&0&1&1&0&0&1&1&0&0&1&1 \\ 0&1&0&1&0&1&0&1&0&1&0&1&0&1&0&1 \end{bmatrix}.$$

Matrix $\mathbf{Q}_r$ is obtained simply by considering all the combinations of r rows of $\mathbf{G}^{(1,m)}$ and by obtaining the product of these vectors, component by component. The result of this multiplication constitutes a row of $\mathbf{Q}_r$. For example, for the combination having the rows of $\mathbf{G}^{(1,m)}$ with indices i_1, i_2, $\ldots$, i_r, the j-th coefficient of the row thus obtained is equal to $G^{(1,m)}_{i_1,j} G^{(1,m)}_{i_2,j} \cdots G^{(1,m)}_{i_r,j}$, the multiplication being carried out in the field $\mathbf{F}_2$. For example, for $r = 2$, we obtain:

$$\mathbf{Q}_2 = \begin{bmatrix} 0&0&0&0&0&0&0&0&0&0&0&0&1&1&1&1 \\ 0&0&0&0&0&0&0&0&0&0&1&1&0&0&1&1 \\ 0&0&0&0&0&0&0&0&0&1&0&1&0&1&0&1 \\ 0&0&0&0&0&0&1&1&0&0&0&0&0&0&1&1 \\ 0&0&0&0&0&1&0&1&0&0&0&0&0&1&0&1 \\ 0&0&0&1&0&0&0&1&0&0&0&1&0&0&0&1 \end{bmatrix}$$

We can show that the code $\mathrm{RM}_{m-r-1,m}$ is the dual code of the code $\mathrm{RM}_{r,m}$, that is, the generator matrix of code $\mathrm{RM}_{m-r-1,m}$ is the parity check matrix of code $\text{RM}_{r,m}$. For some values of r and m, the generator matrix of code $\text{RM}_{r,m}$ is also its parity check matrix. We then say that code $\text{RM}_{r,m}$ is *self dual*. Code $\text{RM}_{1,3}$, for example, is a *self dual* code.

4.1.7 Cyclic codes

Cyclic codes are the largest class of linear block codes. Their relatively easy implementation, from shift registers and logical operators, has made them attractive and widely-used codes.

Definition and polynomial representation

A linear block code $C(n,k)$ is cyclic if, for any codeword $\mathbf{c} = \begin{bmatrix} c_0 & c_1 & \cdots & c_{n-1} \end{bmatrix}$, $\mathbf{c}_1 = \begin{bmatrix} c_{n-1} & c_0 & \cdots & c_{n-2} \end{bmatrix}$, obtained by circular shift to the right of a symbol of $\mathbf{c}$, is also a codeword. This definition of cyclic codes means that any circular shift to the right of j symbols of a codeword, gives another codeword.
For cyclic codes, we use a polynomial representation of the codewords and of the datawords. Thus, with codeword $\mathbf{c}$ we associate the polynomial $c(x)$ of degree $n-1$.

$$c(x) = c_0 + c_1 x + \cdots + c_j x^j + \cdots + c_{n-1} x^{n-1}$$

and with dataword $\mathbf{d}$ the polynomial $d(x)$ of degree $k-1$.

$$d(x) = d_0 + d_1 x + \cdots + d_j x^j + \cdots + d_{k-1} x^{k-1}$$

where d_j and c_j take their values in $\mathbf{F}_2$.
Multiplying $c(x)$ by x,

$$xc(x) = c_0 x + c_1 x^2 + \cdots + c_j x^{j+1} + \cdots + c_{n-1} x^n$$

then dividing $xc(x)$ by x^n+1, we obtain:

$$xc(x) = (x^n+1)c_{n-1} + c_1(x)$$

where $c_1(x)$ is the remainder of the division of $xc(x)$ by x^n+1 with:

$$c_1(x) = c_{n-1} + c_0 x + \cdots c_j x^{j+1} + \cdots c_{n-2} x^{n-1}$$

We can note that $c_1(x)$ corresponds to the codeword $\mathbf{c}_1 = (c_{n-1} c_0 \ldots c_j \ldots c_{n-2})$.
Using the same method as above, we obtain:

$$x^j c(x) = (x^n+1)q(x) + c_j(x) \tag{4.14}$$

where $c_j(x)$ is also a codeword obtained by j circular shifts to the right of the symbols of $c(x)$.

The codewords of a cyclic code are multiples of a normalized polynomial $g(x)$ of degree $(n-k)$ called a generator polynomial .

$$g(x) = g_0 + g_1 x + \cdots + g_j x^j + \cdots + x^{n-k}$$

where g_j takes its values in $\mathbf{F}_2$. The generator polynomial of a cyclic code is a divisor of $x^n + 1$. There exists a polynomial $h(x)$ of degree k such that equation (4.15) is satisfied.

$$g(x)h(x) = x^n + 1 \tag{4.15}$$

The product $d(x)g(x)$ is a polynomial of degree lower than or equal to $n-1$, so it can represent a codeword. The polynomial $d(x)$ having 2^k realizations, $d(x)g(x)$ enables 2^k codewords to be generated. Let us denote $d_l(x)$ the l-th realization of $d(x)$ and $c_l(x)$ the polynomial representation of the associated codeword. We can write:

$$c_l(x) = d_l(x)g(x) \tag{4.16}$$

We will now show that the codewords satisfying relation (4.16) satisfy the properties of cyclic codes. To do so, we re-write relation (4.14) in the form:

$$c_j(x) = (x^n + 1)q(x) + x^j c(x) \tag{4.17}$$

Since $c(x)$ represents a codeword, there exists a polynomial $d(x)$ of degree at most $k-1$, such that $c(x) = d(x)g(x)$. Using (4.15), we can therefore express (4.17) in another way:

$$c_j(x) = g(x)[h(x)q(x) + x^j d(x)] \tag{4.18}$$

The codewords $c_j(x)$ are therefore multiples of the generator polynomial, and they can be generated from the $d_j(x)$ by applying relation (4.16).

- *Generator polynomial of the dual code of $C(n,k)$*

The dual code of a cyclic block code is also cyclic. Polynomial $h(x)$ of degree k can be used to build the dual code of $C(n,k)$. The reciprocal polynomial $\tilde{h}(x)$ of $h(x)$ is defined as follows:

$$\tilde{h}(x) = x^k h(x^{-1}) = 1 + h_{k-1}x + h_{k-2}x^2 + \cdots + h_1 x^{k-1} + x^k$$

We can write (4.15) differently:

$$[x^{n-k} g(x^{-1})][x^k h(x^{-1})] = x^n + 1 \tag{4.19}$$

The polynomial $\tilde{h}(x)$ is also a divisor of $x^n + 1$; it is the generator polynomial of a $C^\perp = C(n, n-k)$ code that is the dual code of $C(n,k)$.

Note: the code of generator polynomial $h(x)$ is equivalent to dual code $C^\perp$. The vector representation of the codewords generated by $h(x)$ corresponds to the reversed vector representation of the codewords of $C^\perp$.

$$\begin{array}{ccc} C^\perp\text{, generated by } \tilde{h}(x) & \leftrightarrow & \text{Code generated by } h(x) \\ \tilde{\mathbf{c}} = \begin{bmatrix} c_0 & c_1 & \cdots & c_{n-1} \end{bmatrix} & \leftrightarrow & \mathbf{c} = \begin{bmatrix} c_{n-1} & \cdots & c_1 & c_0 \end{bmatrix} \end{array}$$

• *Generator matrix of a cyclic code*

From the generator polynomial $g(x)$ it is possible to build a generator matrix $\mathbf{G}$ of code $C(n,k)$. We recall that the k rows of the matrix $\mathbf{G}$ are made up of k linearly independent codewords. These k codewords can be obtained from a set of k independent polynomials of the form:

$$x^j g(x) \qquad j = k-1, k-2, \ldots, 1, 0.$$

Let $d(x)$ be the polynomial representation of any dataword. The k codewords generated by the polynomials $x^j g(x)$ have the expression:

$$\mathbf{c}_j(x) = x^j g(x) d(x) \qquad j = k-1, k-2, \cdots, 1, 0$$

and the k rows of the matrix $\mathbf{G}$ have for their elements the binary coefficients of the monomials of $\mathbf{c}_j(x)$.

Example 4.7

Let $C(7,4)$ be the generator polynomial code $g(x) = 1 + x^2 + x^3$. Let us take $d(x) = 1$ for the dataword. The 4 rows of the generator matrix $\mathbf{G}$ are obtained from the 4 codewords $\mathbf{c}_j(x)$.

$$\begin{aligned}
\mathbf{c}_3(x) &= x^3 + x^5 + x^6 \\
\mathbf{c}_2(x) &= x^2 + x^4 + x^5 \\
\mathbf{c}_1(x) &= x + x^3 + x^4 \\
\mathbf{c}_0(x) &= 1 + x^2 + x^3
\end{aligned}$$

A generator matrix of the code $C(7,4)$ is equal to:

$$\mathbf{G} = \begin{bmatrix} 0 & 0 & 0 & 1 & 0 & 1 & 1 \\ 0 & 0 & 1 & 0 & 1 & 1 & 0 \\ 0 & 1 & 0 & 1 & 1 & 0 & 0 \\ 1 & 0 & 1 & 1 & 0 & 0 & 0 \end{bmatrix}$$

Cyclic code in systematic form

When the codewords are in systematic form, the data coming from the information source are separated from the redundancy symbols. The codeword $c(x)$ associated with dataword $d(x)$ is then of the form:

$$c(x) = x^{n-k} d(x) + v(x) \tag{4.20}$$

where $v(x)$ is the polynomial of degree at most equal to $n - k - 1$ associated with the redundancy symbols.

Taking into account the fact that $c(x)$ is a multiple of the generator polynomial and that the addition and the subtraction can be merged in $\mathbf{F}_2$, we can then write:

$$x^{n-k}d(x) = q(x)g(x) + v(x)$$

$v(x)$ is therefore the remainder of the division of $x^{n-k}d(x)$ by the generator polynomial $g(x)$. The codeword associated with dataword $d(x)$ is equal to $x^{n-k}d(x)$ increased by the remainder of the division of $x^{n-k}d(x)$ by the generator polynomial.

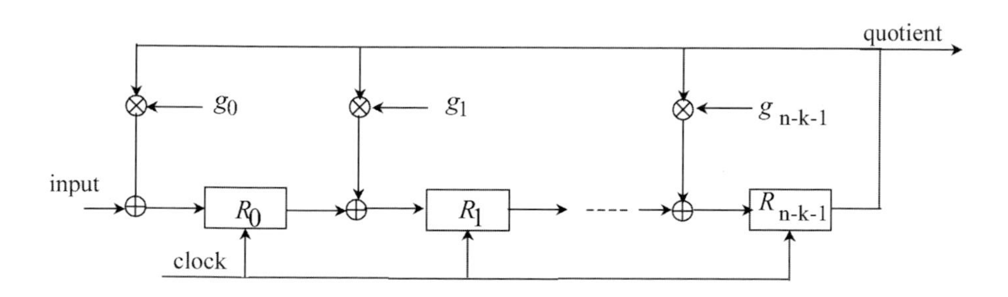

Figure 4.2 – Schematic diagram of a circuit divisor by $g(x)$.

Example 4.8

To illustrate the computation of a codeword written in systematic form, let us take the example of a $C(7,4)$ code of generator polynomial $g(x) = 1+x+x^3$ and let us determine the codeword $c(x)$ associated with message $d(x) = 1+x^2+x^3$, that is:

$$c(x) = x^3 d(x) + v(x)$$

The remainder of the division of $x^3d(x)$ by $g(x) = 1 + x + x^3$ being equal to 1, codeword $\mathbf{c}(x)$ associated with dataword $d(x)$ is:

$$c(x) = 1 + x^3 + x^5 + x^6$$

Thus, with data block $\mathbf{d}$, made up of 4 binary information symbols, is associated codeword $\mathbf{c}$ with:

$$\mathbf{d} = \begin{bmatrix} 1 & 0 & 1 & 1 \end{bmatrix} \rightarrow \mathbf{c} = \begin{bmatrix} 1 & 0 & 0 & 1 & 0 & 1 & 1 \end{bmatrix}$$

To obtain the generator matrix, it suffices to encode

$$d(x) = \quad 1, \quad x, \quad x^2, \quad x^3.$$

We obtain:

$d(x)$	$c(x)$
1	$1+x+x^3$
x	$x+x^2+x^4$
x^2	$1+x+x^2+x^5$
x^3	$1+x^2+x^6$

and thus the generator matrix in a systematic form:

$$\mathbf{G} = \begin{bmatrix} 1 & 1 & 0 & 1 & 0 & 0 & 0 \\ 0 & 1 & 1 & 0 & 1 & 0 & 0 \\ 1 & 1 & 1 & 0 & 0 & 1 & 0 \\ 1 & 0 & 1 & 0 & 0 & 0 & 1 \end{bmatrix}$$

We can verify that for

$$\mathbf{d} = \begin{bmatrix} 1 & 0 & 1 & 1 \end{bmatrix},$$

the matrix product $\mathbf{dG}$ does give

$$\mathbf{c} = \begin{bmatrix} 1 & 0 & 0 & 1 & 0 & 1 & 1 \end{bmatrix}.$$

Implementation of an encoder

We have just seen that the encoder must carry out the division of $x^{n-k}d(x)$ by the generator polynomial $g(x)$ then add the remainder $v(x)$ of this division to $x^{n-k}d(x)$. This operation can be done using only shift registers and adders in field $\mathbf{F}_2$. As the most difficult operation to carry out is the division of $x^{n-k}d(x)$ by $g(x)$, let us first examine the schematic diagram of a divisor by $g(x)$ shown in Figure 4.2. The circuit divisor is realized from a shift register with $(n-k)$ memories denoted R_i and the same number of adders. The shift register is initialized to zero and the k coefficients of the polynomial $x^{n-k}d(x)$ are introduced sequentially into the circuit divisor. After k clock pulses, we can verify that the result of the division is available at the output of the circuit divisor, as well as the remainder $v(x)$ which is in the shift register memories.

The schematic diagram of the encoder shown in Figure 4.3, uses the circuit divisor of Figure 4.2. The multiplication of $d(x)$ by x^{n-k}, corresponding to a simple shift, is realized by introducing polynomial $d(x)$ at the output of the shift register of the divisor.

The k data coming from the information source are introduced sequentially into the encoder (switch I in position 1) that carries out the division of $x^{n-k}d(x)$ by $g(x)$. Simultaneously, the k data coming from the information source are also transmitted. Once this operation is finished, the remainder $v(x)$ of the division is in the $(n-k)$ shift register memories. Switch I then moves to position 2, and the $(n-k)$ redundancy symbols are sent to the output of the encoder.

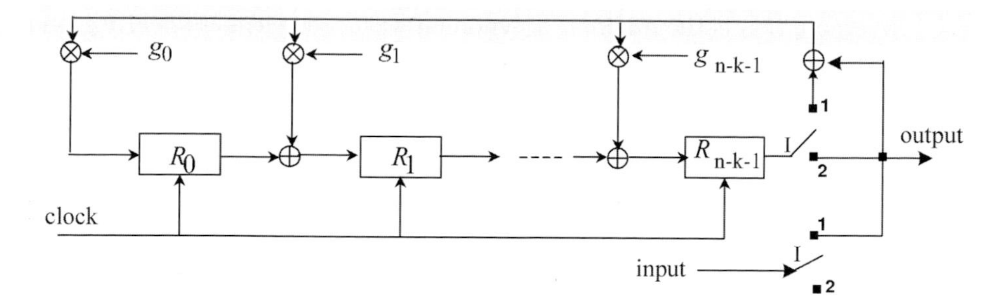

Figure 4.3 – Schematic diagram of an encoder for a cyclic code.

BCH codes

Bose-Chaudhuri-Hocquenghem codes, called BCH codes, enable cyclic codes to be built systematically correcting at least t errors in a block of n symbols, that is, codes whose minimum distance $d_{\min}$ is at least equal to $2t+1$.

To build a BCH code, we set t or equivalently d, called the constructed distance of the code and we determine its generator polynomial $g(x)$. The code obtained has a minimum distance $d_{\min}$ that is always higher than or equal to the constructed distance.

Primitive BCH code

The generator polynomial $g(x)$ of a primitive BCH code constructed over a Galois field $\mathbf{F}_q$ with $q = 2^m$ elements, with a constructed distance d has $(d-1)$ roots of the form: $\alpha^l, \cdots, \alpha^{l+j}, \cdots, \alpha^{l+d-2}$, where $2t+1$ is a primitive element of Galois field $\mathbf{F}_q$ and l an integer. The BCH code is said to be primitive since the roots of its generator polynomial are powers of α, a primitive element of $\mathbf{F}_q$. We will see later that it is possible to build non-primitive BCH codes.

Generally, parameter l is set to 0 or 1 and we show that for a primitive BCH code exponent $(l+d-2)$ of root α^{j+d-2} must be even. When $l=0$, the constructed distance is therefore necessarily even, that is, equal to $2t+2$ for a code correcting t errors. When $l=1$, the constructed distance is odd, that is, equal to $2t+1$ for a code correcting t errors.

- *Primitive BCH code with $l = 1$*

  The generator polynomial of a primitive BCH code correcting at least t errors (constructed distance $2t+1$) therefore has $\alpha, \cdots, \alpha^j, \cdots, \alpha^{2t}$ as roots. We show that the generator polynomial $g(x)$ of a primitive BCH code is equal to:

$$g(x) = \text{S.C.M.}\left(m_\alpha(x), \cdots, m_{\alpha^i}(x), \cdots, m_{\alpha^{2t}}(x)\right)$$

where $m_{\alpha^i}(x)$ is the minimal polynomial with coefficients in field $\mathbf{F}_2$ associated with α^j, and S.C.M. is the Smallest Common Multiple.

It is shown in the appendix that a polynomial with coefficients in $\mathbf{F}_2$ having α^j as its root also has α^{2j} as its root. Thus, the minimal polynomials $m_{\alpha^i}(x)$ and $m_{\alpha^{2i}}(x)$ have the same roots. This remark enables us to simplify the writing of generator polynomial $g(x)$.

$$g(x) = \text{S.C.M.}\left(m_\alpha(x), m_{\alpha^3}(x), \cdots, m_{\alpha^{2t-1}}(x)\right) \tag{4.21}$$

The degree of a minimal polynomial being lower than or equal to m, degree $(n-k)$ of the generator polynomial of a primitive BCH code correcting at least t errors, is therefore lower than or equal to mt. Indeed, $g(x)$ is at most equal to the product of t polynomials of degree lower than or equal to m.

The parameters of a primitive BCH code constructed over a Galois field $\mathbf{F}_q$ with a constructed distance $d = 2t+1$ are therefore the following:

$$n = 2^m - 1; k \geq 2^m - 1 - mt; d_{\min} \geq 2t+1$$

When $t = 1$ a primitive BCH code is a Hamming code. The generator polynomial of a Hamming code, equal to $m_\alpha(x)$, is therefore a primitive polynomial.

Example 4.9

Let us determine the generator polynomial of a BCH code having parameters $m = 4$ and $n = 15$, $t = 2$ and $l = 1$. To do this, we will use a Galois field with $q = 2^4$ elements built from a primitive polynomial of degree $m = 4(\alpha^4 + \alpha + 1)$. The elements of this field are given in the appendix.

We must first determine the minimal polynomials $m_\alpha(x)$ and $m_{\alpha^3}(x)$ associated with elements α and α^3 respectively of field $\mathbf{F}_{16}$.

We have seen in the appendix that if α is a root of polynomial $m_\alpha(x)$ then $\alpha^2, \alpha^4, \alpha^8$ are also roots of this polynomial (raising α to the powers of 16, 32 etc. gives, modulo $\alpha^4+\alpha+1$, elements $\alpha, \alpha^2, \alpha^4, \alpha^8$). We can therefore write:

$$m_\alpha(x) = (x+\alpha)(x+\alpha^2)(x+\alpha^4)(x+\alpha^8)$$

Developing the expression of $m_\alpha(x)$ we obtain:

$$m_\alpha(x) = [x^2 + x(\alpha^2+\alpha) + \alpha^3][x^2 + x(\alpha^8+\alpha^4) + \alpha^{12}]$$

Using the binary representations of the elements of field $\mathbf{F}_{16}$, we can show that $\alpha^2+\alpha = \alpha^5$ and that $\alpha^4+\alpha^8 = \alpha^5$ (we recall that the binary additions are done modulo 2 in the Galois field). We then continue the development of $m_\alpha(x)$ and finally we have:

$$m_\alpha(x) = x^4 + x + 1$$

For the computation of $m_{\alpha^3}(x)$, the roots to take into account are $\alpha^3, \alpha^6, \alpha^{12}, \alpha^{24} = \alpha^9$ $(\alpha^{15} = 1)$, and the other powers of α^3 $(\alpha^{48}, \alpha^{96}, \cdots)$ give the previous roots again. The minimal polynomial $m_{\alpha^3}(x)$ is therefore equal to:

$$m_{\alpha^3}(x) = (x + \alpha^3)(x + \alpha^6)(x + \alpha^{12})(x + \alpha^9)$$

which after development and simplification gives:

$$m_{\alpha^3}(x) = x^4 + x^3 + x^2 + x + 1$$

The S.C.M. of polynomials $m_\alpha(x)$ and $m_{\alpha^3}(x)$ is obviously equal to the product of these two polynomials since they are irreducible and thus, the polynomial generator is equal to:

$$g(x) = (x^4 + x + 1)(x^4 + x^3 + x^2 + x + 1)$$

Developing this, we obtain:

$$g(x) = x^8 + x^7 + x^6 + x^4 + 1$$

Finally the parameters of this BCH code are:

$$m = 4; n = 15; n - k = 8; k = 7; t = 2$$

The numerical values of parameters (n, k, t) of the main BCH codes and the associated generator polynomials have been put table form and can be found in [4.2]. As an example, we give in Table 4.2 the parameters and the generator polynomials, expressed in octals, of some BCH codes with error correction capability $t = 1$ (Hamming codes).

Note : $g(x) = 13$ in octals gives 1011 in binary, that is, $g(x) = x^3 + x + 1$

- *Primitive BCH code with $l = 0$*

The generator polynomial of a primitive BCH code correcting at least t errors (constructed distance $d = 2t + 2$) has $(2t + 1)$ roots of the form: $\alpha^0, \alpha^1, \cdots, \alpha^j, \cdots, \alpha^{2t}$; that is, one root more (α^0) than when $l = 1$. Taking into account the fact that the minimal polynomials $m_{\alpha^j}(x)$ and $m_{\alpha^{2j}}(x)$ have the same roots, generator polynomial $g(x)$ is equal to:

$$g(x) = \text{S.C.M.}(m_{\alpha^0}(x), m_{\alpha^1}(x), m_{\alpha^3}(x), \cdots, m_{\alpha^{2t-1}}(x))$$

n	k	t	g(x)
7	4	1	13
15	11	1	23
31	26	1	45
63	57	1	103
127	120	1	211
255	247	1	435
511	502	1	1021
1023	1013	1	2011
2047	2036	1	4005
4095	4083	1	10123

Table 4.2 – Parameters of some Hamming codes.

1. Parity check code

   Let us consider a BCH code with $l = 0$ and $t = 0$. Its generator polynomial, $g(x) = (x + 1)$ has only one root $\alpha^0 = 1$. This code uses only one redundancy symbol and the $c(x)$ words of this code satisfy the condition:

   $$c(\alpha^0) = c(1) = 0$$

   This code, which is cyclic since $(x + 1)$ divides $(x^n + 1)$, is a parity check code with parameters $n = k + 1, k, t = 0$. Thus, every time we build a BCH code by selecting $l = 0$, we introduce into the generator polynomial a term in $(x+1)$ and the codewords are of even weight.

2. Cyclic Redundancy Code (CRC)

   Another example of a BCH code for which $l = 0$, is the CRC used for detecting errors. A CRC has a constructed distance of 4 ($t = 1$) and its generator polynomial, from above, is therefore equal to:

   $$g(x) = (x + 1)m_\alpha(x)$$

   α being a primitive element, $m_\alpha(x)$ is a primitive polynomial and thus the generator polynomial of a CRC is a code equal to the product of $(x + 1)$ by the generator polynomial of a Hamming code.

   $$g_{\text{CRC}}(x) = (x + 1)g_{\text{Hamming}}(x)$$

   The parameters of a CRC are therefore:

   $$n = 2^m - 1; (n - k) = m + 1; k = 2^m - m - 2$$

Code	m	$g(x)$
CRC-12	12	14017
CRC-16	16	300005
CRC-CCITT	16	210041
CRC-32	32	40460216667

Table 4.3 – generator polynomials of some codes CRC.

The most widely-used CRC codes have the parameters $m = 12$, 16, 32 and their generator polynomials are given, in octals, in Table 4.3. Note: $g(x) = 14017$ in octals corresponds to 1 100 000 001 111 in binary, that is:

$$g(x) = x^{12} + x^{11} + x^3 + x^2 + x + 1$$

Non-primitive BCH code

The generator polynomial of a non-primitive BCH code (with $l = 1$) correcting at least t errors (constructed distance $d = 2t + 1$) has $2t$ roots of the form: $\beta, \beta^2, \beta^3, \ldots, \beta^{2t}$ where β is a non-primitive element of a Galois field $\mathbf{F}_q$. Taking into account the fact that the minimal polynomials $m_{\beta^j}(x)$ and $m_{\beta^{2j}}(x)$ have the same roots, the generator polynomial of a non-primitive BCH code is equal to:

$$g(x) = \text{S.C.M.}(m_\beta(x), m_{\beta^3}(x).....m_{\beta^{2t-1}}(x))$$

We can show that length n of the words of a non-primitive BCH code is no longer of the form $2^m - 1$ but is equal to p, where p is the exponent of β such that $\beta^p = 1$ (p is the order of β). A Galois field $\mathbf{F}_q$ has non-primitive elements if $2^m - 1$ is not prime. The non-primitive elements are then of the form $\beta = \alpha^\lambda$ where λ is a divisor of $2^m - 1$ and α is a primitive element of the field.

Example 4.10

Let there be a Galois field $\mathbf{F}_q$ with $m = 6$ and $q = 64$. The quantity $2^m - 1 = 63$ is not equal to a prime number; it is divisible by 3, 7, 9, 21 and 63. The non-primitive elements of this field are therefore $\alpha^3, \alpha^7, \alpha^9, \alpha^{21}, \alpha^{63} = 1$. Let us build, for example, a non-primitive BCH code having an error correction capability at least equal to $t = 2$ on field $\mathbf{F}_{64}$ and let us take $\beta = \alpha^3$ as the non-primitive element. We have two minimal polynomials to calculate $m_\beta(x)$ and $m_{\beta^3}(x)$. Taking into account the fact that $\beta^{21} = \alpha^{63} = 1$, the roots of these polynomials are:

$$m_\beta(x) : \text{roots } \beta, \beta^2, \beta^4, \beta^8, \beta^{16}, \beta^{32} = \beta^{11}$$
$$m_{\beta^3}(x) : \text{roots } \beta^3, \beta^6, \beta^{12}$$

The generator polynomial of this code is equal to:

$$g(x) = m_{\beta}(x)m_{\beta^3}(x)$$

which, after development and simplification, gives:

$$g(x) = x^9 + x^8 + x^7 + x^5 + x^4 + x + 1$$

The parameters of this non-primitive BCH code are:

$$n = 21; (n - k) = 9; k = 12$$

- *Golay code*

  Among non-primitive BCH codes, the most well-known is certainly the Golay code constructed over a Galois field $\mathbf{F}_q$ with $m = 11, q = 2048$. Noting that $2^m - 1 = 2047 = 23 \times 89$, the non-primitive element used to build a Golay code is $\beta = \alpha^{89}$. The computation of the generator polynomial of this code constructed on field $\mathbf{F}_{2048}$ leads to the following expression:

  $$g(x) = x^{11} + x^9 + x^7 + x^6 + x^5 + x + 1$$

  We can show that the minimum distance $d_{\min}$ of a Golay code is 7 and thus, its correction capability is 3 errors in a block of 23 binary symbols ($\beta^{23} = \alpha^{2047} = 1$). The parameters of a Golay code are therefore:

  $$n = 23; (n - k) = 11; k = 12; t = 3$$

  Note that the reciprocal polynomial of $g(x)$, equal to $\tilde{g}(x) = x^{11}g(x^{-1})$ also enables a Golay code to be produced.

  $$\tilde{g}(x = x^{11} + x^{10} + x^6 + x^5 + x^4 + x^2 + 1$$

4.2 Block codes with non-binary symbols

4.2.1 Reed-Solomon codes

Reed-Solomon or RS codes are the most well-known and the most widely-used codes having non-binary symbols. For codes with non-binary symbols the coefficients c_j of the codewords and d_j of the datawords take their value in a Galois field $\mathbf{F}_q$ with $q = 2^m$ elements. Thus, each symbol of these codes can be encoded on m binary symbols. Reed-Solomon codes being cyclic codes, they are generated by a generator polynomial $g(x)$ divisor of $x^n + 1$ whose coefficients g_j $j = 0, 1, \cdots, n - k - 1$ also take their value in the Galois field $\mathbf{F}_q$.

The generator polynomial of a Reed-Solomon code, with a constructed distance d has $(d-1)$ roots $\alpha^l, \cdots, \alpha^{l+j}, \cdots, \alpha^{l+d-2}$ where α is a primitive element of Galois field $\mathbf{F}_q$. It therefore has the expression:

$$g(x) = \text{S.C.M.}(m_{\alpha^l}(x), \cdots, m_{\alpha^{l+j}}(x), \cdots, m_{\alpha^{l+d-2}}(x))$$

where $m_{\alpha^{l+j}}$ is the minimal polynomial associated with the α^{l+j} element of field $\mathbf{F}_q$.

Using the results of the appendix on the minimal polynomials with coefficients in $\mathbf{F}_q$, the minimal polynomial $m_{\alpha^{l+j}}$ has only one root α^{l+j}.

$$m_{\alpha^{j+i}}(x) = (x + \alpha^{j+i})$$

The generator polynomial of a Reed-Solomon code is therefore of the form:

$$g(x) = (x + \alpha^j)(x + \alpha^{j+1})...(x + \alpha^{j+i})...(x + \alpha^{j+d-2})$$

In general, parameter j is set to 0 or 1 like for binary BCH codes. The generator polynomial of a Reed-Solomon code, of degree $(n-k)$, has $(d-1)$ roots, that is $n-k = d-1$. Its constructed distance is therefore equal to:

$$d = n - k + 1$$

For a block code k $C(n, k)$ the minimum distance $d_{\min}$ being lower than or equal to $n-k+1$, the minimum distance of a Reed-Solomon code is therefore always equal to its constructed distance. A code whose minimum distance is equal to $n-k+1$ is called a maximum distance code.
The parameters of a Reed-Solomon code correcting t errors in a block of n q-ary symbols are therefore:

$$n = q - 1; \quad n - k = d_{\min} - 1 = 2t; \quad k = n - 2t$$

Example 4.11

Let us determine the generator polynomial of a Reed-Solomon code built from a Galois field with 16 elements having a correction capability of $t = 2$ errors. The minimum distance of this code is therefore $d_{\min} = 5$. Taking for example $l = 1$, the generator polynomial of this code is therefore of the form:

$$g(x) = (x + \alpha)(x + \alpha^2)(x + \alpha^3)(x + \alpha^4)$$

Developing the expression above, we obtain:

$$g(x) = \lfloor x^2 + x(\alpha + \alpha^2) + \alpha^3 \rfloor \lfloor x^2 + x(\alpha^3 + \alpha^4) + \alpha^7 \rfloor$$

Using the binary representations of the elements of field $\mathbf{F}_{16}$ (Appendix), the polynomial $g(x)$ after development and simplification is equal to:

$$g(x) = x^4 + \alpha^3 x^3 + \alpha^6 x^2 + \alpha^3 x + \alpha^{10}$$

4.2.2 Implementing the encoder

The schematic diagram of an encoder for Reed-Solomon codes is quite similar to that of an encoder for cyclic codes with binary symbols, but the encoder must now carry out multiplications between q-ary symbols and memorize q-ary symbols.

As an example, we have shown in Figure 4.4 the schematic diagram of the encoder for the Reed-Solomon code treated in the example above.

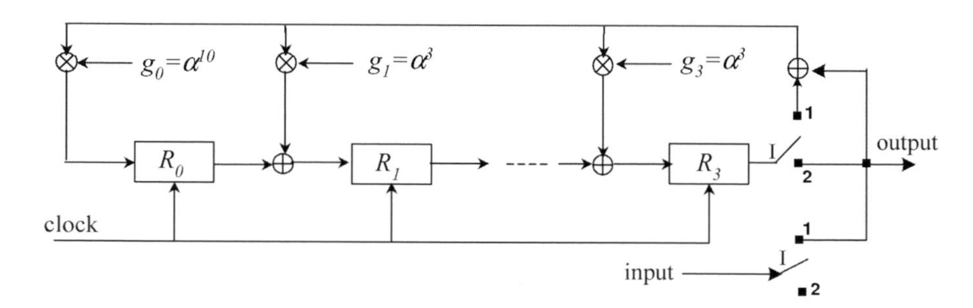

Figure 4.4 – Schematic diagram of the encoder for the RS code (15,11).

4.3 Decoding and performance of codes with binary symbols

4.3.1 Error detection

Considering a binary symmetric transmission channel, the decoder receives binary symbols assumed to be perfectly synchronized with the encoder. This means that the splitting into words having n symbols at the input of the decoder corresponds to the splitting used by the encoder. Thus, in the absence of errors, the decoder sees codewords at its input.

Let us assume that codeword $\mathbf{c}$ is transmitted by the encoder and let $\mathbf{r}$ be the word of n symbols received at the input of the decoder. Word $\mathbf{r}$ can always be written in the form:

$$\mathbf{r} = \mathbf{c} + \mathbf{e}$$

where $\mathbf{e}$ is a word whose non-null symbols represent the errors. A non-null symbol of $\mathbf{e}$ indicates the presence of an error in the corresponding position of $\mathbf{c}$.

Errors are detected by using the orthogonality property of the parity check matrix with the codewords and calculating the quantity $\mathbf{s}$ called the error syndrome.

$$\mathbf{s} = \mathbf{r}\mathbf{H}^\mathbf{T} = (\mathbf{c} + \mathbf{e})\mathbf{H}^\mathbf{T} = \mathbf{e}\mathbf{H}^\mathbf{T}$$

Syndrome $\mathbf{s}$ is null if, and only if, $\mathbf{r}$ is a codeword. A non-null syndrome implies the presence of errors. However, it should be noted that a null syndrome does not necessarily mean absence of errors since $\mathbf{r}$ can belong to the set of codewords even though it is different from $\mathbf{c}$. For this to occur, it suffices for word $\mathbf{e}$ to be a codeword. Indeed, for a linear block code, the sum of two codewords is another codeword.

Finally, let us note that for any linear block code, there are configurations of non-detectable errors.

Detection capability

Let $\mathbf{c}_j$ be the transmitted codeword and $\mathbf{c}_l$ its nearest neighbour. We have the following inequality:

$$d_H(\mathbf{c}_j, \mathbf{c}_l) \geqslant d_{\min}$$

Introducing the received word $\mathbf{r}$, we can write:

$$d_{\min} \leq d_H(\mathbf{c}_j, \mathbf{c}_l) \leq d_H(\mathbf{c}_j, \mathbf{r}) + d_H(\mathbf{c}_l, \mathbf{r})$$

and thus all the errors can be detected if the Hamming distance between $\mathbf{r}$ and $\mathbf{c}_l$ is higher than or equal to 1, that is, if $\mathbf{r}$ is not merged with $\mathbf{c}_l$.

The detection capability of a $C(n,k)$ code with minimum distance $d_{\min}$ is therefore equal to $d_{\min} - 1$.

Probability of non-detection of errors

Considering a block code $C(n,k)$ and a binary symmetric channel with error probability p, the probability of non-detection of the errors P_{nd} is equal to:

$$P_{nd} = \sum_{j=d_{\min}}^{n} A_j p^j (1-p)^{n-j} \tag{4.22}$$

where A_j is the number of codewords with weight j.

Examining the hypothesis of a completely degraded transmission, that is, of an error probability of $p = 1/2$ on the channel, and taking into account the fact that for any block code we have:

$$\sum_{j=d_{\min}}^{n} A_j = 2^k - 1$$

(the -1 in the above expression corresponds to the null codeword), probability P_{nd} is equal to:

$$P_{nd} = \frac{2^k - 1}{2^n} \cong 2^{-(n-k)}$$

The detection of errors therefore remains efficient whatever the error probability on the transmission channel if the number of redundancy symbols $(n-k)$ is large enough. The detection of errors is therefore not very sensitive to error statistics.

When erroneous symbols are detected, the receiver generally asks the source to send them again. To transmit this re-transmission request, it is then necessary to have a receiver source link, called a return channel. The data rate on the return channel being low (a priori, requests for retransmission are short and few in number), we can always arrange it so that the error probability on this channel is much lower than the error probability on the transmission channel. Thus, the performance of a transmission system using error detection and repetition does not greatly depend on the return channel.

In case of error detection, the emission of the source can be interrupted to enable the retransmission of the corrupted information. The data rate is therefore not constant, which can present problems in some cases.

4.3.2 Error correction

Error correction involves looking for the transmitted codeword $\mathbf{c}$ given the received word $\mathbf{r}$. Two strategies are possible. The first one corresponds to a received word $\mathbf{r}$ at the input of the decoder made up of binary symbols (the case of a binary symmetric channel) and the second, to a received word $\mathbf{r}$ made up of analogue symbols (the case of a Gaussian channel). In the first case, we speak of hard input decoding whereas in the second case we speak of soft input decoding. We will now examine these two types of decoding, already mentioned in Chapter 1.

Hard decoding

- *Maximum a posteriori likelihood decoding*

  For hard decoding the received word $\mathbf{r}$ is of the form:

$$\mathbf{r} = \mathbf{c} + \mathbf{e}$$

where $\mathbf{c}$ and $\mathbf{e}$ are words with binary symbols.

Maximum *a posteriori* likelihood decoding involves looking for the codeword $\hat{\mathbf{c}}$ such that:

$$\Pr\{\hat{\mathbf{c}}\,|\mathbf{r}\} > \Pr\{\mathbf{c}_i\,|\mathbf{r}\} \quad \forall\, \mathbf{c}_i \neq \hat{\mathbf{c}} \in C(n,k)$$

Using Bayes' rule and assuming that all the codewords are equiprobable, the above decision rule can also be written:

$$\hat{\mathbf{c}} = \mathbf{c}_i \Leftrightarrow \Pr\left(\mathbf{r}\,|\mathbf{c} = \mathbf{c}_i\right) > \Pr\left(\mathbf{r}\,|\mathbf{c} = \mathbf{c}_j\right), \forall \mathbf{c}_j \neq \mathbf{c}_i \in C\left(n,k\right)$$

Again taking the example of a binary symmetric channel with error probability p and denoting $d_H(\mathbf{r},\hat{\mathbf{c}})$ the Hamming distance between $\mathbf{r}$ and $\hat{\mathbf{c}}$, the decision rule is:

$$\hat{\mathbf{c}} = \mathbf{c}_i \Leftrightarrow p^{d_H(\mathbf{c}_i,\mathbf{r})} (1-p)^{n-d_H(\mathbf{c}_i,\mathbf{r})} > p^{d_H(\mathbf{c}_j,\mathbf{r})} (1-p)^{n-d_H(\mathbf{c}_j,\mathbf{r})}, \ \forall \mathbf{c}_j \neq \mathbf{c}_i$$

Taking the logarithm of the two parts of the above inequality and considering $p < 0.5$, the decision rule of the maximum *a posteriori* likelihood can finally be written:

$$\hat{\mathbf{c}} = \mathbf{c}_i \Leftrightarrow d_H(\mathbf{r}, \mathbf{c}_i) \leq d_H(\mathbf{r}, \mathbf{c}_j), \quad \forall \mathbf{c}_j \neq \mathbf{c}_i \in C(n,k)$$

If two or several codewords are the same distance from $\mathbf{r}$, the codeword $\hat{\mathbf{c}}$ is chosen arbitrarily among the codewords equidistant from $\mathbf{r}$.

This decoding procedure which is optimal, that is, which minimizes the probability of erroneous decoding, becomes difficult to implement when the number of codewords becomes large, which is often the case for the widely-used block codes.

- *Decoding from the syndrome*

To get around this difficulty, it is possible to perform the decoding using syndrome $\mathbf{s}$. We recall that the syndrome is a vector of dimension $(n-k)$ that depends solely on the error configuration $\mathbf{e}$. For a binary symbol block code, the syndrome has 2^{n-k} configurations, which is generally much lower than the 2^k codewords.

To decode from the syndrome, we use a table with n rows and two columns. We write respectively in each row of the first column the null syndrome $\mathbf{s}$ (all the symbols are at zero, no errors) then the syndromes $\mathbf{s}$ corresponding to the configuration of an error then two errors etc. until the n rows are filled. All the configurations of the syndromes of the first column must be different. In the second column, we write the error configuration associated with each syndrome of the first column.

For a received word $\mathbf{r}$ we calculate the syndrome $\mathbf{s}$ then, using the table, we deduce the error word $\mathbf{e}$. Finally, we add word $\mathbf{e}$ to $\mathbf{r}$ and we obtain the most likely codeword.

Example 4.12

Let us consider a code $C(7,4)$ with a parity check matrix $\mathbf{H}$ with:

$$\mathbf{H} = \begin{bmatrix} 1 & 1 & 1 & 0 & 1 & 0 & 0 \\ 1 & 1 & 0 & 1 & 0 & 1 & 0 \\ 1 & 0 & 1 & 1 & 0 & 0 & 1 \end{bmatrix}$$

This code has 16 codewords but only 8 configurations for the syndrome as indicated in Table 4.4.

Syndrome $\mathbf{s}$	Error word $\mathbf{e}$
000	0000000
001	0000001
010	0000100
011	0001000
100	0000100
101	0010000
110	0100000
111	1000000

Table 4.4: Syndromes and corresponding error words for a $C(7,4)$ code.

Let us assume that the codeword transmitted is $\mathbf{c} = [0101101]$ and that the received word $\mathbf{r} = [0111101]$ has an error in position 3. The syndrome is then equal to $\mathbf{s} = [101]$ and, according to the table, $\mathbf{e} = [0010000]$. The decoded codeword is $\hat{\mathbf{c}} = \mathbf{r} + \mathbf{e} = [0101101]$ and the error is corrected.

If the number of configurations of the syndrome is still too high to apply this decoding procedure, we use decoding algorithms specific to certain classes of codes but that, unfortunately, do not always exploit the whole correction capability of the code. These algorithms will be presented below.

- *Correction power*

Let $\mathbf{c}_j$ be the codeword transmitted and $\mathbf{c}_l$ its nearest neighbour. We have the following inequality:

$$d_H(\mathbf{c}_j, \mathbf{c}_l) \geqslant d_{\min}$$

Introducing the received word $\mathbf{r}$ and assuming that the minimum distance $d_{\min}$ is equal to $2t+1$ (integer t), we can write:

$$2t + 1 \leq d_H(\mathbf{c}_j, \mathbf{c}_l) \leq d_H(\mathbf{c}_j, \mathbf{r}) + d_H(\mathbf{c}_l, \mathbf{r})$$

We see that if the number of errors is lower than or equal to t, $\mathbf{c}_j$ is the most likely codeword since it is nearer to $\mathbf{r}$ than to $\mathbf{c}_l$ and thus the t errors can be corrected. If the minimum distance is now $(2t+2)$, using the same reasoning, we arrive at the same error correction capability. In conclusion, the correction capability of a linear block code with minimum distance $d_{\min}$ with hard decoding is equal to:

$$t = \left\lfloor \frac{d_{\min} - 1}{2} \right\rfloor \tag{4.23}$$

where $\lfloor x \rfloor$ is the whole part of x rounded down (for example $\lfloor 2.5 \rfloor = 2$).

• *Probability of erroneous decoding of a codeword*

For a linear block code $C(n,k)$ of error correction capability t, the codeword transmitted will be wrongly decoded if there are $t+j$ errors, $j = 1, 2, \cdots, n-t$, in the received word $\mathbf{r}$. For a binary symmetric channel of probability p, the probability $P_{e,\text{word}}$ of performing an erroneous decoding of the transmitted codeword is upper bounded by:

$$P_{e,\text{word}} < \sum_{j=t+1}^{n} \binom{n}{j} p^j (1-p)^{n-j} \tag{4.24}$$

We can also determine the binary error probability $P_{e,\text{bit}}$ on the information data after decoding. In presence of erroneous decoding, the maximum *a posteriori* likelihood decoder adds at most t errors by choosing the codeword with the minimum distance from the received word. The error probability is therefore bounded by:

$$P_{e,\text{bit}} < \frac{1}{n} \sum_{j=t+1}^{n} (j+t) \binom{n}{j} p^j (1-p)^{n-j} \tag{4.25}$$

If the transmission is performed with binary phase modulation (2-PSK, 4-PSK), probability p is equal to:

$$p = \frac{1}{2}\text{erfc}\sqrt{\frac{RE_b}{N_0}}$$

where R is the coding rate, E_b the energy received per transmitted information bit and N_0 the unilateral power spectral density of the noise. Figure 4.5 shows the binary error probability and word error probability after algebraic decoding for the (15,7) BCH code. The modulation is 4-PSK and the channel is Gaussian. The higher bounds expressed by (4.24) and (4.25) respectively are also plotted.

Soft decoding

Considering a channel with additive white Gaussian noise and binary phase modulation transmission (2-PSK or 4-PSK), the components $r_j, \quad j = 0, 1, \cdots, n-1$ of the received word $\mathbf{r}$ have the form:

$$r_j = \sqrt{E_s}\tilde{c}_j + b_j, \quad \tilde{c}_j = 2c_j - 1$$

where $c_j = 0, 1$ is the symbol in position j of codeword $\mathbf{c}$, $\tilde{c}_j$ is the binary symbol associated with c_j, E_s is the energy received per transmitted symbol and b_j is white Gaussian noise, with zero mean and variance equal to σ_b^2.

• *Maximum a posteriori likelihood decoding*

Decoding using the maximum *a posteriori* likelihood criterion means searching for codeword $\hat{\mathbf{c}}$ such that:

$$\hat{\mathbf{c}} = \mathbf{c} \Leftrightarrow \Pr\{\mathbf{c}\,|\mathbf{r}\} > \Pr\{\mathbf{c}'\,|\mathbf{r}\}, \quad \forall \mathbf{c} \neq \mathbf{c}' \in C(n,k)$$

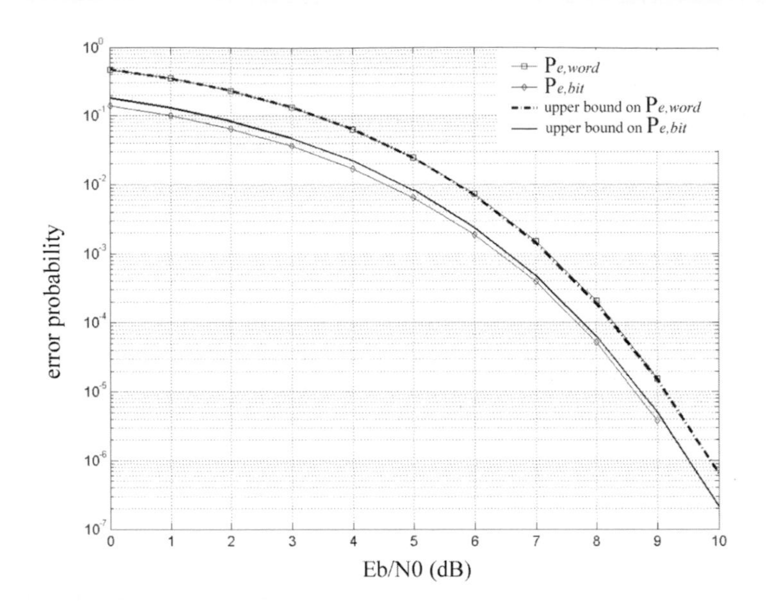

Figure 4.5 – Performance of the algebraic decoding of the (15,7) BCH code. 4-PSK transmission on a Gaussian channel.

Using the Bayes' rule and assuming all the codewords equiprobable, the above inequality can also be written:

$$\hat{\mathbf{c}} = \mathbf{c} \text{ if } p(\mathbf{r}\,|\mathbf{c}\,) > p(\mathbf{r}\,|\mathbf{c}'\,), \qquad \forall\, \mathbf{c} \neq \mathbf{c}' \in C(n,k) \tag{4.26}$$

where $p(\mathbf{r}\,|\mathbf{c}\,)$ is the probability density function of observation $\mathbf{r}$ conditionally to codeword $\mathbf{c}$.

For a Gaussian channel, probability density function $p(\mathbf{r}\,|\mathbf{c}\,)$ is equal to:

$$p(\mathbf{r}\,|\mathbf{c}\,) = \left(\frac{1}{\sqrt{2\pi}\sigma_b}\right)^n \exp\left(-\frac{1}{2\sigma_b^2}\sum_{j=0}^{n-1}(r_j - \sqrt{E_s}\tilde{c}_j)^2\right)$$

where σ_b^2 is the variance of the noise.

Replacing the two probability density functions by their respective expressions in inequality (4.26) and after some basic computation, we obtain:

$$\hat{\mathbf{c}} = \mathbf{c} \Leftrightarrow \sum_{j=0}^{n-1} r_j c_j > \sum_{j=0}^{n-1} r_j {c'}_j, \quad \forall \mathbf{c} \neq \mathbf{c}' \in C(n,k)$$

The decoded codeword is the one that maximizes the scalar product $\langle \mathbf{r}, \mathbf{c}\rangle$. We could also show that the decoded codeword is the one that minimizes the square of the Euclidean distance $\left\|\mathbf{r} - \sqrt{E_s}\tilde{\mathbf{c}}\right\|^2$.

This decoding procedure is applicable when the number of codewords is not too high. In the presence of a large number of codewords we can use a Chase algorithm whose principle is to apply the above decoding procedure and restrict the search space to a subset of codewords.

- *Chase algorithm*

The Chase algorithm is a sub-optimal decoding procedure that uses the maximum *a posteriori* likelihood criterion but considers a very reduced subset of codewords. To determine this subset of codewords, the Chase algorithm works in the following way.

- Step 1: The received word $\mathbf{r}$, made up of analogue symbols, is transformed into a word with binary symbols $\mathbf{z}_0 = (z_{00} \cdots z_{0j} \cdots z_{0n-1})$ by thresholding,

$$z_{0j} = \text{sgn}(r_j)$$

  with the following convention:

$$\begin{aligned} \text{sgn}(x) &= 1 \text{ if } x \geqslant 0 \\ &= 0 \text{ if } x < 0 \end{aligned}$$

  The binary word $\mathbf{z}_0$ is then decoded by a hard decision algorithm other than the maximum *a posteriori* likelihood algorithm (we will present algorithms for decoding block codes later). Let $\mathbf{c}_0$ be the codeword obtained.

- Step 2: Let $j_1, j_2, \cdots, j_t$ be the positions of the least reliable symbols, that is, such that the $|r_j|$ amplitudes are the smallest.

  $2^t - 1$ words $\mathbf{e}_i$ are built by forming all the non-null binary combinations possible on positions $j_1, j_2, \cdots, j_t$. On the positions other than $j_1, j_2, \cdots, j_t$, the symbols of $\mathbf{e}_i$ are set to zero. Recall that t is the correction capability of the code.

- Step 3: Each of the $2^t - 1$ words $\mathbf{e}_i$ is used to define the words $\mathbf{z}_i$ with:

$$\mathbf{z}_i = \mathbf{z}_0 + \mathbf{e}_i$$

  A hard decoder processes the words $\mathbf{z}_i$ to obtain at most $2^t - 1$ codewords $\mathbf{c}_i$. Note that the word at the output of the algebraic decoder is not always a codeword and only codewords will be considered when applying the decision criterion.

- Step 4: The maximum *a posteriori* likelihood rule is applied to the subset of the codewords $\mathbf{c}_i$ created in the previous step.

Example 4.13

Let there be a code $C(n, k)$ with correction capability $t = 3$. The subset of the codewords is made up of 8 codewords, 7 of which are elaborated from the words $\mathbf{e}_i$. Words $\mathbf{e}_i$ are words of length n whose components are null except possibly those with indices j_1, j_2 and j_3 (see the table below).

i	e_{i,j_1} e_{i,j_2} e_{i,j_3}
1	0 0 1
2	0 1 0
3	0 1 1
4	1 0 0
5	1 0 1
6	1 1 0
7	1 1 1

- *Probability of erroneous decoding of a codeword*

Let us assume that the transmitted codeword is $\mathbf{c}_0 = (c_{01} \cdots c_{0j} \cdots c_{0n-1})$ and let $\mathbf{r}_0 = (r_0 \cdots r_j \cdots r_{n-1})$ be the received word with:

$$r_j = \sqrt{E_s}\tilde{c}_{0j} + b_j$$

Codeword $\mathbf{c}_0$ will be wrongly decoded if:

$$\sum_{j=0}^{n-1} r_j c_{0,j} < \sum_{j=0}^{n-1} r_j c_{l,j} \qquad \forall\, \mathbf{c}_l \neq \mathbf{c}_0 \in C(n, k)$$

The code being linear, we can, without loss of generality, assume that the codeword transmitted is the null word, that is, $c_{0,j} = 0$ for $j = 0, 1, \cdots, n-1$.

The probability of erroneous decoding $P_{\text{e,word}}$ of a codeword is then equal to:

$$P_{\text{e,word}} = \Pr\left(\sum_{j=0}^{n-1} r_j c_{1,j} > 0 \text{ or } \ldots \sum_{j=0}^{n-1} r_j c_{l,j} > 0 \text{ or } \ldots\right)$$

Probability $P_{\text{e,word}}$ can be upper bounded by a sum of probabilities and, after some standard computation, it can be written in the form:

$$P_{\text{e,word}} \leq \frac{1}{2}\sum_{j=2}^{2^k} \operatorname{erfc}\sqrt{w_j \frac{E_s}{N_0}}$$

where w_j is the Hamming weight of the j-th codeword.

Assuming that code $C(n,k)$ has A_w codewords of weight w, probability $P_{e,\text{word}}$ can again be written in the form:

$$P_{e,\text{word}} < \frac{1}{2} \sum_{w=d_{\min}}^{n} A_w \text{erfc}\sqrt{w\frac{E_s}{N_0}} \tag{4.27}$$

Introducing the energy E_b received per bit of information transmitted, probability $P_{e,\text{word}}$ can finally be upper bounded by:

$$P_{e,\text{word}} < \frac{1}{2} \sum_{w=d_{\min}}^{n} A_w \text{erfc}\sqrt{w\frac{RE_b}{N_0}} \tag{4.28}$$

where R is the coding rate.
We can also establish an upper bound of the binary error probability on the information symbols after decoding.

$$P_{e,\text{bit}} < \frac{1}{2} \sum_{w=d_{\min}}^{n} \frac{w}{n} A_w \text{erfc}\sqrt{w\frac{RE_b}{N_0}} \tag{4.29}$$

To calculate probabilities $P_{e,\text{word}}$ and $P_{e,\text{bit}}$ we must know the number A_w of codewords of weight w. For extended BCH codes the quantities A_w are given in [4.1].

As an example, Table 4.5 gives the A_w quantities for three extended Hamming codes.

n	k	$d_{\min}$	A_4	A_6	A_8	A_{10}	A_{12}	A_{14}	A_{16}
8	4	4	14	-	1	-	-	-	-
16	11	4	140	448	870	448	140	-	1
32	26	4	1240	27776	330460	2011776	7063784	14721280	18796230

Table 4.5 – A_w for three extended Hamming codes.

For the code (32,26) the missing A_w quantities are obtained from the relation $A_w = A_{n-w}$ for $0 \leq w \leq n/2$, $n/2$ even.

The A_w quantities for non-extended Hamming codes can be deduced from those of extended codes by resolving the following system of equations:

$$(n+1)A_{w-1} = wA_w^{\text{extended}}$$
$$wA_w = (n+1-w)A_{w-1}$$

where n is the length of the words of the non-extended code.

For the Hamming code (7,4), for example, the A_w quantities are:

$$\begin{aligned} 8A_3 &= 4A_4^{\text{extended}} & A_3 = 7 \\ 4A_4 &= 4A_3 & A_4 = 7 \\ 8A_7 &= 8A_8^{\text{extended}} & A_7 = 1 \end{aligned}$$

Weight	A_w (23,12)	A_w (24,12)
0	1	1
7	253	0
8	506	759
11	1288	0
12	1288	2576
15	506	0
16	253	759
23	1	0
24	0	1

Table 4.6 – A_w for extended Golay and Golay codes.

For Golay and extended Golay codes, the A_w quantities are given in Table 4.6.

With a high signal to noise ratio, error probability $P_{e,\text{word}}$ is well approximated by the first term of the series:

$$P_{e,\text{word}} \cong \frac{1}{2} A_{d_{\min}} \operatorname{erfc}\sqrt{\frac{Rd_{\min}E_b}{N_0}} \quad \text{if} \quad \frac{E_b}{N_0} >> 1 \tag{4.30}$$

The same goes for error probability $P_{e,\text{bit}}$ on the information symbols.

$$P_{e,\text{bit}} \cong \frac{d_{\min}}{n} P_{e,\text{word}} \quad \text{if} \quad \frac{E_b}{N_0} >> 1 \tag{4.31}$$

In the absence of coding, the error probability on the binary symbols is equal to:

$$p = \frac{1}{2} \operatorname{erfc}\sqrt{\frac{E_b}{N_0}}$$

As seen in Section 1.5, comparing the two expressions of the binary error probability with and without coding, we observe that the signal to noise ratio E_b/N_0 is multiplied by $Rd_{\min}$ in the presence of coding. If this multiplying coefficient is higher than 1, the coding acts as an amplifier of the signal to noise ratio whose asymptotic gain is approximated by

$$G_a = 10 \log(Rd_{\min})(\text{dB})$$

To illustrate these bounds, let us again take the example of the (15,7) BCH code transmitted on a Gaussian channel with 4-PSK modulation. In Figure 4.6, we show the evolution of the binary error probability and word error probability obtained by simulation from the sub-optimal Chase algorithm (4 non-reliable positions). We also show the first two terms of the sums appearing in the bounds given by (4.28) and (4.29). As a reference, we have also plotted the binary error probability curve of a 4-PSK modulation without coding.

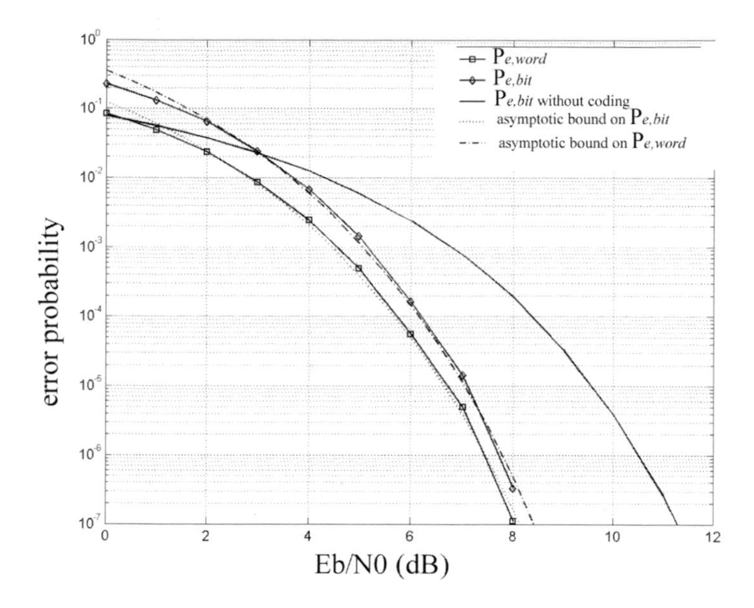

Figure 4.6 – Performance of the soft input decoding of the (15,7) BCH code. 4-PSK transmission on a Gaussian channel.

4.4 Decoding and performance of codes with non-binary symbols

4.4.1 Hard input decoding of Reed-Solomon codes

Hard input decoding algorithms make it possible to decode Reed-Solomon (RS) codes and BCH codes with binary symbols. We begin by presenting the principle of decoding RS codes then we treat the case of BCH codes using binary symbols as a particular case of decoding RS codes.

Assuming that $c(x)$ is the transmitted codeword, then for a channel with discrete input and output, the received word can always be written in the form:

$$r(x) = c(x) + e(x)$$

with:

$$e(x) = e_0 + e_1 x + \cdots + e_j x^j + \cdots + e_{n-1} x^{n-1}, \quad e_j \in \mathbf{F}_q \quad \forall j$$

When $e_j \neq 0$ there is an error in position j.

It was seen above that the generator polynomial of an RS code or of a BCH code (with $l = 1$) correcting t errors had the roots $\alpha, \cdots, \alpha^j, \cdots, \alpha^{2t}$ and that the codewords were multiples of the generator polynomial. Thus, for any codeword, we can write:

$$c(\alpha^i) = 0; \forall i = 1, 2, \cdots, 2t$$

Decoding RS codes and binary BCH codes can be performed from a vector with $2t$ components $\mathbf{S} = [S_1 \cdots S_j \cdots S_{2t}]$, called a *syndrome*.

$$S_j = r(\alpha^j) = e(\alpha^j), \quad j = 1, 2, \cdots, 2t \tag{4.32}$$

When the components of vector $\mathbf{S}$ are all null, there are no errors or, at least, no detectable errors. When some components of the vector $\mathbf{S}$ are non-null, errors are present that, in certain conditions, can be corrected.

In the presence of t transmission errors, the error polynomial $e(x)$ is of the form:

$$e(x) = e_{n_1}x^{n_1} + e_{n_2}x^{n_2} + \cdots + e_{n_t}x^{n_t}$$

where the e_{n_l} are non-null coefficients taking their value in the field $\mathbf{F}_q$. The components S_j of syndrome $\mathbf{S}$ are equal to:

$$S_j = e_{n_1}(\alpha^j)^{n_1} + \cdots + e_{n_l}(\alpha^j)^{n_l} + \cdots + e_{n_t}(\alpha^j)^{n_t}$$

Putting $Z_l = \alpha^{n_l}$ and, to simplify the notations $e_{n_l} = e_l$, the component S_j of the syndrome is again equal to:

$$S_j = e_1 Z_1^j + \cdots + e_l Z_l^j + \cdots + e_t Z_t^j \tag{4.33}$$

To determine the position of the transmission errors it is therefore sufficient to know the value of quantities $Z_l; j = 1, 2, \cdots, t$ then, in order to correct the errors, to evaluate coefficients $e_l; l = 1, 2, \cdots, t$.

The main difficulty in decoding RS codes or binary BCH codes is determining the position of the errors. Two methods are mainly used to decode RS codes or binary BCH codes: Peterson's direct method and the iterative method using the Berlekamp-Massey algorithm or Euclid algorithm .

4.4.2 Peterson's direct method

Description of the algorithm for codes with non-binary symbols

This method is well adapted for decoding RS codes or binary BCH codes correcting a low number of errors, typically 1 to 3. Indeed, the complexity of this method increases as the square of the correction capability of the code, whereas for the iterative method, the complexity increases only linearly with the correction capability of the code.

To determine the position of the errors let us introduce a polynomial $\sigma_d(x)$ called the *error locator* polynomial whose roots are exactly the quantities Z_l.

$$\sigma_d(x) = \prod_{l=1}^{t} (x + Z_l)$$

Developing this expression, the polynomial $\sigma_d(x)$ is again equal to:

$$\sigma_d(x) = x^t + \sigma_1 x^{t-1} + \cdots + \sigma_j x^{t-j} + \cdots + \sigma_t$$

where the coefficients σ_j are functions of the quantities Z_l.

From the expression of S_j we can build a non-linear system of $2t$ equations.

$$S_j = \sum_{i=1}^{t} e_i Z_i^j, \quad j = 1, 2, \cdots, 2t$$

The quantities $Z_l, l = 1, \cdots, t$ being the roots of the error locator polynomial $\sigma_d(x)$, we can write:

$$\sigma_d(Z_l) = Z_l^t + \sum_{j=1}^{t} \sigma_j Z_l^{t-j} = 0, \quad l = 1, 2, \cdots, t \tag{4.34}$$

Multiplying the two parts of this expression by the same term $e_l Z_l^q$, we obtain:

$$e_l Z_l^{t+q} + \sum_{j=1}^{t} \sigma_j e_l Z_l^{t+q-j} = 0, \quad l = 1, 2, \cdots, t \tag{4.35}$$

Summing relations (4.35) for l from 1 to t and taking into account the definition of component S_j of syndrome $\mathbf{S}$, we can write:

$$S_{t+q} + \sigma_1 S_{t+q-1} + \cdots + \sigma_j S_{t+q-j} + \cdots + \sigma_t S_q = 0, \quad \forall q \tag{4.36}$$

For an RS code correcting one error ($t = 1$) in a block of n symbols, syndrome $\mathbf{S}$ has two components S_1 and S_2. Coefficient σ_1 of the error locator polynomial is determined from relation (4.36) by making $t = 1$ and $q = 1$.

$$S_2 + \sigma_1 S_1 = 0 \rightarrow \sigma_1 = \frac{S_2}{S_1} \tag{4.37}$$

In the same way, for an RS code correcting two errors ($t = 2$) in a block of n symbols, the syndrome has four components S_1, S_2, S_3, S_4. Using relation (4.36) with $t = 2$ and $q = 1, 2$ we obtain the following system with two equations:

$$\begin{aligned} \sigma_1 S_2 + \sigma_2 S_1 &= S_3 \\ \sigma_1 S_3 + \sigma_2 S_2 &= S_4 \end{aligned}$$

Resolving this system of two equations enables us to determine coefficients σ_1 and σ_2 of the error locator polynomial.

$$\begin{aligned} \sigma_1 &= \tfrac{1}{\Delta_2} \left[S_1 S_4 + S_2 S_3 \right] \\ \sigma_2 &= \tfrac{1}{\Delta_2} \left[S_2 S_4 + S_3^2 \right] \end{aligned} \tag{4.38}$$

where Δ_2 is the determinant of the system with two equations.

$$\Delta_2 = S_2^2 + S_1S_3$$

Finally, for an RS code correcting three errors ($t = 3$), the relation (4.36) with $t = 3$ and $q = 1, 2, 3$ leads to the following system of three equations:

$$\begin{aligned}\sigma_1 S_3 + \sigma_2 S_2 + \sigma_3 S_1 &= S_4\\ \sigma_1 S_4 + \sigma_2 S_3 + \sigma_3 S_2 &= S_5\\ \sigma_1 S_5 + \sigma_2 S_4 + \sigma_3 S_3 &= S_6\end{aligned}$$

The resolution of this system enables us to determine coefficients σ_1, σ_2 and σ_3 of the error locator polynomial.

$$\begin{aligned}\sigma_1 &= \tfrac{1}{\Delta_3}\left[S_1S_3S_6 + S_1S_4S_5 + S_2^2S_6 + S_2S_3S_5 + S_2S_4^2 + S_3^2S_4\right]\\ \sigma_2 &= \tfrac{1}{\Delta_3}\left[S_1S_4S_6 + S_1S_5^2 + S_2S_3S_6 + S_2S_4S_5 + S_3^2S_5 + S_3S_4^2\right]\\ \sigma_3 &= \tfrac{1}{\Delta_3}\left[S_2S_4S_6 + S_2S_5^2 + S_3^2S_6 + S_4^3\right]\end{aligned} \quad (4.39)$$

where Δ_3 is the determinant of the system with three equations.

$$\Delta_3 = S_1S_3S_5 + S_1S_4^2 + S_2^2S_5 + S_3^3$$

Implementation of Peterson's decoder for an RS code with parameter $t = 3$

1. Calculate the $2t$ syndromes S_j: $S_j = r(\alpha^j)$
2. Determine the number of errors:
   - Case (a) $S_j = 0, \forall j$: no detectable error.
   - Case (b) $\Delta_3 \neq 0$: presence of three errors.
   - Case (c) $\Delta_3 = 0$ and $\Delta_2 \neq 0$: presence of two errors.
   - Case (d) $\Delta_3 = \Delta_2 = 0$ and $S_1 \neq 0$: presence of one error.
3. Calculate the error locator polynomial $\sigma_d(x)$
   - Case (b) Use (4.39)
   - Case (c) Use (4.38)
   - Case (d) Use (4.37)
4. Look for the roots of $\sigma_d(x)$ in field $\mathbf{F}_q$
5. Calculate the error coefficients e_i

- Case (b)

$$e_i = \frac{1}{\Delta}\left[S_1(Z_k^2 Z_p^3 + Z_k^3 Z_p^2) + S_2(Z_k^3 Z_p + Z_k Z_p^3) + S_3(Z_k^2 Z_p + Z_k Z_p^2)\right],$$
$$k \neq p \neq i, (i,k,p) \in \{1,2,3\}^3$$

$$\Delta = \sum_{\substack{1 \leq i_1, i_2, i_3 \leq 3 \\ i_1 + i_2 + i_3 = 6 \\ i_1 \neq i_2 \neq i_3}} Z_1^{i_1} Z_2^{i_2} Z_3^{i_3}$$

- Case (c)

$$e_i = \frac{S_1 Z_p + S_2}{Z_i(Z_1 + Z_2)},\ p \neq i,\ (i,p) \in \{1,2\}^2$$

- Case (d)

$$e_1 = \frac{S_1^2}{S_2}$$

6. Correct the errors: $\hat{c}(x) = r(x) + e(x)$

Example 4.14

To illustrate the decoding of an RS code using the direct method, we now present an example considering an RS code correcting up to three errors ($t = 3$) and having the following parameters:

$$m = 4 \quad q = 16 \quad n = 15 \quad n - k = 6$$

Let us assume, for example, that the transmitted codeword is $c(x) = 0$ and that the received word has two errors.

$$r(x) = \alpha^7 x^3 + \alpha^3 x^6$$

1. Calculate the components of the syndrome

$$\begin{array}{ll} S_1 = \alpha^{10} + \alpha^9 = \alpha^{13} & S_4 = \alpha^{19} + \alpha^{27} = \alpha^6 \\ S_2 = \alpha^{13} + \alpha^{15} = \alpha^6 & S_5 = \alpha^{22} + \alpha^{33} = \alpha^4 \\ S_3 = \alpha^{16} + \alpha^{21} = \alpha^{11} & S_6 = \alpha^{25} + \alpha^{39} = \alpha^{13} \end{array}$$

2. Determine the number of errors

$$\Delta_3 = 0 \qquad \Delta_2 = \alpha^8$$

Δ_3 being null and $\Delta_2 \neq 0$, we have two errors.

3. Calculate coefficients σ_1 and σ_2 of the error locator polynomial.

$$\begin{aligned} \sigma_1 &= \tfrac{1}{\Delta_2}\left[S_1S_4 + S_2S_3\right] = \tfrac{\alpha^{19}+\alpha^{17}}{\alpha^8} = \alpha^{11} + \alpha^9 = \alpha^2 \\ \sigma_2 &= \tfrac{1}{\Delta_2}\left[S_2S_4 + S_3^2\right] = \tfrac{\alpha^{12}+\alpha^{22}}{\alpha^8} = \alpha^4 + \alpha^{14} = \alpha^9 \end{aligned}$$

The error locator polynomial is therefore equal to:

$$\sigma_d(x) = x^2 + \alpha^2 x + \alpha^9$$

4. Look for the two roots of the error locator polynomial.

   Looking through the elements of field $\mathbf{F}_{16}$ we find that α^3 and α^6 cancel the polynomial $\Lambda(x)$. The errors therefore concern the terms in x^3 and in x^6 of word $r(x)$.

5. Calculate the error coefficients e_1 and e_2.

$$\begin{aligned} e_1 &= \tfrac{S_1Z_2+S_2}{Z_1Z_2+Z_1^2} = \tfrac{\alpha^{19}+\alpha^6}{\alpha^9+\alpha^6} = \tfrac{\alpha^{12}}{\alpha^5} = \alpha^7 \\ e_2 &= \tfrac{S_1Z_1+S_2}{Z_1Z_2+Z_2^2} = \tfrac{\alpha^{16}+\alpha^6}{\alpha^9+\alpha^{12}} = \tfrac{\alpha^{11}}{\alpha^8} = \alpha^3 \end{aligned}$$

6. Correct the errors

$$c(x) = (\alpha^7x^3 + \alpha^3x^6) + (\alpha^7x^3 + \alpha^3x^6) = 0$$

   The transmitted codeword is the null word; the two errors have therefore been corrected.

Simplification of Peterson's algorithm for binary codes

For BCH codes with binary symbols it is not necessary to calculate the coefficients e_j. Indeed, as these coefficients are binary, they are necessarily equal to 1 in the presence of an error in position j. The computation of coefficients σ_j can also be simplified by taking into account the fact that for a code with binary symbols we have:

$$S_{2j} = e(\alpha^{2j}) = \left[e(\alpha^j)\right]^2 = S_j^2$$

For a BCH code with binary symbols correcting up to $t = 3$ errors, taking into account the previous remark and using the expressions of the three coefficients σ_j of the error locator polynomial, we obtain:

$$\begin{aligned} \sigma_1 &= S_1 \\ \sigma_2 &= \tfrac{S_1^2S_3+S_5}{S_1^3+S_3} \\ \sigma_3 &= (S_1^3 + S_3) + S_1\tfrac{S_1^2S_3+S_5}{S_1^3+S_3} \end{aligned} \quad (4.40)$$

For a BCH code with binary symbols correcting up to $t = 2$ errors, also taking into account the previous remark and using the expressions of the two coefficients σ_j of the error locator polynomial, we obtain:

$$\begin{aligned} \sigma_1 &= S_1 \\ \sigma_2 &= \frac{S_3+S_1^3}{S_1} \end{aligned} \quad (4.41)$$

Finally, in the presence of an error $\sigma_2 = \sigma_3 = 0$ and $\sigma_1 = S_1$.

Example 4.15

Let us consider a BCH code correcting two errors ($t = 2$) in a block of $n = 15$ symbols of a generator polynomial equal to:

$$g(x) = x^8 + x^7 + x^6 + x^4 + 1$$

Let us assume that the transmitted codeword is $c(x) = 0$ and that the received word $r(x)$ has two errors.

$$r(x) = x^8 + x^3$$

There are three steps to the decoding: calculate syndrome $\mathbf{S}$, determine the coefficients σ_l of the error locator polynomial and search for its roots in field $\mathbf{F}_{16}$.

1. Calculate syndrome $\mathbf{S}$: we only need to calculate the odd index components S_1 and S_3 of syndrome $\mathbf{S}$. Using the binary representations of the elements of field $\mathbf{F}_{16}$ given in the appendix, and taking into account the fact that $\alpha^{15} = 1$, we have:

$$\begin{aligned} S_1 &= r(\alpha) = \alpha^8 + \alpha^3 = \alpha^{13} \\ S_3 &= r(\alpha^3) = \alpha^{24} + \alpha^9 = \alpha^9 + \alpha^9 = 0 \end{aligned}$$

2. Determine the coefficients σ_1 and σ_2 of the error locator polynomial. Using the expressions of coefficients σ_1 and σ_2, we obtain:

$$\begin{aligned} \sigma_1 &= S_1 = \alpha^{13} \\ \sigma_2 &= \frac{S_3+S_1^3}{S_1} = S_1^2 = \alpha^{26} = \alpha^{11} \quad (\alpha^{15} = 1) \end{aligned}$$

and the error locator polynomial is equal to:

$$\sigma_d(x) = x^2 + \alpha^{13}x + \alpha^{11}$$

3. Search for the roots of the error locator polynomial in field $\mathbf{F}_{16}$. By trying all the elements of field $\mathbf{F}_{16}$, we can verify that the roots of the error locator polynomial are α^3 and α^8. Indeed, we have

$$\sigma(\alpha^3) = \alpha^6 + \alpha^{16} + \alpha^{11} = \alpha^6 + \alpha + \alpha^{11} = 1100 + 0010 + 1110 = 0000$$

$$\sigma(\alpha^8) = \alpha^{16} + \alpha^{21} + \alpha^{11} = \alpha + \alpha^6 + \alpha^{11} = 0010 + 1100 + 1110 = 0000$$

The transmission errors concern the terms x^8 and x^3 of received word $r(x)$. The transmitted codeword is therefore $c(x) = 0$ and the two errors have been corrected.

The reader can verify that in the presence of a single error, $r(x) = x^j$; $0 \leq j \leq (n-1)$, the correction is still performed correctly since:

$$S_1 = \alpha^j;\ S_3 = \alpha^{3j};\ \sigma_1 = \alpha^j;\ \sigma_2 = 0;\ \sigma_d(x) = x(x + \sigma_1)$$

and the error locator polynomial has one sole root $\sigma_1 = \alpha^j$.

Chien algorithm

To search for the error locator polynomial roots in the case of codes with binary symbols, we can avoid going through all the elements of field $\mathbf{F}_q$ by using Chien's iterative algorithm.
Dividing polynomial $\sigma_d(x)$ by x^t, we obtain:

$$\tilde{\sigma}_d(x) = \frac{\sigma_d(x)}{x^t} = 1 + \sigma_1 x^{-1} + \cdots + \sigma_j x^{-j} + \cdots + \sigma_t x^{-t}$$

The roots of polynomial $\sigma_d(x)$ that are also the roots of $\tilde{\sigma}_d(x)$ have the form α^{n-j} where $j = 1, 2, \ldots, n-1$ and $n = q - 1$.
Thus α^{n-j} is a root of $\tilde{\sigma}_d(x)$ if:

$$\sigma_1 \alpha^{-n+j} + \cdots + \sigma_p x^{-np+jp} + \cdots + \sigma_t x^{-nt+jt} = 1$$

Taking into account the fact that $\alpha^n = 1$, the condition to satisfy in order for α^{n-j} to be a root of the error locator polynomial is:

$$\sum_{p=1}^{t} \sigma_p \alpha^{jp} = 1; \quad j = 1, 2, \cdots, (n-1) \tag{4.42}$$

Chien's algorithm has just tested whether condition (4.42) is satisfied using the circuit shown in Figure 4.7.
This circuit has a register with t memories initialized with the t coefficients σ_j of the error locator polynomial and a register with n memories that stocks symbols r_j; $j = 0, 1, \cdots, (n-1)$ of word $r(x)$. At the first clock pulse, the circuit performs the computation of the left-hand part of expression (4.42) for $j = 1$. If the result of this computation is equal to 1, α^{n-1} is a root of the error locator polynomial and the error that concerned symbol r_{n-1} is then corrected. If the result of this computation is equal to 0, no correction is performed. At the end of this first phase, the σ_j coefficients contained in the t memories of the register are replaced by $\sigma_j \alpha^j$. At the second clock pulse the circuit again

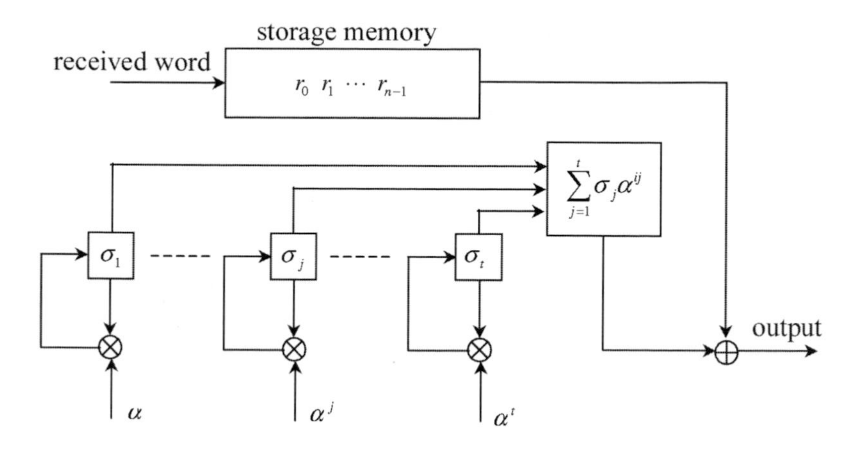

Figure 4.7 – Schematic diagram of the circuit implementing the Chien algorithm.

performs the computation of the left-hand part of expression (4.42) for $j = 2$. If the result of this computation is equal to 1, α^{n-2} is a root of the error locator polynomial and the error that concerned symbol r_{n-2} is then corrected. The algorithm continues in the same way for the following clock pulses.

4.4.3 Iterative method

Decoding RS codes or binary BCH codes with the iterative method uses two polynomials, error locator polynomial $\Lambda(x)$ and error evaluator polynomial $\Gamma(x)$. These two polynomials are defined respectively by:

$$\Lambda(x) = \prod_{j=1}^{t} (1 + Z_j x) \tag{4.43}$$

$$\Gamma(x) = \sum_{i=1}^{t} e_i Z_i x \frac{\Lambda(x)}{1 + Z_i x} \tag{4.44}$$

The error locator polynomial whose roots are Z_j^{-1} enables the position of the errors to be determined and the error evaluator polynomial enables the value of the error e_j to be determined. Indeed, taking into account the fact that $\Lambda(Z_j^{-1}) = 0$, the polynomial $\Gamma(x)$ taken in Z_j^{-1} is equal to:

$$\begin{aligned} \Gamma(Z_j^{-1}) &= e_j \prod_{p \neq j} (1 + Z_p Z_j^{-1}) \\ &= e_j Z_j^{-1} \Lambda'(Z_j^{-1}) \end{aligned}$$

where $\Lambda'(x) = \frac{d\Lambda}{dx}(x)$.

The value of error e_j is then given by the Forney algorithm:

$$e_j = Z_j \frac{\Gamma(Z_j^{-1})}{\Lambda'(Z_j^{-1})} \tag{4.45}$$

Introducing the polynomial $S(x)$ defined by:

$$S(x) = \sum_{j=1}^{2t} S_j x^j \tag{4.46}$$

we can show that:

$$\Lambda(x)S(x) \equiv \Gamma(x) \text{ modulo } x^{2t+1} \tag{4.47}$$

This relation is called *the key equation* for decoding a cyclic code.
To determine polynomials $\Lambda(x)$ and $\Gamma(x)$ two iterative algorithms are mainly used, the Berlekamp-Massey algorithm and Euclid's algorithm.

Berlekamp-Massey algorithm for codes with non-binary symbols

Computation of polynomials $\Lambda(x)$ and $\Gamma(x)$ using the Berlekamp-Massey algorithm is performed iteratively. It requires two intermediate polynomials denoted $\Theta(x)$ and $\Omega(x)$. The algorithm has $2t$ iterations. Once the algorithm has terminated, the Chien algorithm must be implemented to determine the roots Z_j^{-1} of $\Lambda(x)$ and consequently the position of the errors. Next, the Forney algorithm expressed by (4.45) enables the value of the errors e_j to be calculated.

Initial conditions:

$$\begin{aligned} L_0 &= 0 & & \\ \Lambda^{(0)}(x) &= 1 & \Theta^{(0)}(x) &= 1 \\ \Gamma^{(0)}(x) &= 0 & \Omega^{(0)}(x) &= 1 \end{aligned}$$

Recursion: $1 \leq p \leq 2t$

$$\Delta_p = \sum_j \Lambda_j^{(p-1)} S_{p-j}$$

$$\begin{aligned} \delta_p &= 1 \text{ if } \Delta_p \neq 0 \text{ and } 2L_{p-1} \leq p-1 \\ &= 0 \text{ otherwise} \end{aligned}$$

$$L_p = \delta_p(p - L_{p-1}) + (1-\delta_p)L_{p-1}$$

$$\begin{bmatrix} \Lambda^{(p)} & \Gamma^{(p)} \\ \Theta^{(p)} & \Omega^{(p)} \end{bmatrix} = \begin{bmatrix} 1 & \Delta_p x \\ \Delta_p^{-1}\delta_p & (1-\delta_p)x \end{bmatrix} \begin{bmatrix} \Lambda^{(p-1)} & \Gamma^{(p-1)} \\ \Theta^{(p-1)} & \Omega^{(p-1)} \end{bmatrix}$$

Termination :

$$\Lambda(x) = \Lambda^{(2t)}(x)$$
$$\Gamma(x) = \Gamma^{(2t)}(x)$$

Example 4.16

To illustrate the decoding of an RS code using the Berlekamp-Massey algorithm, let us consider an RS code correcting up to two errors ($t = 2$) and having the following parameters:

$$m = 4; \quad q = 16; \quad n = 15; \quad n - k = 4$$

Let us assume, for example, that the transmitted codeword is $c(x) = 0$ and that the received word has two errors.

$$r(x) = \alpha^7 x^3 + \alpha^3 x^6$$

The set of calculations performed to decode this RS code will be done in field $\mathbf{F}_{16}$ whose elements are given in the appendix.

1. Calculate syndrome $\mathbf{S} = (S_1, S_2, S_3, S_4)$

$$S_1 = \alpha^{10} + \alpha^9 = \alpha^{13} \quad S_3 = \alpha^{16} + \alpha^{21} = \alpha^{11}$$
$$S_2 = \alpha^{13} + \alpha^{15} = \alpha^6 \quad S_4 = \alpha^{19} + \alpha^{27} = \alpha^6$$

The polynomial $S(x)$ is therefore equal to:

$$S(x) = \alpha^{13} x + \alpha^6 x^2 + \alpha^{11} x^3 + \alpha^6 x^4$$

2. Calculate polynomials $\Lambda(x)$ and $\Gamma(x)$ from the Berlekamp-Massey algorithm

p	Δ_p	δ_p	L_p	$\Lambda^p(x)$	$\Theta^p(x)$	$\Gamma^p(x)$	$\Omega^p(x)$
0			0	1	1	0	1
1	α^{13}	1	1	$1 + \alpha^{13} x$	α^2	$\alpha^{13} x$	0
2	α	0	1	$1 + \alpha^8 x$	$\alpha^2 x$	$\alpha^{13} x$	0
3	α^{10}	1	2	$1 + \alpha^8 x + \alpha^{12} x^2$	$\alpha^5 + \alpha^{13} x$	$\alpha^{13} x$	$\alpha^3 x$
4	α^{10}	0	2	$1 + \alpha^2 x + \alpha^9 x^2$	$\alpha^5 x + \alpha^{13} x^2$	$\alpha^{13} x + \alpha^{13} x^2$	$\alpha^3 x^2$

In the table above, all the calculations are done in field $\mathbf{F}_{16}$ and take into account the fact that $\alpha^{15} = 1$.

The error locator and error evaluator polynomials are:

$$\Lambda(x) = 1 + \alpha^2 x + \alpha^9 x^2$$
$$\Gamma(x) = \alpha^{13} x + \alpha^{13} x^2$$

We can verify that the key equation for the decoding has been satisfied. Indeed, we do have:

$$\Lambda(x) S(x) = \alpha^{13} x + \alpha^{13} x^2 + \alpha^4 x^5 + x^6 \equiv \alpha^{13} x + \alpha^{13} x^2 = \Gamma(x) \text{ modulo } x^5$$

3. Search for the roots of the error locator polynomial

   By looking through all the elements of field $\mathbf{F_{16}}$ we find that α^{12} and α^9 are roots of polynomial $\Lambda(x)$. The errors are therefore in position $x^3(\alpha^{-12} = \alpha^3)$ and $x^6(\alpha^{-9} = \alpha^6)$ and error polynomial $e(x)$ is equal to:

$$e(x) = e_3 x^3 + e_6 x^6$$

4. Calculate error coefficients e_j (4.45).

$$\begin{aligned} e_3 &= \alpha^3 \frac{\alpha^6}{\alpha^2} = \alpha^7 \\ e_6 &= \alpha^6 \frac{\alpha^{14}}{\alpha^2} = \alpha^3 \end{aligned}$$

   Error polynomial $e(x)$ is therefore equal to:

$$e(x) = \alpha^7 x^3 + \alpha^3 x^6$$

   and the estimated codeword is $\hat{c}(x) = r(x) + e(x) = 0$. The two transmission errors are corrected.

Euclid's algorithm

Euclid's algorithm enables us to solve the key equation for decoding, that is, to determine polynomials $\Lambda(x)$ and $\Gamma(x)$.

Initial conditions:

$$R_{-1}(x) = x^{2t};\ R_0(x) = S(x);\ U_{-1}(x) = 0;\ U_0(x) = 1$$

Recursion:

calculate $Q_j(x)$, $R_{j+1}(x)$ and $U_{j+1}(x)$ from the two following expressions:

$$\begin{aligned} \frac{R_{j-1}(x)}{R_j(x)} &= Q_j(x) + \frac{R_{j+1}(x)}{R_j(x)} \\ U_{j+1}(x) &= Q_j(x)U_j(x) + U_{j-1}(x) \end{aligned}$$

When $\deg(U_j) \leq t$ and $\deg(R_j) \leq t$ then:

$$\begin{aligned} \Lambda(x) &= U_{j+1}(x) \\ \Gamma(x) &= R_{j+1}(x) \end{aligned}$$

Example 4.17

Let us again take the RS code used to illustrate the Berlekamp-Massey algorithm. Assuming that the received word is always $r(x) = \alpha^7 x^3 + \alpha^3 x^6$ when the transmitted codeword is $c(x) = 0$, the decoding algorithm is the following:

1. Calculate syndrome $\mathbf{S} = (S_1, S_2, S_3, S_4)$

$$\begin{array}{ll} S_1 = \alpha^{10} + \alpha^9 = \alpha^{13} & S_3 = \alpha^{16} + \alpha^{21} = \alpha^{11} \\ S_2 = \alpha^{13} + \alpha^{15} = \alpha^6 & S_4 = \alpha^{19} + \alpha^{27} = \alpha^6 \end{array}$$

Polynomial $S(x)$ is therefore equal to:

$$S(x) = \alpha^{13}x + \alpha^6 x^2 + \alpha^{11}x^3 + \alpha^6 x^4$$

2. Calculate polynomials $\Lambda(x)$ and $\Gamma(x)$ from Euclid's algorithm (the calculations are performed in field $\mathbf{F_{16}}$ whose elements are given in the appendix).

$$\begin{array}{rcl} j & = & 0 \\ R_{-1}(x) & = & x^5 \\ R_0(x) & = & S(x) \\ Q_0(x) & = & \alpha^9 x + \alpha^{14} \\ R_1(x) & = & \alpha^5 x^3 + \alpha^{13}x^2 + \alpha^{12}x \\ U_1(x) & = & \alpha^9 x + \alpha^{14} \end{array} \qquad \begin{array}{rcl} j & = & 1 \\ R_0(x) & = & S(x) \\ R_1(x) & = & \alpha^5 x^3 + \alpha^{13}x^2 + \alpha^{12}x \\ Q_1(x) & = & \alpha x + \alpha^5 \\ R_2(x) & = & \alpha^{14}x^2 + \alpha^{14}x \\ U_2(x) & = & \alpha^{10}x^2 + \alpha^3 x + \alpha \end{array}$$

We can verify that $\deg(U_2(x)) = 2$ is lower than or equal to t ($t = 2$) and that $\deg(R_2(x)) = 2$ is lower than or equal to t. The algorithm is therefore terminated and polynomials $\Lambda(x)$ and $\Gamma(x)$ respectively have the expression:

$$\begin{array}{l} \Lambda(x) = U_2(x) = \alpha + \alpha^3 x + \alpha^{10}x^2 = \alpha(1 + \alpha^2 x + \alpha^9 x^2) \\ \Gamma(x) = R_2(x) = \alpha^{14}x + \alpha^{14}x^2 = \alpha(\alpha^{13}x + \alpha^{13}x^2) \end{array}$$

We can verify that the key equation for the decoding is satisfied and that the two polynomials obtained are identical, to within one coefficient α, to those determined using the Berlekamp-Massey algorithm.

The roots of the polynomial $\Lambda(x)$ are therefore $1/\alpha^3$ and $1/\alpha^6$, and error polynomial $e(x)$ is equal to:

$$e(x) = \alpha^7 x^3 + \alpha^3 x^6$$

Calculating coefficients e_j by a transform

It is possible to calculate the coefficients e_j; $j = 0, 1, \cdots, (n-1)$ of error polynomial $e(x)$ without determining the roots of the error locator polynomial $\Lambda(x)$. To do this, we introduce the *extended syndrome* $S^*(x)$ defined by:

$$S^*(x) = \Gamma(x)\frac{1 + x^n}{\Lambda(x)} = \sum_{j=1}^{n} S_j x^j \tag{4.48}$$

Coefficient e_j is null (no errors) if α^{-j} is not a root of error locator polynomial $\Lambda(x)$. In this case, we have $S^*(\alpha^{-j}) = 0$ since $\alpha^{-jn} = 1$ (recall that $n = q - 1$ and $\alpha^{q-1} = 1$).
A contrario if α^{-j} is a root of the locator polynomial, coefficient e_j is non-null (presence of an error) and $S^*(\alpha^{-j})$ is of the form $0/0$. This indetermination can be removed by calculating the derivation of the numerator and the denominator of expression (4.48).

$$S^*(\alpha^{-i}) = \Gamma(\alpha^{-i})\frac{n\alpha^{-j(n-1)}}{\Lambda'(\alpha^{-j})}$$

Using Equation (4.45) and taking into account the fact that $\alpha^{-j(n-1)} = \alpha^j$ and that $na = a$ for n odd in a Galois field, coefficient e_j is equal to:

$$e_j = S^*(\alpha^{-j}) \tag{4.49}$$

The extended syndrome can be computed from polynomials $\Lambda(x)$ and $\Gamma(x)$ using the following relation deduced from expression (4.48).

$$\Lambda(x)S^*(x) = \Gamma(x)(1 + x^n) \tag{4.50}$$

Coefficients S_j of the extended syndrome are identical to those of syndrome $S(x)$ for j from 1 to $2t$ and are determined by cancelling the coefficients of the x^j terms in the product $\Lambda(x)S^*(x)$, for j from $2t + 1$ to n.

Example 4.18

Again taking the example of the RS code ($q = 16$; $n = 15$; $k = 11$; $t = 2$) used to illustrate the Berlekamp-Massey algorithm let us determine the extended syndrome.

$$S^*(x) = \sum_{j=1}^{15} S_j x^j$$

with:

$$\begin{array}{ll} S_1 = \alpha^{13} & S_3 = \alpha^{11} \\ S_2 = \alpha^6 & S_4 = \alpha^6 \end{array}$$

Equation (4.50) provides us with the following relation:

$$S(x) + \alpha^2 x S(x) + \alpha^9 x^2 S(x) = \alpha^{13}(x + x^2 + x^{16} + x^{17})$$

$$\begin{array}{l} S_1 x + (\alpha^2 S_1 + S_2)x^2 \\ + \sum\limits_{k=3}^{15} (\alpha^9 S_{k-2} + \alpha^2 S_{k-1} + S_k)x^k \\ +(\alpha^2 S_{15} + \alpha^9 S_{14})x^{16} + \alpha^9 S_{15} x^{17} \quad = \quad \alpha^{13}(x + x^2 + x^{16} + x^{17}) \end{array}$$

From this there results the recurrence relation:

$$S_k = \alpha^2 S_{k-1} + \alpha^9 S_{k-2}, \quad k = 3, 4, \cdots, 15$$

We thus obtain the coefficients of the extended syndrome:

$$\begin{array}{l} S_5 = \alpha^4, S_6 = \alpha^{13}, S_7 = \alpha^6, S_8 = \alpha^{11}, S_9 = \alpha^6, S_{10} = \alpha^4 \\ S_{11} = \alpha^{13}, S_{12} = \alpha^6, S_{13} = \alpha^{11}, S_{14} = \alpha^6, S_{15} = \alpha^4 \end{array}$$

Another way to obtain the extended polynomial involves dividing $\Gamma(x)(1+x^n)$ by $\Lambda(x)$ by increasing power orders.
The errors being with monomials x^3 and x^6, let us calculate coefficients e_3 and e_6.

$$\begin{array}{rcl} e_3 & = & S^*(\alpha^{12}) = \alpha^2 + \alpha^4 + \alpha^{10} + \alpha^7 = \alpha^7 \\ e_6 & = & S^*(\alpha^9) = \alpha^4 + \alpha^7 = \alpha^3 \end{array}$$

The values found for coefficients e_3 and e_6 are obviously identical to those obtained in example 4.16. We can verify that the other e_j coefficients are all null.

Berlekamp-Massey algorithm for binary cyclic codes

For binary BCH codes the Berlekamp-Massey algorithm can be simplified since it is no longer necessary to determine the error evaluator polynomial, and since it is possible to show that the Δ_j terms are null for j even. This implies:

$$\begin{array}{l} \delta_{2p} = 0 \\ L_{2p} = L_{2p-1} \\ \Lambda^{(2p)}(x) = \Lambda^{(2p-1)}(x) \\ \Theta^{(2p)}(x) = x\Theta^{(2p-1)}(x) \end{array}$$

Hence the algorithm in t iterations:
Initial conditions:

$$\begin{array}{rcl} L_{-1} & = & 0 \\ \Lambda^{(-1)}(x) & = & 1 \quad \Theta^{(-1)}(x) = x^{-1} \end{array}$$

Recursion: $0 \leq p \leq t-1$

$$\begin{array}{rcl} \Delta_{2p+1} & = & \sum_j \Lambda_j^{(2p-1)} S_{2p+1-j} \\ \delta_{2p+1} & = & 1 \quad \text{if } \Delta_{2p+1} \neq 0 \text{ and } L_{2p-1} \leq p \\ & = & 0 \quad \text{if not} \\ L_{2p+1} & = & \delta_{2p+1}(2p+1-L_{2p-1}) + (1-\delta_{2p+1})L_{2p-1} \end{array}$$

$$\begin{bmatrix} \Lambda^{(2p+1)} \\ \Theta^{(2p+1)} \end{bmatrix} = \begin{bmatrix} 1 & \Delta_{2p+1}x^2 \\ \Delta_{2p+1}^{-1}\delta_{2p+1} & (1-\delta_{2p+1})x^2 \end{bmatrix} \begin{bmatrix} \Lambda^{(2p-1)} \\ \Theta^{(2p-1)} \end{bmatrix}$$

Termination:

$$\Lambda(x) = \Lambda^{(2t-1)}(x)$$

Example 4.19

Again taking the BCH code that was used to illustrate the computation of the error locator polynomial with the direct method, let us assume that the received word is $r(x) = x^8 + x^3$ when the transmitted codeword is $c(x) = 0$.

1. Syndrome **S** has four components.

$$\begin{aligned} S_1 &= r(\alpha) = \alpha^8 + \alpha^3 = \alpha^{13} \\ S_3 &= r(\alpha^3) = \alpha^{24} + \alpha^9 = 0 \\ S_2 &= S_1^2 = \alpha^{26} = \alpha^{11} \\ S_4 &= S_2^2 = \alpha^{22} = \alpha^7 \end{aligned}$$

Polynomial $S(x)$ is equal to:

$$S(x) = \alpha^{13}x + \alpha^{11}x^2 + \alpha^7 x^4$$

2. Calculate polynomial $\Lambda(x)$ from the Berlekamp-Massey algorithm

p	Δ_{2p+1}	δ_{2p+1}	L_{2p+1}	$\Lambda^{2p+1}(x)$	$\Theta^{2p+1}(x)$
-1			0	1	x^{-1}
0	α^{13}	1	1	$1+\alpha^{13}x$	α^2
1	α^9	1	2	$1+\alpha^{13}x+\alpha^{11}x^2$	$\alpha^6+\alpha^4 x$

Note that polynomial $\Lambda(x)$ obtained is identical to that determined using the direct method. The roots of $\Lambda(x)$ are $1/\alpha^3$ and $1/\alpha^8$, and the errors therefore concern terms x^3 and x^8. The estimated codeword is $\hat{c}(x) = 0$.

Euclid's algorithm for binary codes

Example 4.20

Let us again take the decoding of the (15,7) BCH code. The received word is $r(x) = x^8 + x^3$.

$$\begin{array}{rclrcl} j &=& 0 & j &=& 1 \\ R_{-1}(x) &=& x^5 & R_0(x) &=& S(x) \\ R_0(x) &=& S(x) & R_1(x) &=& \alpha^4x^3+\alpha^6x^2 \\ Q_0(x) &=& \alpha^8x & Q_1(x) &=& \alpha^3x+\alpha^5 \\ R_1(x) &=& \alpha^4x^3+\alpha^6x^2 & R_2(x) &=& \alpha^{13}x \\ U_1(x) &=& \alpha^8x & U_2(x) &=& \alpha^{11}x^2+\alpha^{13}x+1 \end{array}$$

We can verify that $\deg(U_2(x)) = 2$ is lower than or equal to t ($t = 2$) and that the degree of R_2 is lower than or equal to t. The algorithm is therefore terminated and polynomial $\Lambda(x)$ has the expression:

$$\Lambda(x) = U_2(x) = 1 + \alpha^{13}x + \alpha^{11}x^2$$
$$\Gamma(x) = R_2(x) = \alpha^{13}x$$

For a binary BCH code it is not necessary to use the error evaluator polynomial to determine the value of coefficients e_3 and e_8. However, we can verify that:

$$\begin{aligned} e_3 &= \alpha^3 \frac{\Gamma(\alpha^{-3})}{\Lambda'(\alpha^{-3})} = 1 \\ e_8 &= \alpha^8 \frac{\Gamma(\alpha^{-8})}{\Lambda'(\alpha^{-8})} = 1 \end{aligned}$$

The decoded word is therefore $\hat{c}(x) = r(x) + e(x) = 0$ and the two errors have been corrected.

4.4.4 Hard input decoding performance of Reed-Solomon codes

Recall that for a Reed-Solomon code, the blocks of information to encode and the codewords are made up of k and $n = q - 1$ ($q = 2^m$) q-ary symbols respectively. The probability $P_{e,\text{word}}$ of having a wrong codeword after hard decoding can be upper bounded by:

$$P_{e,\text{word}} \leq \sum_{j=t+1}^{n} \binom{n}{j} p_s^j (1 - p_s)^{n-j} \tag{4.51}$$

where p_s is the error probability per q-ary symbol on the transmission channel and t is the code correction capability in number of q-ary symbols.
When a codeword is wrongly decoded, the error probability per corresponding $P_{e,\text{symbol}}$ symbol after decoding is upper bounded by:

$$P_{e,\text{symbol}} \leq \frac{1}{n} \sum_{j=t+1}^{n} (j + t) \binom{n}{j} p_s^j (1 - p_s)^{n-j} \tag{4.52}$$

The binary error probability after decoding is obtained from the error probability per symbol, taking into account that a symbol is represented by m bits:

$$P_{e,\text{bit}} = 1 - (1 - P_{e,\text{symbol}})^{\frac{1}{m}}$$

At high signal to noise ratio, we can approximate the binary error probability after decoding:

$$P_{e,\text{bit}} \cong \frac{1}{m} P_{e,\text{symbol}} \qquad \frac{E_b}{N_0} >> 1 .$$

Bibliography

[4.1] R.H. Morelos-Zaragoza. *The Art of Error Correcting Coding.* John Wiley & sons, 2005.

[4.2] J. G. Proakis. *Digital Communications.* McGraw-Hill, New-York, 4th edition, 2000.

Appendix

Notions about Galois fields and minimal polynomials

Definition

A Galois field with $q = 2^m$ elements denoted $\mathbf{F}_q$, where m is a positive integer is defined as a polynomial extension of the field with two elements $(0, 1)$ denoted $\mathbf{F}_2$. The polynomial $\varphi(x)$ used to build field $\mathbf{F}_q$ must be

- irreducible, that is, non factorizable in $\mathbf{F}_2$ (in other words, 0 and 1 are not roots of $\varphi(x)$),
- of degree m,
- and with coefficients in $\mathbf{F}_2$.

The elements of a Galois field $\mathbf{F}_q$ are defined modulo $\varphi(x)$ and thus, each element of this field can be represented by a polynomial with degree at most equal to $(m-1)$ and with coefficients in $\mathbf{F}_2$.

Example 1

Consider an irreducible polynomial $\varphi(x)$ in the field $\mathbf{F}_2$ of degree $m = 2$.

$$\varphi(x) = x^2 + x + 1$$

This polynomial enables a Galois field to be built with 4 elements. The elements of this field $\mathbf{F}_4$ are of the form:

$$a\alpha + b \quad \text{where} \quad a, b \in \mathbf{F_2}$$

that is:

$$\mathbf{F}_4 : \{0,\, 1,\, \alpha,\, \alpha + 1\}$$

We can see that if we raise element α to successive powers 0, 1 and 2 we obtain all the elements of field $\mathbf{F}_4$ with the exception of element 0. Indeed, α^2 is still equal to $(\alpha+1)$ modulo $\varphi(\alpha)$. Element α is called the *primitive element* of field $\mathbf{F}_4$.

The elements of field $\mathbf{F}_4$ can also be represented in binary form:

$$\mathbf{F}_4 : \{00,\ 01,\ 10,\ 11\}$$

The binary couples correspond to the four values taken by coefficients a and b.

Primitive element of a Galois field

We call the primitive element of a Galois field $\mathbf{F}_q$, an element of this field that, when it is raised to successive powers $0, 1, 2, \cdots, (q-2)$; $q = 2^m$, makes it possible to retrieve all the elements of the field except element 0. Every Galois field has at least one primitive element. If α is a primitive element of field $\mathbf{F}_q$ then, the elements of this field are:

$$\mathbf{F}_q = \left\{0, \alpha^0, \alpha^1, \cdots, \alpha^{q-2}\right\} \text{ with } \alpha^{q-1} = 1$$

Note that in such a Galois field the "-" sign is equivalent to the "+" sign, that is:

$$-\alpha^j = \alpha^j \quad \forall j \in \{0, 1, \cdots, (q-2)\}$$

Observing that $2\alpha^j = 0$ modulo 2, we can always add the zero quantity $2\alpha^j$ to $-\alpha^j$ and we thus obtain the above equality.

For example, for field $\mathbf{F}_4$ let us give the rules that govern the addition and multiplication operations. All the operations are done modulo 2 and modulo $\alpha^2 + \alpha + 1$.

$+$	**0**	**1**	α	α^2
0	0	1	α	α^2
1	1	0	$1+\alpha=\alpha^2$	$1+\alpha^2=\alpha$
α	α	$1+\alpha=\alpha^2$	0	$\alpha+\alpha^2=1$
α^2	α^2	$1+\alpha^2=\alpha$	$\alpha+\alpha^2=1$	0

Table 4.7 – Addition in field $\mathbf{F}_4$.

Minimal polynomial with coefficients in $\mathbf{F}_2$ associated with an element of a Galois field $\mathbf{F}_q$

The minimal polynomial $m_\beta(x)$ with coefficients in $\mathbf{F}_2$ associated with any element β of a Galois field $\mathbf{F}_q$, is a polynomial of degree at most equal to

$\times$	**0**	**1**	α	α^2
0	0	0	0	0
1	0	1	α	α^2
α	0	α	α^2	$\alpha^3 = 1$
α^2	0	α^2	$\alpha^3 = 1$	$\alpha^4 = \alpha$

Table 4.8 – Multiplication in field $\mathbf{F}_4$.

$m = \log_2(q)$, having β as a root. This polynomial is unique and irreducible in $\mathbf{F_2}$. If β is a primitive element of Galois field $\mathbf{F}_q$ then polynomial $m_\beta(x)$ is exactly of degree m. Note that a polynomial with coefficients in $\mathbf{F_2}$ satisfies the following property:

$$[f(x)]^2 = f(x^2) \Rightarrow [f(x)]^{2^p} = f(x^{2^p})$$

So, if β is a root of polynomial $f(x)$ then $\beta^2, \beta^4, \cdots$ are also roots of this polynomial. The minimal polynomial with coefficients in $\mathbf{F}_2$ having β as a root can then be written in the form:

$$m_\beta(x) = (x + \beta)(x + \beta^2)(x + \beta^4)\cdots$$

If β is a primitive element of $\mathbf{F}_q$, the minimal polynomial with coefficients in $\mathbf{F_2}$ being of degree m, it can also be written:

$$m_\beta(x) = (x + \beta)(x + \beta^2)(x + \beta^4)\cdots(x + \beta^{2^{m-1}})$$

Example 2

Let us calculate the minimal polynomial associated with the primitive element α of Galois field $\mathbf{F}_4$.

$$\mathbf{F}_4 : \{0,\ 1,\ \alpha,\ \alpha^2\}$$

The minimal polynomial associated with element α therefore has α and α^2 ($m = 2$) as roots, and can be expressed:

$$m_\alpha(x) = (x + \alpha)(x + \alpha^2) = x^2 + x(\alpha + \alpha^2) + \alpha^3$$

Taking into account the fact that $\alpha^3 = 1$ and that $\alpha + \alpha^2 = 1$ in field $\mathbf{F}_4$, the polynomial $m_\alpha(x)$ is thus equal to:

$$m_\alpha(x) = x^2 + x + 1$$

Minimal polynomial with coefficients in $\mathbf{F}_q$ associated with an element in a Galois field $\mathbf{F}_q$

The minimal polynomial $m_\beta(x)$, with coefficients in the Galois field $\mathbf{F}_q$ associated with an element $\beta = \alpha^j$ (α a primitive element of field $\mathbf{F}_q$) of this field, is the lowest degree polynomial having β as a root.
Recalling that for a polynomial with coefficients in $\mathbf{F}_q$, we can write:

$$[f(x)]^q = f(x^q) \Rightarrow [f(x)]^{q^p} = f(x^{q^p})$$

Then if β is a root of polynomial $f(x)$, $\beta^q, \beta^{q^2}, \cdots$ are also roots of this polynomial.
Since in field $\mathbf{F}_q$ $\alpha^{q-1} = 1$, then $\beta^{q^p} = (\alpha^j)^{q^p} = \alpha^j = \beta$ and, thus, minimal polynomial $m_\beta(x)$ is simply equal to:

$$m_\beta(x) = x + \beta$$

These results on minimal polynomials are used to determine the generator polynomials of particular cyclic codes (BCH and Reed-Solomon).

Primitive polynomials

A polynomial with coefficients in $\mathbf{F}_2$ is primitive if it is the minimal polynomial associated with a primitive element of a Galois field. A primitive polynomial is thus irreducible in $\mathbf{F}_2$ and consequently can be used to build a Galois field. When a primitive polynomial is used to build a Galois field, all the elements of the field are obtained by raising the primitive element, the root of the primitive polynomial, to successively increasing powers. As the main primitive polynomials are listed in the literature, the construction of a Galois field with $q = 2^m$ elements can then be done simply by using a primitive polynomial of degree m. Table 4.9 gives some primitive polynomials.

To end this introduction to Galois fields and minimal polynomials, let us give an example of a Galois field with $q = 16$ ($m = 4$) elements built from the primitive polynomial x^4+x+1. This field is used to build generator polynomials of BCH and Reed-Solomon codes and to decode them. The elements of this field are:

$$\mathbf{F}_{16} = \left\{0,\, 1,\, \alpha,\, \alpha^2,\, \alpha^3 \cdots \alpha^{14}\right\}$$

where α is a primitive element of $\mathbf{F}_{16}$. With these 16 elements, we can also associate a polynomial representation and a binary representation. The polynomial representation of an element of this field is of the form:

$$a\alpha^3 + b\alpha^2 + c\alpha + d$$

where a, b, c and d are binary coefficients belonging to $\mathbf{F}_2$.

Degree of the polynomial	**Primitive polynomial**
2	$\alpha^2 + \alpha + 1$
3	$\alpha^3 + \alpha + 1$
4	$\alpha^4 + \alpha + 1$
5	$\alpha^5 + \alpha^2 + 1$
6	$\alpha^6 + \alpha + 1$
7	$\alpha^7 + \alpha^3 + 1$
8	$\alpha^8 + \alpha^4 + \alpha^3 + \alpha^2 + 1$
9	$\alpha^9 + \alpha^4 + 1$
10	$\alpha^{10} + \alpha^3 + 1$

Table 4.9 – Examples of primitive polynomials

Galois field $\mathbf{F}_{16}$ being made up of 16 elements, the binary representation of an element of this field is done with the help of 4 binary symbols belonging to $\mathbf{F}_2$. These 4 symbols are equal to the values taken by coefficients a, b, c and d respectively.

Elements of the field	**Polynomial representation**	**Binary representation**
0	0	0 0 0 0
1	1	0 0 0 1
α	α	0 0 1 0
α^2	α^2	0 1 0 0
α^3	α^3	1 0 0 0
α^4	$\alpha + 1$	0 0 1 1
α^5	$\alpha^2 + \alpha$	0 1 1 0
α^6	$\alpha^3 + \alpha^2$	1 1 0 0
α^7	$\alpha^3 + \alpha + 1$	1 0 1 1
α^8	$\alpha^2 + 1$	0 1 0 1
α^9	$\alpha^3 + \alpha$	1 0 1 0
α^{10}	$\alpha^2 + \alpha + 1$	0 1 1 1
α^{11}	$\alpha^3 + \alpha^2 + \alpha$	1 1 1 0
α^{12}	$\alpha^3 + \alpha^2 + \alpha + 1$	1 1 1 1
α^{13}	$\alpha^3 + \alpha^2 + 1$	1 1 0 1
α^{14}	$\alpha^3 + 1$	1 0 0 1

Table 4.10 – Different representations of the elements of Galois field $\mathbf{F}_{16}$

Example 3

Some calculations in field $\mathbf{F}_{16}$ are given in table 4.11 for addition, in table 4.12 for multiplication and in table 4.13 for division.

$+$	α^2	α^4
α^8	$0100 + 0101 = 0001 = 1$	$0011 + 0101 = 0110 = \alpha^5$
α^{10}	$0100 + 0111 = 0011 = \alpha^4$	$0011 + 0111 = 0100 = \alpha^2$

Table 4.11 – Addition in $\mathbf{F}_{16}$

$\times$	α^2	α^6
α^8	α^{10}	α^{14}
α^{14}	$\alpha^{16} = \alpha$ as $\alpha^{15} = 1$	$\alpha^{20} = \alpha^5$ as $\alpha^{15} = 1$

Table 4.12 – Multiplication in $\mathbf{F}_{16}$

$\div$	α^2	α^{12}
α^8	$\alpha^{-6} = \alpha^9$ as $\alpha^{15} = 1$	α^4
α^{14}	$\alpha^{-12} = \alpha^3$ as $\alpha^{15} = 1$	$\alpha^{-2} = \alpha^{13}$ as $\alpha^{15} = 1$

Table 4.13 – Division in $\mathbf{F}_{16}$

Chapter 5

Convolutional codes and their decoding

5.1 History

It was in 1955 that Peter Elias introduced the notion of convolutional code [5.5]. The example of an encoder described is illustrated in Figure 5.1. It is a systematic encoder, that is, the coded message contains the message to be transmitted, to which redundant information is added. The message is of infinite length, which at first sight limits the field of application of this type of code. It is however easy to adapt it for packet transmissions thanks to *tail-biting* techniques.

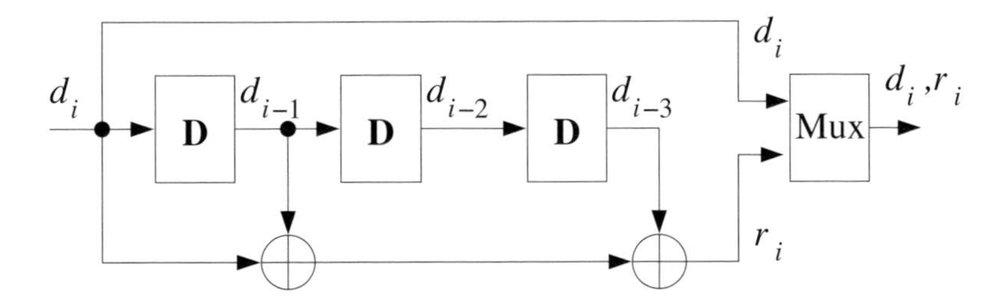

Figure 5.1 – Example of a convolutional encoder.

The encoder presented in Figure 5.1 is designed around a shift register with three memory elements. The redundancy bit at instant i, denoted r_i is constructed with the help of a modulo 2 sum of the information at instant i, d_i and the data present at instants $i-1$ and $i-3$ (d_{i-1} and d_{i-3}). A multiplexer plays the role of a parallel to serial converter and provides the result of the encoding at a rate twice that of the rate at the input. The coding rate of this encoder is

1/2 since, at each instant i, it receives data d_i and delivers two elements at the output: d_i (systematic part) and r_i (redundant part).

It was not until 1957 that the first algorithm capable of decoding such codes appeared. Invented by Wozencraft [5.15], this algorithm, called sequential decoding, was then improved by Fano [5.6] in 1963. Four years later, Viterbi introduced a new algorithm that was particularly interesting when the length of the shift register of the encoder is not too large [5.14]. Indeed, the complexity of the Viterbi algorithm increases exponentially with the size of this register whereas the complexity of the Fano algorithm is almost independent of it.

In 1974, Bahl, Cocke, Jelinek and Raviv presented a new algorithm [5.1] capable of associating a probability with the binary decision. This property is very widely used in the decoding of concatenated codes and more particularly turbocodes, which have brought this algorithm back into favour. It is now referred to in the literature in one of these three ways: BCJR (initials of the inventors), MAP (Maximum *A Posteriori*) or APP (*A Posteriori* Probability). The MAP algorithm is rather complex to implement in its initial version, and it exists in simplified versions, the most common ones being presented in Chapter 7.

In parallel with these advances in decoding algorithms, a number of works have treated the construction of convolutional encoders. The aim of these studies has not been to decrease the complexity of the encoder, since its implantation is trivial. The challenge is to find codes with the highest possible error correction capability. In 1970, Forney wrote a reference paper on the algebra of convolutional codes [5.7]. It showed that a *good* convolutional code is not necessarily systematic and suggested a construction different from that of Figure 5.1. For a short time, that paper took systematic convolutional codes away from the field of research on channel coding.

Figure 5.2 gives an example of a non-systematic convolutional encoder. Unlike the encoder in Figure 5.1, the data are not present at the output of the encoder and are replaced by a modulo 2 sum of the data at instant i, d_i, and of the data present at instants $i-2$ and $i-3$ (d_{i-2} and d_{i-3}). The rate of the encoder remains unchanged at 1/2 since the encoder always provides two elements at the output: $r_i^{(1)}$ and $r_i^{(2)}$, at instant i.

When Berrou *et al.* presented their work on turbocodes [5.4], they rehabilitated systematic convolutional codes by using them in a recursive form. The interest of recursive codes is presented in Sections 5.2 and 5.3. Figure 5.3 gives an example of an encoder for recursive systematic convolutional codes. The original message being transmitted (d_i), the code is therefore truly systematic. A feedback loop appears, the structure of the encoder now being similar to that of pseudo-random sequence generators.

This brief overview has allowed us to present the three most commonly used families of convolutional codes: systematic, non-systematic, and recursive systematic codes. The next two sections tackle the representation and performance

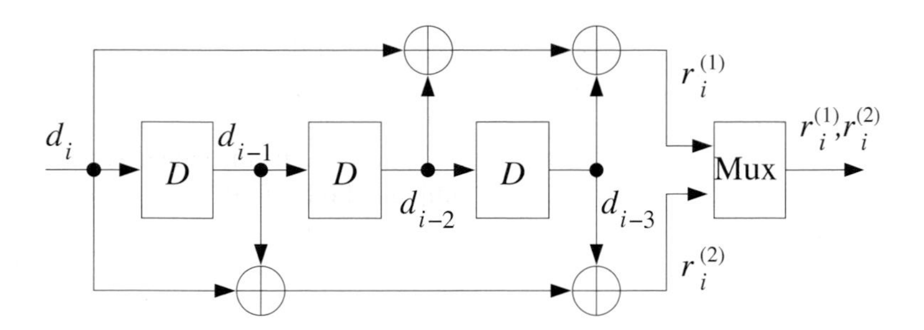

Figure 5.2 – Example of an encoder for non-systematic convolutional codes.

of convolutional codes. They give us the opportunity to compare the properties of these three families. The decoding algorithms most commonly used in current systems are presented in Section 5.4. Finally, Section 5.5 tackles the main *tail-biting* and *puncturing* techniques.

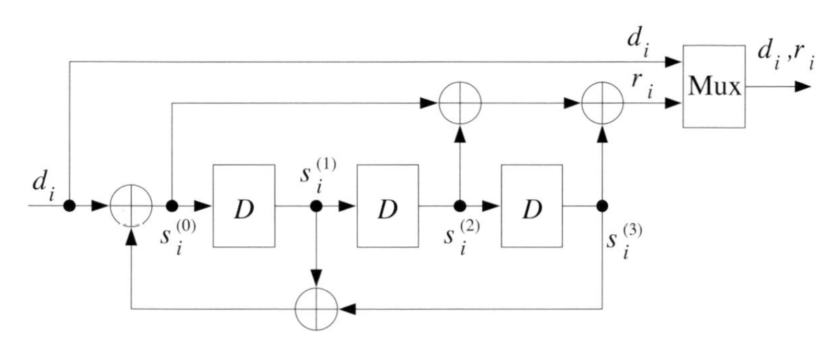

Figure 5.3 – Example of a encoder for recursive systematic convolutional codes.

5.2 Representations of convolutional codes

This chapter makes no claim to tackle the topic of convolutional codes exhaustively. Non-binary or non-linear codes are not treated, nor are encoders with several registers. Only the most commonly used codes, in particular for the construction of turbocodes, are introduced. The reader wishing to go further into the topic can, for example, refer to [5.11].

5.2.1 Generic representation of a convolutional encoder

Figure 5.4 gives a sufficiently general model for us to represent all the convolutional codes studied in this chapter. At each instant i, it receives a vector $\mathbf{d}_i$ of m bits at its input. The code thus generated is a binary code. However,

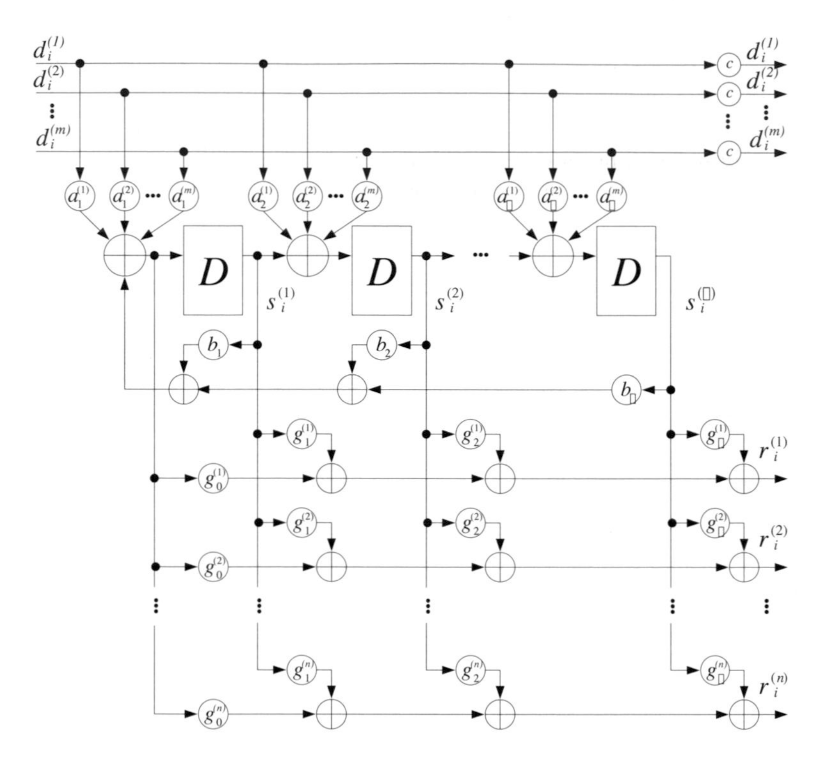

Figure 5.4 – Generic representation of an encoder for convolutional codes.

to simplify the writing, we will call it an *m-binary* and *double-binary* code if $m = 2$ (as in the example presented in Figure 5.5). When $c = 1$, the code generated is systematic, since $\mathbf{d}_i$ is transmitted at the output of the encoder. The code is thus made up of the systematic part $\mathbf{d}_i$ on m bits and the redundancy $\mathbf{r}_i$ on n bits. The coding rate is then $R = m/(m+n)$. If $c = 0$, the code is non-systematic and the rate becomes $R = m/n$.

The non-systematic part is constructed with the help of a shift register made up of ν flip-flops and of binary adders, in other words, of *XOR* gates. We then define an important characteristic of convolutional codes: the *constraint length*, here equal to $\nu + 1$ (some authors denote it ν, which implies that the register is then made up of $\nu - 1$ flip-flops). The register at instant i is characterized by the ν bits $s_i^{(1)}, s_i^{(2)}, \ldots, s_i^{\nu}$ memorized: they define its *state*, that we can thus code on ν bits and represent in the form of a vector $\mathbf{s}_i = (s_i^{(1)}, s_i^{(2)}, \cdots, s_i^{\nu})$. This type of convolutional encoder thus has 2^{ν} possible state values, that we often denote in natural binary or binary decimal form. Thus the state of an encoder made up of three flip-flops can take $2^3 = 8$ values. If $s_1 = 1$, $s_2 = 1$ and $s_3 = 0$, the encoder is in state 110 in natural binary, that is, 6 in decimal.

Using the coefficients $a_j^{(l)}$, each of the m components of the vector $\mathbf{d}_i$ is selected or not as the term of an addition with the content of a previous flip-flop (except in the case of the first flip-flop) to provide the value to be stored in the following flip-flop. The new content of a flip-flop thus depends on the current input and on the content of the previous flip-flop. The case of the first flip-flop has to be considered differently. If all the b_j coefficients are null, the input is the result of the sum of the only components selected of $\mathbf{d}_i$. In the opposite case, the contents of the flip-flops selected by the non-null b_j coefficients are added to the sum of the components selected of $\mathbf{d}_i$. The code thus generated is recursive. Thus, the succession of states of the register depends on the departure state and on the succession of data at the input. The components of redundancy $\mathbf{r}_i$ are finally produced by summing the content of the flip-flops selected by the coefficients g.
Let us consider some examples.

- The encoder represented in Figure 5.1 is systematic binary, therefore $m = 1$ and $c = 1$. Moreover, all the $a_j^{(l)}$ coefficients are null except $a_1^{(1)} = 1$. This encoder is not recursive since all the coefficients b_j are null. The redundancy (or parity) bit is defined by $g_0^{(1)} = 1$, $g_1^{(1)} = 1$, $g_2^{(1)} = 0$ and $g_3^{(1)} = 1$.

- In the case of the non-systematic non-recursive binary (here called "classical") encoder in Figure 5.2, $m = 1$, $c = 0$; among the $a_j^{(l)}$, only $a_1^{(1)} = 1$ is non-null and $b_j = 0 \quad \forall j$. Two parity bits come from the encoder and are defined by $g_0^{(1)} = 1$, $g_1^{(1)} = 1$, $g_2^{(1)} = 0$, $g_3^{(1)} = 1$ and $g_0^{(2)} = 1$, $g_1^{(2)} = 0$, $g_2^{(2)} = 1$, $g_3^{(2)} = 1$.

- Figure 5.3 presents a recursive systematic binary encoder ($m = 1$, $c = 1$ and $a_1^{(1)} = 1$). The coefficients of the recursivity loop are then $b_1 = 1$, $b_2 = 0$, $b_3 = 1$ and those of the redundancy are $g_0^{(1)} = 1$, $g_1^{(1)} = 0$, $g_2^{(1)} = 1$, $g_3^{(1)} = 1$.

- Figure 5.5 represents a recursive systematic double-binary encoder. The only coefficients that differ from the previous case are the $a_j^{(l)}$: coefficients $a_1^{(1)}$, $a_1^{(2)}$, $a_2^{(2)}$ and $a_3^{(2)}$ are equal to 1, the other $a_j^{(l)}$ are null.

To define an encoder, it is not however necessary to make a graphic representation since knowledge of the parameters presented in Figure 5.4 is sufficient. A condensed representation of these parameters is known as generator polynomials. This notation is presented in the following paragraph.

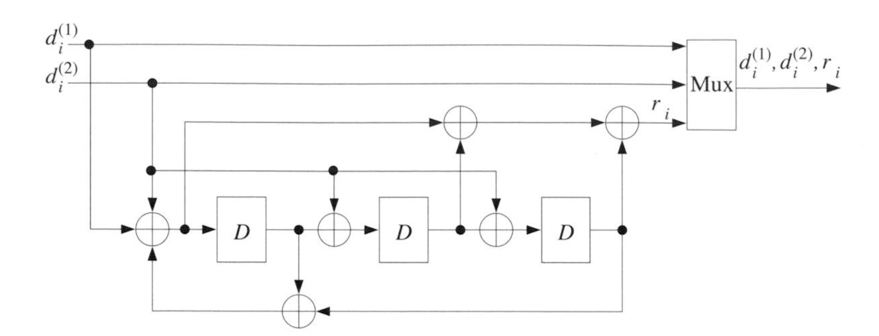

Figure 5.5 – Example of a recursive systematic double-binary convolutional encoder.

5.2.2 Polynomial representation

Let us first consider a classical (non-systematic and non-recursive) binary code: $c = 0$, all the coefficients b_j are null and $m = 1$. The knowledge of $\left[g_j^{(k)}\right]_{j=0..\nu}^{k=1..n}$ then suffices to describe the code. The coefficients $\left[g_j^{(k)}\right]_{j=0..\nu}$ therefore define n generator polynomials $G^{(k)}$ in D (*Delay*) algebra:

$$G^{(k)}(D) = \sum_{j=0\ldots\nu} g_j^{(k)} D^j \tag{5.1}$$

Let us take the case of the encoder defined in Figure 5.2. The outputs $r_i^{(1)}$ and $r_i^{(2)}$ are expressed as functions of the successive data d as follows:

$$r_i^{(1)} = d_i + d_{i-2} + d_{i-3} \tag{5.2}$$

which can also be written, via the transform in D:

$$r^{(1)}(D) = G^{(1)}(D)\ d(D) \tag{5.3}$$

with $G^{(1)}(D) = 1 + D^2 + D^3$, the first generator polynomial of the code and $d(D)$ the transform in D of the message to be encoded. Likewise, the second generator polynomial is $G^2(D) = 1 + D + D^3$.

These generator polynomials can also be resumed by the series of their coefficients, (1011) and (1101) respectively, generally denoted in octal representation, $(13)_{octal}$ and $(15)_{octal}$ respectively. In the case of a non-recursive systematic code , like the example in Figure 5.1, the generator polynomials are expressed according to the same principle. In this example, the encoder has generator polynomials $G^{(1)}(D) = 1$ and $G^{(2)}(D) = 1 + D + D^3$.

To define the generator polynomials of a recursive systematic code is not straightforward. Let us consider the example of Figure 5.3. The first generator polynomial is trivial since the code is systematic. To identify the second, we must note that

$$s_i^{(0)} = d_i + s_i^{(1)} + s_i^{(3)} = d_i + s_{i-1}^{(0)} + s_{i-3}^{(0)} \tag{5.4}$$

and that:

$$r_i = s_i^{(0)} + s_i^{(2)} + s_i^{(3)} = s_i^{(0)} + s_{i-2}^{(0)} + s_{i-3}^{(0)} \tag{5.5}$$

which is equivalent to:

$$\begin{aligned} d_i &= s_i^{(0)} + s_{i-1}^{(0)} + s_{i-3}^{(0)} \\ r_i &= s_i^{(0)} + s_{i-2}^{(0)} + s_{i-3}^{(0)} \end{aligned} \tag{5.6}$$

This result can be reformulated by introducing the transform in D:

$$\begin{aligned} d(D) &= G^{(2)}(D)s(D) \\ r(D) &= G^{(1)}(D)s(D) \end{aligned} \tag{5.7}$$

where $G^{(1)}(D)$ and $G^{(2)}(D)$ are the generator polynomials of the code shown in Figure 5.2, which leads to

$$\begin{aligned} s(D) &= \frac{d(D)}{G^{(2)}(D)} \\ r(D) &= \frac{G^{(1)}(D)}{G^{(2)}(D)}d(D) \end{aligned} \tag{5.8}$$

Thus, a recursive systematic code can easily be derived from a non-systematic non-recursive code. The codes generated by such encoders can be represented graphically according to three models: the tree, the trellis and the state machine.

5.2.3 Tree of a code

The first graphic representation of a code and certainly the least pertinent for the rest of the chapter is the tree representation. It enables all the sequences of possible states to be presented. The root is associated with the initial state of the encoder. From the root, we derive all the possible successive states as a function of input d_i of the encoder. The branch linking a father state to a son state is labelled with the value of the outputs of the encoder during the associated transition. This principle is iterated for each of the strata and so forth. The tree diagram associated with the systematic encoder of Figure 5.1 is illustrated in Figure 5.6. This type of diagram will not be used in what follows and the only use that is made of it concerns a sequential decoding algorithm (Fano's algorithm) not treated in this book.

5.2.4 Trellis of a code

The most common representation of a convolutional code is the trellis diagram. It is of major importance both for defining the properties of a code and for decoding it, as we shall see in the next part of Chapter 5.

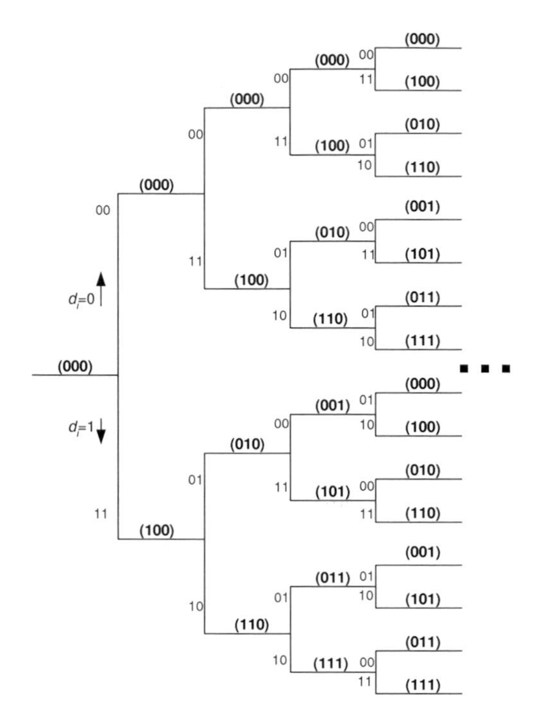

Figure 5.6 – Tree diagram of the code of polynomials $[1, 1 + D + D^3]$. The binary pairs indicate the outputs of the encoder and the values in brackets are the future states.

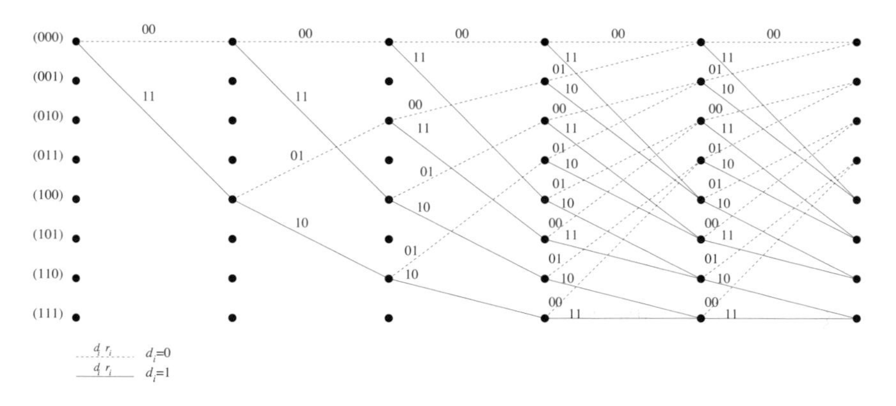

Figure 5.7 – Trellis diagram of a code with generator polynomials $[1, 1 + D + D^3]$.

At an instant i, the state of a convolutional encoder can take 2^ν values. Each of these possible values is represented by a node. With each instant i is associated a states-nodes column and according to input d_i, the encoder transits from a state $\mathbf{s}_i$ to $\mathbf{s}_{i+1}$ while delivering the coded bits. This transition between

two states is represented by an arc between the two associated nodes and labelled with the outputs of the encoder. In the case of a binary code, the transition on an input at 0 (resp. 1) is represented by a dotted (resp. solid) line. The succession of $\mathbf{s}_i$ states up to instant t is represented by the different paths between the initial state and the different possible states at instant t.
Let us show this with the example of the systematic encoder of Figure 5.1. Hypothesizing that the initial state $\mathbf{s}_0$ is state (000) :

- If $d_1 = 0$ then the following state, $\mathbf{s}_1$, is also (000). The transition is represented by a dotted line and labelled in this first case 00, the value of the encoder outputs;
- If $d_1 = 1$, then the following state, $\mathbf{s}_1$, is (100). The transition is represented by a solid line and here labelled 11.
- We must next envisage the four possible transitions: $\mathbf{s}_1 = (000)$ if $d_2 = 0$ or $d_2 = 1$ and $\mathbf{s}_1 = (100)$ if $d_2 = 0$ or $d_2 = 1$.

Iterating this construction, we reach the representation, in Figure 5.7, of all the possible successions of states from the initial state to instant 5, without the unlimited increase of the tree diagram.

A complete section of the trellis suffices to characterize the code. The trellis section of the previous code is thus shown in Figure 5.8(a). Likewise, the encoders presented in Figures 5.2 and 5.3 are associated with trellis sections, illustrated in Figures 5.8(b) and 5.8(c) respectively.

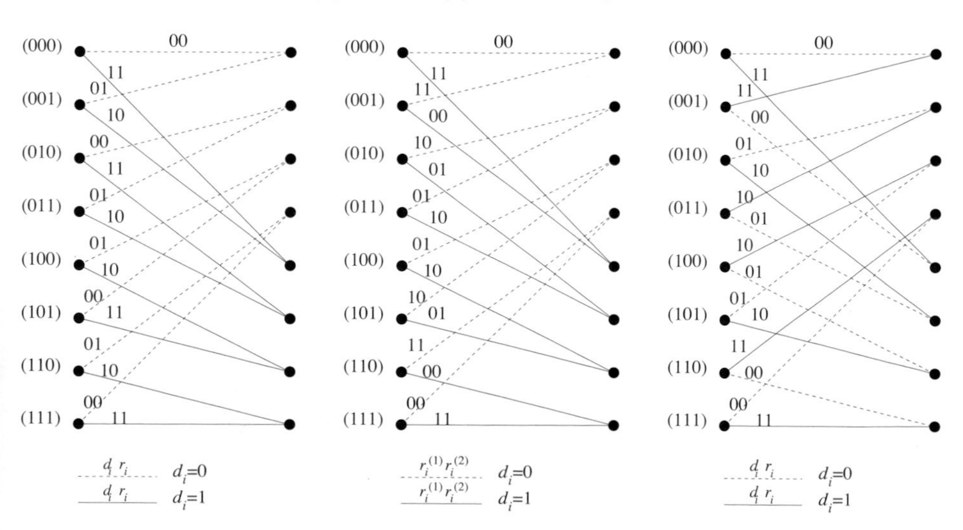

Figure 5.8 – Trellis sections of the codes with generator polynomials $[1, 1 + D + D^3]$ (a), $[1 + D^2 + D^3, 1 + D + D^3]$ (b) and $[1, (1 + D^2 + D^3)/(1 + D + D^3)]$ (c)

Such a representation shows the basic pattern of these trellises: the butterfly, so called because of its shape. Each of the sections of Figures 5.8 is thus made up of 4 butterflies (the transitions of states 0 and 1 towards states 0 and 4 make up one). The butterfly structure of the three trellises illustrated is identical but the sequences coded differ. It should be noted in particular that all the transitions arriving at the same node of a trellis of a non-recursive code are due to the same value at the input of the encoder. Thus, among the two non-recursive examples treated (Figures 5.8(a) and 5.8(b)), a transition associated with a 0 at the input necessarily arrives at one of the states between 0 and 3 and a transition with a 1 arrives at one of the states between 4 and 7. It is different in the case of a recursive code (like the one presented in Figure 5.8(c)): each state allows one incident transition associated with an input with a 0, and another one associated with 1. We shall see the consequences of this in Section 5.3.

5.2.5 State machine of a code

To represent the different transitions between the states of an encoder, there is a final representation, that of a state machine. The convention for defining the transition branches is identical to that used in the previous section. Only 2^{ν} nodes are represented, independently of instant i, the previous representation being translated into a trellis section. The encoders of Figures 5.1, 5.2 and 5.3 thus allow a representation in the form of a state machine illustrated in Figures 5.9, 5.10 and 5.11 respectively.

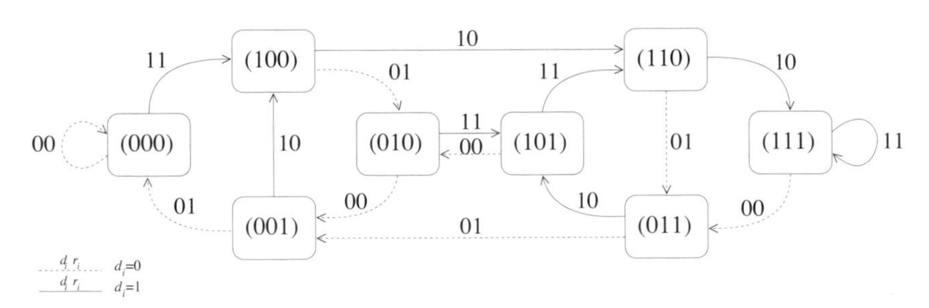

Figure 5.9 – State machine for a code with generator polynomials $[1, 1 + D + D^3]$.

This representation is particularly useful for determining the transfer function and the distance spectrum of a convolutional code (see Section 5.3). It is now possible to see a notable difference between a state machine of a recursive code and that of a non-recursive code: the existence and the number of cycles[1] on an all-zero sequence at the input.

In the case of two non-recursive state machines, there is a single cycle on a null sequence at the input: the loop on state 0.

[1] A cycle is a succession of states such that the initial state is also the final state.

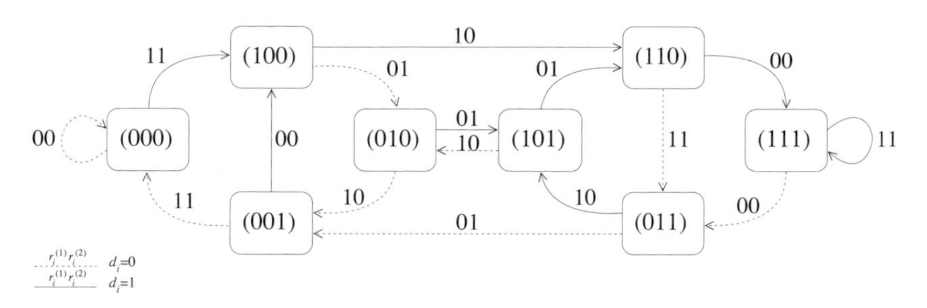

Figure 5.10 – State machine for a code with generator polynomials $[1 + D^2 + D^3, 1 + D + D^3]$.

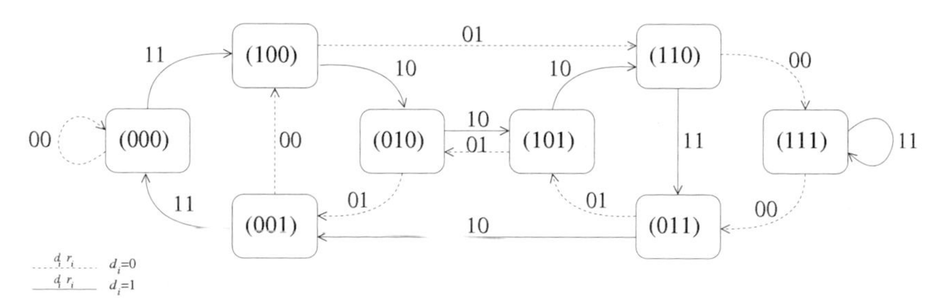

Figure 5.11 – State machine for a code with generator polynomials $[1, (1+D^2+D^3)/(1+D+D^3)]$.

However, the recursive state machine allows another cycle on a null sequence at the input: state 4 → state 6 → state 7 → state 3 → state 5 → state 2 → state 1 → state 4.
Moreover, this cycle is linked to the loop on state 0 by two transitions associated with inputs at 1 (transitions 0 → 4 and 1 → 0). There therefore exists an infinite number of input sequences with Hamming weight [2] equal to 2 producing a cycle on state 0. This weight 2 is the minimum weight of any sequence that makes the recursive encoder leave state 0 and return to zero. Because of the linearity of the code (see Chapter 1), this value of 2 is also the smallest distance that can separate two sequences with different inputs that make the encoder leave the same state and return to the same state.

In the case of non-recursive codes, the Hamming weight of the input sequences allowing a cycle on state 0 can only be 1 (state 0 → state 4 → state 2 → state 1 → state 0). This distinction is essential for understanding the interest of recursive codes used alone (see Section 5.3) or in a turbocode structure (see Chapter 7).

[2] The Hamming weight of a binary sequence is equal to the number of bits equal to 1.

5.3 Code distances and performance

5.3.1 Choosing a good code

As a code is exploited for its correcting capacities, we have to be able to estimate these in order to make a judicious choice of one code rather than another, according to the application targeted. Among the bad possible choices, *catastrophic codes* are such that a finite number of errors at the input of the decoder can produce an infinite number of errors at the output of the decoder, which explains their name. One main property of these codes is that there exists at least one input sequence of infinite weight that generates a coded sequence of finite weight: systematic codes therefore cannot be catastrophic. These codes can be identified very simply if they have a rate of the form $R = 1/N$. We can then show that the code is catastrophic if the largest common divisor (L.C.D.) of its generator polynomials is different from unity. Thus, the code with generator polynomials $G^{(1)}(D) = 1 + D + D^2 + D^3$ and $G^{(2)}(D) = 1 + D^3$ is catastrophic since the L.C.D. is $1 + D$.

However, choosing a convolutional code cannot be limited to the question "Is it catastrophic?". By exploiting the graphic representations introduced above, the properties and performance of codes can be compared.

5.3.2 *RTZ* sequences

Since convolutional codes are linear, to determine the distances between the different coded sequences amounts to determining the distances between the non-null coded sequences and the "all zero" sequence. Therefore, it suffices to calculate the Hamming weight of all the coded sequences that leave from state 0 and that return to it. These sequences are called *Return To Zero* (RTZ) sequences. The smallest Hamming weight thus obtained is called the *free distance* of the code. The minimum Hamming distance of a convolutional code is equal to its free distance from a certain length of coded sequence. In addition, the number of RTZ sequences that have the same weight is called the multiplicity of this weight.

Let us consider the codes that have been used as examples so far. Each RTZ sequence of minimum weight is shown in bold in Figures 5.12, 5.13 and 5.14.

The non-recursive systematic code has an RTZ sequence with minimum Hamming weight equal to 4. The free distance of this code is therefore equal to 4. On the other hand, as the classical code and the recursive systematic code each possess two RTZ sequences with minimum weight 6, their free distance is therefore 6. The correction capacity of non-recursive non-systematic and recursive systematic codes is therefore better than that of the non-recursive systematic code.

It is interesting, in addition, to compare the weights of the sequences at the input associated with the RTZ sequences with minimum weight. In the case of

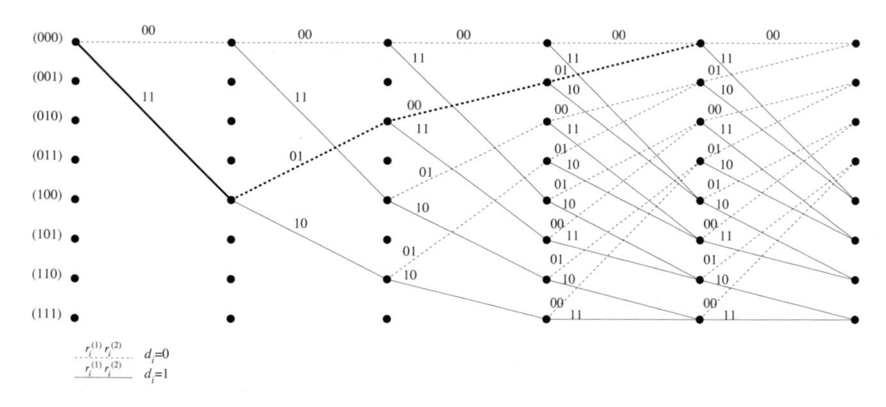

Figure 5.12 – RTZ sequence (in bold) defining the free distance of the code with generator polynomials $[1, 1 + D + D^3]$.

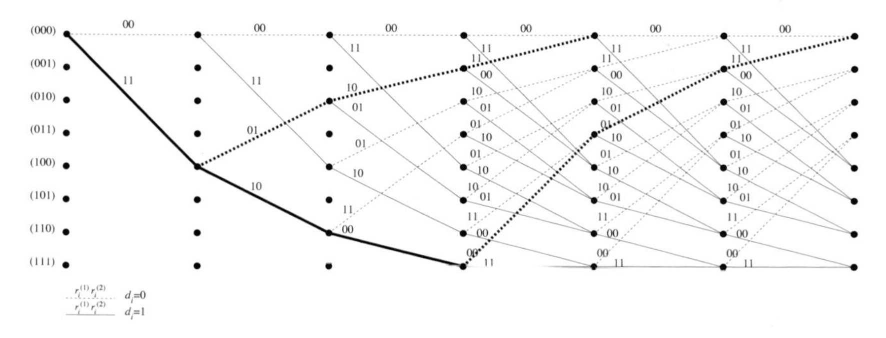

Figure 5.13 – RTZ sequences (in bold) defining the free distance of the code with generator polynomials $[1 + D^2 + D^3, 1 + D + D^3]$.

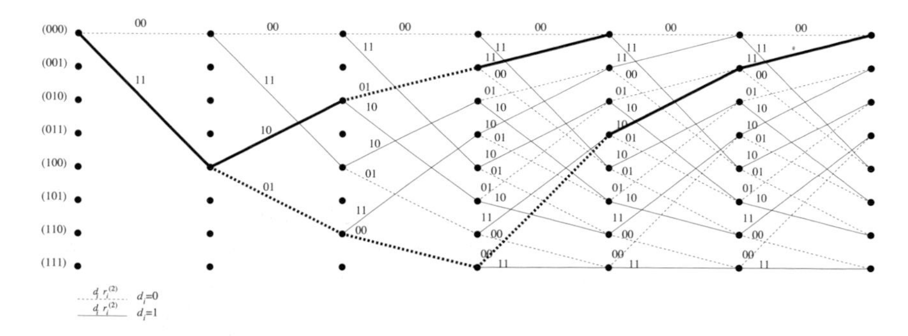

Figure 5.14 – RTZ sequences (in bold) defining the free distance of the code with generator polynomials $[1, (1 + D^2 + D^3)/(1 + D + D^3)]$.

the non-recursive systematic code, the only sequence of this type has a weight equal to 1, which means that if the RTZ sequence is decided instead of the transmitted "all zero" sequence, only one bit is erroneous. In the case of the classical code, one sequence at the input has a weight of 1 and another a weight of 3: one or three bits are therefore wrong if such an RTZ sequence is decoded. In the case of the recursive systematic code, the RTZ sequences with minimum weight have an input weight of 3.

Knowledge of the minimum Hamming distance and of the input weight associated with it is not sufficient to closely evaluate the error probability at the output of the decoder of a simple convolutional code. It is necessary to compute the distances, beyond the minimum Hamming distance, and their weight in order to make this evaluation. This computation is called the *distance spectrum*.

5.3.3 Transfer function and distance spectrum

The error correction capability of a code depends on all the RTZ sequences, which we will consider in the increasing order of their weight. Rather than computing them by reading the graphs, it is possible to establish the transfer function of the code. The latter is obtained from the state transition diagram in which the initial state (000) is cut into two states a_e and a_s, which are no other than the initial state and the arrival state of any RTZ sequence.

Let us illustrate the computation of the transfer function with the example of the systematic code of Figure 5.1, whose state transition diagram is again represented in Figure 5.15.

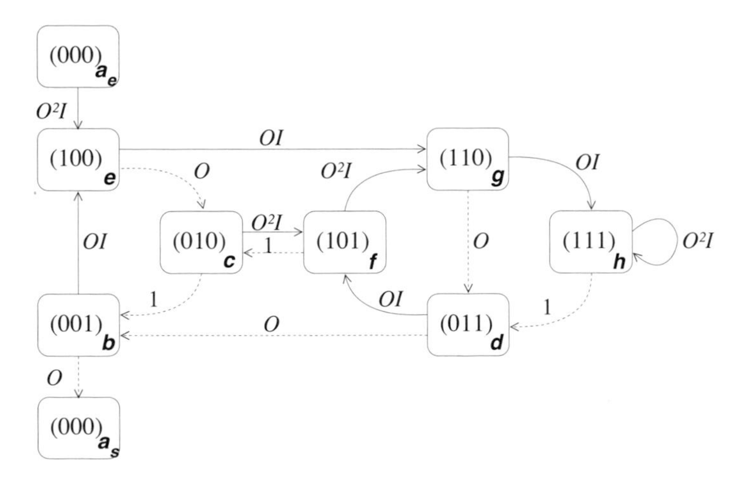

Figure 5.15 – Machine state of the code $[1, 1 + D + D^3]$, modified for the computation of the associated transfer function.

Each transition has a label $O^i I^j$, where i is the weight of the sequence coded and j that of the sequence at the input of the encoder. In our example, j can take the value 0 or 1 according to the level of the bit at the input of the encoder at each transition and i varies between 0 and 2, since 4 coded symbols are possible (00, 01, 10, 11), with weights between 0 and 2.

The transfer function of the code $T(O, I)$ is then defined by:

$$T(O,I) = \frac{a_s}{a_e} \tag{5.9}$$

To establish this function, we have to solve the system of equations coming from the relations between the 9 states (ae, b, c... h and as):

$$\begin{aligned}
b &= c + Od \\
c &= Oe + f \\
d &= h + Dg \\
e &= O^2 I a_e + OIb \\
f &= O^2 Ic + OId \\
g &= OIe + O^2 If \\
h &= O^2 Ih + OIg \\
a_s &= Ob
\end{aligned} \tag{5.10}$$

Using a formal computation tool, it is easy to arrive at the following result:

$$T(O,I) = \frac{-I^4O^{12} + (3I^4 + I^3)O^{10} + (-3I^4 - 3I^3)O^8 + (I^4 + 2I^3)O^6 + IO^4}{I^4O^{10} + (-3I^4 - I^3)O^8 + (3I^4 + 4I^3)O^6 + (-I^4 - 3I^3)O^4 - 3IO^2 + 1}$$

$T(O, I)$ can then be developed as a series:

$$\begin{aligned}
T(O,I) = \; & IO^4 \\
& +(I^4 + 2I^3 + 3I^2)O^6 \\
& +(4I^5 + 6I^4 + 6I^3)O^8 \\
& +(I^8 + 5I^7 + 21I^6 + 24I^5 + 17I^4 + I^3)O^{10} \\
& +(7I^9 + 30I^8 + 77I^7 + 73I^6 + 42I^5 + 3I^4)O^{12} \\
& +\cdots
\end{aligned} \tag{5.11}$$

This enables us to observe that an RTZ sequence with weight 4 is produced by an input sequence with weight 1, that the RTZ sequences with weight 6 are produced by a sequence with weight 4, by two sequences with weight 3 and by three sequences with weight 2, etc.

In the case of the classical code mentioned above, the transfer function is:

$$\begin{aligned}
T(O,I) = \; & (I^3 + I)O^6 \\
& +(2I^6 + 5I^4 + 3I^2)O^8 \\
& +(4I^9 + 16I^7 + 21I^5 + 8I^3)O^{10} \\
& +(8I^{12} + 44I^{10} + 90I^8 + 77I^6 + 22I^4)O^{12} \\
& +(16I^{15} + 112I^{13} + 312I^{11} + 420I^9 + 265I^7 + 60I^5)O^{14} \\
& +\cdots
\end{aligned} \tag{5.12}$$

Likewise, the recursive systematic code already studied has as its transfer function:

$$\begin{aligned} T(O,I) = \; & 2I^3O^6 \\ & +(I^6+8I^4+I^2)O^8 \\ & +(8I^7+33I^5+8I^3)O^{10} \\ & +(I^{10}+47I^8+145I^6+47I^4+I^2)O^{12} \\ & +(14I^{11}+254I^9+649I^7+254I^5+14I^3)O^{14} \\ & +\cdots \end{aligned} \tag{5.13}$$

Comparing the transfer functions from the point of view of the monomial with the smallest degree allows us to appreciate the error correction capability at very high signal to noise ratio (asymptotic behaviour). Thus, the non-recursive systematic code is weaker than its rivals since it has a lower minimum distance. A classical code and its equivalent recursive systematic code have the same free distance, but their monomials of minimal degree differ. The first is in $(I^3+I)O^6$ and the second in $2I^3O^6$. This means that with the classical code an input sequence with weight 3 and another with weight 1 produce an RTZ sequence with weight 6 whereas with the recursive systematic code two sequences with weight 3 produce an RTZ sequence with weight 6. Thus, if an RTZ sequence with minimum weight is introduced by the noise, the classical code will introduce one or three errors, whereas its recursive systematic code will introduce three or three other errors. In conclusion, the probability of a binary error on such a sequence is lower with a classical code than with a recursive systematic code, which explains that the former will be slightly better at high signal to noise ratio. Things are generally different when the codes are punctured (see Section 5.5) in order to have higher rates [5.13].

To compare the performance of codes with low signal to noise ratio, we must consider all the monomials. Let us take the example of the monomial in O^{12} for the non-recursive systematic code, the classical code and the recursive systematic code, respectively:

$$\begin{aligned} & (7I^9+30I^8+77I^7+73I^6+42I^5+3I^4)O^{12} \\ & (8I^{12}+44I^{10}+90I^8+77I^6+22I^4)O^{12} \\ & (I^{10}+47I^8+145I^6+47I^4+I^2)O^{12} \end{aligned}$$

If 12 errors are introduced by the noise on the channel, 232 RTZ sequences are "available" as errors for the first code, 241 for the second and 241 again for the third. It is therefore (a little) less probable that an RTZ sequence will appear if the code used is the non-recursive systematic code. Moreover, the error expectancy per RTZ sequence of the three codes is 6.47, 7.49 and 6.00, respectively: the recursive systematic code therefore introduces, on average, fewer decoding errors than the classical code on RTZ sequences with 12 errors on the frame coded. This is also true for higher degree monomials. Recursive and non-recursive systematic codes are therefore more efficient at low signal to

d	6	8	10	12	14	...
$\omega(d)$	6	40	245	1446	8295	...

Table 5.1 – First terms of the spectrum of the recursive systematic code with generator polynomials $[1, (1 + D^2 + D^3)/(1 + D + D^3)]$.

noise ratio than the classical code. Moreover, we find the monomials I^2O^{8+4c}, where c is an integer, in the transfer function of the recursive code. The infinite number of monomials of this type is due to the existence of the cycle on a null input sequence different from the loop on state 0. Moreover, such a code does not provide any monomials of the form IO^c, unlike non-recursive codes. These conclusions concur with those drawn from the study of state machines in Section 5.2.

This notion of transfer function is therefore efficient for studying the performance of a convolutional code. A derived version is moreover essential for the classification of codes according to their performance. This is the distance spectrum $\omega(d)$ whose definition is as follows:

$$(\frac{\partial T(O,I)}{\partial I})_{I=1} = \sum_{d=d_f}^{\infty} \omega(d) O^d \tag{5.14}$$

For example, the first terms of the spectrum of the recursive systematic code, obtained from (5.13), are presented in Table 5.1. This spectrum is essential for estimating the performance of codes in terms of calculating their error probability, as illustrated in the vast literature on this subject [5.9].

The codes used in the above examples have a rate of 1/2. By increasing the number of redundancy bits n the rate becomes lower. In this case, the powers of O associated with the branches of the state machines will be higher than or equal to those of the figures above. This leads to higher transfer functions with powers of O, that is, to RTZ sequences with a greater Hamming weight. The codes with lower rates therefore have a higher error correction capability.

5.3.4 Performance

The performance of a code is defined by the decoding error probability after transmission on a noisy channel. The previous section allows us to intuitively compare non-recursive non-systematic, non-recursive systematic and recursive systematic codes with the same constraint length. However, to estimate the absolute performance of a code, we must be able to estimate the decoding error probability as a function of the noise, or at least to limit it. The literature, for example [5.9], thus defines many bounds that are not described here and we will limit ourselves to comparing the three categories of convolutional codes. To do this, a transmission on a Gaussian channel of blocks of 53 then 200 bytes

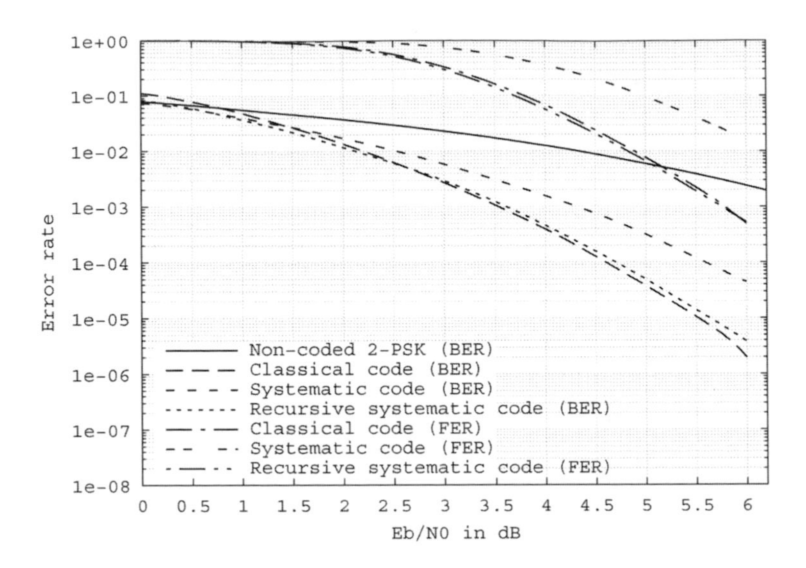

Figure 5.16 – Comparison of simulated performance (Binary Error Rate and Packet Error Rate) of three categories of convolutional codes after transmission of packets of 53 bytes on a Gaussian channel (decoding using the MAP algorithm).

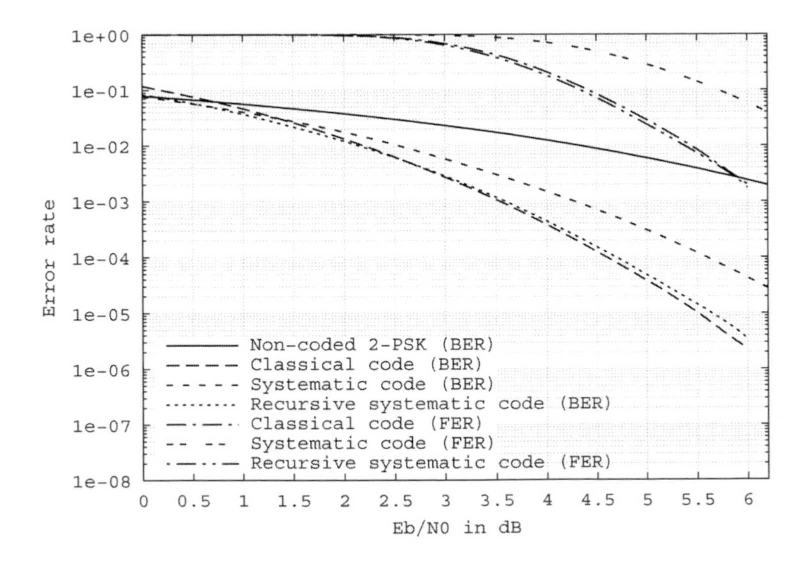

Figure 5.17 – Comparison of simulated performance of three categories of convolutional codes after transmission of blocks of 200 bytes on a Gaussian channel.

coded according to different schemes was simulated (Figures 5.16 and 5.17): classical (non-recursive non-systematic), non-recursive systematic and recursive systematic.

The blocks were constructed following the classical trellis termination technique for non-recursive codes whereas the recursive code is circular tail-biting (see Section 5.5). The decoding algorithm used is the MAP algorithm.

The BER curves are in perfect agreement with the conclusions drawn during the analysis of the free distance of codes and of their transfer function: the systematic code is not as good as the others at high signal to noise ratio and the classical code is then slightly better than the recursive code. At low signal to noise ratios, the hierarchy is different: the recursive code and the systematic code are equivalent and better than the classical code.

Comparing performance as a function of the size of the frame (53 and 200 bytes) shows that the performance hierarchy of the codes is not modified. Moreover, the bit error rates are almost identical. This was predictable as the sizes of the frames are large enough for the transfer functions of the codes not to be affected by edge effects. However, the packet error rate is affected by the length of the blocks since although the bit error probability is constant, the packet error probability increases with size.

The comparisons above only concern codes with 8 states. It is, however, easy to see that the performance of a convolutional code is linked with its capacity to provide information on the succession of data transmitted: the more the code can integrate successive data into its output symbols, the more it improves the quality of protection these data. In other words, the greater the number of states (therefore the size of the register of the encoder), the more efficient a convolutional code is (within its category). Let us compare three recursive systematic codes:

- 4 states $[1, (1 + D^2)/(1 + D + D^2)]$,
- 8 states $[1, (1 + D^2 + D^3)/(1 + D + D^3)]$
- and 16 states $[1, (1 + D + D^2 + D^4)/(1 + D^3 + D^4)]$.

Their performance in terms of BER and PER were simulated on a Gaussian channel and are presented in Figure 5.18.

The higher the number of states of the code, the lower the residual error rates are. For a BER of 10^{-4}, 0.6 dB are thus gained when passing from 4 states to 8 states and 0.5 dB when passing from 8 states to 16 states. This remark is coherent with the qualitative justification of the interest of a large number of states. It would therefore seem logical to choose a convolutional code with a large number of states to ensure the desired protection, especially since such codes offer the possibility of producing redundancy on far more than two components, and therefore of providing even higher protection. Thus, the *Big Viterbi Decoder* project at *NASA*'s *Jet Propulsion Laboratory* used for transmissions with space

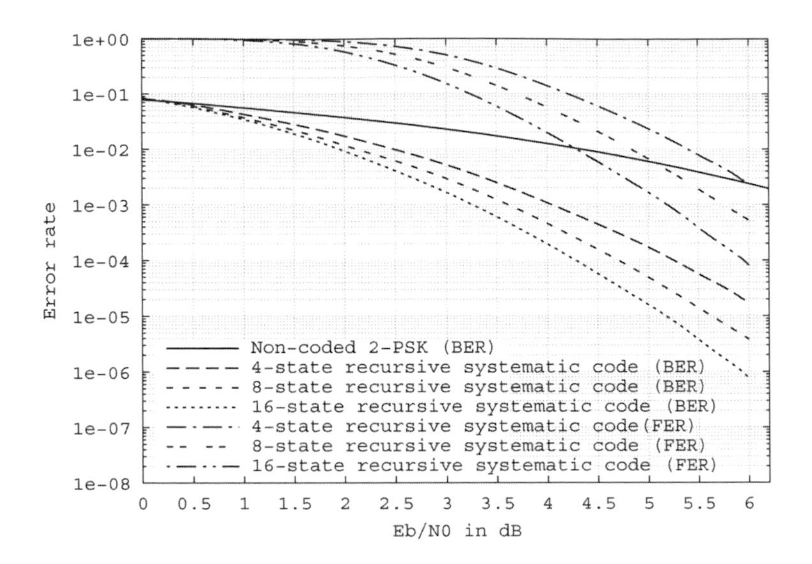

Figure 5.18 – Comparison of simulated performance of recursive systematic convolutional codes with 4, 8 and 16 states after transmitting packets of 53 bytes on a Gaussian channel (decoding according to the MAP algorithm)

probes was designed to process frames encoded with convolutional codes with 2 to 16384 states and rates much lower than 1/2 (16384 states and $R = 1/6$ for the *Cassini* probe to Saturn and the *Mars Pathfinder* probe). Why not use such codes for terrestrial radio-mobile transmissions for the general public? Because the complexity of the decoding would become unacceptable for current terrestrial transmissions using a reasonably-sized terminal operating in real time, and fitting into a pocket.

5.4 Decoding convolutional codes

There exist several algorithms for decoding convolutional codes. The most famous is probably the Viterbi algorithm which relies on the trellis representation of codes [5.14][5.8]. It enables us to find the most probable sequence of states in the trellises from the received symbol sequence.

The original Viterbi algorithm performs *hard output* decoding, that is, it provides a binary estimation of each of the symbols transmitted. It is therefore not directly adapted to iterative systems that require information about the *trustworthiness* of the decisions. Adaptations of the Viterbi algorithm such as those proposed in [5.2], [5.10] or [5.3] have led to versions with weighted output called *Soft-Output Viterbi Algorithm* (SOVA). SOVA algorithms are not described in this book since, for turbo decoding, we prefer another family of algorithms relying on the minimization of the error probability of each symbol transmitted. Thus, the Maximum *A Posteriori* (MAP) algorithm enables the calculation of the exact value of the *a posteriori* probability associated with each symbol transmitted using the received sequence [5.1]. The MAP algorithm and its variants are described in Chapter 7.

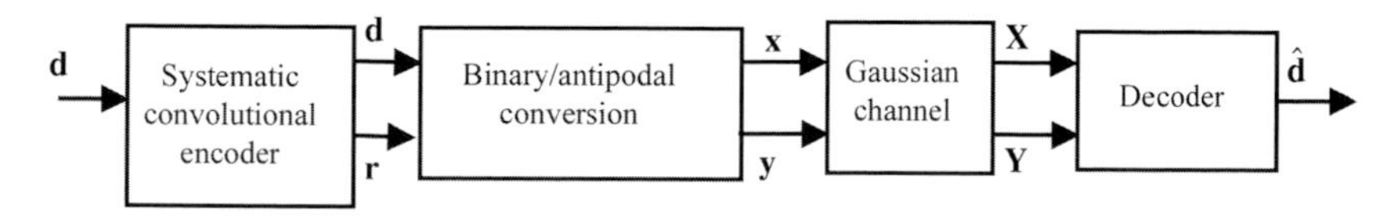

Figure 5.19 – Model of the transmission chain studied.

5.4.1 Model of the transmission chain and notations

Viterbi and MAP algorithms are used in the transmission chain shown in Figure 5.19. The convolutional code considered is systematic and, for each data sequence of information $\mathbf{d} = \mathbf{d}_1^N = \{\mathbf{d}_1, \cdots, \mathbf{d}_N\}$, calculates a redundant sequence $\mathbf{r} = \mathbf{r}_1^N = \{\mathbf{r}_1, \cdots, \mathbf{r}_N\}$. The code is m-binary with rate $R = m/(m+n)$: each vector of data $\mathbf{d}_i$ at the input of the encoder is thus made up of m bits, $\mathbf{d}_i = (d_i^{(1)}, d_i^{(2)}, \cdots, d_i^{(m)})$, and the corresponding redundancy vector at the output is written $\mathbf{r}_i = (r_i^{(1)}, r_i^{(2)}, \cdots, r_i^{(n)})$. The value of $\mathbf{d}_i$ can also be represented by the scalar integer variable $j = \sum_{l=1}^{m} 2^{l-1} d_i^{(l)}$, between 0 and $2^m - 1$, and we can then write $\mathbf{d}_i \equiv j$.

The systematic $\mathbf{d}$ and redundant $\mathbf{r}$ data sequences are transmitted after a binary/antipodal conversion making an antipodal value (-1 or +1) transmitted towards the channel correspond to each value binary (0 or 1) coming from the encoder. $\mathbf{X}$ and $\mathbf{Y}$ represent the noisy systematic and redundant symbol sequences received at the input of the decoder and $\hat{\mathbf{d}}$ the decoded sequence that can denote either a binary sequence in the case of the Viterbi algorithm, or a sequence of weighted decisions associated with the $\mathbf{d}_i$ at the output of the MAP algorithm.

5.4.2 The Viterbi algorithm

The Viterbi algorithm is the most widely used method for the maximum-likelihood (ML) decoding of convolutional codes with low constraint length (typically $\nu \leq 8$). Beyond this limit, its complexity of implementation means that we have to resort to a sequential decoding algorithm, like Fano's [5.6].

ML decoding is based on a search for the codeword $\mathbf{c}$ that is the shortest distance away from the received word. In the case of a channel with binary decisions (binary symmetric channel), ML decoding relies on the Hamming distance, whereas in the case of a Gaussian channel, it relies on the Euclidean distance (see Chapter 2). An exhaustive search for codewords associated with the different paths in the trellis leads to $2^{\nu+k}$ paths being taken into account. In practice, searching for the path with the minimum distance on a working window with a width l lower than k, limits the search to $2^{\nu+l}$paths.

The Viterbi algorithm enables a notable reduction in the complexity of the computation. It is based on the idea that, among the set of paths of the trellis that converge in a node at a given instant, only the most probable path can be retained for the following search steps. Let us denote $\mathbf{s}_i = (s_i^{(1)}, s_i^{(2)}, \ldots, s_i^{(\nu)})$ the state of the encoder at instant i and $T(i, \mathbf{s}_{i-1}, \mathbf{s}_i)$ the branch of the trellis corresponding to the emission of data $\mathbf{d}_i$ and associated with the transition between nodes $\mathbf{s}_{i-1}$ and $\mathbf{s}_i$. Applying the Viterbi algorithm involves performing the set of operations described below.

- *At each instant i, for i ranging from 1 to k:*

  - Calculate for each branch of a *branch metric*, $d\left(T(i, \mathbf{s}_{i-1}, \mathbf{s}_i)\right)$. For a binary output channel, this metric is defined as the Hamming distance between the symbol carried by the branch of the trellis and the received symbol, $d\left(T(i, \mathbf{s}_{i-1}, \mathbf{s}_i)\right) = d_H\left(T(i, \mathbf{s}_{i-1}, \mathbf{s}_i)\right)$.
  For a Gaussian channel, the metric is equal to the square of the Euclidean distance between the branch considered and the observation at the input of the decoder (see also Section 1.3):

$$\begin{aligned} d\left(T(i, \mathbf{s}_{i-1}, \mathbf{s}_i)\right) &= \left\|\mathbf{X}_i - \mathbf{x}_i\right\|^2 + \left\|\mathbf{Y}_i - \mathbf{y}_i\right\|^2 \\ &= \sum_{j=1}^{m} \left(x_i^{(j)} - X_i^{(j)}\right)^2 + \sum_{j=1}^{n} \left(y_i^{(j)} - Y_i^{(j)}\right)^2 \end{aligned}$$

  - Calculate the *accumulated metric* associated with each branch $T(i, \mathbf{s}_{i-1}, \mathbf{s}_i)$ defined by:

$$\lambda(T(i, \mathbf{s}_{i-1}, \mathbf{s}_i)) = \mu(i - 1, \mathbf{s}_{i-1}) + d(T(i, \mathbf{s}_{i-1}, \mathbf{s}_i))$$

  where $\mu(i-1, \mathbf{s}_{i-1})$ is the accumulated metric associated with node $\mathbf{s}_{i-1}$.

  - For each node $\mathbf{s}_i$, select the branch of the trellis corresponding to the minimum accumulated metric and memorize this branch in memory (in practice, it is the value of d_i associated with the branch that is stored). The path in the trellis made up of the branches successively memorized at the instants between 0 and i is the *survivor path* arriving in $\mathbf{s}_i$. If the two paths that converge in $\mathbf{s}_i$ have identical accumulated metrics, the survivor is then chosen arbitrarily between these two paths.

- Calculate the accumulated metric associated with each node $\mathbf{s}_i$, $\mu(i, \mathbf{s}_i)$. It is equal to the accumulated metric associated with the survivor path arriving in $\mathbf{s}_i$:

$$\mu(i, \mathbf{s}_i) = \min_{\mathbf{s}_{i-1}} \left(\lambda \left(T(i, \mathbf{s}_{i-1}, \mathbf{s}_i) \right) \right).$$

- *Initialization (instant i =0):*

The initialization values of the metrics μ when commencing the algorithm depend on the initial state $\mathbf{s}$ of the encoder: $\mu(0, \mathbf{s}) = +\infty$ if $\mathbf{s} \neq \mathbf{s}_0$ and $\mu(0, \mathbf{s}_0) = 0$. If this state is not known, all the metrics are initialized to the same value, typically 0. In this case, the decoding of the beginning of the frame is less efficient since the accumulated metrics associated with each branch at instant 1 depend only on the branch itself. The past cannot be taken into account since it is not known: $\lambda(T(1, \mathbf{s}_0, \mathbf{s}_1)) = d(T(1, \mathbf{s}_0, \mathbf{s}_1))$.

- *Calculating the decisions (instant i=k):*

At instant k, if the final state of the encoder is known, the maximum likelihood path is the survivor path coming from the node corresponding to the final state of the encoder. The decoded sequence is given by the series of values of d_i, i ranging from 1 to k, stored in the memory associated with the maximum likelihood path. This operation is called *trellis traceback.*

If the final state is not known, the maximum likelihood path is the survivor path coming from the node with minimum accumulated metric. In this case, the problem is similar to the one mentioned for initialization: the decoding of the end of the frame is less efficient.

When the sequence transmitted is long, or even infinite in length, it is not possible to wait for the whole transmitted binary sequence to be received to begin the decoding operation. To limit the decoding latency and the size of memory necessary to memorize the survivor paths, the trellis must be truncated. Observing the algorithm unfold, we can note that by tracing back sufficiently in time from instant i, the survivor paths coming from the different nodes of the trellis nearly always converge towards a same path. In practice, memorizing the survivors can therefore be limited to a time interval of duration l. It is then sufficient to do a trellis traceback at each instant i over a length l in order to take the decision on the data d_{i-l}. To decrease the complexity, the survivors are sometimes memorized on an interval higher than l (for example l+3). The number of trellis traceback operations is then decreased (divided by 3 in our example) but each of the tracebacks provides several decisions (d_{i-l}, d_{i-l-1} and d_{i-l-2}, in our example).

The higher the code memory and coding rate are, the greater the value of l must be. We observe that, for a systematic code, values of l corresponding to the production by the encoder of a number of redundancy symbols equal to 5 times the constraint length of the code are sufficient. As an example, for coding rate $R = 1/2$, we typically take l equal to $5(\nu + 1)$.

From the point of view of complexity, the Viterbi algorithm requires the calculation of $2^{\nu+1}$ accumulated metrics at each instant i and its complexity varies linearly with the length of sequence k or of decoding window l.

Example of applying the Viterbi algorithm

Let us illustrate the different steps of the Viterbi algorithm described above by applying it to decode the binary recursive systematic convolutional code (7,5) with 4 states and encoding rate $R = 1/2$ ($m = n = 1$). The structure of the encoder and the trellis are shown in Figure 5.20.

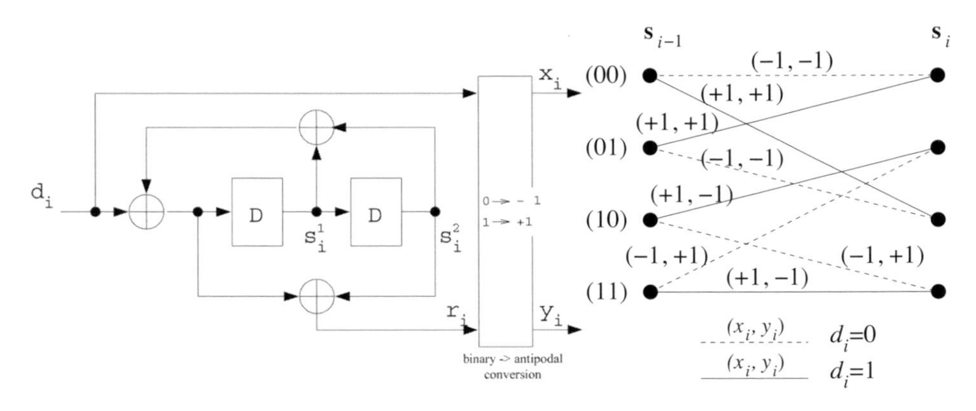

Figure 5.20 – Structure of the recursive systematic convolutional code (7,5) and associated trellis.

Calculate the branch metrics at instant i:

$$\begin{aligned}
d(T(i,\mathbf{0},\mathbf{0})) = d(T(i,\mathbf{1},\mathbf{2})) &= (X_i+1)^2 + (Y_i+1)^2 \\
d(T(i,\mathbf{0},\mathbf{2})) = d(T(i,\mathbf{1},\mathbf{0})) &= (X_i-1)^2 + (Y_i-1)^2 \\
d(T(i,\mathbf{2},\mathbf{1})) = d(T(i,\mathbf{3},\mathbf{3})) &= (X_i-1)^2 + (Y_i+1)^2 \\
d(T(i,\mathbf{2},\mathbf{3})) = d(T(i,\mathbf{3},\mathbf{1})) &= (X_i+1)^2 + (Y_i-1)^2
\end{aligned}$$

Calculate the accumulated metrics of the branches at instant i:

$$\begin{aligned}
\lambda(T(i,\mathbf{0},\mathbf{0})) &= \mu(i-1,\mathbf{0}) + d(T(i,\mathbf{0},\mathbf{0})) = \mu(i-1,\mathbf{0}) + (X_i+1)^2 + (X_i+1)^2 \\
\lambda(T(i,\mathbf{1},\mathbf{2})) &= \mu(i-1,\mathbf{1}) + d(T(i,\mathbf{1},\mathbf{2})) = \mu(i-1,\mathbf{1}) + (X_i+1)^2 + (X_i+1)^2 \\
\lambda(T(i,\mathbf{0},\mathbf{2})) &= \mu(i-1,\mathbf{0}) + d(T(i,\mathbf{0},\mathbf{2})) = \mu(i-1,\mathbf{0}) + (X_i-1)^2 + (X_i-1)^2 \\
\lambda(T(i,\mathbf{1},\mathbf{0})) &= \mu(i-1,\mathbf{1}) + d(T(i,\mathbf{1},\mathbf{0})) = \mu(i-1,\mathbf{1}) + (X_i-1)^2 + (X_i-1)^2 \\
\lambda(T(i,\mathbf{2},\mathbf{1})) &= \mu(i-1,\mathbf{2}) + d(T(i,\mathbf{2},\mathbf{1})) = \mu(i-1,\mathbf{2}) + (X_i-1)^2 + (X_i+1)^2 \\
\lambda(T(i,\mathbf{3},\mathbf{3})) &= \mu(i-1,\mathbf{3}) + d(T(i,\mathbf{3},\mathbf{3})) = \mu(i-1,\mathbf{3}) + (X_i-1)^2 + (X_i+1)^2 \\
\lambda(T(i,\mathbf{2},\mathbf{3})) &= \mu(i-1,\mathbf{2}) + d(T(i,\mathbf{2},\mathbf{3})) = \mu(i-1,\mathbf{2}) + (X_i+1)^2 + (X_i-1)^2 \\
\lambda(T(i,\mathbf{3},\mathbf{1})) &= \mu(i-1,\mathbf{3}) + d(T(i,\mathbf{3},\mathbf{1})) = \mu(i-1,\mathbf{3}) + (X_i+1)^2 + (X_i-1)^2
\end{aligned}$$

Calculate the accumulated metrics of the nodes at instant i:

$$\begin{aligned}
\mu(i,\mathbf{0}) &= \min\left(\lambda(T(i,\mathbf{0},\mathbf{0}), \lambda(T(i,\mathbf{1},\mathbf{0})\right)\\
\mu(i,\mathbf{1}) &= \min\left(\lambda(T(i,\mathbf{3},\mathbf{1}), \lambda(T(i,\mathbf{2},\mathbf{1})\right)\\
\mu(i,\mathbf{2}) &= \min\left(\lambda(T(i,\mathbf{0},\mathbf{2}), \lambda(T(i,\mathbf{1},\mathbf{2})\right)\\
\mu(i,\mathbf{3}) &= \min\left(\lambda(T(i,\mathbf{3},\mathbf{3}), \lambda(T(i,\mathbf{2},\mathbf{3})\right)
\end{aligned}$$

For each of the four nodes of the trellis, the value of d_i corresponding to the transition of minimum accumulated metric λ is stored in memory.

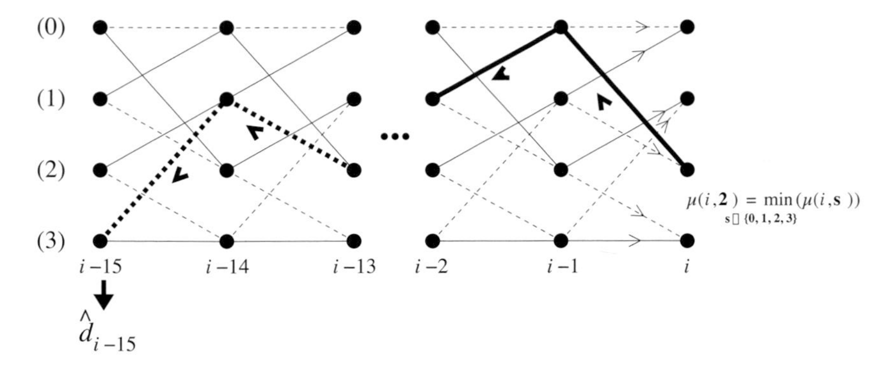

Figure 5.21 – Survivor path traceback operation (in bold) in the trellis from instant i and determining the binary decision at instant $i-15$.

After selecting the node with minimum accumulated metric, denoted $\mathbf{s}$ (in the example of Figure 5.21, $\mathbf{s}=\mathbf{3}$), we trace back in the trellis along the survivor path to a depth $l=15$. At instant $i-15$, the binary decision $\hat{d}_{i-15}$ is equal to the value of d_{i-15} stored in the memory associated with the survivor path.

The aim of applying the Viterbi algorithm with weighted inputs is to search for codeword $\mathbf{c}$ that is the shortest Euclidean distance between two codewords. Equivalently (see Chapter 1), this also means looking for the codeword that maximizes the scalar product $\langle \mathbf{x},\mathbf{X}\rangle + \langle \mathbf{y},\mathbf{Y}\rangle = \sum\limits_{i=1}^{k}\left(\sum\limits_{l=1}^{m} x_i^{(l)}X_i^{(l)} + \sum\limits_{l=1}^{n} y_i^{(l)}Y_i^{(l)}\right)$. In this case, applying the Viterbi algorithm uses branch metrics of the form $d\left(T(i,\mathbf{s}_{i-1},\mathbf{s}_i)\right) = \sum\limits_{l=1}^{m} x_i^{(l)}X_i^{(l)} + \sum\limits_{l=1}^{n} y_i^{(l)}Y_i^{(l)}$ and the survivor path then corresponds to the path with maximum accumulated metric.

Figure 5.22 provides the performance of the two variants, with hard and weighted inputs, of a decoder using the Viterbi algorithm for the code (7,5) RSC for a transmission on a channel with additive white Gaussian noise. In practice, we observe a gain of around 2 dB when we substitute weighted input decoding for hard input decoding.

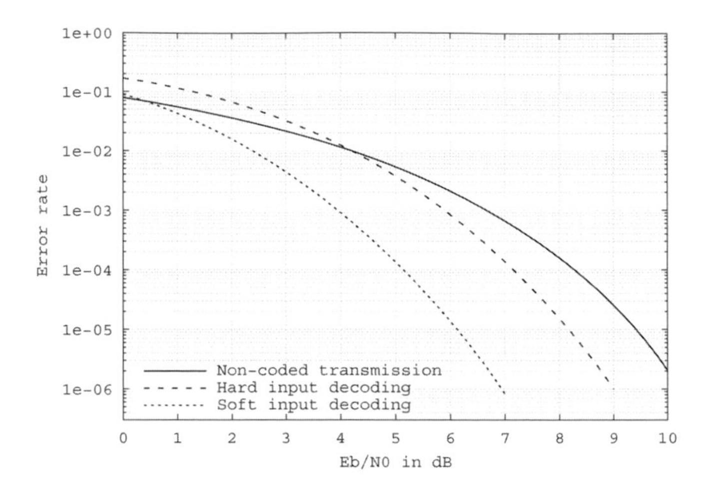

Figure 5.22 – Example of correction performance of the Viterbi algorithm with hard inputs and with weighted inputs on a Gaussian channel. Recursive systematic convolutional code (RSC) with generator polynomials 7 (recursivity) and 5 (redundancy). Coding rate $R = 1/2$.

5.4.3 The Maximum *A Posteriori* algorithm or MAP algorithm

The Viterbi algorithm determines the codeword closest to the received word. However, it does not necessarily minimize the error probability of the bits or symbols transmitted. The MAP algorithm enables us to calculate the *a posteriori* probability of each bit or each symbol transmitted and, at each instant, the corresponding decoder selects the most probable bit or symbol. This algorithm was published in 1974 by Bahl, Cocke, Jelinek and Raviv [5.1]. The impact of this decoding method remained little known until the discovery of turbocodes since it does not provide any notable improvement in performance compared to the Viterbi algorithm for decoding convolutional codes and it turned out to be more complex to implement. However, the situation changed in 1993 as decoding turbocodes uses elementary decoders with weighted or soft outputs and the MAP algorithm, unlike the Viterbi algorithm, enables us to associate a weighting naturally with each decision. The MAP algorithm is presented in Chapter 7.

5.5 Convolutional block codes

Convolutional codes are naturally adapted to transmission applications where the message transmitted is of infinite length. However, most telecommunications systems use independent frame transmissions. Paragraph 5.4.2 showed the importance of knowing the initial and final states of the encoder during the decoding of a frame. In order to know these states, the technique used is usually called trellis termination. This generally involves forcing the initial and final states to values known by the decoder (in general zero).

5.5.1 Trellis termination

Classical trellis termination

As the encoder is constructed around a register, it is easy to initialize it by using reset inputs before beginning the encoding of a frame. This operation has no consequence on the coding rate. But termination at the end of a frame is not so simple.

When the k bits of the frame have been coded, the register of the encoder is in any of the 2^{ν} possible states. The aim of termination is to lead the encoder towards state zero by following one of the paths in the trellis so that the decoding algorithm can use this knowledge of the final state. In the case of non-recursive codes, the final state is forced to zero by injecting ν zero bits at the end of the frame. It is as if the coded frame were of length $k+\nu$ with $d_{k+1} = d_{k+2} = \ldots = d_{k+\nu} = 0$. The encoding rate is slightly decreased by the transmission of the *termination bits*. However, taking into account the size of the frames generally transmitted, this degradation in rate is very often negligible. In the case of recursive codes, it is also possible to inject a zero at the input of the register. Figure 5.23 shows a simple way to solve this question.

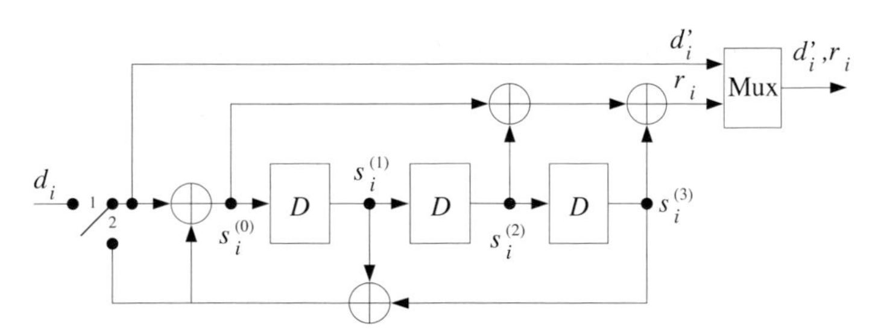

Figure 5.23 – Example of an encoder of recursive systematic convolutional codes allowing termination at state 0 in ν instants.

After initializing the register to zero, switch I is kept in position 1 and data d_1 to d_k are coded. At the end of this encoding operation, instants k to $k+\nu$, switch I is placed in position 2 and d_i takes the value coming from the feedback of the register, that is, a value that forces one register input to zero. Indeed, $S_i^{(0)}$ is the result of a modulo-2 sum of two identical members. As for the encoder, it continues to produce the associated redundancies r_i.

This classical termination has one main drawback: the protection of the data is not independent of their position in the frame. In particular, this can lead to edge effects in the construction of a turbocode (see Chapter 7).

Tail-biting

A technique was introduced in the 70s and 80s [5.12] to terminate the trellis of convolutional codes without edge effects: tail-biting. This involves making the decoding trellis circular, that is, ensuring that the departure and the final states of the encoder are identical. This state is then called the circulation state. This technique is trivial for non-recursive codes as the circulation state is merely the last ν bits of the sequence to encode. As for RSC codes, tail-biting requires operations that are described in the following. The trellis of such a code, called circular recursive systematic codes (CRSC), is shown in Figure 5.24.

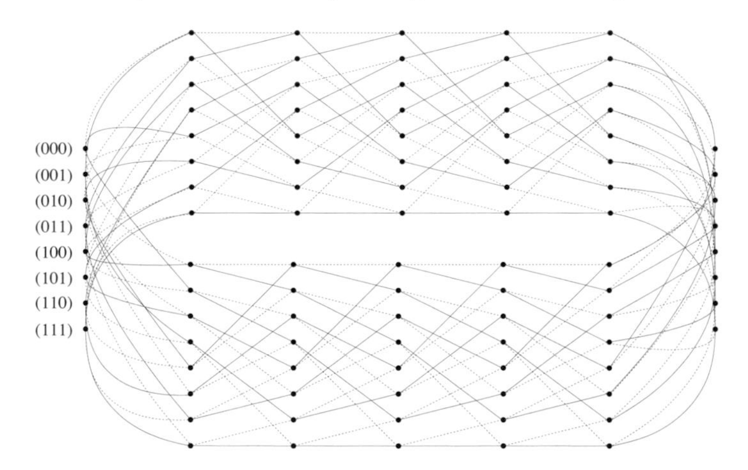

Figure 5.24 – Trellis of a CRSC code with 8 states.

It is possible to establish a relation between state $\mathbf{s}_{i+1}$ of the encoder at an instant $i+1$, its state $\mathbf{s}_i$ and the input data $\mathbf{d}_i$ at the previous time:

$$\mathbf{s}_{i+1} = \mathbf{A}\mathbf{s}_i + \mathbf{B}\mathbf{d}_i \tag{5.15}$$

where $\mathbf{A}$ is the state matrix and $\mathbf{B}$ the input matrix. In the case of the recursive systematic code of generator polynomials $[1, (1+D^2 + D^3)/(1+D + D^3)]$ mentioned above, these matrices are

$$\mathbf{A} = \begin{bmatrix} 1 & 0 & 1 \\ 1 & 0 & 0 \\ 0 & 1 & 0 \end{bmatrix} \text{ and } \mathbf{B} = \begin{bmatrix} 1 \\ 0 \\ 0 \end{bmatrix}.$$

If the encoder is initialized to state 0 ($\mathbf{s}_0 = 0$), the final state $\mathbf{s}_k^0$ obtained at the end of a frame of length k is:

$$\mathbf{s}_k^0 = \sum_{j=1}^{k} \mathbf{A}^{j-1}\mathbf{B}\mathbf{d}_{k-j} \tag{5.16}$$

When it is initialized in any state $\mathbf{s}_c$, the final state $\mathbf{s}'_k$ is expressed as follows:

$$\mathbf{s}'_k = \mathbf{A}^k \mathbf{s}_c + \sum_{j=1}^{k} \mathbf{A}^{j-1}\mathbf{B}\mathbf{d}_{k-j} \tag{5.17}$$

For this state $\mathbf{s}'_k$ to be equal to the departure state $\mathbf{s}_c$ and for the latter therefore to become the circulation state, it is necessary and sufficient that:

$$\left(\mathbf{I} - \mathbf{A}^k\right) \mathbf{s}_c = \sum_{j=1}^{k} \mathbf{A}^{j-1}\mathbf{B}\mathbf{d}_{k-j} \tag{5.18}$$

where $\mathbf{I}$ is the identity matrix of dimension $\nu \times \nu$.
Thus, by introducing state $\mathbf{s}_k^0$ of the encoder after initialization to 0

$$\mathbf{s}_c = \left(\mathbf{I} - \mathbf{A}^k\right)^{-1} \mathbf{s}_k^0 \tag{5.19}$$

This is only possible on condition that the matrix $(\mathbf{I} - \mathbf{A}^k)$ is *invertible* : this is the condition for the existence of the circulation state.

In conclusion, if the circulation state exists, it can be obtained in two steps. The first involves coding the frame of k bits at the input after initializing the encoder to state zero and keeping the termination state. The second is to simply deduce the circulation state from the previous termination state and from a table (obtained by inverting $\mathbf{I} - \mathbf{A}^k$).

Take for example the recursive systematic code used above. Since it is a binary code, addition and subtraction are equivalent: the circulation state exists if $\mathbf{I} + \mathbf{A}^k$ is invertible. Matrix $\mathbf{A}$ is such that $\mathbf{A}^7$ is equal to $\mathbf{I}$. Thus, if k is a multiple of $7 (= 2^\nu - 1)$, $\mathbf{I} + \mathbf{A}^k$ is null and therefore non-invertible: this case should be avoided. Another consequence is that $\mathbf{A}^k = \mathbf{A}^{k \bmod 7}$: it suffices to calculate once and for the 6 state transformation tables associated with the 6 possible values of $(\mathbf{I} - \mathbf{A}^k)^{-1}$, to store them and to read the right table, after calculating $k \mod 7$. The table of the CRSC code with 8 states of generator polynomials $[1, (1 + D^2 + D^3)/(1 + D + D^3)]$ is given in Table 5.2.

A simple method for encoding according to a circular trellis can be summarized in five steps, after checking the existence of the circulation state:

1. Initialize the encoder to state 0
2. Code the frame to obtain the final state $\mathbf{s}_k^0$;

$\mathbf{s}_k^0$ \ k mod 7	1	2	3	4	5	6
0	0	0	0	0	0	0
1	6	3	5	4	2	7
2	4	7	3	1	5	6
3	2	4	6	5	7	1
4	7	5	2	6	1	3
5	1	6	7	2	3	4
6	3	2	1	7	4	5
7	5	1	4	3	6	2

Table 5.2 – Table of the CRSC code with generator polynomials $[1, (1+D^2+D^3)/(1+D+D^3)]$ providing the circulation state as a function of $k \mod 7$ (k being the length of the frame at the input) and of the terminal state $\mathbf{s}_k^0$ obtained after encoding initialized to state 0.

3. Calculate the circulation state $\mathbf{s}_c$ from the tables already calculated and stored;
4. Initialize the encoder to state $\mathbf{s}_c$;
5. Code the frame and transmit the redundancies calculated.

5.5.2 Puncturing

Some applications can only allocate a small space for the redundant part of the codewords. But, by construction, the natural rate of a systematic convolutional code is $m/(m+n)$, where m is the number of input bits $\mathbf{d}_i$ of the encoder and n is the number of output bits. It is therefore maximum when $n = 1$ and becomes $R = m/(m+1)$. High rates can therefore only be obtained with high values of m. Unfortunately, the number of transitions leaving any one node of the trellis is 2^m. In other words, the complexity of the trellis, and therefore of the decoding, increases exponentially with the number of input bits of the encoder. Therefore, this solution is generally not satisfactory. It is often avoided in favour of a technique with a slightly lower error correction capability, but easier to implement: *puncturing*.

The puncturing technique is commonly used to obtain high rates. It involves using an encoder with a low value of m (1 or 2 for example), to keep a reasonable decoding complexity, but transmitting only part of the bits coded. An example is proposed in Figure 5.25. In this example, a 1/2 rate encoder produces outputs d_i and r_i at each instant i. Only 3 bits out of 4 are transmitted, which leads to a global rate of 2/3. The pattern in which the bits are punctured is called the puncturing mask.

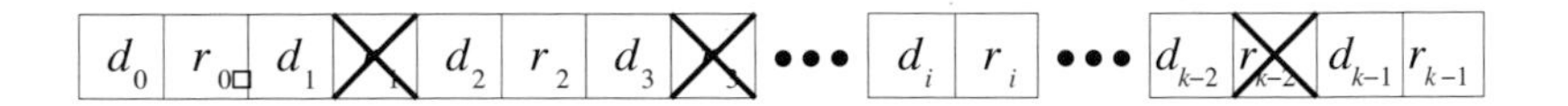

Figure 5.25 – Puncturing a systematic code to obtain a rate 2/3.

In the case of systematic codes, it is generally the redundancy that is punctured. Figure 5.26 shows the trellis of code $[1, (1 + D^2 + D^3)/(1 + D + D^3)]$ resulting from a puncturing operation according to the mask of Figure 5.25. The "X"s mark the bits that are not transmitted and that therefore cannot be used for the decoding.

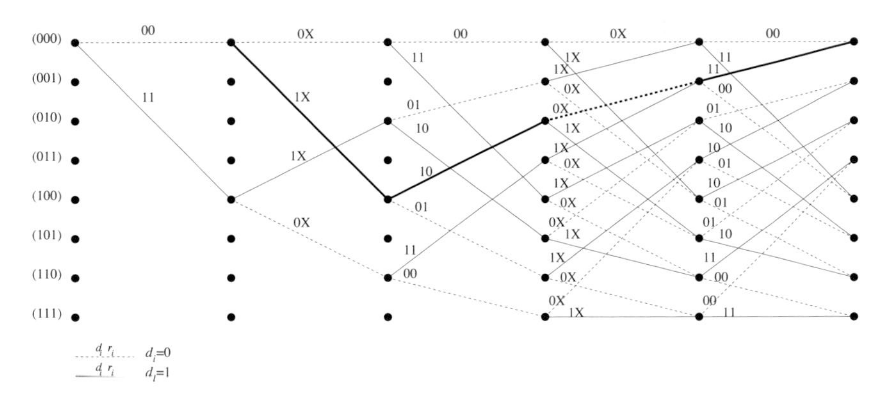

Figure 5.26 – Trellis diagram of the punctured recursive code for a rate 2/3.

The most widely used decoding technique involves taking the decoder of the original code and inserting *neutral* values in the place of the punctured elements. The neutral values are values representing information that is *a priori* not known. In the usual case of a transmission using antipodal signalling (+1 for the logical '1', -1 for the logical '0'), the null value (analogue 0) is taken as the neutral value.

The introduction of puncturing increases the coding rate but, of course, decreases its correction capability. Thus, in the example of Figure 5.26, the free distance of the code is reduced from 6 to 4 (an associated RTZ sequence is shown in the figure). Likewise Figure 5.27, in which we present the error rate curves of code $[1, (1 + D^2 + D^3)/(1 + D + D^3)]$ for rates 1/2, 2/3, 3/4 and 6/7, shows a decrease in error correction capability with the increase in coding rate.

The choice of puncturing mask obviously influences the performance of the code. It is thus possible to favour one part of the frame, transporting sensitive data, by slightly puncturing it to the detriment of another part that is more highly punctured. A regular mask is, however, often chosen as it is simple to implement.

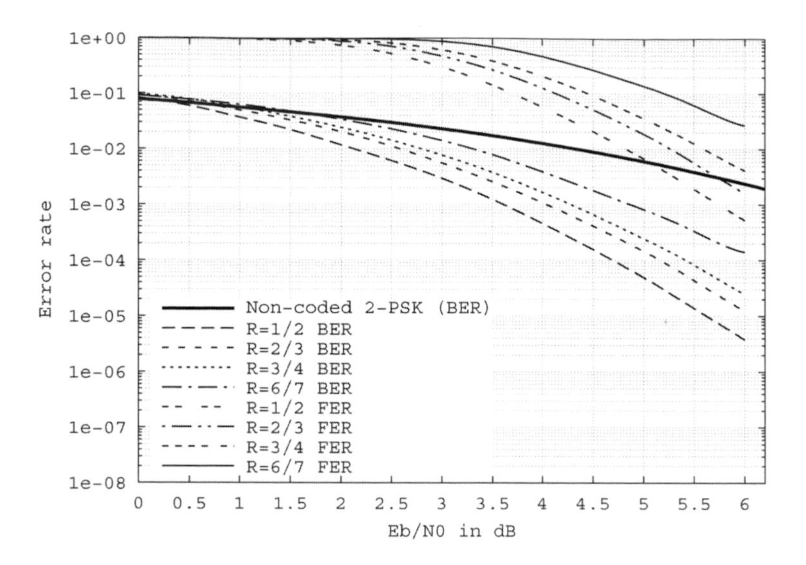

Figure 5.27 – Simulated performance of a CRSC code with 8 punctured states, as a function of its rate.

Puncturing is therefore a flexible technique, and easy to implement. As the encoder and the decoder remain identical whatever the puncturing applied, it is possible to modify the encoding rate at any moment. Some applications use this flexibility to adapt the rate, as they go along, to the channel and/or to the importance of the data transmitted.

Bibliography

[5.1] L. R. Bahl, J. Cocke, F. Jelinek, and J. Raviv. Optimal decoding of linear codes for minimizing symbol error rate. *IEEE Transactions on Information Theory*, IT-20:284–287, March 1974.

[5.2] G. Battail. Weighting of the symbols decoded by the viterbi algorithm. *Annals of Telecommunications*, 42(1-2):31–38, Jan.-Feb. 1987.

[5.3] C. Berrou, P. Adde, E. Angui, and S. Faudeuil. A low complexity soft-output viterbi decoder architecture. In *Proceedings of IEEE International Conference on Communications (ICC'93)*, pages 737–740, GENEVA, May 1993.

[5.4] C. Berrou, A. Glavieux, and P. Thitimajshima. Near shannon limit error-correcting coding and decoding: turbo-codes. In *Proceedings of IEEE In-*

ternational Conference on Communications (ICC'93), pages 1064–1070, GENEVA, May 1993.

[5.5] P. Elias. Coding for noisy channels. *IRE Convention Records*, 3(4):37–46, 1955.

[5.6] R. M. Fano. A heuristic discussion of probabilistic decoding. *IEEE Transactions on Information Theory*, IT-9:64–74, Apr. 1963.

[5.7] G. D. Forney. Convolutional codes i: Algebraic structure. *IEEE Transactions on Information Theory*, IT-16:720–738, Nov. 1970.

[5.8] G. D. Forney. The viterbi algorithm. *Proceedings of the IEEE*, 61(3):268–278, March 1973.

[5.9] A. Glavieux. *Codage de canal, des bases théoriques aux turbocodes.* Hermès-Science, 2005.

[5.10] J. Hagenauer and P. Hoeher. A viterbi algorithm with soft-decision outputs and its applications. In *Proceedings of IEEE Global Communications Conference (Globecom'89)*, pages 1680–1686, Dallas, Texas, USA, Nov. 1989.

[5.11] R. Johannesson and K. Sh. Zigangirov. *Fundamentals of Convolutional Coding.* Piscataway, IEEE Press, 1999.

[5.12] H. H. Ma and J. K. Wolf. On tail-biting convolutional codes. *IEEE Transactions on Communications*, COM-34:104–111, Feb. 1986.

[5.13] P. Thitimajshima. *Les codes convolutifs récursifs systématiques et leur application à la concaténation parallèle.* Phd thesis, Université de Bretagne Occidentale, Dec. 1993.

[5.14] A. J. Viterbi. Coding for noisy channels. *IEEE Transactions on Information Theory*, IT-13:260–269, Apr. 1967.

[5.15] J. M. Wozencraft. Sequential decoding for reliable communication. *IRE National Convention Records*, 5 part 2(2):11–25, 1957.

Chapter 6

Concatenated codes

The previous chapters presented the elementary laws of encoding like BCH, Reed-Solomon or CRSC codes. Most of these elementary codes are asymptotically good, in the sense that their minimum Hamming distances (MHD) can be made as large as we want, by sufficiently increasing the degree of the generator polynomials. The complexity of the decoders is unfortunately unacceptable for the degrees of polynomials that would guarantee the MHD required by practical applications.

A simple means of having codes with a large MHD and nevertheless easily decodable is to combine several reasonably-sized elementary codes, in such a way that the resulting global code has a high error correction capability. The decoding is performed in steps, each of them corresponding to one of the elementary encoding steps. The first composite encoding scheme was proposed by Forney during work on his thesis in 1965, called concatenated codes [6.4]. In this scheme, a first encoder, called the outer encoder, provides a codeword that is then re-encoded by a second encoder, called the inner encoder. If the two codes are systematic, the concatenated code is itself systematic. In the rest of this chapter, only systematic codes will be considered.

Figure 6.1(a) shows a concatenated code, as imagined by Forney, and the corresponding step decoder. The most judicious choice of constituent code is an algebraic code, typically a Reed-Solomon code, for the outer code, and a convolutional code for the inner code. The inner decoder is then the Viterbi decoder, which easily takes advantage of the soft values provided by the demodulator, and the outer decoder, which works on symbols with several bits (for example, 8 bits), can handle errors in bursts at the output of the first decoder. A permutation or interleaving function inserted between the two encoders, and its inverse function placed between the two decoders, can greatly increase the robustness of the concatenated code (Figure 6.1(b)). Such an encoding scheme has worked very successfully in applications as varied as deep space transmissions and digital, satellite and terrestrial television broadcasting. In particular, it is the encoding scheme adopted in many countries for digital terrestrial television [6.1].

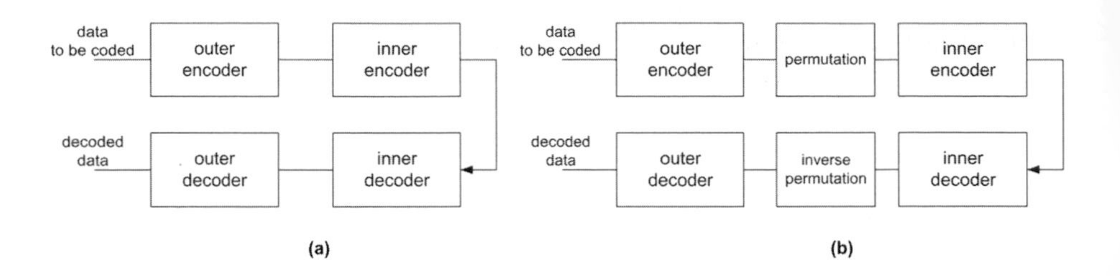

Figure 6.1 – serial concatenated code, (a) without and (b) with permutation. In both cases, the output of the outer encoder is entirely recoded by the inner encoder.

Nowadays, this first version of concatenated codes is called *serial concatenation* (SC). Its decoding, presented in Figures 6.1 is not optimal. Indeed, even if, locally, the two elementary decoders are optimal, the simple sequencing of these two decodings is not globally optimal as the inner decoder does not take any advantage of the redundancy produced by the outer code. It is this observation, that occurred fairly late in the history of information theory, that led to the development of new decoding principles, beginning with turbo decoding. We now know how to decode, quasi-optimally, all sorts of concatenated schemes, with the sole condition that the decoders of elementary codes are of the SISO (*soft-in/soft-out*) type. In this sense, we can note that the concept of concatenation has greatly evolved in the last few years, moving towards a wider notion of multi-dimensional encoding. Here, the *dimension of a code*, which should not be confused with the length (k) of the information message that we also call dimension, is the number of elementary codes used in the production of the final codeword.

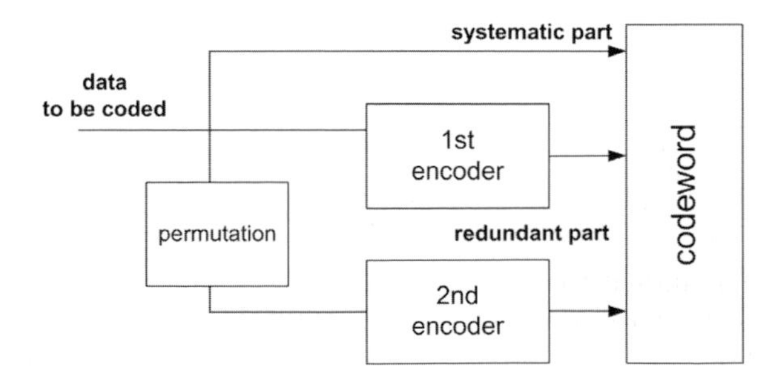

Figure 6.2 – Parallel concatenation of systematic encoders.

A new form of concatenation, called *parallel concatenation* (PC), was introduced at the beginning of the 1990s to elaborate turbo codes [6.3]. Figure 6.2 presents a PC with dimension 2, which is the classical dimension for turbo codes. In this scheme, the message is coded twice, in its natural order and in a permuted order. The redundant part of the codeword is formed by concatenating the redundant outputs of the two encoders. PCs differ from SCs in several ways, described in the next section.

6.1 Parallel concatenation and serial concatenation

Limiting ourselves to dimension 2, the PC, which associates two elementary codes with rates R_1 (code C_1) and R_2 (code C_2), has a global encoding rate:

$$R_p = \frac{R_1 R_2}{R_1 + R_2 - R_1 R_2} = \frac{R_1 R_2}{1 - (1 - R_1)(1 - R_2)} \tag{6.1}$$

This rate is higher than the global rate R_s of a serial concatenated code ($R_s = R_1 R_2$), for identical values of R_1 and R_2, and the lower the encoding rates the greater the difference. We can deduce from this that with the same error correction capability of component codes, parallel concatenation offers a better encoding rate, but this advantage diminishes when the rates considered tend towards 1. When the dimension of the composite code increases, the gap between R_p and R_s also increases. For example, three component codes of rate 1/2 form a concatenated code with global rate 1/4 for parallel, and 1/8 for serial concatenation. That is the reason why it does not seem to be useful to increase the dimension of a serial concatenated code beyond 2, except for rates very close to unity.

However, with SC, the redundant part of a word processed by the outer decoder has benefited from the correction of the decoder(s) that precede(s) it. Therefore, at first sight, the correction capability of a serial concatenated code seems to be greater than that of a parallel concatenated code, in which the values representing the redundant part are never corrected. In other terms, the MHD of a serial concatenated code must normally be higher than that of a parallel concatenated code. We therefore find ourselves faced with the dilemma given in Chapter 1: PC performs better in the convergence zone (near the theoretical limit) since the encoding rate is more favourable, and the SC behaves better at low error rates thanks to a larger MHD. Encoding solutions based on the SC of convolutional codes have been studied [6.3], which can be an interesting alternative to classical turbo codes, when low error rates are required. Serial convolutional concatenated codes will not, however, be described in the rest of this book.

When the redundant parts of the inner and outer codewords both undergo supplementary encoding, the concatenation is said to be double serial concatenation. The most well-known example of this type of encoding structure is the product code, which implements BCH codes (see Chapters 4 and 8). Mixed structures, combining parallel and serial concatenations have also been proposed [6.6]. Moreover, elementary concatenated codes can be of a different nature, for example a convolutional code and a BCH code [6.2]. We then speak of hybrid concatenated codes. From the moment elementary decoders accept and produce weighted values, all sorts of mixed and/or hybrid schemes can be imagined.

Whilst SC can use systematic or non-systematic codes indifferently, parallel concatenation uses systematic codes. If they are convolutional codes, at least one of these codes must be recursive, for a fundamental reason to do with the minimum input weight w_{min}, which is only 1 for non-recursive codes but is 2 for recursive codes (see Chapter 5). To show this, see Figure 6.3 which presents two non-recursive systematic codes, concatenated in parallel. The input sequence is "all zero" (reference sequence) except in one position. This single "1" perturbs the output of the encoder C_1 for a short length of time, equal to the constraint length 4 of the encoder. The redundant information Y_1 is poor, in relation to this particular sequence, as it contains only 3 values different from 0. After permutation, of whatever type, the sequence is still "all zero", except in one single position. Again, this "1" perturbs the output of the encoder C_2 for a length of time equal to the constraint length, and redundancy Y_2 provided by the second code is as poor in information as redundancy Y_1. In fact, the minimum distance of this two-dimensional code is not higher than that of a single code, with the same rate as that of the concatenated code. If we replace at least one of the two non-recursive encoders by a recursive encoder, the "all zero" sequence except in one position is no longer a "Return to Zero" (RTZ, see Section 5.3.2) sequence for this recursive encoder, and the redundancy that it produces is thus of much higher weight.

What we have explained above about the PC of non-recursive convolutional codes suggests that the choice of elementary codes for the PC in general is limited. As another example, let us build a parallel concatenated code from the extended Hamming code defined by Figure 1.1 and the encoding Table 1.1. The information message contains 16 bits, arranged in a 4x4 square (Figure 6.4(a)). Each line and each column is encoded by the elementary Hamming code. The horizontal and vertical parity bits are denoted $r_{i,j}$ and $r'_{i,j}$, respectively. The global coding rate is 1/3. Decoding this type of code can be performed using the principles of turbo decoding (optimal local decoding according to the maximum likelihood and continuous exchanges of extrinsic information).

The MHD of the code is given by the pattern of errors of input weight 1 (Figure 6.4(b)). Whatever the position of the 1 in the information message, the weight is 7. The figure of merit Rd_{min} (see Section 1.5) is therefore equal to 7x(1/3), compared with the figure of merit of the elementary code 4x(1/2).

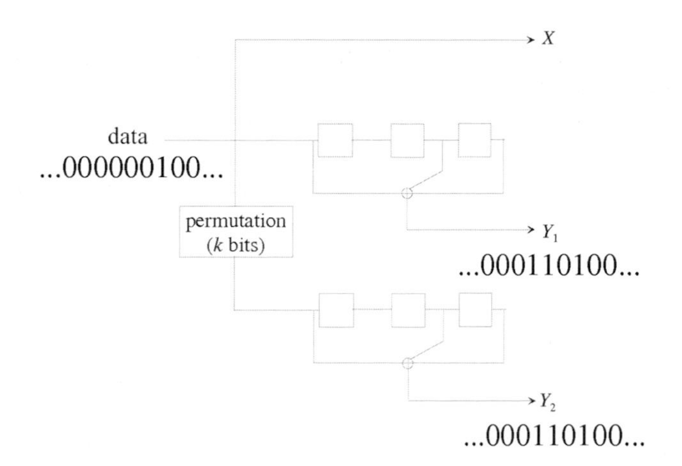

Figure 6.3 – The parallel concatenation of non-recursive systematic codes is a poor code concerning the information sequences of weight 1. In this example, the redundancy symbols Y_1 and Y_2 each contain only three 1.

$d_{0,0}$	$d_{0,1}$	$d_{0,2}$	$d_{0,3}$	$r_{0,0}$	$r_{0,1}$	$r_{0,2}$	$r_{0,3}$
$d_{1,0}$	$d_{1,1}$	$d_{1,2}$	$d_{1,3}$	$r_{1,0}$	$r_{1,1}$	$r_{1,2}$	$r_{1,3}$
$d_{2,0}$	$d_{2,1}$	$d_{2,2}$	$d_{2,3}$	$r_{2,0}$	$r_{2,1}$	$r_{2,2}$	$r_{2,3}$
$d_{3,0}$	$d_{3,1}$	$d_{3,2}$	$d_{3,3}$	$r_{3,0}$	$r_{3,1}$	$r_{3,2}$	$r_{3,3}$
$r'_{0,0}$	$r'_{0,1}$	$r'_{0,2}$	$r'_{0,3}$				
$r'_{1,0}$	$r'_{1,1}$	$r'_{1,2}$	$r'_{1,3}$				
$r'_{2,0}$	$r'_{2,1}$	$r'_{2,2}$	$r'_{2,3}$				
$r'_{3,0}$	$r'_{3,1}$	$r'_{3,2}$	$r'_{3,3}$				

(a)

1	0	0	0	0	1	1	1
0	0	0	0	0	0	0	0
0	0	0	0	0	0	0	0
0	0	0	0	0	0	0	0
0	0	0	0				
1	0	0	0				
1	0	0	0				
1	0	0	0				

(b)

Figure 6.4 – Parallel concatenation of extended Hamming codes (global rate: 1/3). On the right: a pattern of errors of input weight 1 and total weight 7.

The asymptotic gain has therefore not been extraordinarily increased by means of the concatenation (0,67 dB precisely), and a great reduction in the coding rate has occurred. If we wish to keep the same global rate of 1/2, a part of the redundancy must be punctured. We can choose, for example, not to transmit the 16 symbols present in the last two columns and the last two lines of the table of Figure 6.4(a). The MHD then drops to the value 3, that is, less than the MHD of the elementary code. The PC is therefore of no interest in this case.

Again from the extended Hamming code, a double serial concatenation can be elaborated in the form of a product code (Figure 6.5(a)). In this scheme, the redundant parts of the horizontal and vertical codewords are themselves re-encoded by elementary codes, which produce redundancy symbols denoted $w_{i,j}$. One useful algebraic property of this product code is the identity of the redundancy symbols coming from the second level of encoding, in the horizontal and vertical directions. The MHD of the code, which has a global rate $1/4$, is again given by the patterns of errors of input weight 1 and is equal to 16, that is, the square of the MHD of the elementary code (Figure 6.5(b)). The figure of merit $Rd_{min} = 4$ has therefore been greatly increased compared to parallel concatenation. To attempt to increase the rate of this code by puncturing the redundancy symbols while keeping a good MHD is bound to fail.

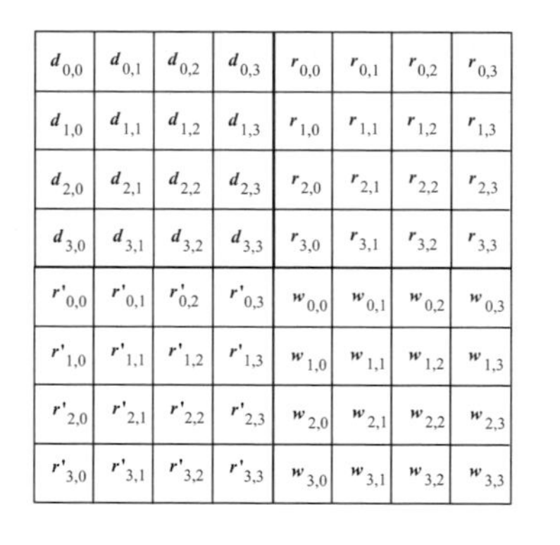

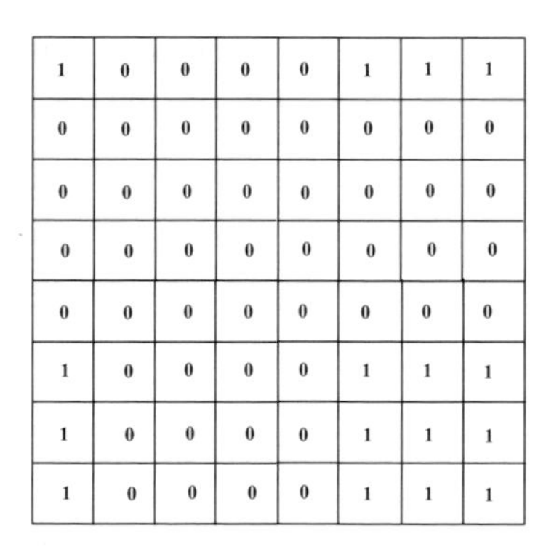

Figure 6.5 – Double serial concatenation (product code) of extended Hamming codes (global rate: $1/4$). On the right: a pattern of errors of input weight 1 and total weight 16.

In conclusion, parallel concatenation cannot be used with just any elementary code. Today, only convolutional recursive systematic codes are used in this type of concatenation, with 2 dimensions. Serial concatenation can offer large MHD. The choice of codes is greater: convolutional codes, recursive or not, BCH codes or Reed-Solomon codes. However, with the same coding rates of elementary codes, serial concatenation has a lower global rate than parallel concatenation.

6.2 Parallel concatenation and LDPC codes

LDPC codes, which are described in Chapter 9, are codes where the lines and columns of the parity check matrix contain few 1s. LDPC codes can be seen as a multiple concatenation of $n - k$ parity relations containing few variables.

Here it is not a concatenation in the sense that we defined above, since the parity relations contain several redundancy variables and these variables appear in several relations. We cannot therefore assimilate LDPC codes to standard serial or parallel concatenation schemes. However, we can, like MacKay [6.5], observe that a turbo code is an LDPC code. An RSC code with generator polynomials $G_X(D)$ (recursivity) and $G_Y(D)$ (redundancy), whose input is X and redundant output Y, is characterized by the sliding parity relation:

$$G_Y(D)X(D) = G_X(D)Y(D) \tag{6.2}$$

Using the tail-biting technique (see CRSC, Section 5.5.1, the parity check matrix takes a very regular form, such as the one presented in Figure 6.6 for a coding rate 1/2, and choosing $G_X(D) = 1 + D + D^3$ and $G_Y(D) = 1 + D^2 + D^3$. A CRSC code is therefore an LDPC code since the check matrix is sparse. This is certainly not a good LDPC code, as the check matrix does not respect certain properties about the positions of the 1s. In particular, the 1s on a same line are very close to each other, which is not favourable to the belief propagation decoding method.

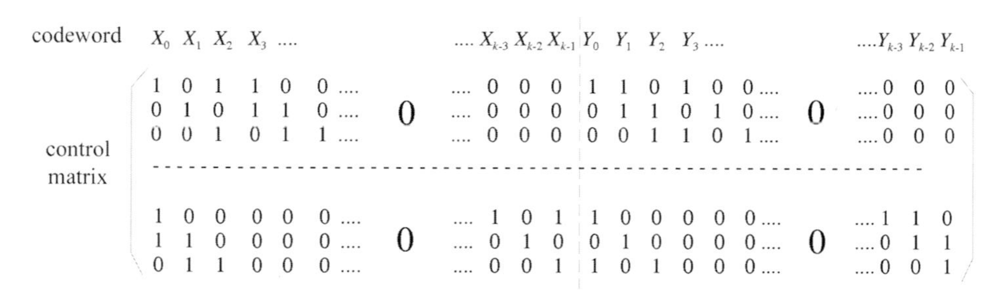

Figure 6.6 – Check matrix of a tail-biting convolutional code. A convolutional code, particularly of the CRSC type, can be seen as an LDPC code since the check matrix is sparse.

A parallel concatenation of CRSC codes, that is a turbo code, is also an LDPC code since it associates elementary codes that are of the LDPC type. Of course, there are more degrees of freedom in the construction of an LDPC code, as each 1 of the check matrix can be positioned independently of the others. On the other hand, decoding a convolutional code, via an algorithm based on the trellis, does not encounter the problem of correlation between successive symbols that a belief propagation type of decoding would encounter, if it was applied to a simple convolutional code. A turbo code cannot therefore be decoded like an LDPC code.

6.3 Permutations

The functions of permutation or interleaving, used between elementary encoders in a concatenated scheme, have a twofold role. On the one hand, they ensure, at the output of each component decoder, a time spreading of the errors that can be produced by it in bursts. These packets of errors then become isolated errors for the following decoder, with far lower correlation effects. This technique for the spreading of errors is used in a wider context than that of channel coding. We can use it profitably, for example, to reduce the effects of more or less long attenuation in transmissions affected by fading, and more generally in situations where perturbations can alter consecutive symbols. On the other hand, in close liaison with the characteristics of constituent codes, the permutation is designed so that the MHD of the concatenated code is as large as possible. This is a problem of pure mathematics associating geometry, algebra and combinatory logic which, in most cases, has not yet found a definitive answer. Sections 7.3.2 and 9.1.6 develop the topic of permutation for turbo codes and graphs for LDPC codes, respectively.

6.4 Turbo crossword

To end this chapter, here is an example of parallel concatenation that is familiar to everyone: crosswords. The content of a grid has been altered during its retranscription, as can be seen in Figure 6.7. Fortunately, we have a correct clue for each line and for each column and we have at our disposal a dictionary of synonyms.

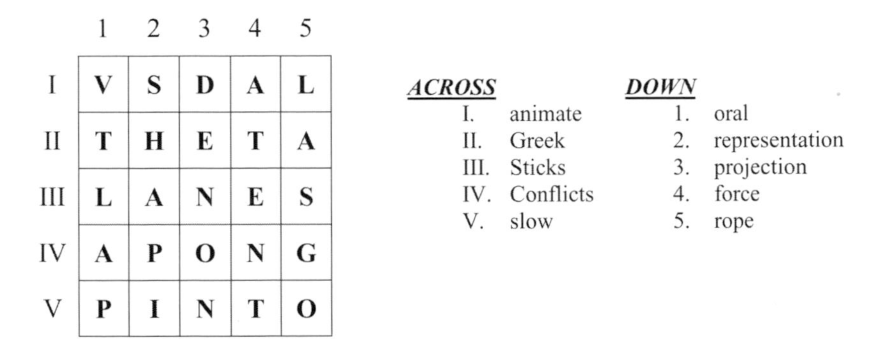

Figure 6.7 – Crossword grid with wrong answers but correct clues.

To correct (or decode) this grid, we must operate iteratively by line and by column. The basic decoding rule is the following: "If there is a word in the dictionary, a synonym or an equivalent to the definition given that differs from the word in the grid by at most one letter, then this synonym is adopted".

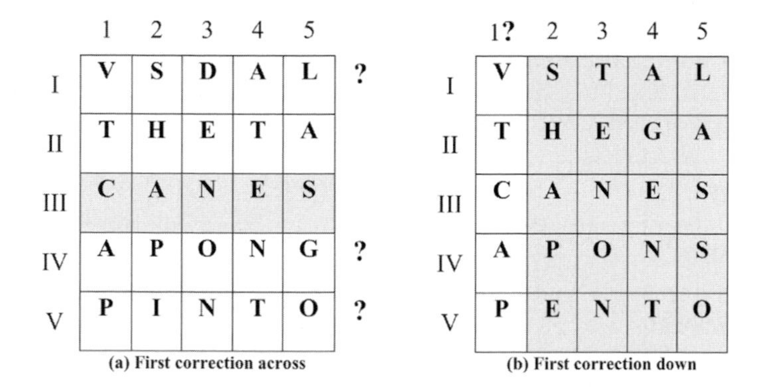

Figure 6.8 – First iteration of the line - column decoding process.

The horizontal definitions allow us to begin correcting the lines in the grid (Figure 6.8(a)):

I. There must be more than one wrong letter.

II. THETA is a Greek letter.

III. We can replacc thc L with a C, getting CANES (sticks).

IV. There must be more than one wrong letter.

V. There must be more than one wrong letter

After decoding this line, two words are correct or have been corrected, and three are still to be found. Using the vertical definitions, we can now decode the columns (Figure 6.8(b)):

1. There must be more than one wrong letter. No correction is possible.
2. Replacing the I with an E, we get SHAPE (representation).
3. A TENON is a projection (of wood).
4. Replacing the T with a G, we get AGENT (force).
5. A LASSO is a kind of rope.

After decoding the columns, there are still some unknown words and we have to perform a second iteration of the line - column decoding process (Figure 6.9). Line decoding leads to the following result (Figure 6.9(a)):

I. Replacing the S with an I, we get VITAL (animate).

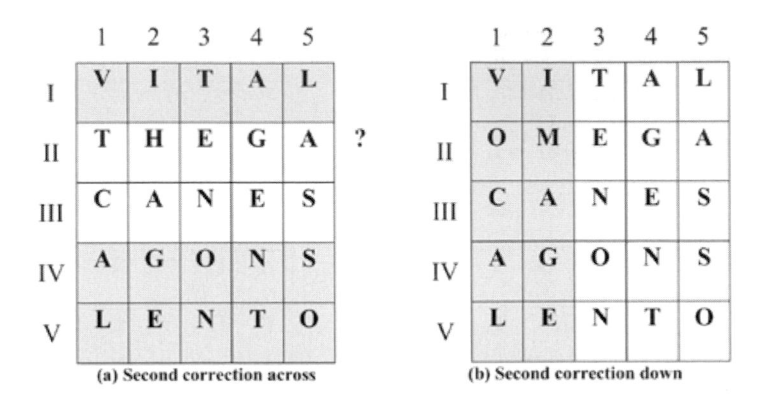

Figure 6.9 – Second iteration of the line - column decoding process.

II. There must be at least 2 wrong letters.

III. CANES is correct

IV. Replacing the P with a G, we get AGONS (conflicts)

V. We can replace the P with an L, getting LENTO (slow).

After this step, there is still one wrong line. It is possible to correct it by decoding the columns again (Figure 6.9(b)).

1. VOCAL is a synonym of oral.
2. IMAGE is also a kind of representation.
3. TENON is correct.
4. AGENT is correct.
5. LASSO is correct.

After this final decoding step, the wrong word on line II is identified: it is OMEGA, another Greek letter.

A certain number of remarks can be made after this experience of decoding a parallel concatenated code.

- To arrive at the right result, two iterations of the line and column decoding were necessary. It would have been a pity to stop after just one iteration. But that is what we do in the case of a classical concatenated code such as that of Figure 6.1.

- A word that was correct at the beginning (THETA is indeed a Greek letter) turned out to be wrong. Likewise, a correction made during the second step (SHAPE) turned out to be wrong. So the intermediate results must be considered with some caution and we must avoid any hasty decisions. In modern iterative decoders, this caution is measured by a probability that is never exactly 0 or 1.

Bibliography

[6.1] Dvb-terrestrial. ETSI EN 302 296 V1.1.1 (2004-04).

[6.2] P. Adde, R. Pyndiah, and C. Berrou. Performance of hybrid turbo codes. *Elect. Letters*, 32(24):2209–2210, Nov. 1996.

[6.3] S. Benedetto, D. Divsalar, G. Montorsi, and F. Pollara. Serial concatenation of interleaved codes: performance analysis, design, and iterative decoding. *IEEE Trans. Info. Theory*, 44(3):909–926, May 1998.

[6.4] Jr. G. D. Forney. Performance of concatenated codes. In E. R. Berlekamp, editor, *Key papers in the development of coding theory*, pages 90–94. IEEE Press, 1974.

[6.5] D. J. C. MacKay. Good error-correcting codes based on very sparse matrices. *IEEE Transactions on Information Theory*, 45(2):399–431, March 1999.

[6.6] K. R. Narayanan and G. L. Stüber. Selective serial concatenation of turbo codes. *IEEE Comm. Letters*, 1(5):136–139, Sept. 1997.

Chapter 7

Convolutional turbo codes

The error correction capability of a convolutional code increases when the length of the encoding register increases. This is shown in Figure 7.1, which provides the performance of four RSC codes with respective memories $\nu = 2, 4, 6$ and 8, for rates 1/2, 2/3, 3/4 and 4/5, decoded according to the MAP algorithm. For each of the rates, the error correction capability improves with the increase in ν, above a certain signal to noise ratio that we can assimilate almost perfectly with the theoretical limit calculated in Chapter 3 and identified here by an arrow. To satisfy the most common applications of channel coding, a memory of the order of 30 or 40 would be necessary (from a certain length of register and for a coding rate 1/2, the minimum Hamming distance of a convolutional code with memory ν is of the order of ν). If we knew how to easily decode a convolutional code with over a billion states, we would no longer speak much about channel coding and this book would not exist.

A turbo code is a coding trick, aiming to imitate a convolutional code with a large memory ν. It is built on the principle of the saying *divide and rule*, that is, by associating several small RSC codes whose particular decodings are of reasonable complexity. A judicious exchange of information between the elementary decoders enables the composite decoder to approximate the performance of maximum likelihood decoding.

7.1 The history of turbo codes

The invention of turbo codes is not the outcome of a mathematical development. It is the result of an intuitive experimental approach whose origin can be found in the work of several European researchers: Gerard Battail, Joachim Hagenauer and Peter Hoeher who, at the end of the 80s [7.8, 7.7, 7.31, 7.30] highlighted the interest of probabilistic processing in receivers. Others before them, mainly in the United States: Peter Elias [7.25], Michael Tanner [7.45], Robert Gallager

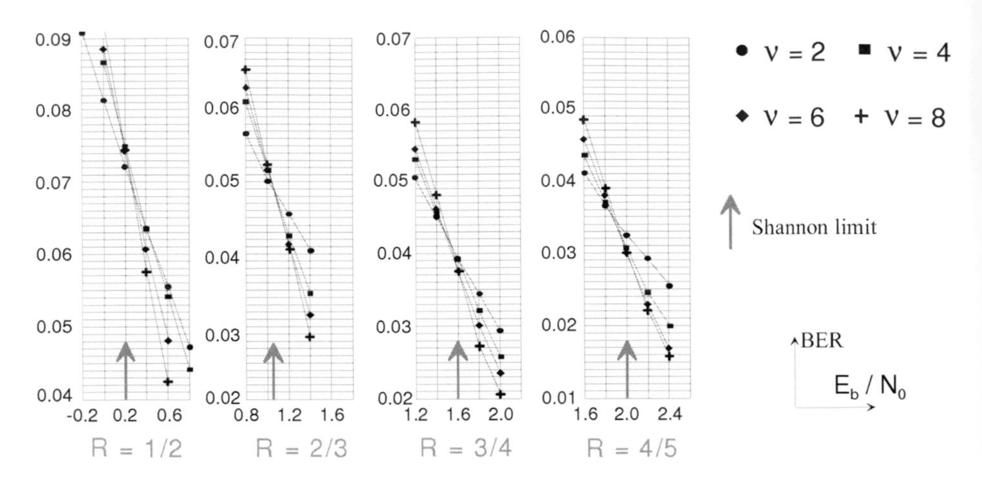

Figure 7.1 – Performance of recursive systematic convolutional codes (RSC) for different rates and four values of the memory of code ν. Comparison with Shannon limits.

[7.26], etc. had earlier imagined procedures for coding and decoding that were the forerunners of turbo codes.

In a laboratory at École Nationale Supérieure des Télécommunications de Bretagne (Telecom Bretagne), Claude Berrou and Patrick Adde were attempting to transcribe the Viterbi algorithm with weighted input (SOVA: *Soft-Output Viterbi Algorithm*) proposed in [7.7], into MOS transistors, in the simplest possible way. A suitable solution [7.10] was found after two years which enabled these researchers to form an opinion about probabilistic decoding. Claude Berrou, then Alain Glavieux, pursued the study and observed, after Gerard Battail, that a decoder with weighted input and output could be considered as a signal to noise ratio amplifier. This encouraged them to implement the concepts commonly used in amplifiers, mainly feedback. Perfecting turbo codes involved many very pragmatic stages and also the introduction of neologisms, like "parallel concatenation" or "extrinsic information", nowadays common in information theory jargon. The publication in 1993 of the first results [7.14], with a performance 0,5 dB from the Shannon limit, shook the coding community. A gain of almost 3 dB, compared to solutions existing at that time, had been found by a small team that was not only unknown, but also French (France, a country known for its mathematical rigour, *versus* turbo codes, an empirical invention to say the least). There followed a very distinct evolution in habits, as underlined by A. R. Calderbank in [7.20] (p. 2573): "*It is interesting to observe that the search for theoretical understanding of turbo codes has transformed coding theorists into experimental scientists*"

[7.13] presents a chronology describing the successive ideas that appeared in the search to perfect turbo codes. This new coding and decoding technique was first baptized turbo-code, with a hyphen to show that it was a code decoded in a turbo way (by analogy with the turbo engine that uses exhaust gas to increase its power). As the hyphen is not used much in English, it became *turbo code*, that is, a "turbo" code, which does not mean very much. In French today, turbo code is written as a single word: *turbocode*.

7.2 Multiple concatenation of RSC codes

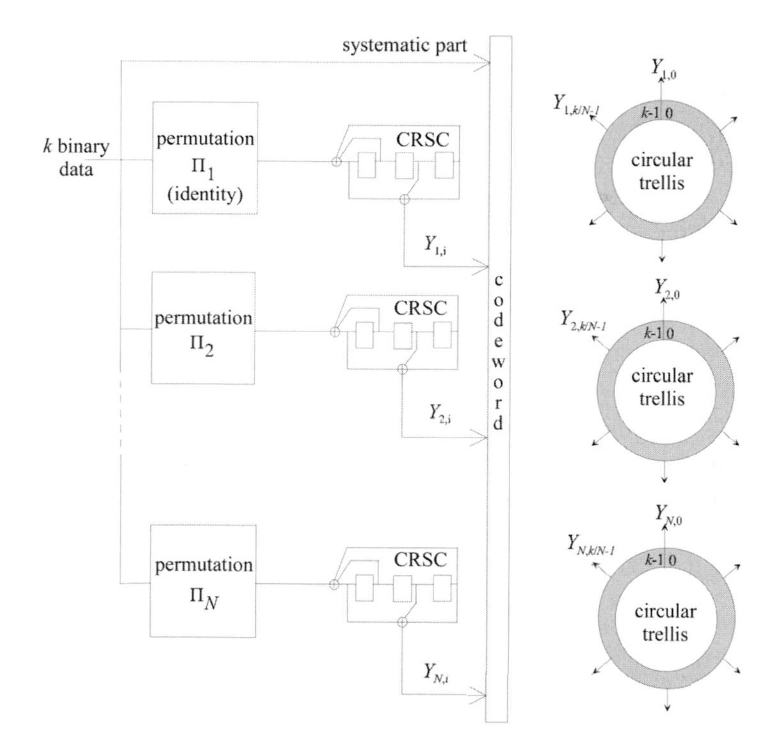

Figure 7.2 – Multiple parallel concatenation of circular recursive systematic convolutional (CRSC) codes. Each encoder produces k/N redundancy symbols uniformly distributed on the circular trellis. Global coding rate: 1/2.

Since the seminal work of Shannon, random codes have always been a reference for error correction coding (see Section 3.1.5). The systematic random coding of a block of k information bits, leading to a codeword of length n, can, as the first step and once for all, involve drawing at random and memorizing k binary markers containing $n - k$ bits, whose memorization address is denoted i $(0 \leq i \leq k-1)$. The redundancy associated with any block of information is then formed by the modulo 2 sum of all the markers whose address i is such that the i-th information bit equals 1. In other words, the k markers are the bases of a vector space of dimension k. The codeword is finally made up of the concatenation of the k information bits and of the $n-k$ redundancy bits. The rate R of the code is k/n. This very simple construction of the codeword relies on the linearity property of the addition and leads to high minimum distances for sufficiently large values of $n-k$. Because two codewords are different by at least one information bit and the redundancy is drawn at random, the average minimum distance is $1 + \frac{n-k}{2}$. However, the minimum distance of this code being a random variable, its different realizations can be lower than this value. A simple realistic approximation of the effective minimum distance is $\frac{n-k}{4}$.

A way to build an almost random encoder is presented in Figure 7.2. It is a multiple parallel concatenation of circular recursive systematic convolutional codes (CRSC, see Chapter 5) [7.12]. The sequence of k binary data is coded N times by N CRSC encoders, in a different order each time. The permutations Π_j are drawn at random, except the first one that can be the identity permutation. Each elementary encoder produces $\frac{k}{N}$ redundancy symbols (N being a divisor of k), the global rate of the concatenated code being 1/2.

The proportion of input sequences of a recursive encoder built from a pseudo-random generator with memory ν, initially positioned in state 0, which return the register back to the same state at the end of the coding, is:

$$p_1 = 2^{-\nu} \tag{7.1}$$

since there are 2^ν possible return states, with the same probability. These sequences, called *Return To Zero* (RTZ) sequences,(see Chapter 5), are linear combinations of the minimum RTZ sequence, which is given by the recursivity polynomial of the generator ($1 + D + D^3$ in the case of Figure 7.2).

The proportion of RTZ sequences for the multi-dimensional encoder is lowered to:

$$p_N = 2^{-N\nu} \tag{7.2}$$

since the sequence must, after each permutation, remain RTZ for the N encoders.

The other sequences, with proportion $1 - p_N$, produce codewords that have a distance d satisfying:

$$d > \frac{k}{2N} \tag{7.3}$$

This worst case value assumes that a single permuted sequence is not RTZ and that redundancy Y takes the value 1, every other time on average, on the corresponding circle. If we take $N = 8$ and $\nu = 3$ for example, we obtain $p_8 \approx 10^{-7}$ and, for sequences to encode of length $k = 1024$, we have $d_{min} = 64$, which is a sufficient minimum distance if we refer to the curves of Figure 3.6.

Random coding can thus be approximated by using small codes and random permutations. The decoding can be performed following the turbo principle, described in Section 7.4 for $N = 2$. The scheme of Figure 7.2 is, however, not used in practice, for reasons linked to the performance and complexity of the decoding. First, the convergence threshold of the turbo decoder, that is, the signal to noise ratio from which the turbo decoder can begin to correct most of the errors, degrades when the dimension of the concatenation increases. Indeed, the very principle of turbo decoding means considering the elementary codes one after the other, iteratively. As their redundancy rate decreases when the dimension of the composite code increases, the first steps in the decoding are penalized compared to a concatenated code with a simple dimension 2. Then, the complexity and the latency of the decoder are proportional to the number of elementary encoders.

7.3 Turbo codes

Fortunately, concerning the above, it is not necessary to carry dimension N to a high value. By replacing the random permutation Π_2 by a judiciously elaborated permutation, good performance can be obtained by limiting ourselves to a dimension $N = 2$. That is the principle of turbo codes.

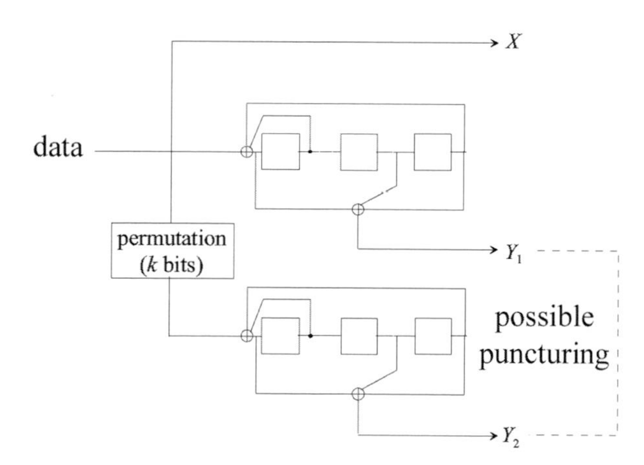

Figure 7.3 – A binary turbo code with memory $\nu = 3$ using identical elementary RSC encoders (polynomials 15, 13). The natural coding rate of the turbo code, without puncturing, is 1/3.

Figure 7.3 presents a turbo code in its most classical version [7.14]. The binary input message, of length k, is encoded in its natural order and in a permuted order by two RSC encoders called C_1 and C_2, which can be terminated or not. In this example, the two elementary encoders are identical (generator polynomials 15 for the recursivity and 13 for the construction of the redundancy) but this is not a necessity. The natural coding rate, without puncturing, is 1/3. To obtain higher rates, redundancy symbols Y_1 and Y_2 are punctured. Another way to have higher rates is to adopt m-binary codes (see 7.5.2).

As the permutation function (Π) concerns a message of finite size k, the turbo code is by construction a block code. However, to distinguish it from concatenated algebraic codes decoded in a "turbo" way, like product codes which were later called *block turbo codes*, this turbo coding scheme is called a *convolutional* code or, more technically, a *Parallel Concatenated Convolutional Code* (PCCC).

Arguments in favour of this coding scheme (some of which have already been introduced in Chapter 6) are the following:

1. A decoder for convolutional codes is vulnerable to errors arriving in packets. Coding the message twice, following two different orders (before and after permutation), makes fairly improbable the simultaneous appearance of error packets at the input of the decoders of C_1 and C_2. If there are errors grouped at the input of the decoder of C_1, the permutation disperses them over time and they become isolated errors that are easy for the decoder of C_2 to correct. This reasoning also holds for error packets at the input of this second decoder, which correspond, before permutation, to isolated errors. Thus two-dimensional coding, on at least one of the two dimensions, greatly reduces the vulnerability of convolutional coding concerning grouped perturbations. But which of the two decoders should be relied on to take the final decision? No criterion allows us to be more confident about one or the other. The answer is given by the "turbo" algorithm that avoids having to make this choice. This algorithm implements exchanges of probabilities between the two decoders and constrains them to converge, during these exchanges, towards the same decisions.

2. As we saw in Section 6.1, parallel concatenation leads to a higher coding rate than that of serial concatenation. Parallel concatenation is therefore more favourable when signal to noise ratios close to the theoretical limits are considered, with average error rates targeted. It can be different when very low error rate are sought for since the MHD of a serial concatenated code can be larger.

3. Parallel concatenation uses systematic codes and at least one of these codes must be recursive, for reasons also described in Section 6.1

4. Elementary codes are small codes: codes with 16, 8, or even 4 states. Even if the decoding implements repeated probabilistic processing, it remains of reasonable complexity.

Figure 7.4 presents turbo codes used in practice and Table 7.2 lists some industrial applications. For a detailed overview of the applications of turbo and LDPC codes see [7.29]. The parameters defining a particular turbo code are the following:

a– m is the number of bits in the symbols applied to the turbo encoder. The applications known to this day consider binary ($m = 1$) or double-binary ($m = 2$) symbols.

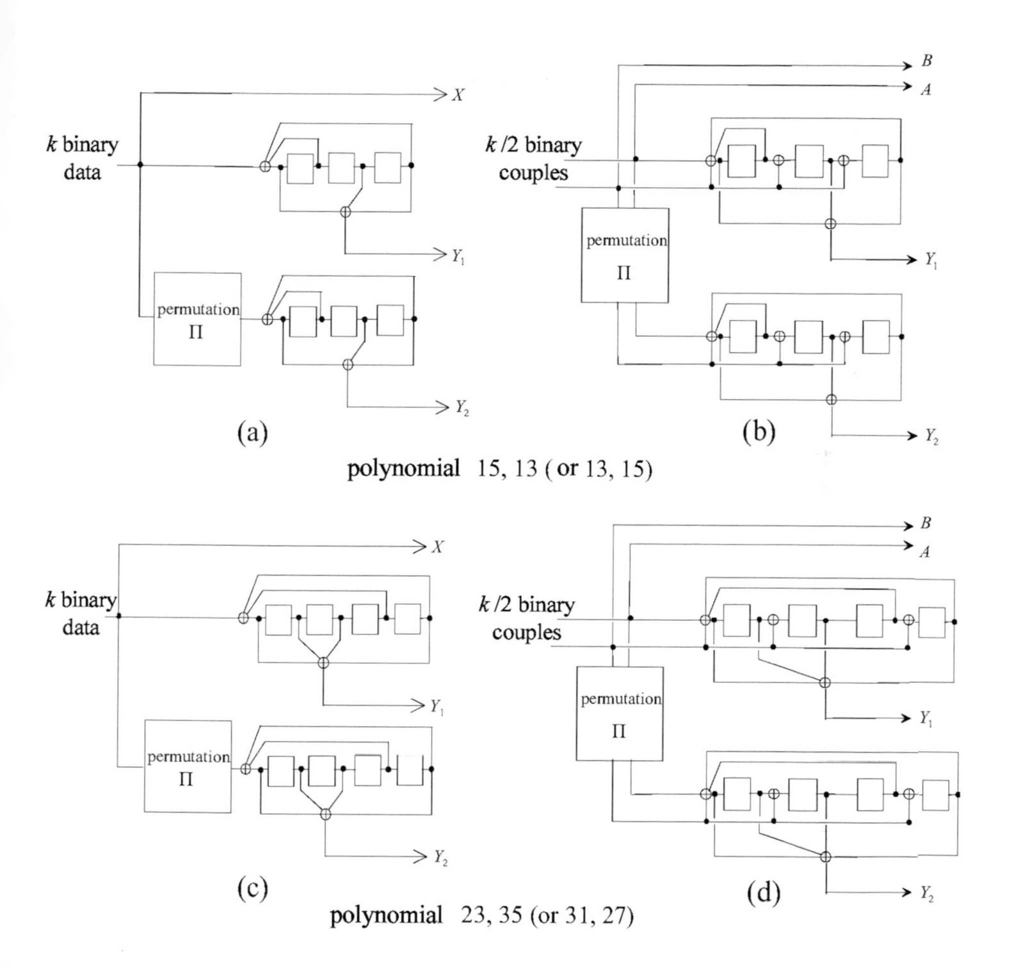

Figure 7.4 – Turbo codes used in practice.

b– Each of the two elementary encoders C_1 and C_2 is characterized by

- its code memory ν
- its generator polynomials for recursivity and redundancy
- its rate

The values of ν are in practice lower than or equal to 4. The generator polynomials are generally those that we use for classical convolutional codes and that were the subject of much literature in the 1980s and 1990s.

c– The way in which we perform the permutation is important when the targeted binary error rate is lower than around 10^{-5}. Above this value, per-

Application	Turbo code	Termination	Polynomials	Rates
CCSDS (deep space)	binary, 16 states	tail bits	23, 33, 25, 37	1/6, 1/4, 1/3, 1/2
UMTS, CDMA2000 (mobile 3G)	binary, 8 states	tail bits	13, 15, 17	1/4, 1/3, 1/2
DVB-RCS (Return Channel on Satellite)	double-binary, 8 states	circular	15, 13	from 1/3 to 6/7
DVB-RCT (Return Channel on Terrestrial)	double-binary, 8 states	circular	15, 13	1/2, 3/4
Inmarsat (M4)	binary, 16 states	none	23, 35	1/2
Eutelsat (skyplex)	double-binary, 8 states	circular	15, 13	4/5, 6/7
IEEE 802.16 (WiMAX)	double-binary, 8 states	circular	15, 13	from 1/2 to 7/8

Table 7.2 – Standardized applications of convolutional turbo codes.

formance is not very sensitive to permutation, on condition of course that it respects at least the principle of dispersion (which, for example, can be a regular permutation). For a low or very low targeted error rate, performance is dictated by the minimum distance of the code and the latter is highly dependent on the permutation Π.

d– The puncturing pattern must be the as regular as possible, in the same way as for classical convolutional codes. However, it can be advantageous to have a slightly irregular puncturing pattern when we are looking for very low error rates and when the puncturing period is a divisor of the period of the generator polynomial of recursivity or parity.

Puncturing is performed classically on the redundancy symbols. It can be envisaged instead to puncture the information symbols, in order to increase the minimum distance of the code. This is done to the detriment of the convergence threshold of the turbo decoder. From this point of view, in fact, puncturing data shared by the two decoders is more penalizing than puncturing data that are only useful to one of the decoders.

What must be closely considered when building a turbo code and decoding it, are the RTZ sequences, whose output weights limit the minimum distance of the code and fix its asymptotic performance. In what follows it will be assumed that the error patterns that are not RTZ do not contribute to the MHD of the turbo code and will therefore not have to be considered.

7.3.1 Termination of constituent codes

For a turbo code, there are two trellises to be terminated and the solutions presented in Section 5.5.1 can be envisaged:

Doing nothing in particular concerning the terminal states: the data situated at the end of the block, in either the natural order or in the permuted order, are thus less well protected. This leads to a decrease in the asymptotic gain, but this degradation, which is a function of the size of the block, may be compatible with some applications. It should be noted that non-termination of the trellis penalizes the PER (Packet Error Rates) more greatly than the BER.

Terminating the trellis of one or both elementary codes using tail bits: CCSDS [7.4] and UMTS [7.3] standards use this technique. The bits ensuring the termination of one of the two trellises are not used in the other encoder. These bits are therefore not turbo encoded, which leads, but to a lesser degree, to the same drawbacks as those presented in the previous case. Moreover, the transmission of the tail bits causes a decrease in the coding rate and therefore in the spectral efficiency.

Using interleaving enables the automatic termination of the trellis: it is possible to close the trellis of a turbo code automatically, without adding any tail bits, by slightly transforming the coding scheme (self-concatenation) and by using interleaving that respects certain periodicity rules. This solution described in [7.15] does not decrease the spectral efficiency but imposes constraints on the interleaving which makes it difficult to control performance at low error rates.

Adopting circular encoding: a circular encoder for convolutional codes guarantees that the initial state and the final state of the register are identical. The trellis then takes the form of a circle, which, from the point of view of the decoder, can be considered as a trellis with infinite length [7.32, 7.48]. This termination process, already known as *tail-biting* for non-recursive codes, offers two main advantages:

- Unlike the other techniques, circular termination does not present any edge effects: all the bits of the message are protected in the same way and are all doubly encoded by the turbo code. Therefore, during the design of the permutation, there is no need to give special importance to such and such a bit, which leads to simpler permutation models.

- The sequences that are not RTZ have an influence on the whole circle: on average, one parity symbol out of two is modified along the block. For typical values of k (a few hundred or more), the corresponding output weight is therefore very high and these error patterns do not contribute to the MHD of the code, as already mentioned at the end of the previous section. Without termination or with termination using tail bits, only the part of the block after the beginning of the non-RTZ sequence has any effect on the parity symbols.

To these two advantages we can, of course, add the interest of having to transmit no additional information about termination and therefore losing nothing in spectral efficiency.

The circular termination technique was chosen for the DVB-RCS and DVB-RCT [7.2, 7.1] standards, for example.

7.3.2 The permutation function

Whether we call it interleaving or permutation, the technique that involves dispersing the data over time has always been very useful in digital communications. For example, we use it profitably to reduce the effects of more or less long attenuations in transmissions affected by fading and, more generally, in situations where perturbations can alter consecutive symbols. In the case of turbo codes too, permutation allows us to efficiently combat the appearance of error packets, on at least one of the dimensions of the composite code. But its role does not stop there: in close relation with the properties of the constituent codes, it also determines the minimum distance of the concatenated code.

Let us consider the turbo code presented in Figure 7.3. The worst permutation that we can use is, of course, identity permutation, which minimizes the diversity of the coding (we then have $Y_1 = Y_2$). On the other hand, the best permutation that we could imagine, but that probably does not exist [7.42], would enable the concatenated code to be equivalent to a sequential machine of which the number of irreducible states would be 2^{k+6}. There are indeed $k + 6$ binary memorization elements in the structure: k for the permutation memory and 6 for the two convolutional codes. If we could assimilate this sequential machine to a convolutional encoder and for common values of k, the number of corresponding states would be very large, in any case large enough to guarantee a large minimum distance. For example, a convolutional encoder with a code memory of 60 (10^{18} states !) shows a free distance of the order of a hundred (for $R = 1/2$), which is quite sufficient.

Thus, from the worst to the best permutation, there is a wide choice and we have not yet discovered any perfect permutation. Having said that, good permutations have been defined even so in order to elaborate standardized turbo coding schemes.

There are two ways to specify a permutation, the first by equations linking addresses before and after permutation, the second by a *look-up table* providing the correspondence between addresses. The first is preferable from the point of view of simplicity in the specification of the turbo code (standardization committees are sensitive to this aspect) but the second can lead to better results since the degree of freedom is generally larger when designing the permutation.

Regular permutation

The point of departure when designing interleaving is regular permutation, which is described in Figure 7.5 in two different forms. The first assumes that the block containing k bits can be organized as a table of M rows and N columns. The interleaving then involves writing the data in an *ad hoc* memory, row by row, and reading them column by column (Figure 7.5(a)). The second applies without any hypothesis about the value of k. After writing the data in a linear memory (address i, $0 \leq i \leq k-1$), the block is assimilated to a circle, the two extremities ($i = 0$ and $i = k-1$) then being adjacent (Figure 7.5(b)). The binary data are then extracted in such a way that the j-th datum read has been previously written in position i, with value:

$$i = \Pi(j) = Pj + i_0 \text{ mod } k \tag{7.4}$$

where P is a prime integer with k and i_0 is the index of departure[1].

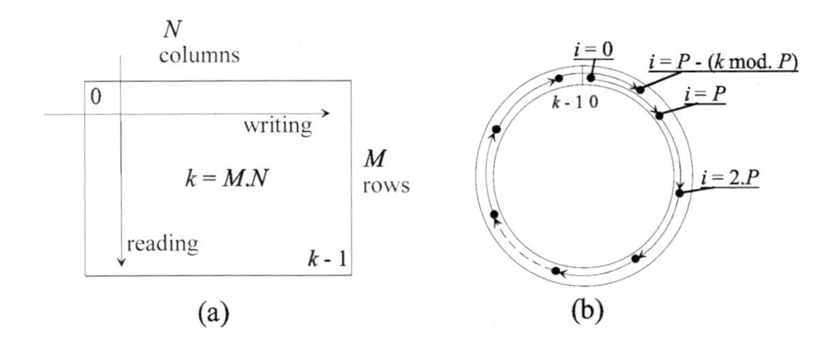

Figure 7.5 – Regular permutation in rectangular (a) and circular (b) form.

For circular permutation, let us define the accumulated spatial distance $S(j_1, j_2)$ as the sum of the two spatial distances separating two bits, before and after permutation, whose reading indices are j_1 and j_2:

$$S(j_1, j_2) = f(j_1, j_2) + f(\Pi(j_1), \Pi(j_2)) \tag{7.5}$$

[1] The permutation can, of course, be defined in a reciprocal form, that is, j function of i. It is a convention that is to be adopted once and for all, and the one that we have chosen is compatible with most standardized turbo codes.

where:

$$f(u,v) = \min\{|u-v|, k-|u-v|\} \tag{7.6}$$

Function f is introduced to take into account the circular nature of the addresses. Finally, we call S_{min} the smallest of the values of $S(j_1, j_2)$, for all the possible pairs j_1 and j_2:

$$S_{\min} = \min_{j_1, j_2}\{S(j_1, j_2)\} \tag{7.7}$$

It is proved in [7.19] that an upper bound of S_{min} is:

$$\sup S_{\min} = \sqrt{2k} \tag{7.8}$$

This upper bound is only reached in the case of a regular permutation and with conditions:

$$P = P_0 = \sqrt{2k} \tag{7.9}$$

and:

$$k = \frac{P_0}{2} \bmod P_0 \tag{7.10}$$

Let us now consider a sequence of any weight that, after permutation, can be written:

$$\tilde{d}(D) = \sum_{j=0}^{k-1} a_j D^j \tag{7.11}$$

where a_j can take the binary value 0 (no error) or 1 (one error) and, before permutation:

$$d(D) = \sum_{i=0}^{k-1} a_i D^i = \sum_{j=0}^{k-1} a_{\Pi(j)} D^{\Pi(j)} \tag{7.12}$$

We denote j_{min} and j_{max} the j indices corresponding to the first and last non-null values a_j in $\tilde{d}(D)$. Similarly, we define i_{min} and i_{max} for sequence $d(D)$. Then, the regular permutation satisfying (7.9) and (7.10) guarantees the property:

$$(j_{\max} - j_{\min}) + (i_{\max} - i_{\min}) > \sqrt{2k} \tag{7.13}$$

This is because $d(D)$ and $\tilde{d}(D)$, both considered between min and max indices, contain at least 2 bits whose accumulated spatial distance, as defined by (7.5), is maximum and equal to $\sqrt{2k}$. We must now consider two cases:

- sequences $d(D)$ and $\tilde{d}(D)$ are both of the *simple* RTZ type, that is, they begin in state 0 of the encoder and return to it once, at the end. The parity bits produced by these sequences are statistically 1s, every other time. Taking into account (7.13), for common values of k ($k > 100$), the redundancy weights are high and these RTZ sequences do not contribute to the MHD of the turbo code.

- at least one of sequences $d(D)$ and $\tilde{d}(D)$ is of the *multiple* RTZ type, that is, it corresponds to the encoder passing several times through state 0. If these passes through state 0 are long, the parity associated with the sequence may have reduced weight and the associated distance may be low. Generally, in this type of situation, the sequences before and after permutation are both multiple RTZ.

The performance of a turbo code, at low error rates, is closely linked with the presence of multiple RTZ patterns and regular permutation is not a good solution for eliminating these patterns.

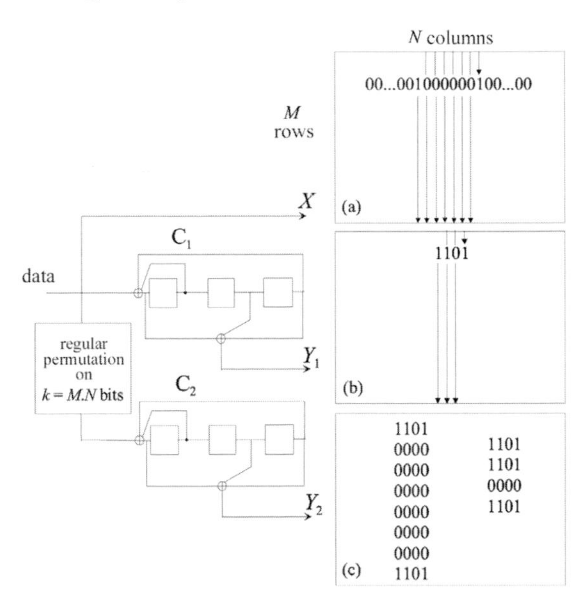

Figure 7.6 – Possible error patterns of weight 2, 3, 6 or 9 with a turbo code whose elementary encoders have a period 7 and with regular permutation.

Necessity for disorder

Again assuming that the error patterns that are not RTZ have weights high enough not to have any effect on performance, an ideal permutation for a turbo code could be defined by the following rule:

If a sequence is of the RTZ type before permutation, then it is no longer so after permutation and vice-versa.

The rule above is impossible to satisfy in practice and a more realistic objective is:

If a sequence is of the RTZ type before permutation, then it is no longer so after permutation or it has become a simple long RTZ sequence and vice-versa.

The dilemma in designing good permutations for turbo codes lies in the need to satisfy this objective for two distinct classes of input sequences that require opposing types of processing: simple RTZ sequences and multiple RTZ sequences, as defined above. To illustrate this problem, consider a rate 1/3 turbo code, with regular rectangular permutation (writing in M rows, lecture in N columns) over blocks of $k = MN$ bits (Figure 7.6). Elementary encoders are encoders with 8 states whose period is 7 (recursivity generator 15).

The first pattern (a) of Figure 7.6 concerns a sequence of possible errors with input weight $w = 2 : 10000001$ for code C_1, that we can also call the horizontal code. This is the RTZ minimum sequence with weight 2 for the encoder considered. The redundancy produced by this encoder is of weight 6 (exactly: 11001111). The redundancy produced by the vertical encoder C_2, for which the sequence considered is also RTZ (its length is a multiple of 7), is much more informative because it is simple RTZ and produced over seven columns. Assuming that Y_2 is equal to 1 every other time on average, the weight of this redundancy is around $w(Y_2) \approx \frac{7M}{2}$. When we make k tend towards infinity via the values of M and N ($M \approx N \approx \sqrt{k}$), the redundancy produced by one of the two codes, for this type of pattern, also tends towards infinity. We then say that the code is *good*.

The second pattern (b) is that of the minimum RTZ sequence of input weight 3. Here again, the redundancy is poor on the first dimension and much more informative on the second. The conclusions are the same as above.

The other two diagrams in (c) present examples of multiple RTZ sequences, made up of short RTZ sequences on each of the two dimensions. The input weights are 6 and 9. The distances associated with these patterns (respectively 30 and 27 for this rate 1/3 code) are not generally sufficient to ensure good performance at low error rates. Moreover, these distances are independent of block size and therefore, in relation to the patterns considered, the code is not *good*.

Regular permutation is therefore a good permutation for the class of simple RTZ error patterns. For multiple RTZ patterns, however, regular permutation is not appropriate. A good permutation must "break" the regularity of rectangular composite patterns like those of Figure 7.6(c), by introducing some disorder. But this must not be done to the detriment of the patterns for which regular permutation is good. The disorder must therefore be managed well! Therein lies the whole problem when looking for a permutation that must lead to a high enough minimum distance. A good permutation cannot be found independently of the properties of elementary codes, of their RTZ patterns, their periodicities, etc.

Intra-symbol disorder

When the elementary codes are m-binary codes, we can introduce a certain disorder into the permutation of a turbo code without however removing its regular nature! To do this, in addition to intersymbol classical permutation, we implement intra-symbol permutation, that is, a non-regular modification of the content of the symbols of m bits, before coding by the second code [7.11]. We briefly develop this idea with the example of double-binary turbo codes ($m = 2$).

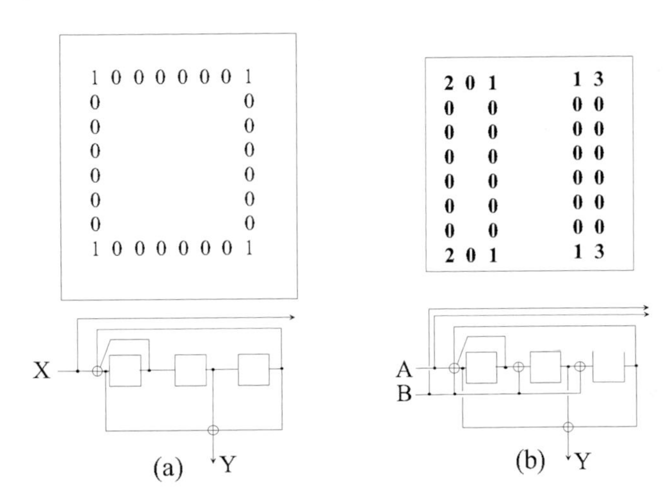

Figure 7.7 – Possible error patterns with binary (a) and double-binary (b) turbo codes and regular permutation.

Figure 7.7(a) presents the minimum pattern of errors with weight $w = 4$, again using the code of Figure 7.6. It is a square pattern whose side is equal to the period of the pseudo-random generator with polynomial 15, that is, 7. It has already been mentioned that some disorder had to be introduced into the permutation to "break" this kind of error pattern but without altering the properties of the regular permutation in relation to patterns with weight 2 and 3, which is not easy. If, as an elementary encoder, we replace the binary encoder by a double-binary encoder, the error patterns to consider are no longer made up of bits but of couples of bits. Figure 7.7(b) gives an example of a double-binary encoder and of possible error patterns, when the permutation is regular. The couples are numbered from **0** to **3**, according to the following correspondence:

$$(0,0) : \mathbf{0}; \quad (0,1) : \mathbf{1}; \quad (1,0) : \mathbf{2}; \quad (1,1) : \mathbf{3}$$

The periodicities of the double-binary encoder are resumed in the diagram of Figure 7.8. There we can find all the combinations of pairs of couples of the RTZ type. For example, if the encoder, initialized in state 0, is fed by the successive couples **1** and **3**, it immediately returns to state 0. It is the same for the sequences **201** or **2003** or **3000001**, for example.

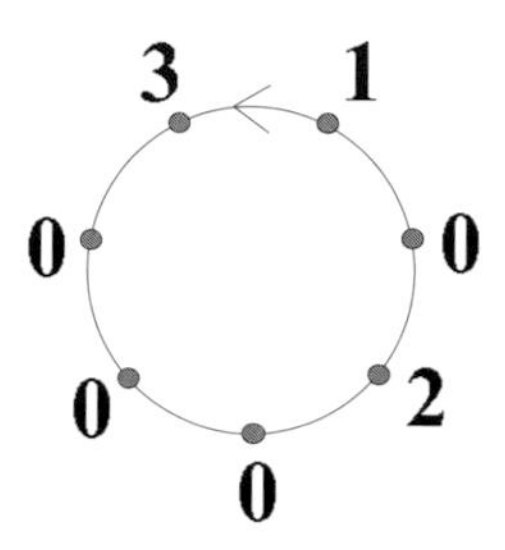

Figure 7.8 – Periodicities of the double-binary encoder of Figure 7.7(b). The four input couples $(0, 0)$, $(0, 1)$, $(1, 0)$ and $(1, 1)$ are denoted 0, 1, 2 and 3, respectively. This diagram gives all the combinations of pairs of couples of the RTZ type.

Figure 7.7(b) gives two examples of rectangular, minimum size error patterns. First note that the perimeter of these patterns is larger than half the perimeter of the square of Figure 7.7(a). Now, for a same coding rate, the redundancy of a double-binary code is twice as dense as that of a binary code. We thus deduce that the distances of the double-binary error patterns will naturally be larger, everything else being equal, than those of binary error patterns. Moreover, there is a simple way to eliminate these elementary patterns.

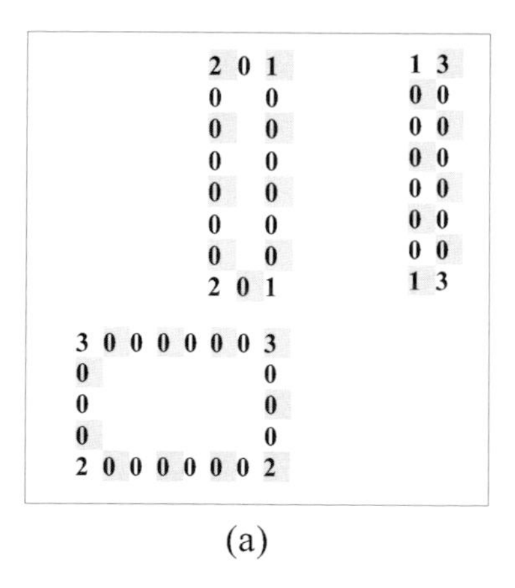

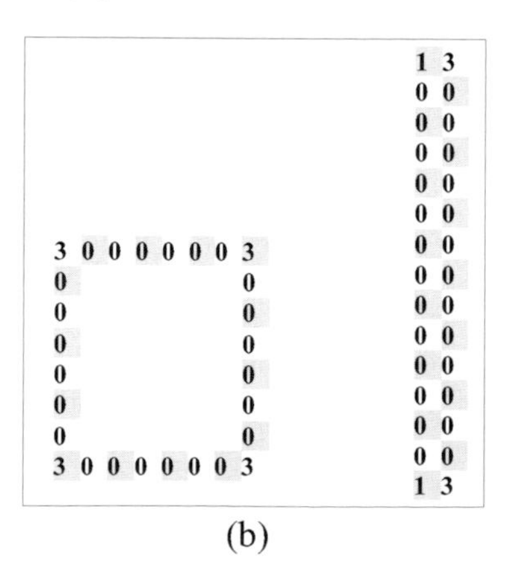

Figure 7.9 – The couples of the grey boxes are inverted before the second (vertical) encoding. 1 becomes 2, 2 becomes 1; 0 and 3 remain unchanged. The patterns of Figure 7.7(b), redrawn in (a), are no longer possible error patterns. But those of (b) are, with distances 24 and 26 for coding rate 1/2.

Assume, for example, that the couples are inverted (**1** becomes **2** and vice-versa), every other time, before being applied to the vertical encoder. Then the error patterns presented in Figure 7.9(a) no longer exist; for example, although **30002** does represent an RTZ sequence for the encoder considered, **30001** no longer does. Thus, many of the error patterns, in particular the smallest, disappear thanks to the disorder introduced inside the symbols. Figure 7.9(b) gives two examples of patterns that the periodic inversion does not modify. The corresponding distances are high enough (24 and 26 for a rate 1/2) not to pose a problem for small or average block sizes. For long blocks (several thousand bits), additional intersymbol disorder, of low intensity, can be added to the intra-symbol non-uniformity, to obtain even higher minimum distances.

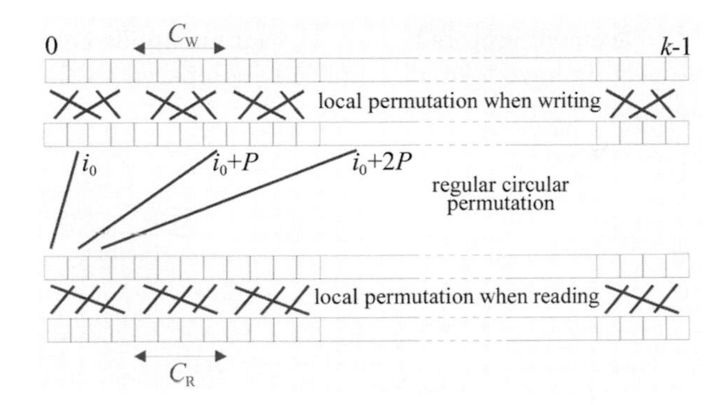

Figure 7.10 – Permutation of the DRP type. This is a circular regular permutation to which local permutations before writing and after reading are added.

Irregular permutations

In this section, we will not describe all the irregular permutations that have been imagined so far, and that have been the subject of numerous publications and several book chapters (see [7.40, 7.34] for example). We prefer to present what seems, for the moment, to be both the simplest and the most efficient type of permutation. These are almost regular circular permutations, called almost regular permutation (ARP)[7.17] or dithered relatively prime (*DRP*) [7.21] permutations, depending on their authors. In all cases, the idea is not to stray too far away from the regular permutation, which is well adapted to simple RTZ error patterns and to instil some small, controlled disorder to counter multiple RTZ error patterns.

Figure 7.10 gives an example, taken from [7.21], of what this small disorder can be. Before the circular regular permutation is performed, the bits undergo local permutation. This permutation is performed in groups of C_W bits. C_W, which is the *writing cycle* disorder, is a divisor of length k of the message. Similarly, a local C_R *reading cycle* permutation is applied before the final reading.

In practice, C_W and C_R can be identical values $C_W = C_R = C$, typically 4 or 8. This way of introducing disorder, in small local fluctuations, does not significantly decrease the accumulated spatial distance, whose maximum value is $\sqrt{2k}$. However, it enables us to suppress the error patterns comparable to those of Figures 7.6(b) and 7.7(c) on condition that the heights and widths of these patterns are not both multiples of C.

Another way to perturb the regular permutation in a controlled way is shown in Figure 7.11. The permutation is represented here in rectangular form, visually more accessible, but it can also be very well applied to circular permutation. One piece of information (bit or symbol) is placed where each row and column cross. With regular permutation, these data are therefore memorized row by row and read column by column. In Figure 7.11, the disorder is introduced by means of four displacement vectors $V_1, \cdots, V_4$ that are applied alternately during reading. These vectors have a small amplitude compared to the dimensions of the permutation matrix.

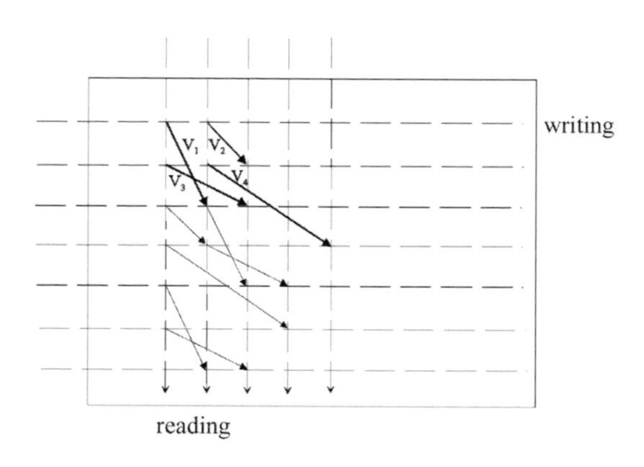

Figure 7.11 – Permutation of the ARP type, following [7.17].

The mathematical model associated with this almost regular permutation, in its circular form, is an extension of (7.4):

$$i = \Pi(j) \equiv Pj + Q(j) + i_0 \bmod k \tag{7.14}$$

If we choose

$$Q(j) = A(j)P + B(j) \tag{7.15}$$

where the positive integers $A(j)$ and $B(j)$ are periodic, with cycle C (divisor of k), then these values correspond to the positive shifts applied respectively before and after regular permutation. That is the difference between the permutation shown in Figure 7.10, in which the writing and reading perturbations are performed inside small groups of data and not by shifts.

For the permutation to really be a bijection, parameters $A(j)$ and $B(j)$ are not just any parameters. To ensure the existence of the permutation, there is one sufficient condition: all the parameters have to be multiples of C. This condition is not very restricting in relation to the efficiency of the permutation. (7.15) can then be rewritten in the form:

$$Q(j) = C(\alpha(j)P + \beta(j)) \tag{7.16}$$

where $\alpha(j)$ and $\beta(j)$ are more often than not small integers, with values 0 to 8. In addition, since the properties of a circular permutation are not modified by a simple rotation, one of the $Q(j)$ values can systematically be 0.

Two typical sets of Q values, with cycle 4 and $\alpha = 0$ or 1, are given below:

$$\begin{array}{ll} \text{if } j = 0 & \text{mod } 4, \text{ then } Q = 0 \\ \text{if } j = 1 & \text{mod } 4, \text{ then } Q = 4P + 4\beta_1 \\ \text{if } j = 2 & \text{mod } 4, \text{ then } Q = 4\beta_2 \\ \text{if } j = 3 & \text{mod } 4, \text{ then } Q = 4P + 4\beta_3 \end{array} \tag{7.17}$$

$$\begin{array}{ll} \text{if } j = 0 & \text{mod } 4, \text{ then } Q = 0 \\ \text{if } j = 1 & \text{mod } 4, \text{ then } Q = 4\beta_1 \\ \text{if } j = 2 & \text{mod } 4, \text{ then } Q = 4P + 4\beta_2 \\ \text{if } j = 3 & \text{mod } 4, \text{ then } Q = 4P + 4\beta_3 \end{array} \tag{7.18}$$

These models require the knowledge of only four parameters (P, β_1, β_2 and β_3), which can be determined using the procedure described in [7.17]. The utilization of m-binary codes (see Section 7.5), instead of binary codes, simply requires k to be replaced by k/m in Equation (7.14). In particular, the permutations defined for double-binary turbo codes ($m = 2$) of the DVB-RCS, DVB-RCT and WiMax standards are inspired by Equations (7.14) and (7.21).

Quadratic Permutation

Recently, Sun and Takeshita [7.41] proposed a new class of deterministic interleavers based on permutation polynomials (PP) over integer rings. The use of PP reduces the design of interleavers to simply a selection of polynomial coefficients. Furthermore, PP-based turbo codes have been shown to have a) good distance properties [7.38] which are desirable for lowering the error floor and b) a maximum contention-free property [7.43] which is desirable for parallel processing to allow high-speed hardware implementation of iterative turbo decoders.

A. PERMUTATION POLYNOMIALS

Before addressing the quadratic PP, we will define the general form of a polynomial and discuss how to verify whether a polynomial is a PP over the ring of integers modulo N, $\mathbb{Z}_N$. Given an integer $N \geq 2$, a polynomial

$$f(x) = a_0 + a_1 + a_2 x2 + \ldots + a_m x_m \text{ modulo } N \quad (7.19)$$

where the coefficients $a_0, a_1, a_2, \ldots, a_m$ and m are non-negative integers, is said to be permutation polynomial over $\mathbb{Z}_N$ when $f(x)$ permutes $\{0, 1, 2, \ldots, N-1\}$ [7.41]. Since we have modulo N operation, it is sufficient for the coefficients $a_0, a_1, a_2, \ldots, a_m$ to be in $\mathbb{Z}_N$. Let us recall that the formal derivative of the polynomial $f(x)$ is given by

$$f'(x) = a_1 + 2a_2 x + 3a_3 x^2 + \ldots + m a_m x^{m-1} \text{ modulo } N \quad (7.20)$$

To verify whether a polynomial is a PP over $\mathbb{Z}_N$, let us discuss the following three cases a) the case $N = 2^n$, where n is an element of the positive integers $\mathbb{Z}_+$, b) the case $N = p^n$ where p is any prime number, and c) the case where N is an arbitrary element of $\mathbb{Z}_+$.

1. Case I ($N = 2^n$): a theorem in [7.36] states that $f(x)$ is a PP over the integer ring $\mathbb{Z}_2^n$ if and only if 1) a_1 is odd, 2) $a_2 + a_4 + a_6 + \ldots$ is even, and 3) $a_3 + a_5 + a_7 + \ldots$ is even.

   Example 1 for $N = 2^3 = 8 : f(x) = 1 + 5x + x^2 + x^3 + x^5 + 3x^6$ is a PP over $\mathbb{Z}_{N=8}$ because it maps the sequence $\{0, 1, 2, 3, 4, 5, 6, 7\}$ to $\{1, 4, 7, 2, 5, 0, 3, 6\}$. Note that $a1 = 5$ is odd, $a_2 + a_4 + a_6 = 1 + 0 + 3 = 4$ is even, and $a_3 + a_5 = 1 + 1 = 2$ is even.

   Example 2 for $N = 2^3 = 8 : f(x) = 1 + 4x + x^2 + x^3 + x^5 + 3x^6$ is not PP over $\mathbb{Z}_{N=8}$ because it maps the sequence $\{0, 1, 2, 3, 4, 5, 6, 7\}$ to $\{1, 3, 5, 7, 1, 3, 5, 7\}$. Note that $a_1 = 4$ is even.

2. Case II ($N = p^n$): a theorem in [7.33] guarantees that $f(x)$ is a PP modulo pn if and only if $f(x)$ is a PP modulo p and $f'(x) \neq 0$ modulo p, for every integer $x \in \mathbb{Z}_p^n$. Note that the Case I is simply a special case of the Case II because $p = 2$ is a prime number.

   Example 1 for $N = 3^n (p = 3) : f(x) = 1+2x+3x^2$ is a PP over $\mathbb{Z}_3^n$ because $f(0) = 1$ modulo $3 = 1, f(1) = 6$ modulo $3 = 0$, $f(2) = 17$ modulo $3 = 2$, and $f'(x) = 2 + 6x = 2$ modulo $3 = 2$ is a non-zero constant for all $xin\mathbb{Z}_3^n$.

   Example 2 for $N = 3^n (p = 3) : f(x) = 1+6x+3x^2$ is not a PP $\mathbb{Z}_3^n$ because $f'(x) = 6+6x = 0$ modulo $3 = 0$ for all $xin\mathbb{Z}_3^n$. For instance, for $N = 3^2 = 9$, $f(x)$ maps the sequence $\{0, 1, 2, 3, 4, 5, 6, 7, 8\}$ to $\{1, 1, 7, 1, 1, 7, 1, 1, 7\}$.

3. Case III (arbitrary N): let $P = \{2, 3, 5, 7, \ldots\}$ be the set of prime numbers. Then, every $N \in \mathbb{Z}_+$ can be factored as $N = \prod_{p \in P} p^{n_{N,p}}$, where all p values are distinct prime numbers, $n_{N,p} \geq 1$ for a certain number of p and $n_{N,p} = 0$ otherwise. For example, if $N = 2500 = 2^2 \times 5^4$, then we have $n_{2500,2} = 2$ and $n_{2500,5} = 4$. A theorem in [7.41] states that for any $N = \prod_{p \in P} p^{n_{N,p}}$, $f(x)$ is a PP modulo N if and only if $f(x)$ is also a PP modulo $p^{n_{N,p}}$,$\forall p$ such that $n_{N,p} \geq 1$. With this theorem, verifying whether a polynomial is a PP modulo N reduces to verifying the polynomial modulo each $p^{n_{N,p}}$ factor of N. For $p = 2$, we use the theorem reported in Case I, which is a simple test on the polynomial coefficients. For $p \neq 2$, we must use the theorem reported in Case II, which cannot be done by simply testing the polynomial coefficients. For an arbitrary N, it is difficult to develop a simple coefficient test to check whether an arbitrary m-degree polynomial $f(x)$ is a PP modulo N. However, for quadratic polynomial ($m = 2$), $f(x) = a_0 + a_1 x + a_2 x^2$, a simple coefficient test have been proposed in [7.39]. Next section will address this coefficient test in details.

B. Quadratic Permutation Polynomials

Since the constant q_0 in the quadratic polynomial $q(x) = q_0 + q_1 x + q_2 x^2$ only causes a "cyclic shif" to the permuted values, we define in this section -without loosing generality- quadratic polynomials as $q(x) = q_1 x + q_2 x^2$. Let us first establish some abbreviations borrowed from [7.43], that we will use throughout this section. We express the fact that b is divisible by a, or a is divisor of b, by $a\,|b$. We also use $a not\,|b$ to express the contrary of $a\,|b$. The greatest common divisor of a and b is denoted by $gcd(a, b)$. Remember that $gcd(a, b) = 1$ indicates that a and b are relatively prime. As we will see in Proposition 1 below, we are mainly interested in the factorization of the coefficient q_2, which can be written according to the previous notation as $q_2 = \prod_{p \in P} p^{n_{q_2,p}}$. The following Proposition 1 provides a necessary and sufficient condition for verifying whether a quadratic polynomial is a PP modulo N.

Proposition 1: Let $N = \prod_{p \in P} p^{n_{N,p}}$. For a quadratic polynomial $q(x) = q_1 x + q_2 x^2$ modulo N to be a PP, the following necessary and sufficient conditions must be satisfied [7.39]

1) Either $2 not\,|N$ or $4\,|N$ (i.e., $n_{N,2} \neq 1$) $gcd(q_1, N) = 1$ and $q_2 = \prod_{p \in P} p^{n_{q_2,p}}$, $n_{q_2,p} \geq 1$, $\forall p$ such that $n_{N,p} \geq 1$.

2) $2\,|N$ or $4not\,|N$ (i.e., $n_{N,2} = 1$), $q_1 + q_2$ is odd, $gcd(q_1, \frac{N}{2}) = 1$ and $q_2 = \prod_{p \in P} p^{n_{q_2,p}}$, $n_{q_2,p} \geq 1$, $\forall p$ such that $p \neq 2$ and $n_{N,p} \geq 1$.

The statement $q_2 = \prod_{p \in P} p^{n_{q_2,p}}$, $n_{q_2,p} \geq 1$, $\forall p$ such that $n_{N,p} \geq 1$ can be expressed in simple words as follows: *each p factor of N must also be a factor of q_2*. It is important to note that this statement still allows q_2 to have prime factors that differ from all p factors of N.

Example 1: if $N = 36$, then we have case 1) of Proposition 1 because $4\,|N$. All possible values of q_1 are simply the set of numbers that are relatively prime to N. Consequently, $q_1 = \{1, 5, 7, 11, 13, 17, 19, 23, 25, 29, 31, 35\}$. Since $36 = 2^2 \times 3^2$ ($p_1 = 2$ and $p_2 = 3$), then 2 and 3 must be a factor of q_2. That is, $q_2 = \{(2 \times 3), (2^2 \times 3), (2^3 \times 3), (2 \times 3^2), 5 \times (2 \times 3)\} = \{6, 12, 24, 18, 30\}$. As mentioned above, the use of the prime number 5 in $5 \times (2 \times 3)$ does not violate the condition in case 1) of Proposition 1. In total, for $N = 36$ there are $12 \times 5 = 60$ possible quadratic PPs (QPP).

The statement $q_2 = \prod_{p \in P} p^{n_{q_2,p}}$, $n_{q_2,p} \geq 1$, $\forall p$ such that $p \neq 2$ and $n_{N,p} \geq 1$ of case 2) can also be expressed in simple words as follows: *each $p6 = 2$ factor of N must also be a factor of q_2*. It is important to note that this statement still allows q_2 to have the prime factor 2 and all prime factors that differ from all p factors of N(q_2 may or may not have 2 as a factor).

Example 2: if $N = 90$, then we have case 2) of Proposition 1 because $2\,|N$ and $4not\,|N$. Since $N = 90 = 2 \times 3^2 \times 5$, all p values that differ from 2 are $p_1 = 3$ and $p_2 = 5$. Therefore, the potential values for q_2 are

$$\{(3 \times 5), (3^2 \times 5), (3 \times 5^2), 2 \times (3 \times 5), 2^2 \times (3 \times 5)\} = \{15, 45, 75, 30, 60\}$$

Under the condition $gcd(q_1, N/2 = 45) = 1$, the potential values for q_1 are have 120 possible QPPs.

$$\{1, 2, 4, 7, 8, 11, 13, 14, 16, 17, 19, 22, 23, 26, 28, 29, 31, 32, 34, 37, 38, 41, 43, 44, 46, \\ 47, 49, 52, 53, 56, 58, 59, 61, 62, 64, 67, 68, 71, 73, 74, 76, 77, 79, 82, 83, 86, 88, 89\}$$

Despite the tight conditions imposed by Proposition 1 on q_1 and q_2, the search space of QPPs is still large, especially for medium to large interleavers. Thus, it is desirable to reduce the search space further. A solution is to consider only QPPs that do have a quadratic inverse (for more details on quadratic inverse for QPP, see [7.39]). This solution for reducing the search space is based on the interesting finding reported by Rosnes and Takeshita [7.38], namely, for $32 \leq N \leq 512$ and $N = 1024$, the class of QPP-based interleavers with quadratic inverses are strictly superior (in term of minimum distance) to the class of QPP-based interleavers with no quadratic inverse. Using exhaustive computer search, Rosnes and Takeshita provided, for turbo codes that use 8 and 16-state constituent codes, a very useful list for the best (in term of minimum distance) QPPs for a wide range of N ($32 \leq N \leq 512$ and $N = 1024$) [7.38]. After discussing a necessary and sufficient condition for verifying whether a quadratic polynomial is a PP modulo N, and providing some examples, let us discuss some properties of QPPs. It is well known that a linear polynomial, $l(x) = l_0 + l_1 x$(or simply $l(x) = l_1 x$), is guaranteed to be a PP modulo N if l_1 is chosen relatively prime to N (i.e., $gcd(l_1, N) = 1$). Consequently, linear permutation polynomials (LPP) always exist for any N, but unfortunately this is not true for QPPs. For example, there are no QPP for $N = 11$ and for $2 \leq N \leq 4096$ there are only 1190 values of N that have QPPs (roughly 29%) [7.44]. A theorem in [7.44] guarantees the existence of QPP for all $N = 8i$, $i \in \mathbb{Z}_+$(i.e., multiples of a typical computer byte size of 8). It is shown in [7.44] that some QPP degenerate to LPP (i.e., there exists an LPP that generates the same permutation over the ring $\mathbb{Z}_N$). A QPP is called reducible if it degenerates to an LPP; otherwise it is called irreducible. For instance, example 1 in Case I of sub-section A could be simply reduced to $f(x) = l + 3x$ modulo 8 to obtain the same permutation. In [7.38], it is shown that some reducible QPPs can achieve better minimum distances than irreducible QPP for some short to medium interleavers. However, for large intereavers, the class of irreducible QPPs are better (in term of minimum distance) than the class of LPP; and if not, that particular length will not have any good minimum distance [7.38].

7.4 Decoding turbo codes

7.4.1 Turbo decoding

Decoding a binary turbo code is based on the schematic diagram of Figure 7.12. The loop allows each decoder to take advantage of all the information available. The values considered at each node of the layout are LLRs, the decoding operations being performed in the logarithmic domain.

The LLR at the output of a decoder of systematic codes can be seen as the sum of two terms: the intrinsic information, coming from the transmission channel, and the extrinsic information, which this decoder adds to the former to perform its correction operation. As the intrinsic information is used by the two decoders (at different instants), it is the extrinsic information produced by each of the decoders that must be transmitted to the other as new information, to ensure joint convergence. Section 7.4.2 details the operations performed to calculate the extrinsic information, by implementing the MAP algorithm or its simplified Max-Log-MAP version.

Because of latency effects, the exchange of extrinsic information, in a digital processing circuit, must be implemented via an iterative process: first decoding

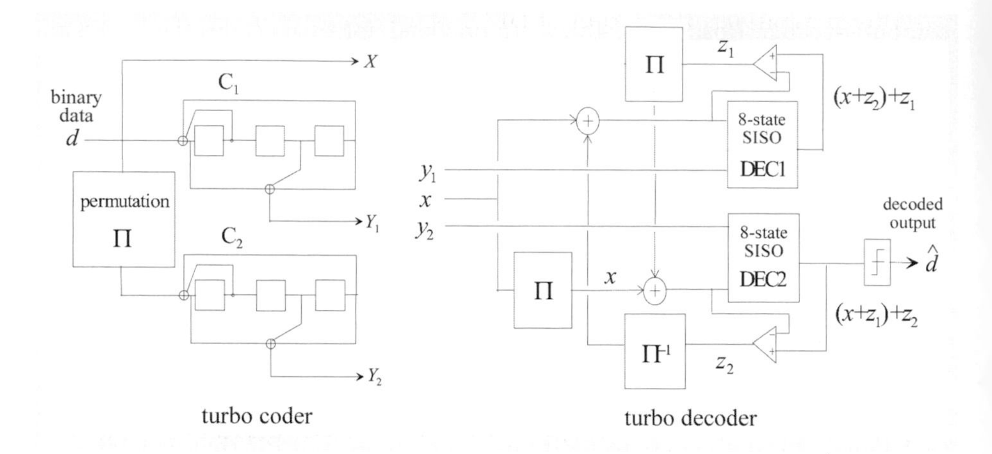

Figure 7.12 – 8-state turbo encoder and schematic structure of the corresponding turbo decoder. The two elementary decoders exchange probabilistic information, called extrinsic information (z)

by DEC_1 and putting extrinsic information z_1 in the memory, second decoding by DEC_2 and putting extrinsic information z_2 in the memory (end of the first iteration), again using DEC_1 and putting z_1 in the memory, etc. Different hardware architectures, with more or less great degrees of parallelism, can be envisaged to accelerate the iterative decoding.

If we wanted to decode the turbo code using a single decoder, which would take into account all the possible states of the encoder, for each element of the message decoded, we would obtain one and only one probability of having a binary value equal to 0 or to 1. As for the composite structure of Figure 7.12, it uses two decoders working jointly. By analogy with the result that the single decoder would provide, they therefore need to *converge towards the same decisions, with the same probabilities*, for each of the data considered. That is the fundamental principle of "turbo" processing, which justifies the structure of the decoder, as the following reasoning shows.

The role of a SISO decoder (see Section 7.4.2), is to process the LLRs at its input to try to make them more reliable, thanks to local redundancy (that is, y_1 for DEC1, y_2 for DEC2). The LLR produced by a decoder of binary codes, relative to data d, can be written simply as

$$\mathrm{LLR}_{\mathrm{output}}(d) = \mathrm{LLR}_{\mathrm{input}(d)+z(d)} \tag{7.21}$$

where $z(d)$ is the extrinsic information specific to d. The LLR is improved when z is negative and d is a 0, or when z is positive and d is a 1.

After p iterations, the output of DEC1 is:

$$\mathrm{LLR}^p_{\mathrm{output},1}(d) = (x + z_2^{p-1}(d)) + z_1^p(d)$$

and the output of DEC2 is

$$\mathrm{LLR}^p_{\mathrm{output},2}(d) = (x + z_1^{p-1}(d)) + z_2^p(d)$$

If the iterative process converges towards a stable solution, $z_1^p(d) - z_1^{p-1}(d)$ and $z_2^p(d) - z_2^{p-1}(d)$ tend towards zero when p tends towards infinity. Consequently, the two LLRs relative to d become identical, thus satisfying the basic criterion of common probability mentioned above. As for proof of the convergence, it is still being studied further and on this topic we can, for example, consult [7.49, 7.24].

Apart from the permutation and inverse permutation functions, Figure 7.13 details the operations performed during turbo decoding:

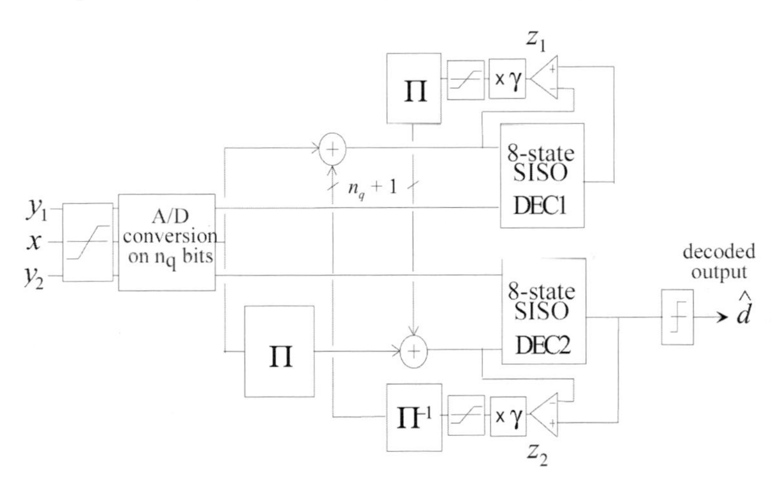

Figure 7.13 – Operations shown (clipping, quantization, attenuation of the extrinsic information) in the turbo decoder of Figure 7.12.

1. Analogue to digital (A/D) conversion transforms the data coming from the demodulator into samples exploitable by the digital decoder. Two parameters are involved in this operation: n_q, the number of quantization bits, and Q, the scale factor, that is, the ratio between the average absolute value of the quantized signal and its maximum absolute value. n_q is fixed to a compromise value between the precision required and the complexity of the decoder. With $n_q = 4$, the performance of the decoder is very close to what we obtain with real samples. The value of Q depends on the modulation, on the coding rate and on the type of channel. It is, for example, larger for a Rayleigh channel than for a Gaussian channel.

2. The role of SISO decoding is to increase the equivalent signal to noise ratio of the LLR, that is, to provide more reliable extrinsic information at output z_{output} than at input (z_{input}). The convergence of the iterative process (see Section 7.6) will depend on the transfer function $\mathrm{SNR}(z_{output}) = G(\mathrm{SNR}(z_{input}))$ of each of the decoders.

   When data is not available at the input of the SISO decoder, due to puncturing for example, a neutral value (analogue zero) is substituted for this missing data.

3. When the elementary decoding algorithm is not the optimal MAP algorithm but a sub-optimal simplified version, the extrinsic information has to undergo some transformations before being used by a decoder:

   - multiplying the extrinsic information by factor γ, lower than 1, guarantees the stability of the looped structure. γ can vary over the iterations, for example, from 0.7 at the beginning of the iterative process, to 1 for the last iteration.
   - clipping the extrinsic information solves both the issue of limiting the size of the memories and that of participating in the stability of the process. A typical value of the maximum dynamics of the extrinsic information is twice the input dynamics of the decoder.

4. Binary decision taking is performed by thresholding at value 0.

   The number of iterations required by turbo decoding depends on the size of the block and on the coding rate. Generally, the larger the decoded block and the slower the convergence, the higher the MHD of the code. The same occurs when the coding rates are low. In practice, we limit the number of iterations to a value between 4 and 10, according to the speed, latency and consumption constraints imposed by the application.

Figure 7.14 gives an example of the performance of a binary turbo code, taken from the UMTS standard [7.3]. We can observe a decrease in packet error rates (PER), just close to the theoretical limit (that is, around 0,5 dB, taking into account the size of the block), but also a fairly pronounced change in slope, due to an MHD that is not very high ($d_{\mathrm{min}} = 26$) for a rate of 1/3.

7.4.2 SISO decoding and extrinsic information

Here we present processing performed in practice in a SISO decoder using the MAP algorithm [7.6] or its simplified version, the Max-log-MAP algorithm, also called the SubMAP algorithm [7.37], to decode RSC m-binary codes and implement iterative decoding. For binary codes and turbo codes, all these equations can be simplified by taking $m = 1$.

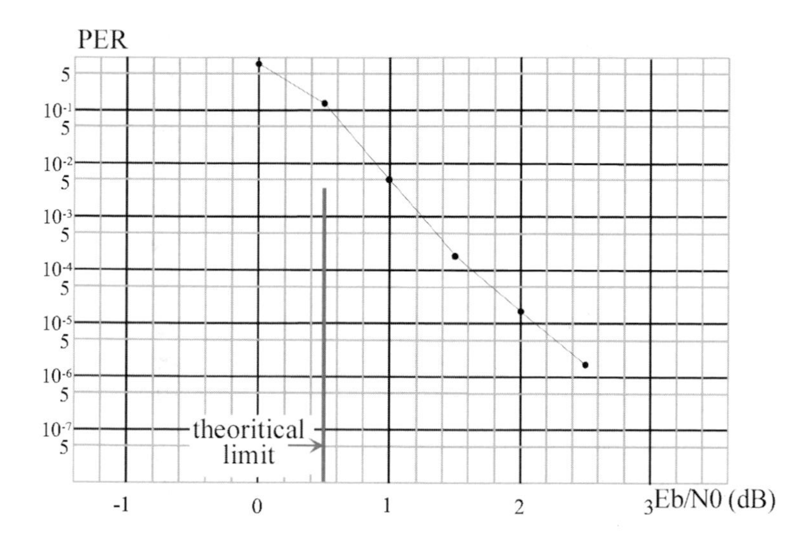

Figure 7.14 – Performance in packet error rates (PER) of the UMTS standard turbo code for $k = 640$ and $R = 1/3$ on a Gaussian channel with 4-PSK modulation. Decoding using the Max-Log-MAP algorithm with 6 iterations.

Notations

A sequence of data $\mathbf{d}$ is defined by $\mathbf{d} \equiv \mathbf{d}_0^{k-1} = (\mathbf{d}_0 \cdots \mathbf{d}_i \cdots \mathbf{d}_{k-1})$, where $\mathbf{d}_i$ is the vector of m-binary data applied at the input of the encoder at instant i: $\mathbf{d}_i = (d_{i,1} \cdots d_{i,l} \cdots d_{i,m})$. The value of $\mathbf{d}_i$ can also be represented by the scalar integer value $j = \sum_{l=1}^{m} 2^{l-1} d_{i,l}$, ranging between 0 and $2^m - 1$ and we can then write $\mathbf{d}_i \equiv j$.

In the case of two or four-phase PSK modulation (2-PSK, 4-PSK), the encoded modulated sequence $\mathbf{u} \equiv \mathbf{u}_0^{k-1} = (\mathbf{u}_0 \cdots \mathbf{u}_i \cdots \mathbf{u}_{k-1})$ is made up of vectors $\mathbf{u}_i$ of size $m + m'$: $\mathbf{u}_i = (u_{i,1} \cdots u_{i,l} \cdots u_{i,m+m'})$, where $u_{i,l} = \pm 1$ for $l = 1 \cdots m + m'$ and m' is the number of redundancy bits added to the m bits of information. The symbol $u_{i,l}$ is therefore representative of a systematic bit for $l \leq m$ and of a redundancy bit for $l > m$.

The sequence observed at the output of the demodulator is denoted $\mathbf{v} \equiv \mathbf{v}_0^{k-1} = (\mathbf{v}_0 \cdots \mathbf{v}_i \cdots \mathbf{v}_{k-1})$, with $\mathbf{v}_i = (v_{i,1} \cdots v_{i,l} \cdots v_{i,m+m'})$. The series of the states of the encoder between instants 0 and k is denoted $\mathbf{S} = \mathbf{S}_0^k = (\mathbf{S}_0 \cdots \mathbf{S}_i \cdots \mathbf{S}_k)$. The following is based on the results presented in the chapter on convolutional codes.

Decoding following the Maximum *A Posteriori* (MAP) criterion

At each instant i, the weighted (probabilistic) estimates provided by the MAP decoder are the 2^m *a posteriori* probabilities (APP) $\Pr(\mathbf{d}_i \equiv j \mid \mathbf{v})$, $j = 0 \cdots 2^m - 1$. The corresponding hard decision, $\hat{\mathbf{d}}_i$, is the binary representation of value j that maximizes the APP.

Each APP can be expressed as a function of the joint likelihoods $p(\mathbf{d}_i \equiv j, \mathbf{v})$:

$$\Pr(\mathbf{d}_i \equiv j \mid \mathbf{v}) = \frac{p(\mathbf{d}_i \equiv j, \mathbf{v})}{p(\mathbf{v})} = \frac{p(\mathbf{d}_i \equiv j, \mathbf{v})}{\sum\limits_{l=0}^{2^m-1} p(\mathbf{d}_i \equiv l, \mathbf{v})} \tag{7.22}$$

In practice, we calculate the joint likelihoods $p(\mathbf{d}_i \equiv j, \mathbf{v})$ for $j = 0 \cdots 2^m - 1$ then each APP is obtained by normalization.

The trellis representative of a code with memory ν has 2^ν states, taking their scalar value s in $(0, 2^\nu - 1)$. The joint likelihoods are calculated from the recurrent *forward* $\alpha_i(s)$ and *backward* probabilities $\beta_i(s)$ and the branch likelihoods $g_i(s', s)$:

$$p(\mathbf{d}_i \equiv j, \mathbf{v}) = \sum_{(s',s)/\mathbf{d}_i(s',s)\equiv j} \beta_{i+1}(s)\, \alpha_i(s')\, g_i(s', s) \tag{7.23}$$

where $(s', s)/\mathbf{d}_i(s', s) \equiv j$ denotes the set of transitions from state to state $s' \rightarrow s$ associated with the m-binary j. This set is, of course, always the same in a trellis that is invariant over time.

The value $g_i(s', s)$ is expressed as:

$$g_i(s', s) = \Pr{}^a(\mathbf{d}_i \equiv j, \mathbf{d}_i(s', s) \equiv j).p(\mathbf{v}_i \mid \mathbf{u}_i) \tag{7.24}$$

where $\mathbf{u}_i$ is the set of systematic and redundant information symbols associated with transition $s' \rightarrow s$ of the trellis at instant i and $\Pr^a(\mathbf{d}_i \equiv j, \mathbf{d}_i(s', s) \equiv j)$ is the *a priori* probability of transmitting the m-tuple of information and that this would correspond to transition $s' \rightarrow s$ at instant i. If transition $s' \rightarrow s$ does not exist in the trellis for $\mathbf{d}_i \equiv j$, then $\Pr^a(\mathbf{d}_i \equiv j, \mathbf{d}_i(s', s) \equiv j) = 0$, otherwise the transition is given by the source statistics (usually uniform, in practice).

In the case of a Gaussian channel with binary inputs, value $p(\mathbf{v}_i \mid \mathbf{u}_i)$ can be written:

$$p(\mathbf{v}_i \mid \mathbf{u}_i) = \prod_{l=1}^{m+m'} \left(\frac{1}{\sigma\sqrt{2\pi}} \exp\left(-\frac{(v_{i,l} - u_{i,l})^2}{2\sigma^2} \right) \right) \tag{7.25}$$

where σ^2 is the variance of the additive white Gaussian noise. In practice, we keep only the terms that are specific to the transition considered and that are not eliminated by division in the expression (7.22):

$$p'(\mathbf{v}_i \mid \mathbf{u}_i) = \exp\left(\frac{\sum_{l=1}^{m+m'} v_{i,l} \cdot u_{i,l}}{\sigma^2}\right) \tag{7.26}$$

The forward and backward recurrent probabilities are calculated as follows:

$$\alpha_i(s) = \sum_{s'=0}^{2^\nu-1} \alpha_{i-1}(s')\, g_{i-1}(s', s) \quad \text{for } i = 1 \cdots k \tag{7.27}$$

and:

$$\beta_i(s) = \sum_{s'=0}^{2^\nu-1} \beta_{i+1}(s')\, g_i(s, s') \quad \text{for } i = k-1 \cdots 0 \tag{7.28}$$

To avoid problems of precision or of overflow in the representation of these values, they have to be normalized regularly. The initialization of the recursions depends on the knowledge or not of the state of the encoder at the beginning and at the end of encoding. If the initial state $\mathbf{S}_0$ of the encoder is known, then $\alpha_0(\mathbf{S}_0) = 1$ and $\alpha_0(s) = 0$ for any other state, otherwise all the $\alpha_0(s)$ are initialized to the same value. The same rule is applied for the final state $\mathbf{S}_k$. For circular codes, initialization is performed automatically after the prologue step, which starts from identical values for all the states of the trellis.

In the context of iterative decoding, the composite decoder uses two elementary decoders exchanging *extrinsic probabilities*. Consequently, the basic decoding brick described above must be reconsidered in order to:

1. take into account an extrinsic probability, $\Pr^{ex}(\mathbf{d}_i \equiv j \,|\, \mathbf{v}')$, in expression (7.24), calculated by the other elementary decoder of the composite decoder, from its own input sequence $\mathbf{v}'$,
2. produce its own extrinsic probability $\Pr^{ex}(\mathbf{d}_i \equiv j \,|\, \mathbf{v}')$ that will be used by the other elementary decoder.

In practice, for each value of j, $j = 0 \cdots 2^m - 1$:

1. in expression (7.24), the *a priori* probability $\Pr^a(\mathbf{d}_i \equiv j, \mathbf{d}_i(s', s) \equiv j)$ is replaced by the modified *a priori* probability $\Pr^{@}(\mathbf{d}_i \equiv j, \mathbf{d}_i(s', s) \equiv j)$, having for its expression, to within one normalization factor:

$$\Pr^{@}(\mathbf{d}_i \equiv j, \mathbf{d}_i(s', s) \equiv j) = \Pr^a(\mathbf{d}_i \equiv j, \mathbf{d}_i(s', s) \equiv j).\Pr^{ex}(\mathbf{d}_i \equiv j \mid \mathbf{v}') \tag{7.29}$$

1. $\Pr^{ex}(\mathbf{d}_i \equiv j \mid \mathbf{v})$ is given by:

$$\Pr^{ex}(\mathbf{d}_i \equiv j \mid \mathbf{v}) = \frac{\sum\limits_{(s',s)/\mathbf{d}_i(s',s)\equiv j} \beta_{i+1}(s)\,\alpha_i(s')\,g_i^*(s',s)}{\sum\limits_{(s',s)} \beta_{i+1}(s)\,\alpha_i(s')\,g_i^*(s',s)} \tag{7.30}$$

The terms $g_i^*(s',s)$ are non-zero if $s' \to s$ corresponds to a transition of the trellis and are then inferred from the expression of $p(\mathbf{v}_i \mid \mathbf{u}_i)$ by eliminating the systematic part of the information. In the case of a transmission over a Gaussian channel with binary inputs and starting from the simplified expression (7.26) of $p'(\mathbf{v}_i \mid \mathbf{u}_i)$, we have:

$$g_i^*(s',s) = \exp\left(\frac{\sum\limits_{l=m+1}^{m+m'} v_{i,l}u_{i,l}}{\sigma^2}\right) \tag{7.31}$$

The simplified Max-Log-MAP or SubMAP algorithm

Decoding following the MAP criterion requires a large number of operations, including calculating exponentials and multiplications. Re-writing the decoding algorithm in the logarithmic domain simplifies the processing. The weighted estimations provided by the decoder are then values proportional to the logarithms of the APPs, called Log-APP logarithms, denoted L:

$$L_i(j) = -\frac{\sigma^2}{2} \ln \Pr(\mathbf{d}_i \equiv j \mid \mathbf{v}), \quad j = 0 \cdots 2^m - 1 \tag{7.32}$$

We define $M_i^{\alpha}(s)$ and $M_i^{\beta}(s)$ the forward and backward metrics relative to node s at instant i, and $M_i(s',s)$, the branch metric relative to the $s' \to s$ transition of the trellis at instant i by:

$$\begin{aligned} M_i^{\alpha}(s) &= -\sigma^2 \ln \alpha_i(s) \\ M_i^{\beta}(s) &= -\sigma^2 \ln \beta_i(s) \\ M_i(s',s) &= -\sigma^2 \ln g_i(s',s) \end{aligned} \tag{7.33}$$

Introduce values $A_i(j)$ and B_i calculated as:

$$A_i(j) = -\sigma^2 \ln\left[\sum_{(s',s)/\mathbf{d}_i(s',s)\equiv j} \beta_{i+1}(s)\alpha_i(s')g_i(s',s)\right] \tag{7.34}$$

$$B_i = -\sigma^2 \ln\left[\sum_{(s',s)} \beta_{i+1}(s)\alpha_i(s')g_i(s',s)\right] \tag{7.35}$$

$L_i(j)$ can then be written, by reference to (7.22) and (7.23), as follows:

$$L_i(j) = \frac{1}{2}\left(A_i(j) - B_i\right) \tag{7.36}$$

Expressions (7.34) and (7.35) can be simplified by applying the so-called Max-Log approximation:

$$\ln(\exp(a) + \exp(b)) \approx \max(a, b) \tag{7.37}$$

For $A_i(j)$ we get:

$$A_i(j) \approx \min_{(s',s)/\mathbf{d}_i(s',s)\equiv j} \left(M_{i+1}^{\beta}(s) + M_i^{\alpha}(s') + M_i(s', s)\right) \tag{7.38}$$

and for B_i:

$$B_i \approx \min_{(s',s)} \left(M_{i+1}^{\beta}(s) + M_i^{\alpha}(s') + M_i(s', s)\right) = \min_{l=0\cdots 2^m-1} A_i(l) \tag{7.39}$$

and finally we get:

$$L_i(j) = \frac{1}{2}\left(A_i(j) - \min_{l=0\cdots 2^m-1} A_i(l)\right) \tag{7.40}$$

Note that these values are always positive or equal to zero.

Introduce the values L^a proportional to the logarithms of the *a priori* probabilities $\Pr^a$:

$$L_i^a(j) = -\frac{\sigma^2}{2} \ln \Pr^a(\mathbf{d}_i \equiv j) \tag{7.41}$$

Branch metrics $M_i(s', s)$ can be written, according to (7.24) and (7.33):

$$M_i(s', s) = 2L_i^a(\mathbf{d}(s', s)) - \sigma^2 \ln p(\mathbf{v}_i \mid \mathbf{u}_i) \tag{7.42}$$

If the statistic of the *a priori* transmission of the m-tuples $\mathbf{d}_i$ is uniform, term $2L_i^a(\mathbf{d}(s', s))$ can be omitted from the above relation since it is the same value that is used in all the branch metrics. In the case of a transmission over a Gaussian channel with binary inputs, we have according to (7.26):

$$M_i(s', s) = 2L_i^a(\mathbf{d}(s', s)) - \sum_{l=1}^{m+m'} v_{i,l} \cdot u_{i,l} \tag{7.43}$$

The forward and backward metrics are then calculated from the following recurrence relations:

$$M_i^{\alpha}(s) = \min_{s'=0\cdots 2^{\nu}-1} \left(M_{i-1}^{\alpha}(s') - \sum_{l=1}^{m+m'} v_{i-1,l} \cdot u_{i-1,l} + 2L_{i-1}^a(\mathbf{d}(s', s))\right) \tag{7.44}$$

$$M_i^\beta(s) = \min_{s'=0\cdots 2^\nu-1}\left(M_{i+1}^\beta(s') - \sum_{l=1}^{m+m'} v_{i,l}\cdot u_{i,l} + 2L_i^a(\mathbf{d}(s,s'))\right) \tag{7.45}$$

Applying the Max-Log-MAP logarithm in fact amounts to performing two Viterbi decodings, in the forward and backward directions. That is the reason why it is also called the *dual Viterbi* algorithm.

If the initial state of the encoder, $\mathbf{S}_0$, is known, then $M_0^\alpha(\mathbf{S}_0) = 0$ and $M_0^\alpha(s) = +\infty$ for any other state, otherwise all the $M_0^\alpha(s)$ are initialized to the same value. The same rule is applied for the final state. For circular codes, all the metrics are initialized to the same value at the beginning of the prologue.

Finally, taking into account (7.38) and replacing $M_i(s', s)$ by its expression (7.43), we obtain:

$$A_i(j) = \min_{(s',s)/\mathbf{d}_i(s',s)\equiv j}\left(M_{i+1}^\beta(s) + M_i^\alpha(s') - \sum_{l=1}^{m+m'} v_{i,l}\cdot u_{i,l}\right) + 2L_i^a(j) \tag{7.46}$$

The hard decision taken by the decoder is the value of j, $j = 0\cdots 2^m - 1$, which minimizes $A_i(j)$. Let us denote this value j_0. According to (7.40), $L_i(j)$ can be written:

$$L_i(j) = \frac{1}{2}\left[A_i(j) - A_i(j_0)\right] \text{ pour } j = 0\cdots 2^m - 1 \tag{7.47}$$

We note that the presence of coefficient σ^2 in definition (7.32) of $L_i(j)$ allows us to ignore the knowledge of this parameter for computing the metrics and hence for all the decoding. This is an important advantage of the Max-Log-MAP method over the MAP method.

In the context of iterative decoding, term $L_i^a(j)$ is modified in order to take into account extrinsic information $L_i^*(j)$ coming from the other elementary decoder:

$$L_i^@(j) = L_i^a(j) + L_i^*(j) \tag{7.48}$$

On the other hand, the extrinsic information produced by the decoder is obtained by eliminating in $L_i(j)$ the terms containing the direct information about $\mathbf{d}_i$, that is, the intrinsic and *a priori* information:

$$\begin{aligned} L_i^*(j) = \tfrac{1}{2}\Bigg[& \min_{(s',s)/\mathbf{d}_i(s',s)\equiv j}\left(M_{i+1}^\beta(s) + M_i^\alpha(s') - \sum_{l=m+1}^{m+m'} v_{i,l}\cdot u_{i,l}\right) \\ & - \min_{(s',s)/\mathbf{d}_i(s',s)\equiv j_0}\left(M_{i+1}^\beta(s) + M_i^\alpha(s') - \sum_{l=m+1}^{m+m'} v_{i,l}\cdot u_{i,l}\right)\Bigg] \end{aligned} \tag{7.49}$$

The expression of $L_i(j)$ can then be formulated as follows:

$$L_i(j) = L_i^*(j) + \frac{1}{2}\sum_{l=1}^{m} v_{i,l}\cdot\left[u_{i,l}|_{\mathbf{d}_i\equiv j} - u_{i,l}|_{\mathbf{d}_i\equiv j_0}\right] + \left[L_i^@(j) - L_i^@(j_0)\right] \tag{7.50}$$

This expression shows that extrinsic information $L_i^*(j)$ can, in practice, be deduced from $L_i(j)$ by simple subtraction. Factor $\frac{1}{2}$ in definition (7.32) of $L_i(j)$ allows us to obtain a weighted decision and extrinsic information $L_i^*(j)$ on the same scale as the noisy samples $v_{i,l}$.

7.4.3 Practical considerations

The simplest way to perform turbo decoding is totally sequential and uses the following operations, here founded on the Max-Log-MAP algorithm and repeated as many times as necessary:

1. Backward recursion for code C_2 (Figure 7.12), calculation and memorization of metrics $M_i^\beta(s)$, $i = k-1, ..., 0$ and $s = 0, ..., 2^\nu - 1$,
2. Forward recursion for code C_2, calculation of metrics $M_i^\alpha(s)$, $i = 0, ..., k-1$ and $s = 0, ..., 2^\nu - 1$. Calculation and memorization of the extrinsic information,
3. Backward recursion for code C_1, calculation and memorization of metrics $M_i^\beta(s)$, $i = k-1, ..., 0$ and $s = 0, ..., 2^\nu - 1$,
4. Forward recursion for code C_1, calculation of metrics $M_i^\alpha(s)$, $i = 0, ..., k-1$ and $s = 0, ..., 2^\nu - 1$. Calculation and memorization of the extrinsic information. Binary decisions (at the last iteration).

The first practical problem lies in the memory necessary to store metrics $M_i^\beta(s)$. Processing the coded messages of $k = 1000$ bits, for example, with 8-state decoders and quantization of the metrics on 6 bits, at first sight requires a storage capacity of 48000 bits for each decoder. In sequential operation (alternate processing of C_1 and C_2), this memory can, of course, be used by the two decoders in turn. The technique used to greatly reduce this memory is that of the *sliding window*. It involves (Figure 7.15) replacing all the backward processing, from $i = k-1$ to 0, by a succession of partial forward processings, from $i = i_F$ to 0, then from $i = 2i_F$ to i_F, from $i = 3i_F$ to $2i_F$ etc., where i_F is an interval of some tens of trellis sections. Each partial backward processing includes a "prologue" (dotted line), that is, a step without memorization whose aim is to estimate as correctly as possible the accumulated backward metrics in positions i_F, $2i_F$, $3i_F$, etc. The parts shown by a solid line correspond to the phases during which these metrics are memorized. The same memory can be used for all the partial backward recursions. The forward recursion is performed without any discontinuity.

The process greatly reduces the storage capacity necessary which, in addition, becomes independent of the length of the messages. The drawback lies in the necessity to perform the additional operations – the prologues – that can increase the total calculation complexity by 10 to 20 %. However, these prologues can be avoided after the first iteration if the estimates of the metrics at the boundary indices are put into memory to be used as departure points for the calculations of the following iteration.

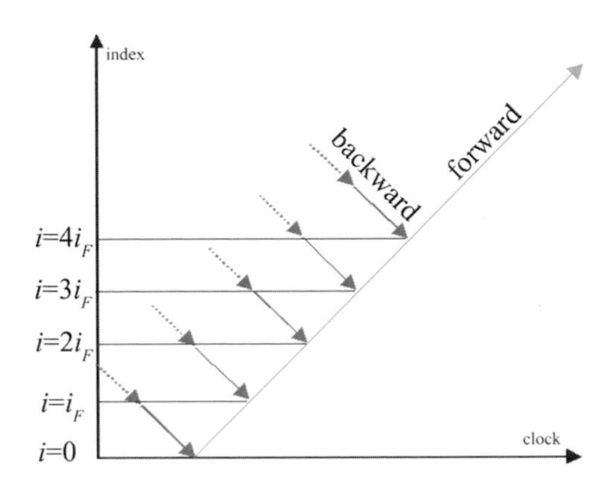

Figure 7.15 – Operation of the forward and backward recursions when implementing the MAP algorithm with a sliding window.

The second practical problem is that of the speed and latency of decoding. The extent of the problem depends of course on the application and on the ratio between the decoding circuit clock and the data rate. If the latter is very high, the operations can be performed by a single machine, in the sequential order presented above. In specialized processors of the DSP (*digital signal processor*) type, cabled co-processors may be available to accelerate the decoding. In dedicated circuits of the ASIC (*application-specific integrated circuit*) type, acceleration of the decoding is obtained by using parallelism, that is, multiplying the number of arithmetical operators, if possible without increasing the capacity of the memories required to the same extent. Then, problems of access to these memories are generally posed.

Note first that only knowledge of permutation $i = \Pi(j)$ is necessary for implementation of the iterative decoding and not that of inverse permutation Π^{-1}, as could be wrongly assumed from the schematic diagrams of Figures 7.12 and 7.13. Consider, for example, two SISO decoders working in parallel to decode the two elementary codes of the turbo code and based on two dual-port memories for the extrinsic information (Figure 7.16). The DEC1 decoder associated with the first code produces and receives the extrinsic information in the natural order i. The DEC2 decoder associated with the second code works according to index j but writes and recovers its data at addresses $i = \Pi(j)$. Knowledge of Π^{-1}, which could pose a problem depending on the permutation model selected, is therefore not required.

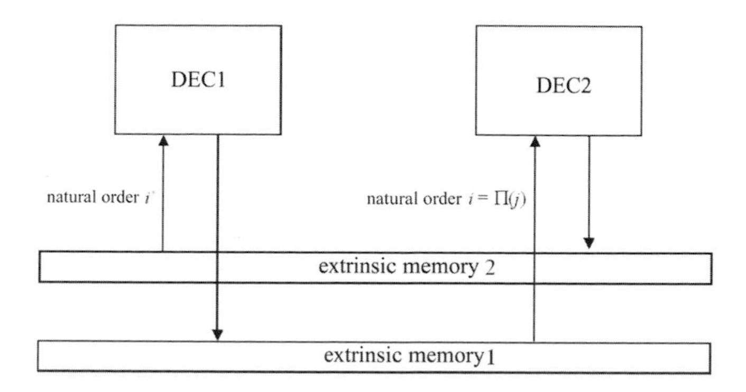

Figure 7.16 – Implementing turbo decoding does not require explicit knowledge of Π^{-1}

In addition, the two extrinsic information memories can be merged into a single one, observing that extrinsic information that has just been read and exploited by a decoder no longer has to be retained. It can thus be replaced immediately afterwards by another datum, which can be the extrinsic information output from the same decoder. Figure 7.17 illustrates this process, which imposes a slight hypothesis: working indices i and j have the same parity and permutation $i = \Pi(j)$ inverses the parity. For example, with the permutation defined by (7.4), this hypothesis is satisfied if the departure index i_0 is odd and the length of the message k is even.

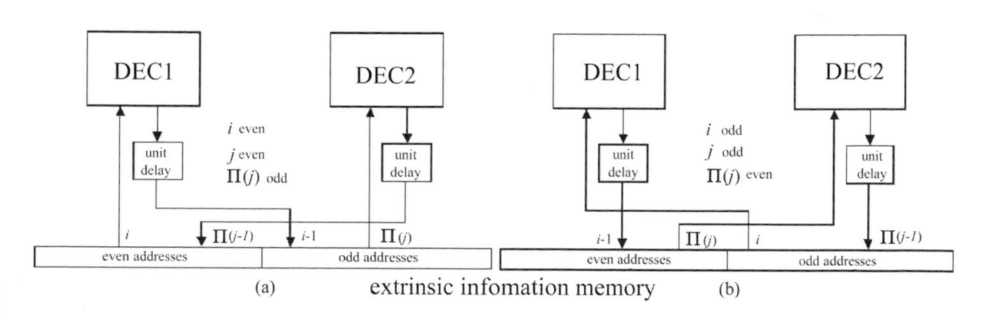

Figure 7.17 – In practice, the storage of the extrinsic information uses only a single memory.

The extrinsic information memory is divided in two pages corresponding to the sub-sets of the even and odd addresses. Access to these two pages, in a dual-port memory, is alternated regularly. In Figure 7.17(a), in the even cycles, DEC1 reads at an even address and writes the extrinsic information produced during the previous cycle, via a buffer memory with unit delay, to an odd address. Meanwhile, DEC2 reads at an even address and writes to an odd address. In

Figure 7.17(b), during the odd cycles, the accesses to the reading-writing pages are exchanged.

To further increase the degree of parallelism in the iterative decoder, the forward and backward recursion operations can also be tackled inside each of the two decoders (DEC1 and DEC2). This can be easily implemented by considering the diagram of Figure 7.15.

Finally, depending on the permutation model used, the number of elementary decoders can be increased beyond two. Consider for example the circular permutation defined by (7.14) and (7.16), with cycle $C = 4$ and k a multiple of 4.

The congruences of j and $\Pi(j)$ modulo 4, are periodic. Parallelism with degree 4 is then possible following the principle described in Figure 7.18 [7.17]. For each forward or backward recursion (these also can be done in parallel), four processors are used. At the same instant, these processors process data whose addresses have different congruences modulo 4. In the example in the figure, the forward recursion is considered and we assume that $k/4$ is also a multiple of 4. Then, we have first processor begin at address 0, the second at address $k/4+1$, the third at address $k/2 + 2$ and finally the fourth at address $3k/4+3$. At each instant, as the processors advance by one place each time, the congruences modulo 4 of the addresses are always different. Addressing conflicts are avoided via a router that directs the four processors towards four memory pages corresponding to the four possible congruences. If $k/4$ is not a multiple of 4, the departure addresses are no longer exactly 0, $k/4+1$, $k/2+2$, $3k/4+3$ but the process is still applicable.

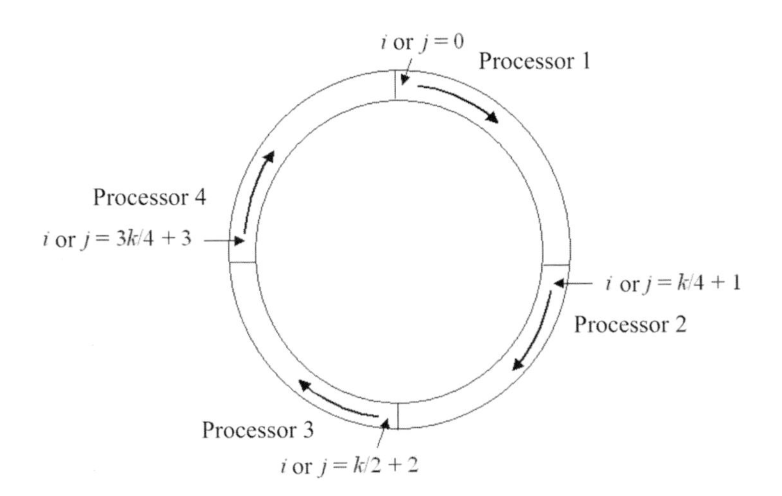

Figure 7.18 – The forward recursion circle is divided into 4 quadrants.

Whatever the value of cycle C, higher degrees of parallelism of value pC, can be implemented. Indeed, any multiple of C, the basic cycle in the permutation, is also a cycle in the permutation, on condition that pC is a divisor of k. That is, j modulo pC and $\Pi(j)$ modulo pC are periodic on the circle of length k, which can then be cut into pC fractions of equal length. For example, a degree 64 parallelism is possible for a value of k equal to 2048.

However, whatever the degree of parallelism, a minimum latency is unavoidable: the time required for receiving a packet and putting it into the buffer memory. While this packet is being put into memory, the decoder works on the information contained in the previous packet. If this decoding is performed in a time at least equal to the memorization time, then the total decoding latency is at maximum twice this memorization time. The level of parallelism in the decoder is adjusted according to this objective, which may be a constraint in certain cases.

For further information about the implementation of turbo decoders, of all the publications on this topic, [7.47] is a good resource.

7.5 m-binary turbo codes

$m-$binary turbo codes are built from recursive systematic convolutional (RSC) codes with m binary inputs ($m \geq 2$). There are at least two ways to build an $m-$binary convolutional code: either from the Galois field $\mathbf{F}_{2^m}$, or from the Cartesian product $(\mathbf{F}_2)^m$. Here, we shall only deal with the latter, which is more convenient. Indeed, a code elaborated in $\mathbf{F}_{2^m}$, with a memory ν, has $2^{\nu m}$ possible states, whereas the number of states of the code defined in $(\mathbf{F}_2)^m$, with the same memory, can be limited to 2^ν.

The advantages of the m-binary construction compared to the classical ($m = 1$) turbo code scheme, are varied: better convergence of the iterative process, larger minimum distances, less puncturing, lower latency, robustness towards the sub-optimality of the decoding algorithm, in particular when the MAP algorithm is simplified into its Max-Log-MAP version [7.23].

The case $m = 2$ has already been adopted in the European standards for the return path in digital video broadcasting via the satellite network and in the terrestrial network [7.2, 7.1] as well as in the IEEE 802.16 standard [7.5]. Combined with the circular trellis technique, these 8-state turbo codes, called double-binary turbo codes, offer good average performance and great flexibility in adapting to different block sizes and different rates, whilst retaining reasonable decoding complexity.

7.5.1 $m-$binary RSC encoders

Figure 7.19 presents the general structure of an $m-$binary RSC encoder. It uses a pseudo-random generator with code memory ν and generator matrix $\mathbf{G}$ (size

$\nu \times \nu$). The input vector $\mathbf{d}$ with m components is connected to the different possible nodes thanks to a grid of interconnections whose binary matrix, size $\nu \times m$, is denoted $\mathbf{C}$. The vector $\mathbf{T}$ applied to the ν possible taps of the register at instant i, is given by:

$$\mathbf{T}_i = \mathbf{C}.\mathbf{d}_i \tag{7.51}$$

with $\mathbf{d}_i = (d_{1,i} \dots d_{m,i})$.

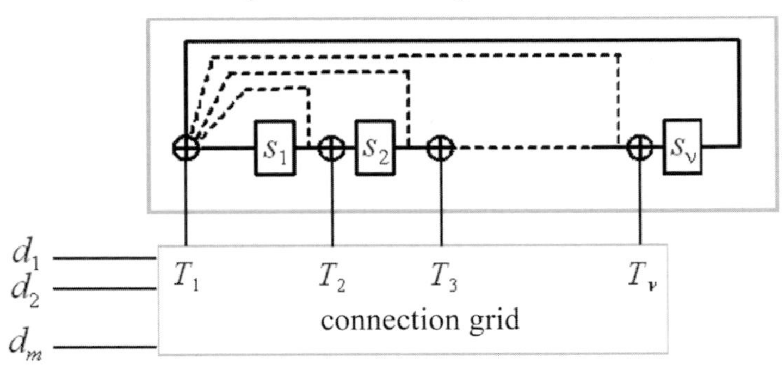

Figure 7.19 – General structure of an m-binary RSC encoder with code memory ν. The time index is not shown.

If we wish to avoid parallel transitions in the trellis of the code, condition $m \leq \nu$ must be respected and matrix $\mathbf{C}$ must be full rank. Except for very particular cases, this encoder is not equivalent to an encoder with a single input on which we would successively present $d_1, d_2, \cdots, d_m$. An $m-$binary encoder is therefore not decomposable generally.

The redundant output of the machine (not shown in the figure) is calculated at instant i according to the expression:

$$y_i = \sum_{j=1\dots m} d_{j,i} + \mathbf{R}^{\mathbf{T}}\mathbf{S}_i \tag{7.52}$$

where $\mathbf{S}_i$ is the state vector at instant i and $\mathbf{R}^{\mathbf{T}}$ is the transposed redundancy vector. The p-th component of $\mathbf{R}$ equals 1 if the p-th component of $\mathbf{S}_i$ is used in the construction of y_i and 0 otherwise. We can show that y_i can also be written as:

$$y_i = \sum_{j=1\dots m} d_{j,i} + \mathbf{R}^{\mathbf{T}}\mathbf{G}^{-1}\mathbf{S}_{i+1} \tag{7.53}$$

on condition that :

$$\mathbf{R}^{\mathbf{T}}\mathbf{G}^{-1}\mathbf{C} \equiv \mathbf{0} \tag{7.54}$$

Expression (7.52) ensures, first, that the Hamming weight of vector $(d_{1,i}, d_{2,i}, \cdots, d_{m,i}, y_i)$ is at least equal to 2 when we leave the reference path ("all zero" path), in the trellis. Indeed, inverting any component of $\mathbf{d}_i$ modifies the value of y_i. Second, expression (7.53) indicates that the Hamming weight of the same vector is also at least equal to 2 when we return to the reference path. In conclusion, relations (7.52) and (7.53) together guarantee that the free distance of the code, whose rate is $R = m/(m+1)$, is at least equal to 4, whatever m.

7.5.2 m-binary turbo codes

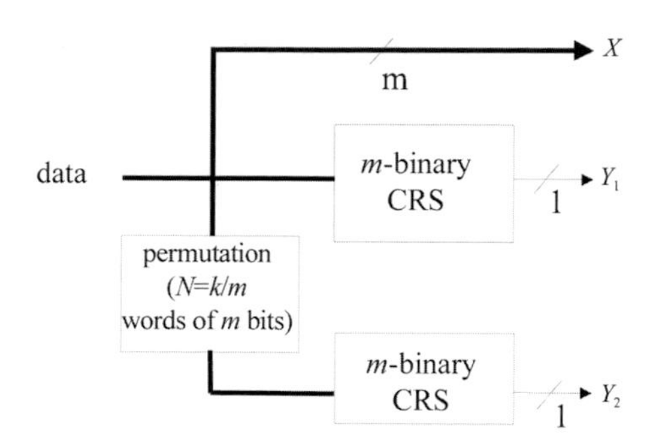

Figure 7.20 – m-binary turbo encoder.

We consider a parallel concatenation of two $m-$binary RSC enoders associated with a permutation as a function of N words of m bits ($k = mN$) (Figure 7.20). The blocks are encoded twice by this two-dimensional code, whose rate is $m/(m+2)$. The circular trellis principle is adopted to enable encoding of the blocks without a termination sequence and without edge effects.

The advantages of this construction compared to classical turbo codes are the following :

- **Better convergence**. This advantage, observed first in [7.9], commented in [7.16] and in a different way in [7.23], can be explained by a lower density of errors on each of the two dimensions of the iterative process. Take relation (7.8) that provides the upper bound of the accumulated spatial distance for a binary code and adapt it to an m-binary code:

$$\sup S_{\min} = \sqrt{\frac{2k}{m}} \tag{7.55}$$

For a coding rate R, the number of parity bits produced by the sequence with accumulated length $\sup S_{\min}$ is:

$$n_{parity}(\sup S_{\min}) = \left(\frac{1-R}{R}\right)\frac{m}{2}\sup S_{\min} = \left(\frac{1-R}{R}\right)\sqrt{\frac{mk}{2}} \tag{7.56}$$

Thus, replacing an ($m = 1$) binary turbo code by an ($m = 2$) double-binary code, the number of parity bits in the sequence considered is multiplied by $\sqrt{2}$, although the accumulated spatial distance has been reduced by the same ratio. Because the parity bits are local information for the two elementary decoders (and are therefore not a source of correlation between them), increasing the number of the former improves convergence. To increase m beyond 2 slightly improves behaviour concerning correlation but the effects are less visible than when passing from $m = 1$ to $m = 2$.

- **Larger minimum distances**. As explained above, the number of parity bits produced by the RTZ sequences of input weight 2 is increased by using m-binary codes. The same is true for all the simple RTZ sequences defined in Section 7.3.2. The number of parity bits for these sequences is at least equal to $n_{parity}(\sup S_{\min})$. The corresponding Hamming distances are therefore even higher than those obtained with binary codes and contribute even less to the MHD of the turbo code. As for the distances associated with multiple RTZ patterns, which are generally those that fix the MHD, they depend on the quality of the permutation implemented (see Section 7.3.2).

- **Less puncturing**. To obtain coding rates greater than $m/(m+1)$ from the encoder of Figure 7.20, it is not necessary to suppress as many redundancy symbols as with a binary encoder. The performance of elementary codes is improved by this, as Figure 7.21 shows. This figure compares the correction capability of convolutional codes of rates 2/3 and 6/7, in the binary ($m = 1$) and double-binary ($m = 2$) versions.

- **Reduced latency**. From the point of view of encoding as well as decoding, the latency (that is, the number of clock cycles necessary for the processing) is divided by m since the data are processed in groups of m bits. However, it may happen that the critical path of the decoding circuit is increased compared to the case $m = 1$ as more data are to be considered in a clock cycle. Parallelism solutions, such as those proposed in [7.35], can help to increase the frequency of the circuit.

- **Robustness of the decoder**. For binary turbo codes, the difference in performance between the MAP algorithm and its simplified versions or between the MAP algorithm and the SOVA algorithm, vary from 0.2 to 0.6 dB, depending on the size of the blocks and the coding rates. This difference is divided by two when we use double-binary turbo codes and can be even lower for $m > 2$. This favourable (and slightly surprising) property can be explained as follows: for a block of a given size (k bits), the lower the number of steps in the trellis, the closer the decoder is to the Maximum Likelihood (ML) decoder, whatever the algorithm on which

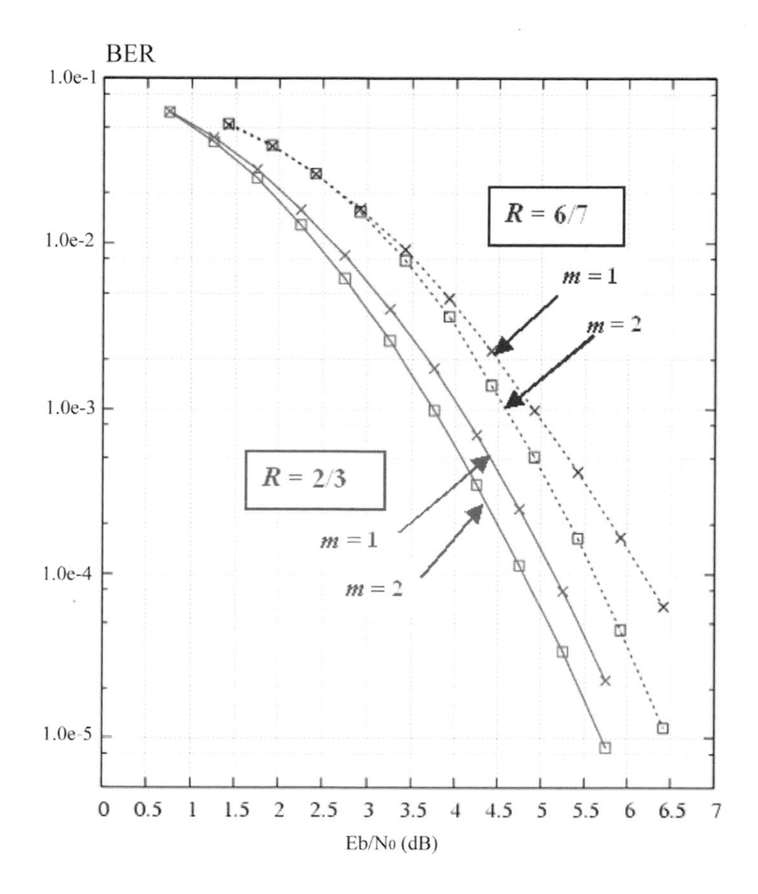

Figure 7.21 – Performance of simple binary ($m = 1$) and double-binary ($m = 2$) convolutional codes, for $R = 2/3$ and $R = 6/7$.

it is based. Ultimately, a trellis reduced to a single step and therefore containing all the possible codewords is equivalent to an ML decoder.

8-state double-binary turbo code

Figure 7.22(a) gives some examples of performance obtained with the turbo code of [7.2], for a rate 2/3. The parameters of the constituent encoders are:

$$\mathbf{G} = \begin{bmatrix} 1 & 0 & 1 \\ 1 & 0 & 0 \\ 0 & 1 & 0 \end{bmatrix}, \mathbf{C} = \begin{bmatrix} 1 & 1 \\ 0 & 1 \\ 0 & 1 \end{bmatrix}, \mathbf{R} = \begin{bmatrix} 1 \\ 1 \\ 0 \end{bmatrix}$$

The permutation function uses both inter- and intrasymbol interference. In particular, we can observe:

- good average performance for this code whose decoding complexity remains very reasonable (around 18,000 gates per iteration plus the memory);
- a certain coherence concerning the variation of performance with block size (in agreement with the curves of Figures 3.6, 3.9, 3.10). The same coherence could also be observed for the variation of performance with coding rate;
- quasi-optimality of decoding with low error rates. The theoretical asymptotic curve for 188 bytes has been calculated from the sole knowledge of the minimum distance of the code (that is, 13 with a relative multiplicity of 0.5) and not from the total spectrum of the distances. In spite of this, the difference between the asymptotic curve and the curve obtained by simulation is only 0.2 dB for a PER of 10^{-7}.

16-state double-binary turbo code

The extension of the previous scheme to 16-state elementary encoders allows the minimum distances to be greatly increased. We can, for example, choose:

$$\mathbf{G} = \begin{bmatrix} 0 & 0 & 1 & 1 \\ 1 & 0 & 0 & 0 \\ 0 & 1 & 0 & 0 \\ 0 & 0 & 1 & 0 \end{bmatrix}, \mathbf{C} = \begin{bmatrix} 1 & 1 \\ 0 & 1 \\ 0 & 0 \\ 0 & 1 \end{bmatrix}, \mathbf{R} = \begin{bmatrix} 1 \\ 1 \\ 1 \\ 0 \end{bmatrix}$$

For the rate 2/3 turbo code, again with blocks of 188 bytes, the minimum distance obtained is equal to 18 (relative multiplicity of 0.75) instead of 13 for the 8-state code. Figure 7.22(b) shows the gain obtained for low error rates: around 1 dB for a PER of 10^{-7} and 1.4 dB asymptotically, considering the respective minimum distances. We can note that the convergence threshold is almost the same for 8-state and 16-state decoders, the curves being practically identical for a PER greater than 10^{-4}. The theoretical limit (TL), for $R = 2/3$ and for a blocksize of 188 bytes, is 1.7 dB. The performance of the decoder in this case is: TL + 0.9 dB for a PER of 10^{-4} and TL + 1.3 dB for a PER of 10^{-7}. These intervals are typical of what we obtain in most rate and blocksize configurations.

Replacing 4-PSK modulation by 8-PSK modulation, in the so-called pragmatic approach, gives the results shown in Figure 7.22(b), for blocks of 188 and 376 bytes. Here again, good performance of the double-binary code can be observed, with losses compared to the theoretical limits (that are around 3.5 and 3.3 dB, respectively) close to those obtained with 4-PSK modulation. Associating turbo codes with different modulations is described in Chapter 10.

For a particular system, the choice between an 8-state or 16-state turbo code depends, apart from the complexity desired for the decoder, on the target error

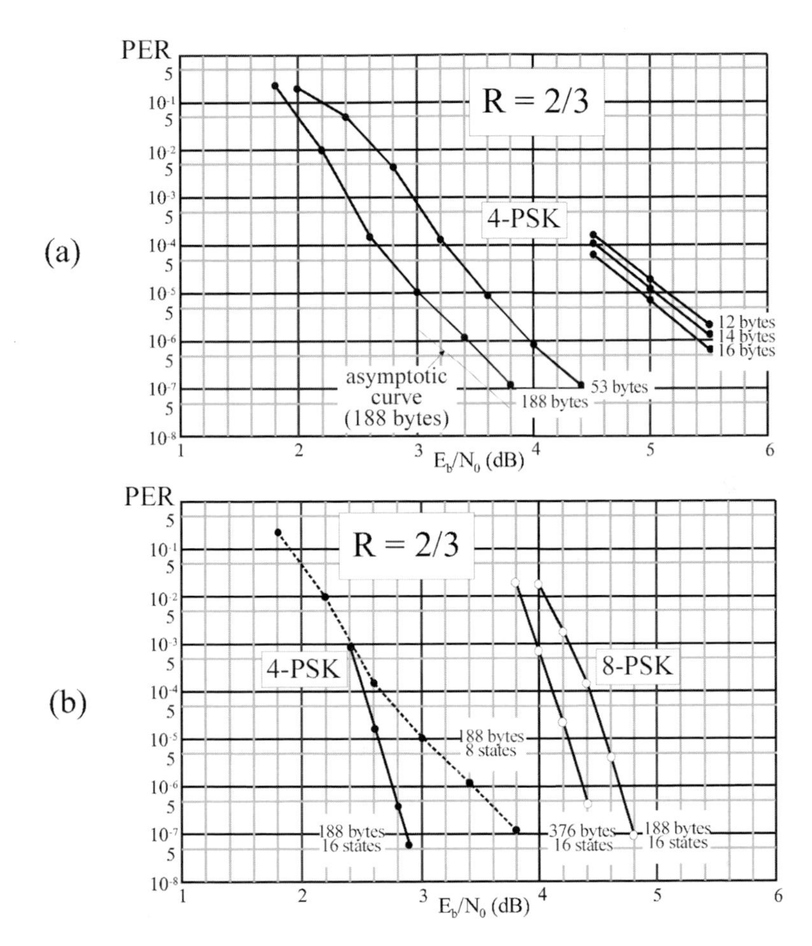

Figure 7.22 – (a) PER performance of a double-binary turbo code with 8 states for blocks of 12, 14, 16, 53 and 188 bytes. 4-PSK, AWGN noise and rate 2/3. Max-Log-MAP decoding with input samples of 4 bits and 8 iterations. (b) PER performance of a double-binary turbo code with 16 states for blocks of 188 bytes (4-PSK and 8-PSK) and 376 bytes (8-PSK), AWGN noise and rate 2/3. Max-Log-MAP decoding with input samples of 4 bits (4-PSK) or 5 bits (8-PSK) and 8 iterations.

rates. To simplify, let us say that an 8-state turbo code suffices for PERs greater than 10^{-4}. This is generally the case for transmissions having the possibility of repetitions (ARQ: Automatic Repeat reQuest). For lower PERs, typical of broadcasting or of mass memory applications, the 16-state code is highly preferable.

7.6 Analysis tools

7.6.1 Theoretical performance

Figure 1.6 shows two essential parameters allowing the performance of an error correcting code and its decoder to be evaluated:

- the *asymptotic gain* measuring the behaviour of the coded system at low error rates. This is mainly dictated by the MHD of the code (see Section 1.5). A low value of the MHD leads to a great change in the slope (*flattening*) in the error rate curve. When the asymptotic gain is reached, the BER(E_b/N_0) curve with coding becomes parallel to the curve without coding.

- the *convergence threshold* defined as the signal to noise ratio from which the coded system becomes more efficient than the non-coded transmission system;

In the case of turbo codes and the iterative process of their decoding, it is not always easy to estimate the performance either of the asymptotic gain or of the convergence. Methods for estimating or determining the minimum distance proposed by Berrou *et al.* [7.18], Garello *et al.* [7.27] and Crozier *et al.* [7.22] are presented in the rest of this chapter. The EXIT diagram method proposed by ten Brink [7.46] to estimate the convergence threshold is also introduced.

7.6.2 Asymptotic behaviour

Determining the performance of error correcting codes with low error rates by simulation requires high calculation power. It is, however, possible to estimate this performance when the MHD $d_{\min}$ and the multiplicity are known (see Section 1.5). Thus, the packet error rate with high signal to noise ratio E_b/N_0 is given by the first term of the *union bound* (UB). The expression of the UB is described by relation (3.21), and estimation of the PER, given by Equation (1.16), is shown again here:

$$\text{PER} \approx \frac{1}{2} N(d_{\min}) \operatorname{erfc}\left(\sqrt{R d_{\min} \frac{E_b}{N_0}}\right) \tag{7.57}$$

where $N(d_{\min})$, the multiplicity, represents the number of codewords at the minimum distance.

The minimum distance of a code is not, in the general case, simple to determine except if the number of codewords is low enough for us to make an exhaustive list of them, or if particular properties of the code enable us to establish an analytical expression of this value (for example, the minimum distance of a product code is equal to the product of the minimum distances of the constituent codes). In the case of convolutional turbo codes, the minimum distance is not obtained analytically; the only methods proposed are based on the total or partial [7.28] enumeration of codewords whose input weight is lower than or equal to the minimum distance. These methods are applicable in practice only for small sizes blocksizes and small minimum distances, which is why they will not be described here.

Error impulse method

This method, proposed by Berrou *et al.* [7.18], is not based on the analysis of the properties of the code but on the correction capacity of the decoder. Its principle, illustrated in Figure 7.23, involves superposing on the input sequence of the decoder an error impulse whose amplitude A_i is increased until the decoder no longer knows how to correct it.

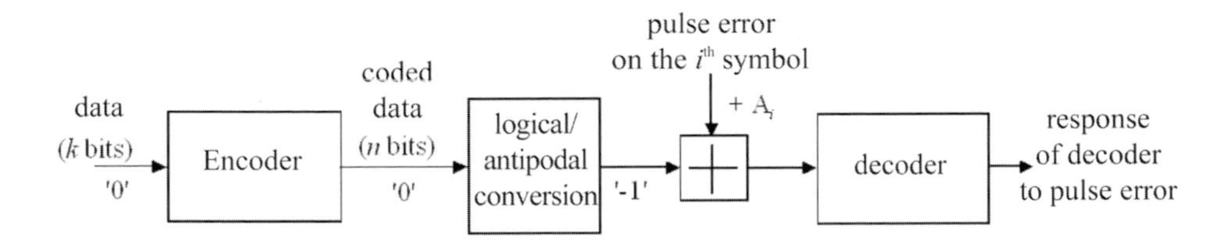

Figure 7.23 – Schematic diagram of the error impulse method.

The code considered being linear, the sequence transmitted is assumed to be the "all zero" sequence. The coding operation then produces codewords that also contain only zeros. These are next converted into real values equal to -1. If this succession of symbols was directly applied at the decoder, the latter would not encounter any difficulty in retrieving the original message since the transmission channel is perfect.

The proposed method involves adding an error impulse to the i-th symbol $(0 \leq i \leq k-1)$ of the information sequence (systematic part), that is, transforming a "-1" symbol into a symbol having a positive value equal to $-1+A_i$. If amplitude A_i is high enough, the decoder does not converge towards the "all zero" word. Let us denote A_i^* the maximum amplitude of the impulse in position i such that the decoded codeword is the "all zero" word. It is shown in [7.18] that, if the decoder performs maximum likelihood decoding, *impulse distance* $d_{imp} = \min_{i=0,\cdots,k-1} (A_i^*)$ is also the minimum distance d_{min} from the code.

It is generally not necessary to test all the positions of the sequence. For a shift invariant code (which is the case of convolutional codes), it suffices to apply the error impulse to just one position of the datablock. For a code presenting a periodicity of period P, it is necessary to test P positions. This method is appli-

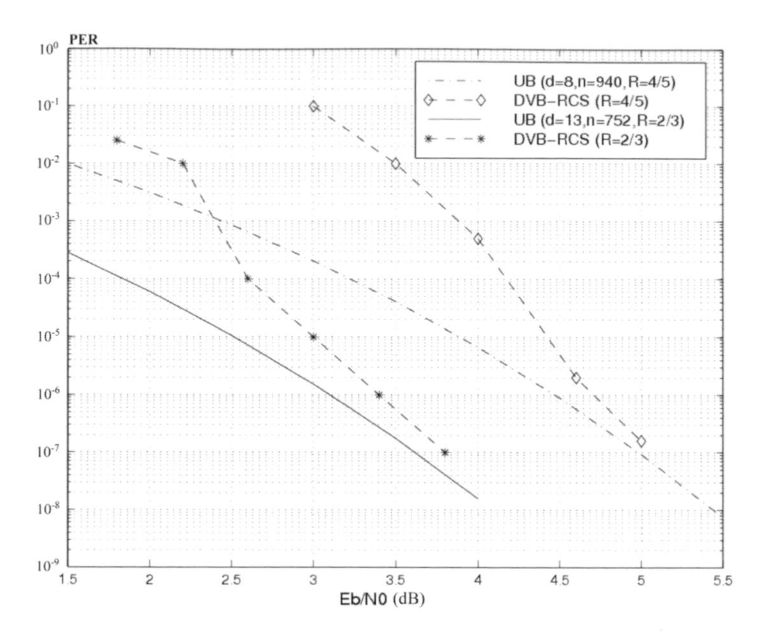

Figure 7.24 – Measured and estimated PER (UB) of the DVB-RCS turbo code for the transmission of MPEG (188 bytes) blocks with coding rates 2/3 and 4/5. 4-PSK modulation and Gaussian channel.

cable to any linear code, for any blocksize and any coding rate, and it requires only a few seconds to a few minutes calculation on an ordinary computer, the calculation time being a linear function of the blocksize or of its period P.

When the decoding is not maximum likelihood, this method is no longer rigorous and produces only an estimation of the minimum distance. In addition, the multiplicity of the codewords at distance $d_{\min}$ is not provided and Equation (7.57) cannot be applied without particular hypotheses about the properties of the code. In the case of turbo codes, two realistic hypotheses can be formulated to estimate multiplicity: a single codeword at distance A_i^* has its i-th information bit at 1 (unicity), and the A_i^* values corresponding to all positions i come from distinct codewords (non-overlapping).

An estimation of the PER is then given by:

$$\text{PER} \approx \frac{1}{2} \sum_{i=0}^{k-1} \text{erfc}(\sqrt{RA_i^* \frac{E_b}{N_0}}) \tag{7.58}$$

The first hypothesis (unicity) under-evaluates the value of the error rate, unlike the second (non-overlapping) that over-evaluates it and, overall, the two effects compensate each other. As an example, Figure 7.24 compares the measured performance of the DVB-RCS turbo code, for two coding rates, with their estimate deduced from (7.58). The parameters obtained by the error impulse method are:

- $d_{\min} = 13$ and $n(d_{\min}) = 752$ for $R = 2/3$
- $d_{\min} = 8$ and $n(d_{\min}) = 940$ for $R = 4/5$

For packet error rates of 10^{-7}, less than 0.2 dB separates the measured and estimated curves.

Modified error impulse method

The approach of Garello *et al.* [7.27] is similar to the error impulse method presented above. It involves placing an impulse in row i in the "all zero" codeword. This time, the amplitude of the impulse is high enough for the decoder not to converge towards the "all zero" codeword but towards another sequence that contains a 1 in position i. In addition, Gaussian noise is added to the input sequence of the decoder, which tends to help the latter converge towards the concurrent word having the lowest weight. This is what often happens when the level of noise is well adjusted. In all cases, the weight of the codeword provided by the decoder is an upper limit of the minimum distance of all the codewords containing a 1 in row i. The minimum distance and the multiplicity are estimated by sweeping all the positions. This algorithm works very well for small and average distances.

Double error impulse method

The method proposed by Crozier *et al.* [7.22] is an improvement of the previous method, at the expense of higher computation time. It involves placing a first high level impulse at row i and a second at row j to the right of i and such that $j - i < r$. The upper limit of r is $2D$ where D is an upper bound of the distance to be evaluated. Then, decoding is applied similar to that described above but with a stronger probability of obtaining a codeword at the minimum distance. The calculation time is increased by a ratio r.

7.6.3 Convergence

A SISO decoder can be seen as a processor that transforms one of its input values, the LLR of the extrinsic information used as *a priori* information, into an output extrinsic LLR. In iterative decoding, the characteristics of the extrinsic information provided by decoder 1 depend on the extrinsic information provided by decoder 2 and vice-versa. The degree of dependency between the input and output extrinsic information can be measured by the mutual information (MI).

The idea implemented by ten Brink [7.46] is to follow the exchange of extrinsic information through the SISO decoders working in parallel on a diagram, called an *EXtrinsic Information Transfer* (EXIT) chart. To elaborate the EXIT chart, it is necessary to know the transfer characteristics of the extrinsic information of each SISO decoder used in the decoding. This section shows how to establish the transfer function of the extrinsic information for a SISO decoder, then construct the EXIT chart, and finally analyse the convergence of the iterative decoder.

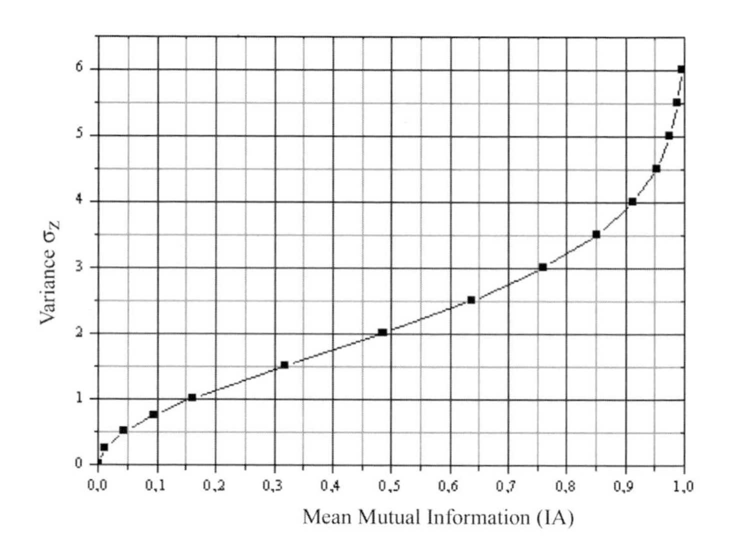

Figure 7.25 – Variation of variance σ_z as a function of the the mutual information IA

Transfer function for a SISO decoder of extrinsic information

a. Definition of the mutual information (MI)

If the weighted extrinsic information z on the coded binary element $x \in \{-1, +1\}$ follows a conditional probability density $f(z|x)$, the MI $I(z, x)$ measures the quantity of information provided on average by z on x and equals

$$\mathrm{I}(z,x) = \frac{1}{2} \sum_{x=-1,+1} \int_{-\infty}^{+\infty} f(z|x) \times \log_2 \left[\frac{2f(z|x)}{f(z|-1) + f(z|+1)} \right] dz \qquad (7.59)$$

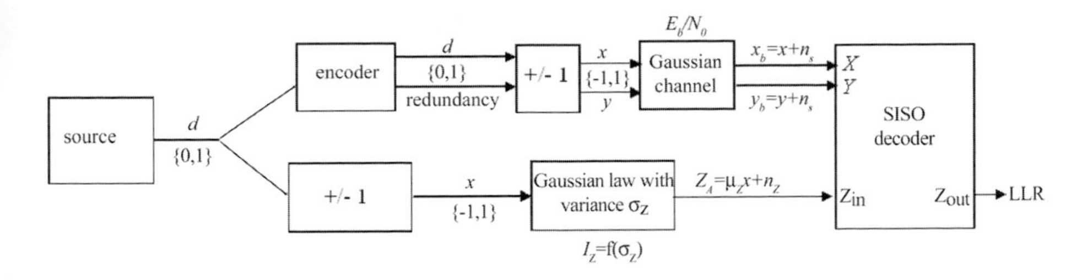

Figure 7.26 – Algorithm for determining the transfer function $IE = T(IA, Eb/N_0)$

b. Definition of the a priori *mutual information*

Hypotheses:

- Hyp. 1: when the interleaving is large enough, the distribution of the input extrinsic information can be approximated by a Gaussian distribution after a few iterations.
- Hyp. 2: probability density $f(z|x)$ satisfies the exponential symmetry condition, that is, $f(z|x) = f(-z|x)exp(-z)$.

The first hypothesis allows the *a priori* LLR Z_A of a SISO decoder to be modelled by a variable having independent Gaussian noise n_z, with variance σ_z and expectation μ_z, applied to the transmitted information symbol x according to the expression

$$Z_A = \mu_z x + n_z$$

The second hypothesis imposes $\sigma_z^2 = 2\mu_z$. The amplitude of the extrinsic information is therefore modelled by the following distribution:

$$f\left(\lambda|\, x\right) = \frac{1}{\sqrt{4\pi\mu_z}} \exp\left[-\frac{\left(\lambda - \mu_z x\right)^2}{4\mu_z}\right] \tag{7.60}$$

From (7.59) and (7.60), observing that $f\left(z|\ 1\right) = f\left(-z| -1\right)$, we deduce the *a priori* mutual information:

$$\text{IA} = \int_{-\infty}^{+\infty} \frac{1}{\sqrt{4\pi\mu_z}} \exp\left[-\frac{\left(\lambda - \mu_z\right)^2}{4\mu_z}\right] \times \log_2\left[\frac{2}{1+\exp\left(-\lambda\right)}\right] d\lambda$$

or again

$$\text{IA} = 1 - \int_{-\infty}^{+\infty} \frac{1}{\sqrt{2\pi}\sigma_z} \exp\left[-\frac{\left(\lambda - \sigma_z^2/2\right)^2}{2\sigma_z^2}\right] \times \log_2\left[1+\exp\left(-\lambda\right)\right] d\lambda \tag{7.61}$$

We can note that $\lim_{\sigma_z \to 0} \mathrm{IA} = 0$ (the extrinsic information does not provide any information about datum x) and that $\lim_{\sigma_z \to +\infty} \mathrm{IA} = 1$ (the extrinsic information perfectly determines datum x).
IA is an increasing monotonous function of σ_z; it is therefore invertible. Function $\sigma_z = f(\mathrm{IA})$ is shown in Figure 7.25.

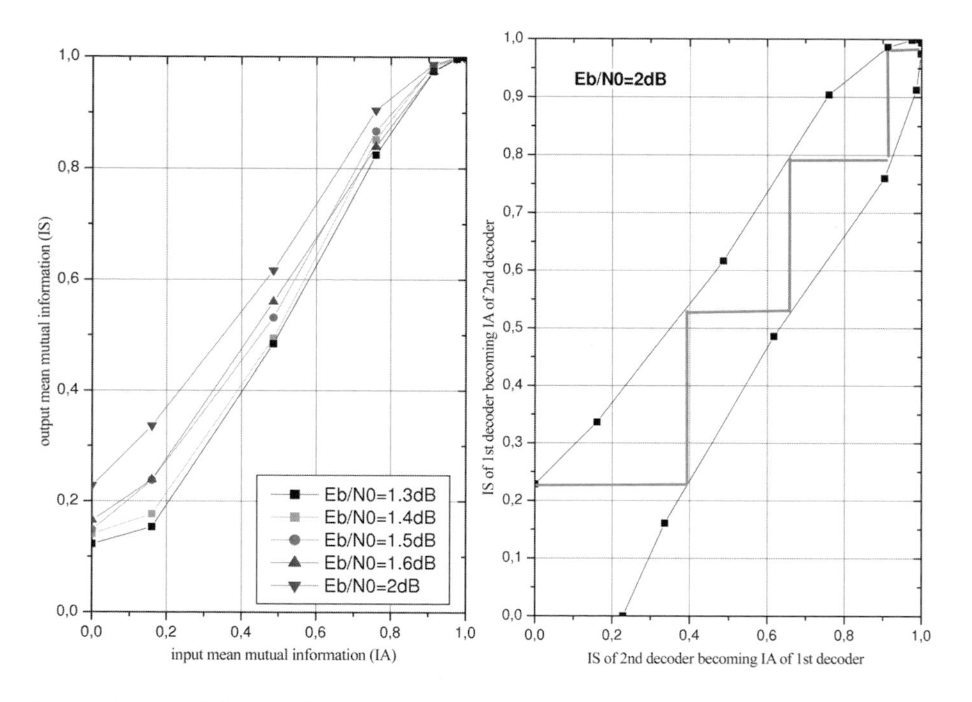

Figure 7.27 – (a) Transfer characteristic of the extrinsic information for a 16-state binary encoder, with rate 2/3 and MAP decoding with different Eb/N0.
(b) EXIT chart and trajectory for the corresponding turbo code, with pseudo-random interleaving on 20,000 bits, for $E_b/N_0 = 2dB$.

c. Calculation of the output mutual information

Relation (7.59) allows the calculation of the mutual information IS linked with the extrinsic information produced by the SISO decoder:

$$\mathrm{IS} = \frac{1}{2} \sum_{x=-1,+1} \int_{-\infty}^{+\infty} f_s\left(z|\,x\right) \times \log_2 \left[\frac{2 f_s\left(z|\,x\right)}{f_s\left(z|-1\right) + f_s\left(z|+1\right)} \right] dz \qquad (7.62)$$

We can note that $\mathrm{IS} \in [0, 1]$.

The distribution f_s is not Gaussian. It is therefore necessary to use a digital calculation tool to determine it, which is the great drawback of this method.

If we view the MI of output IS as a function of IA and of the signal to noise ratio E_b/N_0, the transfer function of the extrinsic information is defined by:

$$IE = T(IA, E_b/N_0) \tag{7.63}$$

d. Practical method to obtain the transfer function of the extrinsic information

Figure 7.26 shows the path taken to establish the transfer characteristic of the extrinsic information of a SISO decoder.

- step 1: Generation of the pseudo random message d to be transmitted; at least 10000 bits are necessary for the statistical properties to be representative.
- step 2: Encoding the data with rate R then 2-PSK modulation of the signal; the systematic and redundancy data both belong to the alphabet {-1,+1}.
- step 3: Application of a Gaussian noise with signal to noise ratio E_b/N_0 (in dB), with variance

$$\sigma = \sqrt{\frac{1}{2} \cdot \frac{10^{-0,1\times E_b/N_0}}{R}}$$

- step 4: Application to the data transmitted (stored in a file) of normal law $N(\mu_z, \sigma_z)$ corresponding to the mutual information IA desired (see Figure 7.25) to obtain the distribution of *a priori* extrinsic information.
- step 5: Initialization of the SISO decoder with the *a priori* LLRs (it might be necessary, depending on the decoding algorithm chosen, to transform the LLRs into probabilities).
- step 6: Recovering the LLRs at the output of the SISO decoder (corresponding to one half-iteration of the decoding process), in a file.
- step 7: Utilization of digital calculation software to evaluate IS (relation (7.62)).
  - Trace the histograms of the LLR distributions output as a function of the bit transmitted (hence the necessity to store this information in two files).
  - Evaluate the integral by the trapeze method.

- The result is the MI of output IS corresponding to the MI of input IA.

e. An example

The simulations were performed on a 16-state binary turbo code with rate 2/3, with a pseudo-random interleaving of 20,000 bits. The decoding algorithm is the MAP algorithm. Figure 7.27(a) shows the relation between IS and IA as a function of the signal to noise ratio of the Gaussian channel.

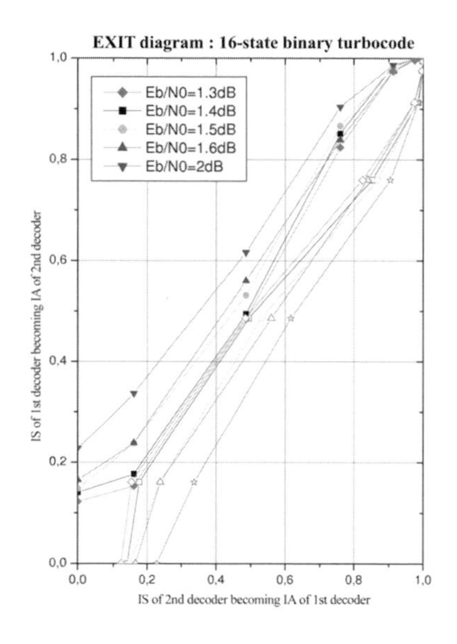

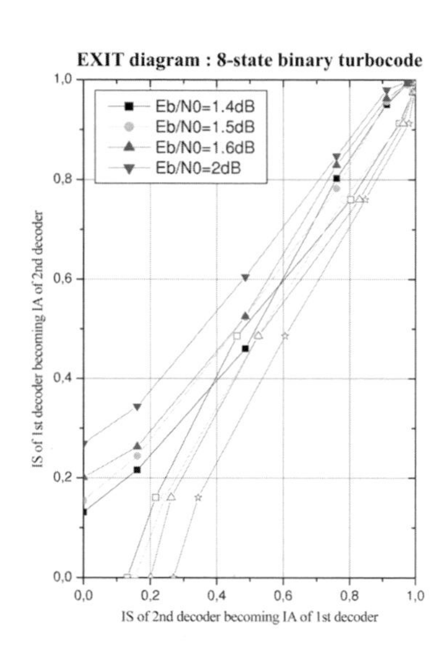

Figure 7.28 – EXIT charts for different E_b/N_0 in the case of binary turbo codes, rate 2/3, pseudo-random interleaving of 20,000 bits, (a) 16-state and (b) 8-state MAP decoding.

EXIT chart

The extrinsic information transfer characteristic is now known for a SISO decoder. In the case of iterative decoding, the output of decoder 1 becomes the input of decoder 2 and vice versa. Curves IS1 $= f(IA1 =$ IS2) and IS2 $= f($IA2 $=$ IS1), identical to one symmetry if the SISO decoders are the same, are placed on the same graph as shown in Figure 7.27(b). In the case of a high enough signal to noise ratio (here 2 dB), the two curves do not have any intersection outside the point of coordinates (1,1) which materializes the knowledge of the received message. Starting from null mutual information, it is then possible to follow the exchange of extrinsic information along the iterations. In the example of Figure 7.27(b), arrival at point (1,1) is performed in 3.5 iterations.

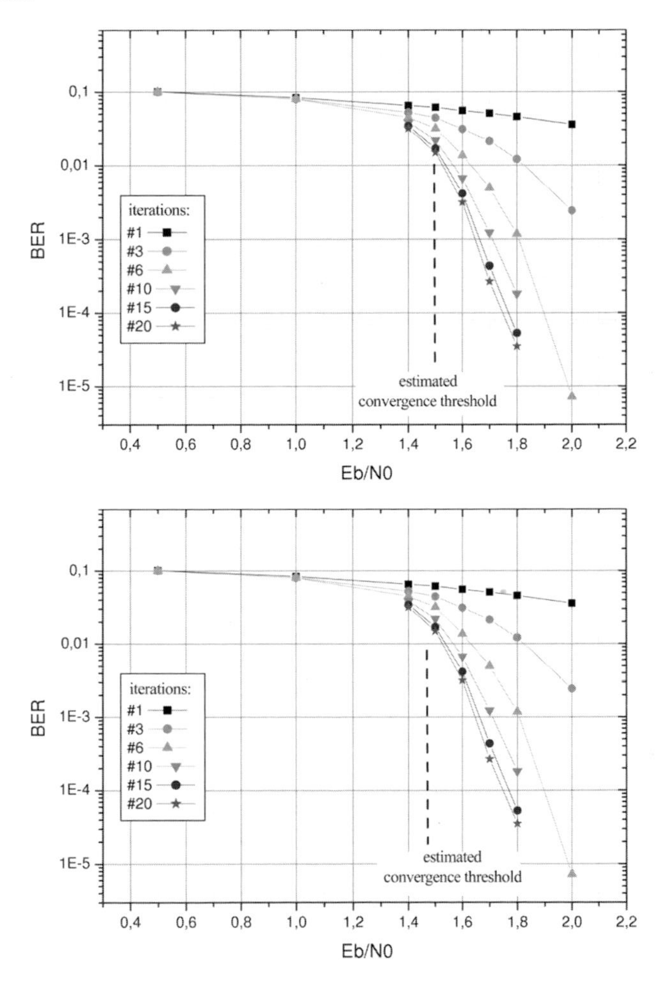

Figure 7.29 – Binary error rates of a 16-state (a) and 8-state (b) binary turbo code with rate 2/3, with pseudo-random interleaving of 20000 bits. MAP decoding with 1, 3, 6, 10, 15 and 20 iterations and comparison with the convergence threshold estimated by the EXIT method.

When the signal to noise ratio is too low, as in case $E_b/N_0 = 1.4$ dB in Figure 7.28(b), the curves have intersection points other than point $(1, 1)$. The iterative process starting from null MI at the input will therefore not be able to lead to a perfectly determined message. The minimum signal to noise ratio for which there is no intersection other than point (1,1) is the convergence threshold of the turbo encoder. In the simulated example, this convergence can be estimated at around 1.4 dB for 16-state (Figure 7.28(a)) and 8-state (Figure 7.28(b)) binary turbo codes.

Figure 7.29 shows the performance of 16-state and 8-state binary turbo codes as a function of the number of iterations, and compares them with the convergence threshold estimated by the EXIT chart method.

Bibliography

[7.1] Interaction channel for digital terrestrial television. DVB, ETSI EN 301 958, V1.1.1, pp. 28-30, Aug. 2001.

[7.2] Interaction channel for satellite distribution systems. DVB, ETSI EN 301 790, V1.2.2, pp. 21-24, Dec. 2000.

[7.3] Multiplexing and channel coding (fdd). 3GPP Technical Specification Group, TS 25.212 v2.0.0, June 1999.

[7.4] Recommendations for space data systems. telemetry channel coding. Consultative Committee for Space Data Systems, BLUE BOOK, May 1998.

[7.5] Ieee standard for local and metropolitan area networks, IEEE Std 802.16a, 2003. Available at http://standards.ieee.org/getieee802/download/802.16a-2003.pdf.

[7.6] L. R. Bahl, J. Cocke, F. Jelinek, and J. Raviv. Optimal decoding of linear codes for minimizing symbol error rate. *IEEE Transactions on Information Theory*, IT-20:284–287, March 1974.

[7.7] G. Battail. Weighting of the symbols decoded by the viterbi algorithm. *Annals of Telecommunications*, 42(1-2):31–38, Jan.-Feb. 1987.

[7.8] G. Battail. Coding for the gaussian channel: the promise of weighted-output decoding. *International Journal of Satellite Communications*, 7:183–192, 1989.

[7.9] C. Berrou. Some clinical aspects of turbo codes. In *Proceedings of 3rd International Symposium on turbo codes & Related Topics*, pages 26–31, Brest, France, Sept. 1997.

[7.10] C. Berrou, P. Adde, E. Angui, and S. Faudeuil. A low complexity soft-output viterbi decoder architecture. In *Proceedings of IEEE International Conference on Communications (ICC'93)*, pages 737–740, GENEVA, May 1993.

[7.11] C. Berrou, C. Douillard, and M. Jézéquel. Designing turbo codes for low error rates. In *IEE colloquium : turbo codes in digital broadcasting – Could it double capacity?*, pages 1–7, London, Nov. 1999.

[7.12] C. Berrou, C. Douillard, and M. Jézéquel. Multiple parallel concatenation of circular recursive convolutional (crsc) codes. *Annals of Telecommunications*, 54(3-4):166–172, March-Apr. 1999.

[7.13] C. Berrou and A. Glavieux. Reflections on the prize paper: "near optimum error correcting coding and decoding: turbo codes". *IEEE IT Society Newsletter*, 48(2):136–139, June 1998. www.ieeeits.org/publications/nltr/98_jun/reflections.html.

[7.14] C. Berrou, A. Glavieux, and P. Thitimajshima. Near shannon limit error-correcting coding and decoding: turbo-codes. In *Proceedings of IEEE International Conference on Communications (ICC'93)*, pages 1064–1070, GENEVA, May 1993.

[7.15] C. Berrou and M. Jézéquel. Frame-oriented convolutional turbo-codes. *Electronics letters*, 32(15):1362–1364, July 1996.

[7.16] C. Berrou and M. Jézéquel. Non binary convolutional codes for turbo coding. *Electronics Letters*, 35(1):39–40, Jan. 1999.

[7.17] C. Berrou, Y. Saouter, C. Douillard, S. Kerouédan, and M. Jézéquel. Designing good permutations for turbo codes: towards a single model. In *Proceedings of IEEE International Conference on Communications (ICC'04)*, Paris, France, June 2004.

[7.18] C. Berrou, S. Vaton, M. Jézéquel, and C. Douillard. Computing the minimum distance of linear codes by the error impulse method. In *Proceedings of IEEE Global Communication Conference (Globecom'2002)*, pages 1017–1020, Taipei, Taiwan, Nov. 2002.

[7.19] E. Boutillon and D. Gnaedig. Maximum spread of d-dimensional multiple turbo codes. *IEEE Transanction on Communications*, 53(8):1237–1242, Aug. 2005.

[7.20] A. R. Calderbank. The art of signaling: fifty years of coding theory. *IEEE Transactions on Information Theory*, 44(6):2561–2595, Oct. 1998.

[7.21] S. Crozier and P. Guinand. Distance upper bounds and true minimum distance results for turbo-codes with drp interleavers. In *Proceedings of 3rd International Symposium on turbo codes & Related Topics*, pages 169–172, Brest, France, Sept. 2003.

[7.22] S. Crozier, P. Guinand, and A. Hunt. Computing the minimum distance of turbo-codes using iterative decoding techniques. In *Proceedings of the 22nd Biennial Symposium on Communications*, pages 306–308, Kingston, Ontario, Canada, 2004,May 31-June3.

[7.23] C. Douillard and C. Berrou. Turbo codes with rate-m/(m+1) constituent convolutional codes. *IEEE Trans. Commun.*, 53(10):1630–1638, Oct. 2005.

[7.24] L. Duan and B. Rimoldi. The iterative turbo decoding algorithm has fixed points. *IEEE Transactions on Information Theory*, 47(7):2993–2995, Nov. 2001.

[7.25] P. Elias. Error-free coding. *IEEE Transactions on Information Theory*, 4(4):29–39, Sept. 1954.

[7.26] R. G. Gallager. Low-density parity-check codes. *IRE Transactions on Information Theory*, IT-8:21–28, Jan. 1962.

[7.27] R. Garello and A. Vila Casado. The all-zero iterative decoding algorithm for turbo code minimum distance computation. In *Proceedings of IEEE International Conference on Communications (ICC'2004)*, pages 361–364, Paris, France, 2004.

[7.28] R. Garello, P. Pierloni, and S. Benedetto. Computing the free distance of turbo codes and serially concatened codes with interleavers: Algorithms and applications. *IEEE Journal on Selected Areas in Communications*, May 2001.

[7.29] K. Gracie and M.-H. Hamon. Turbo and turbo-like codes: Principles and applications in telecommunications. In *Proceedings of the IEEE*, volume 95, pages 1228–1254, June 2007.

[7.30] J. Hagenauer and P. Hoeher. Concatenated viterbi-decoding. In *Proceedings of International Workshop on Information Theory*, pages 136–139, Gotland, Sweden, Aug.-Sept. 1989.

[7.31] J. Hagenauer and P. Hoeher. A viterbi algorithm with soft-decision outputs and its applications. In *Proceedings of IEEE Global Communications Conference (Globecom'89)*, pages 1680–1686, Dallas, Texas, USA, Nov. 1989.

[7.32] J. Hagenauer and P. Hoeher. Turbo decoding with tail-biting trellises. In *Proceedings of Signals, Systems and Electronics, URSI Int'l Symposium*, pages 343–348, Sept.-Oct. 1998.

[7.33] G. H. Hardy and E. M. Wright. *An Introduction to the Theory of Numbers.* Oxford Univ. Press, Oxford, UK, 1979.

[7.34] C. Heegard and S. B. Wicker. Chapter 3. In *Turbo coding.* Kluwer Academic Publishers, 1999.

[7.35] H. Lin and D. G. Messerschmitt. Algorithms and architectures for concurrent viterbi decoding. In *Proceedings of IEEE International Conference on Communications (ICC'89)*, pages 836–840, Boston, June 1989.

[7.36] R. L. Rivest. Permutation polynomials modulo 2w. *Finite Fields Their Applications*, 7:287–292, Apr. 2001.

[7.37] P. Robertson, P. Hoeher, and E. Villebrun. Optimal and suboptimal maximum a posteriori algorithms suitable for turbo decoding. *European Transactions on Telecommunication*, 8:119–125, March-Apr. 1997.

[7.38] E. Rosnes and O. Y. Takeshita. Optimum distance quadratic permutation polynomial-based interleavers for turbo codes. In *Proceedings of IEEE International Symposyum on Information Theory (ISIT'07)*, pages 1988–1992, Seattle, Washington, July 2006.

[7.39] J. Ryu and O. Y. Takeshita. On quadratic inverses for quadratic permutation polynomials over integer rings. *IEEE Transactions On Information Theory*, 52:1254–1260, March 2006.

[7.40] H. R. Sadjadpour, N. J. A. Sloane, M. Salehi, and G. Nebe. Interleaver design for turbo codes. *IEEE Journal on Selected Areas in Commununications*, 19(5):831–837, May 2001.

[7.41] J. Sun and O. Y. Takeshita. Interleavers for turbo codes using permutation polynomials over integer rings. *IEEE Transactions On Information Theory*, 51:101–119, Jan. 2005.

[7.42] Y. V. Svirid. Weight distributions and bounds for turbo-codes. *European Transactions on Telecommunication*, 6(5):543–55, Sept.-Oct. 1995.

[7.43] O. Y. Takeshita. On maximum contention-free interleavers and permutation polynomials over integer rings. *IEEE Transactions On Information Theory*, 52:1249–1253, March 2006.

[7.44] O. Y. Takeshita. Permutation polynomial interleavers: an algebraic-geometric perspective. *IEEE Transactions On Information Theory*, 53:2116–2132, June 2007.

[7.45] R. M. Tanner. A recursive approach to low complexity codes. *IEEE Transactions on Information Theory*, IT-271:533–547, Sept. 1981.

[7.46] S. ten Brink. Convergence behavior of iteratively decoded parallel concatenated codes. *IEEE Transactions On Communications*, 49(10), Oct. 2001.

[7.47] Z. Wang, Z. Chi, and K. K. Parhi. Area-efficient high-speed decoding schemes for turbo decoders. *IEEE Trans. VLSI Systems*, 10(6):902–912, Dec. 2002.

[7.48] C. Weiss, C. Bettstetter, and S. Riedel. Code constuction and decoding of parallel concatenated tail-biting codes. *IEEE Comm. Letters*, 47(1):366–386, Jan. 2001.

[7.49] Y. Weiss and W. T. Freeman. On the optimality of solutions of the max-product belief-propagation algorithm in arbitrary graphs. *IEEE Transactions on Information Theory*, 47(2):736–744, Feb. 2001.

Chapter 8

Turbo product codes

8.1 History

Because of the Gilbert-Varshamov bound, it is necessary to have long codes in order to obtain block codes with a large minimum Hamming distance (MHD) and therefore high error correction capability. But, without a particular structure, it is almost impossible to decode these codes.

The invention of product codes, due to Elias [8.4], can be seen in this context: it means finding a simple way to obtain codes with high error correction capability that are easily decodable from simple elementary codes. These product codes can be seen as a particular realization of the concatenation principle (Chapter 6).

The first decoding algorithm results directly from the construction of these codes. This algorithm alternates the hard decision decoding of elementary codes on the rows and columns. Unfortunately, this algorithm does not allow us to reach the maximum error correction capability of these codes. The Reddy-Robinson algorithm [8.15] does allow us to reach it. But no doubt due to its complexity, it has never been implemented in practical applications.

The aim of this chapter is to give a fairly complete presentation of algorithms for decoding product codes, whether they be algorithms for hard data or soft data.

8.2 Product codes

With conventional constructions, it is theoretically possible to build codes having a high MHD. However, the decoding complexity becomes prohibitive, even for codes having an algebraic structure, like Reed-Solomon codes or BCH (see Chapter 4) codes. For example, for Reed-Solomon codes on $\mathbf{F}_{256}$, the most complex decoder to have been implemented on a circuit has an error correction capability limited to 11 error symbols, which is insufficient for most applications today. The construction of product codes allows this problem to be circumvented: by using simple codes with low correction capability, but whose decoding is not too costly, it is possible to assemble them to obtain a longer code with higher correction capability.

Definition

Let C_1 (resp. C_2) be a linear code of length n_1 (resp. n_2) and with dimension[1] k_1 (resp. k_2). The product code $C = C_1 \otimes C_2$ is the set of matrices M of size $n_1 \times n_2$ such that:

- Each row is a codeword of C_1,
- Each column is a codeword of C_2.

This code is a linear code of length $n_1 \times n_2$ and with dimension $k_1 \times k_2$.

Example 8.1

Let **H** be the Hamming code of length 7 and P be the parity code of length 3. The dimension of **H** is 4 and the dimension of P is 2. The code $C = H \otimes P$ is therefore of length $21 = 7 \times 3$ and dimension $8 = 4 \times 2$. Let the following information word be coded:

$$I = \begin{bmatrix} 0 & 1 & 1 & 0 \\ 1 & 0 & 1 & 0 \end{bmatrix}.$$

Each row of a codeword of C must be a codeword of **H**. Therefore to code I, we begin by multiplying each row of I by the generating matrix of code **H**:

$$\begin{bmatrix} 0 & 1 & 1 & 0 \end{bmatrix} \cdot \begin{bmatrix} 1 & 0 & 0 & 0 & 1 & 1 & 1 \\ 0 & 1 & 0 & 0 & 1 & 1 & 0 \\ 0 & 0 & 1 & 0 & 1 & 0 & 1 \\ 0 & 0 & 0 & 1 & 0 & 1 & 1 \end{bmatrix} = \begin{bmatrix} 0 & 1 & 1 & 0 & 0 & 1 & 1 \end{bmatrix}$$

$$\begin{bmatrix} 1 & 0 & 1 & 0 \end{bmatrix} \cdot \begin{bmatrix} 1 & 0 & 0 & 0 & 1 & 1 & 1 \\ 0 & 1 & 0 & 0 & 1 & 1 & 0 \\ 0 & 0 & 1 & 0 & 1 & 0 & 1 \\ 0 & 0 & 0 & 1 & 0 & 1 & 1 \end{bmatrix} = \begin{bmatrix} 1 & 0 & 1 & 0 & 0 & 1 & 0 \end{bmatrix}$$

[1] or length of message

After encoding the rows, the provisional codeword is therefore:

$$\begin{bmatrix} 0 & 1 & 1 & 0 & 0 & 1 & 1 \\ 1 & 0 & 1 & 0 & 0 & 1 & 0 \end{bmatrix}.$$

Each column of the final codeword must now be a codeword with parity P. The final codeword is therefore obtained by adding a third row made up of the parity bits of each column. The complete codeword is:

$$\begin{bmatrix} 0 & 1 & 1 & 0 & 0 & 1 & 1 \\ 1 & 0 & 1 & 0 & 0 & 1 & 0 \\ 1 & 1 & 0 & 0 & 0 & 0 & 1 \end{bmatrix}.$$

For the codeword to be valid, it must then be verified that the third row of the word is indeed a codeword of **H**. This row vector must therefore be multiplied by the parity control matrix of **H**:

$$\begin{bmatrix} 1 & 1 & 1 & 0 & 1 & 0 & 0 \\ 1 & 1 & 0 & 1 & 0 & 1 & 0 \\ 1 & 0 & 1 & 1 & 0 & 0 & 1 \end{bmatrix} \cdot \begin{bmatrix} 1 \\ 1 \\ 0 \\ 0 \\ 0 \\ 0 \\ 1 \end{bmatrix} = \begin{bmatrix} 0 \\ 0 \\ 0 \end{bmatrix}$$

In fact, it is not worthwhile doing this verification: it is ensured by construction since codes **H** and **P** are linear. In addition, the encoding order is not important: if we first code by columns then by rows, the codeword obtained is the same.

8.3 Hard input decoding of product codes

8.3.1 Row-column decoding

The first decoding algorithm results directly from the construction of the code: we successively alternate decoding the rows by a decoder of code C_1 and decoding the columns by a decoder of code C_2. Let d_1 (resp. d_2) be the minimum distance of code C_1 (resp. C_2). Then, syndrome decoding of C_1 (resp. C_2) is t_1-correcting (resp. t_2-correcting) with $t_1 = \lfloor d_1/2 \rfloor$ (resp. $t_2 = \lfloor d_2/2 \rfloor$).

Property

Row-column decoding is limited by a correction capability of $(t_1 + 1) \cdot (t_2 + 1)$ errors. In other words, row-column decoding decodes any word having at least

$(t_1 + 1) \cdot (t_2 + 1)$ errors (even if it might decode certain patterns having more errors) and there are words having exactly $(t_1 + 1) \cdot (t_2 + 1)$ errors that will not be decoded.

Indeed, assume that we have a pattern with a number of errors strictly lower than $(t_1 + 1) \cdot (t_2 + 1)$. Since the algorithm decoding in rows corrects up to t_1 errors, after the first decoding step, any row with errors contains at least $t_1 + 1$ errors. There are therefore at least t_2 rows with errors after row decoding. Each column therefore contains at least t_2 errors and column decoding then eliminates all the errors.

There are undecodable patterns having exactly $(t_1 + 1) \cdot (t_2 + 1)$ errors. Take a codeword of $C_1 \otimes C_2$ for which we choose $(t_2 + 1)$ rows and $(t_1 + 1)$ columns at random. At each intersection between a row and a column, we insert an error in the initial codeword. By construction, for the word thus obtained, there exists a codeword for the product code at a distance of $(t_1 + 1) \cdot (t_2 + 1)$ errors, but row decoding and column decoding fail.

We can note that row-column decoding is less powerful than syndrome decoding for a product code. Indeed, a product code is a linear code whose minimum distance is $d_1 d_2$. Therefore syndrome decoding allows us to correct all the words having at least t errors with $t = \lfloor (d_1 d_2)/2 \rfloor$. But row-column decoding allows only all the words having less than $(t_1 + 1) \cdot (t_2 + 1) = (\lfloor d_1/2 \rfloor + 1)(\lfloor d_2/2 \rfloor + 1)$ errors to be corrected. We therefore lose around a factor 2 in error correction capability.

Example 8.2

We assume that we have a product code whose row code and column code both have a minimum distance equal to 5. They are therefore both 2-correcting. The row-column decoding of the product code, according to the above, can thus correct any word having at most 8 errors. Figure 8.1 illustrates a word having 10 errors (shown as points) but that can be corrected by row-column decoding. Figure 8.2 shows a pattern having the same number of errors but not correctable.

8.3.2 The Reddy-Robinson algorithm

The Reddy-Robinson algorithm [8.15] is a more sophisticated iterative algorithm which, now, assumes that for codes C_1 and C_2 we have decoders with errors and erasures. An erasure is an unreliable position in the frame received, whose symbol we think might be erroneous. The difference with a standard error is that the position of the erasure is known at the moment of decoding. For an MHD code equal to d, syndrome decoding can be adapted to take into account the erasures and then it is possible to decode any frame with t errors and e erasures as long as we have

$$2t + e < d \tag{8.1}$$

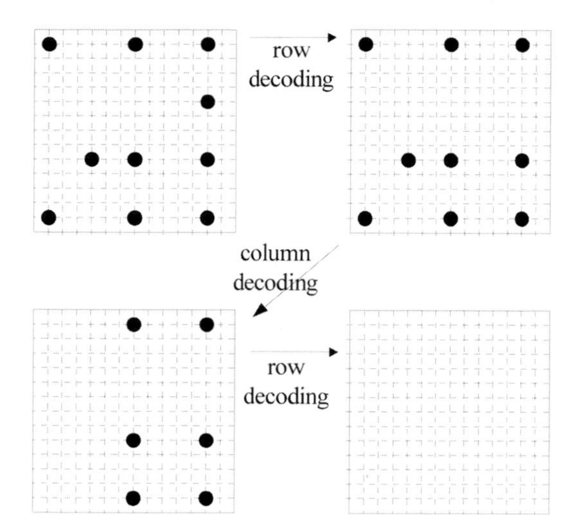

Figure 8.1 – Erroneous word that can be corrected by row-column decoding.

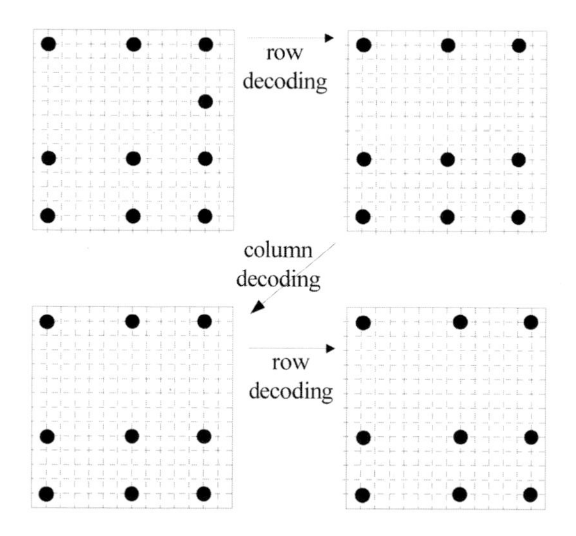

Figure 8.2 – Erroneous word that cannot be corrected by row-column decoding.

The algebraic algorithms of BCH and Reed-Solomon codes can also be adapted to treat erasures.

The Reddy-Robinson algorithm is thus the following:

- Step 1: Find the errors in each row i by applying the decoder of code C_1 without erasures and assign them a weight l_i equal to the number of errors detected. If the decoding fails, the weight assigned is $l_i = d_1/2$. The rows are passed without correction to step 2.
- Step 2: Decode the columns by the decoder of code C_2 with erasures. For each column, we successively perform all the decodings by erasing the least reliable symbols (*i.e.* those that have the highest row weight). We therefore have at least $\lceil d_1/2 \rceil$ decodings per row. At each decoding we assign a weight W such that $W = \sum_{i=1}^{n} w_i$ where $w_i = l_i$ if the symbol of row i is unchanged by the decoding and $w_i = \lceil d_2/2 \rceil$ otherwise.
- Step 3: At the final decoding, for each column we choose the decoded word giving the smallest weight W.

Reddy-Robinson decoding allows any pattern of errors to be corrected whose weight is strictly lower than $d_1 d_2$ [8.15].

Example 8.3

Let us again take the above example with the word of Figure 8.2. During the first step, those rows with 3 errors will not be able to be corrected by the row code since the latter can correct only a maximum of 2 errors (MHD 5). They will therefore be assigned an equal weight 2.5. The row with one error will have a weight equal to 1, while all the remaining rows will have a weight equal to 0. The configuration is then as shown in Figure 8.3.

At the second step, the correction becomes effective. Only three columns have errors. Concerning the most left-hand column with errors, according to the weights provided by step 1, three symbols in this column have a weight equal to 2.5, one symbol with a weight equal to 1 and all the others have a null weight. The second step of the algorithm for this column will therefore involve three successive decodings: decoding without erasure, decoding with three erasures (the symbols having a weight equal to 2.5) and decoding with four erasures (the three previous symbols plus the symbol having a weight of 1). The first decoding fails since the code is only 2-correcting. The second decoding succeeds. Indeed, the column code has a minimum distance of 5. It can thus correct t errors and e erasures when $2t + e < 5$. Now, for this decoding, we have $e = 3$ (since 3 erasures are placed) and $t = 0$ (since there are no additional errors in the column). The weight associated with the second decoding is the sum of the weights of the symbols erased, that is, 7.5. Likewise, the third decoding (with 4 erasures) succeeds and the weight associated with the word decoded is thus 8.5. The algorithm then chooses from among the decodings having succeeded the one whose weight is the lowest, that is, the second decoding in this case.

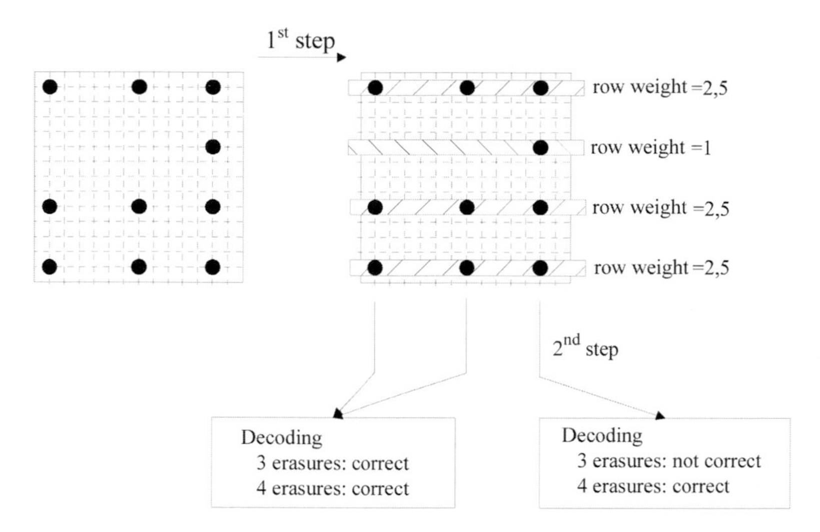

Figure 8.3 – Calculation of the weight of row coming from the first step of the Reddy-Robinson algorithm.

The two other columns with errors are also decoded. However for the most right-hand column, the second decoding fails (since there is one error in the non-erased symbols) and the word decoded for this column is therefore that of the third decoding. Finally, all the errors are corrected.

In this algorithm the role of the rows and of the columns is not symmetric. Thus, if the whole decoding fails in the initial order, it is possible to try again by inverting the role of the rows and the columns.

The Reddy-Robinson decoding algorithm is not iterative in its initial version. We can make it iterative, for example, by starting another decoding with the final word of step 2, if one of the decodings of step 2 succeeded. There are also more sophisticated iterative versions [8.16].

8.4 Soft input decoding of product codes

8.4.1 The Chase algorithm with weighted input

The Chase algorithm, in the case of a block code, enables the values received on the transmission channel to be used to decode in the maximum likelihood sense or, at least approximate its performance. In its basic version, it produces hard decoding. Following the idea of convolutional turbo code decoding [8.2], Pyndiah's improved version [8.14] of the algorithm allows a soft value of the decoded bit to be obtained at the output. With the help of iterative decoding, it is then possible for the row and column decoders of a product code to exchange extrinsic information about the bits. We can thus decode the product codes in a turbo manner.

Let therefore $\mathbf{r} = (r_1, ..., r_n)$ be the received word after encoding and transmission on a Gaussian channel. The Chase-Pyndiah algorithm with t places is decomposed as follows:

- Step 1: Select the t places P_k in the frame containing the least reliable symbols in the frame (i.e. the t places j for which the r_j values are the smallest in absolute value).

- Step 2: Generate the vector of the hard decisions $\mathbf{h_0} = (h_{01}, ..., h_{on})$ such that $h_{0j} = 1$ if $r_j > 0$ and 0 otherwise. Generate the vectors $\mathbf{h_1}, ..., \mathbf{h_{2^t-1}}$ such that $h_{ij} = h_{0j}$ if $j \notin \{P_k\}$ and $h_{iP_k} = h_{0P_k} \oplus Num(i,k)$ where $Num(i,k)$ is the k-th bit in the binary writing of i.

- Step 3: Decode the words $h_0, ..., h_{2^t-1}$ with the hard decoder of the linear code. We thus obtain the concurrent words $c_0, ..., c_{2^t-1}$.

- Step 4: Calculate the metrics of the concurrent words

$$M_i = \sum_{1 \leq j \leq n} r_j(1 - 2c_{ij})$$

- Step 5: Determine the index pp such that

$$M_{pp} = \min \{M_i\}$$

  Codeword c_{pp} is then the most probable codeword.

- Step 6: For each bit j in the frame, calculate the reliability

$$F_j = 1/4(\min \{M_i, c_{ij} \neq c_{pp,j}\} - M_{pp})$$

  If there are no concurrent words for which the j-th bit is different from $c_{pp,j}$, then the reliability F_j is at a fixed value β.

- Step 7: Calculate the extrinsic value for each bit j,

$$E_j = (2 \times c_{pp,j} - 1) \times F_j - r_j$$

The extrinsic values are then exchanged between row decoders and column decoders in an iterative process. The value β, as well as the values of the feedbacks are here more sensitive than in the case of decoding convolutional turbo codes. Inadequate values can greatly degrade the error correction capability. However, it is possible to determine them incrementally. First, for the first iteration we search to see which values give the best performance (for example by dichotomy). Then, these values being fixed, we perform a similar search for the second iteration, and so forth.

Example 8.4

Let $r = (0.5; 0.7; -0.9; 0.2; -0.3; 0.1; 0.6)$ be a received sample of a Hamming codeword $(7, 4, 3)$ and the number of places of the Chase algorithm is equal to $t = 3$. We choose $\beta = 0.6$. The above algorithm thus gives:

- Step 1: $P_1 = 6, P_2 = 4, P_3 = 5$.
- Step 2:

I	h_i
0	(1;1;0;1;0;1;1)
1	(1;1;0;1;0;$\underline{0}$;1)
2	(1;1;0;$\underline{0}$;0;1;1)
3	(1;1;0;$\underline{0}$;0;$\underline{0}$;1)
4	(1;1;0;1;$\underline{1}$;1;1)
5	(1;1;0;1;$\underline{1}$;$\underline{0}$;1)
6	(1;1;0;$\underline{0}$;$\underline{1}$;1;1)
7	(1;1;0;$\underline{0}$;$\underline{1}$;$\underline{0}$;1)

The bits underlined correspond to the inversions performed by the Chase algorithm.

- Step 3:

I	h_i	c_i
0	(1;1;0;1;0;1;1)	(1;1;0;1;0;1;0*)
1	(1;1;0;1;0;$\underline{0}$;1)	(1;1;0;0*;0;0;1)
2	(1;1;0;$\underline{0}$;0;1;1)	(1;1;0;0;0;0*;1)
3	(1;1;0;$\underline{0}$;0;$\underline{0}$;1)	(1;1;0;0;0;0;1)
4	(1;1;0;1;$\underline{1}$;1;1)	(1;1;1*;1;1;1;1)
5	(1;1;0;1;$\underline{1}$;$\underline{0}$;1)	(0*;1;0;1;1;0;1)
6	(1;1;0;$\underline{0}$;$\underline{1}$;1;1)	(1;0*;0;0;1;1;1)
7	(1;1;0;$\underline{0}$;$\underline{1}$;$\underline{0}$;1)	(1;1;0;0;0*;0;1)

The bits with a star in the column of concurrent words c_i correspond to the places corrected by the hard decoder in word h_i.

- Step 4:

I	c_i	M_i
0	(1;1;0;1;0;1;0*)	-(0.5)-(0.7)+(-0.9)-(0.2)+(-0.3)-(0.1)+(0.6)=-2.1
1	(1;1;0;0*;0;0;1)	-(0.5)-(0.7)+(-0.9)+(0.2)+(-0.3)+(0.1)-(0.6)=-2.7
2	(1;1;0;0;0;0*;1)	-(0.5)-(0.7)+(-0.9)+(0.2)+(-0.3)+(0.1)-(0.6)=-2.7
3	(1;1;0;0;0;0;1)	-(0.5)-(0.7)+(-0.9)+(0.2)+(-0.3)+(0.1)-(0.6)=-2.7
4	(1;1;1*;1;1;1;1)	-(0.5)-(0.7)-(-0.9)-(0.2)-(-0.3)-(0.1)-(0.6)=-0.9
5	(0*;1;0;1;1;0;1)	+(0.5)-(0.7)-(-0.9)-(0.2)-(-0.3)+(0.1)-(0.6)=0.3
6	(1;0*;0;0;1;1;1)	-(0.5)+(0.7)+(-0.9)+(0.2)-(-0.3)-(0.1)-(0.6)=-0.9
7	(1;1;0;0;0*;0;1)	-(0.5)-(0.7)+(-0.9)+(0.2)+(-0.3)+(0.1)-(0.6)=-2.7

- Step 5: We have $pp = 1$. The word decoded is $(1; 1; 0; 0; 0; 0; 1)$. Note that we encounter it several times in the list of concurrent words.

- Steps 6 and 7:

j	F_j	E_j
1	((0.3)-(-2.7))/4=0.75	0.75-0.5=0.25
2	((-0.9)-(-2.7))/4=0.475	0.475-0.7=-0.225
3	((-0.9)-(-2.7))/4=0.475	-0.475-(-0.9)=0.525
4	((-0.9)-(-2.7))/4=0.475	-0.475-0.2=-0.675
5	((-2.1)-(-2.7))/4=0.15	-0.15-(-0.3)=0.15
6	((-0.9)-(-2.7))/4=0.475	-0.475-0.1=-0.575
7	((-2.1)-(-2.7))/4=0.15	0.15-0.6=-0.4

8.4.2 Performance of the Chase-Pyndiah algorithm

The Chase-Pyndiah algorithm is the most widespread of weighted decoding algorithms for block codes. Figure 8.4 gives the performance obtained by this algorithm in the context of turbo decoding for different product codes. The curves are circuit oriented, i.e. the data are quantified on $q = 5$ bits. The simulations are done on a Gaussian channel.

Figure 8.5 shows the evolution of the binary error rates during the turbo decoding of the extended BCH code, BCH$(64,57,4)^2$, with the Chase-Pyndiah algorithm. For this code, we can see that most of the gain in correction is obtained during the first 4 iterations.

8.4.3 The Fang-Battail algorithm

Like the Chase-Pyndiah algorithm, the Fang-Battail algorithm enables the reliability of the decoded bits to be calculated from soft data sampled at the output of the transmission channel. We can see this algorithm as a variant of the Chase-Pyndiah algorithm for which only the least reliable bits are modified. Let $r = (r_1, ..., r_n)$ be the received word. The code used is a linear code of length n and dimension k of a generating matrix $\mathbf{G} = \{g_{ij}, 0 \leq i < k, 0 \leq j < n\}$ and parity control matrix $\mathbf{H} = \{h_{ij}, 0 \leq i < n-k, 0 \leq j < n\}$. The algorithm proceeds in the following steps:

- Step 1: Sort the values r_i in the order of decreasing absolute values. We thus obtain a permutation P on all the indices $[1..n]$. We put $r'_l = r_{P(l)}$ and we therefore have $|r'_1| \geq |r'_2| \geq ... \geq |r'_n|$.

- Step 2: Let $\mathbf{H}^*$ be the matrix obtained by permuting the columns of $\mathbf{H}$ by P. Systematize the k most right-hand columns of $\mathbf{H}^*$ by Gauss reduction. Let $\mathbf{H}'$ be the matrix obtained.

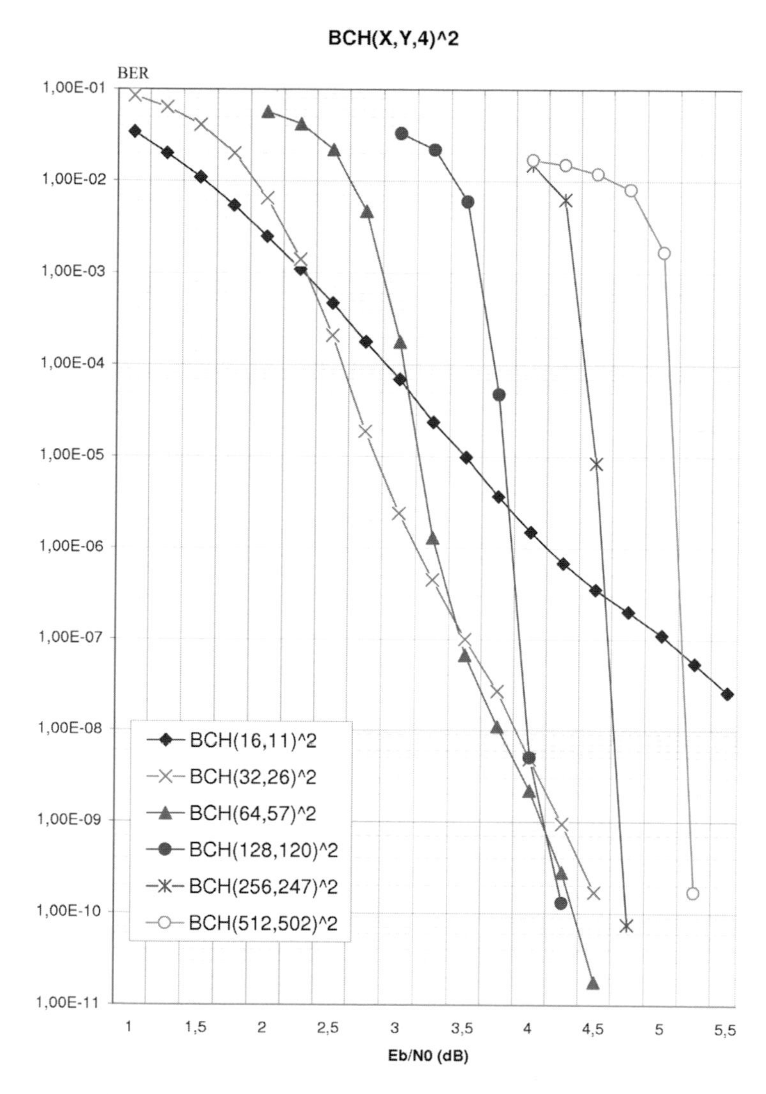

Figure 8.4 – Performance of the Chase-Pyndiah algorithm in the turbo decoding of different 1-correcting codes (2-PSK on a Gaussian channel, 8 iterations, $q = 5$). *(Source P. Adde, Electronics Department, ENST Bretagne).*

- Step 3: Let $\mathbf{s}'$ be the vector of the hard decisions of r'. Let M_1 be the accumulated metrics associated with the $n-k$ first values of r'. That is,

$$\mathbf{s}'' = \mathbf{H}' \begin{bmatrix} s'_1 & s'_2 & \dots & s'_{n-k} & 0 & 0 & \dots & 0 \end{bmatrix}$$

Let M_2 be the accumulated metrics associated with s'' in the k last values of r'. The total accumulated metric is then $M = M_1 + M_2$.

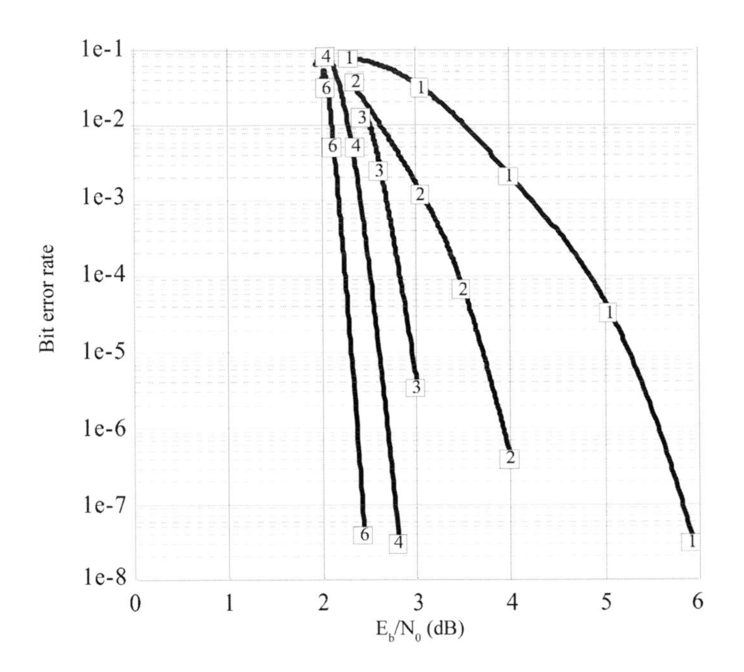

Figure 8.5 – Evolution of the binary error rate during Chase-Pyndiah turbo decoding for 1, 2, 3, 4 and 6 iterations (BCH code(64,51,6), 2-PSK on a Gaussian channel).

- Step 4: List the concurrent words in the order of increasing metrics M_1 (we use a combination generation algorithm with exclusion). Stop as soon as the total metric M begins to increase.

- Step 5: After the inverse permutation of the metrics, calculate the extrinsic values from the concurrent words by using the same equations as in the Chase-Pyndiah algorithm.

Example 8.5

We consider the same example as for the Chase algorithm.

Let $\mathbf{r} = (0.5; 0.7; -0.9; 0.2; -0.3; 0.1; 0.6)$ be a received sample. The parity control matrix of the Hamming code of dimension 4 and length 7 is, as we have already seen,

$$\mathbf{H} = \begin{bmatrix} 1 & 1 & 1 & 0 & 1 & 0 & 0 \\ 1 & 1 & 0 & 1 & 0 & 1 & 0 \\ 1 & 0 & 1 & 1 & 0 & 0 & 1 \end{bmatrix}$$

- Step 1: We have $\mathbf{r}' = (-0.9; 0.7; 0.6; 0.5; -0.3; 0.2; 0.1)$ and $\mathbf{P} = [3, 2, 7, 1, 5, 4, 6]$.

- Step 2:

$$\mathbf{H}^* = \begin{bmatrix} 1 & 1 & 0 & 1 & 1 & 0 & 0 \\ 0 & 1 & 0 & 1 & 0 & 1 & 1 \\ 1 & 0 & 1 & 1 & 0 & 1 & 0 \end{bmatrix}$$

After systematization on the last three columns, we obtain

$$\mathbf{H}' = \begin{bmatrix} 1 & 1 & 0 & 1 & 1 & 0 & 0 \\ 1 & 0 & 1 & 1 & 0 & 1 & 0 \\ 1 & 1 & 1 & 0 & 0 & 0 & 1 \end{bmatrix}$$

- Step 3: The hard decision on the $n - k = 4$ highest received values gives $\mathbf{s}' = (0, 1, 1, 1)$. Matrix $\mathbf{H}'$ without the k last columns makes it possible to find the missing redundancy values:

$$\begin{bmatrix} 1 & 1 & 0 & 1 \\ 1 & 0 & 1 & 1 \\ 1 & 1 & 1 & 0 \end{bmatrix} \cdot \begin{bmatrix} 0 \\ 1 \\ 1 \\ 1 \end{bmatrix} = \begin{bmatrix} 0 \\ 0 \\ 0 \end{bmatrix}$$

The initial decoded vector is therefore $(0, 1, 1, 1, 0, 0, 0)$ which, after inverse permutation, gives the vector $(1, 1, 0, 0, 0, 0, 1)$. The initial metric is $M = M_1 + M_2 = -2.7$ as $M_1 = (-0.9) - 0.7 - 0.6 - 0.5 = -2.7$ and $M_2 = (-0.3) + 0.2 + 0, 1 = 0.0$.

- Step 4: To list the concurrent words, we apply inversion masks to the $n - k$ first bits (we can have from 1 to 4 inversions maximum). Each inversion mask will increase the M_1 part of the metric. For an inversion, the bonus is at minimum 1.0 and at maximum 1.8. The minimum bonus for two inversions is at minimum $2 \times (0.6 + 0.5) = 2.2$. The first concurrent words to consider are therefore all those corresponding to a single inversion. Moreover, modifications on the M_2 part of the metric could decrease it. However, the decrease cannot exceed $2 \times (0.2 + 0.1) = 0.6$ compared to the initial metric. We therefore already know that the found previously word is the most likely word. However, if we decide to list them anyway, we find the following concurrent words which, unlike the Chase-Pyndiah algorithm, are all different from each other:

s_i	C'_i	M
(0,1,1,1)	(0,1,1,1,0,0,0)	-2.7
(0,1,1,0)	(0,1,1,0,1,1,0)	-2.7+1.0+0.2=-1.5
(0,1,0,1)	(0,1,0,1,0,1,1)	-2.7+1.2-0.6=-2.1
(0,0,1,1)	(0,0,1,1,1,0,1)	-2.7+1.4+0.2=-1.1
(1,1,1,1)	(1,1,1,1,1,1,1)	-2.7+1.8+0.0=-0.9
(0,1,0,0)	(0,1,0,0,1,0,1)	-2.7+2.2+0.2=-0.3
(0,0,1,0)	(0,0,1,0,0,1,1)	-2.7+2.4-0.6=-0.9
(0,0,0,1)	(0,0,0,1,1,1,0)	-2.7+2.6+0.2=0.1
(1,1,1,0)	(1,1,1,0,0,0,1)	-2.7+2.8-0.2=-0.1
(1,1,0,1)	(1,1,0,1,1,0,0)	-2.7+3.0+0.6=0.9
(1,0,1,1)	(1,0,1,1,0,1,0)	-2.7+3.2-0.2=0.3
(0,0,0,0)	(0,0,0,0,0,0,0)	-2.7+3.6+0.0=0.9
(1,1,0,0)	(1,1,0,0,0,1,0)	-2.7+4.0-0.2=1.1
(1,0,1,0)	(1,0,1,0,1,0,0)	-2.7+4.2+0.6=2.1
(1,0,0,1)	(1,0,0,1,0,0,1)	-2.7+4.4-0.2=1.5
(1,0,0,0)	(1,0,0,0,1,1,1)	-2.7+5.4+0.0=2.7

- Step 5:

j	$F_{P(j)}$
1	((-0.9)-(-2.7))/4=0.475
2	((-1.1)-(-2.7))/4=0.4
3	((-2.1)-(-2.7))/4=0.15
4	((-1.5)-(-2.7))/4=0.3
5	((-1.5)-(-2.7))/4=0.3
6	((-1.5)-(-2.7))/4=0.3
7	((-2.1)-(-2.7))/4=0.15

J	E_J
1	0.3-0.5=-0.2
2	0.4-0.7=-0.3
3	-0.475-(-0.9)=0.525
4	-0.3-0.2=-0.5
5	-0.3-(-0.3)=0.0
6	-0.15-0.1=-0.25
7	0.15-0.6=-0.4

The Fang-Battail algorithm is theoretically maximum likelihood. However, that will not necessarily suffice for the algorithm applied for decoding product codes to be maximum likelihood itself.

8.4.4 The Hartmann-Nazarov algorithm

Hartmann et al. [8.8] used the properties of duality to describe a maximum likelihood decoding method for linear codes. Initially, only a hard decision was provided by the algorithm. Hagenauer [8.7] then took up these ideas in order to make available the extrinsic values necessary for turbo decoding. Finally, Nazarov and Smolyaninov [8.13] showed how to reduce the complexity cost by using the fast Hadamard transform. This paragraph summarizes these three reference articles.

Let $\mathbf{r} = (r_1, ..., r_n)$ be the received word. The code used is a linear code of length n and dimension k of a generating matrix $\mathbf{G} = \{g_{ij}, 0 \leq i < k, 0 \leq j < n\}$ and parity control matrix $\mathbf{H} = \{h_{ij}, 0 \leq i < n-k, 0 \leq j < n\}$. We assume that the transmission has been done on a channel perturbed by a Gaussian noise with null mean and variance equal to σ^2.

Putting $\rho_l = -\tanh(r_l/\sigma^2)$, we show that the LLR maximum likelihood of the l-th bit in the frame is:

$$LLR_l = \ln\left(\frac{1-C(l)}{1+C(l)}\right) \tag{8.2}$$

with:

$$C(l) = \frac{\rho_l}{2}\left(1 + \frac{F_D(h_l)}{F_D(0)}\right) + \frac{1}{2\rho_l}\left(1 - \frac{F_D(h_l)}{F_D(0)}\right)$$

where h_l is the l-th column of the parity control matrix and function F_D is such that

$$F_D(h_l) = \sum_{\nu=0}^{2^{n-k}-1} D_\nu(l) \exp\left\{ j\pi \sum_{m=0}^{n-k-1} \nu_m h_{ml} \right\}$$

ν_m is the m-th bit of the binary representation of integer ν ($\nu = \sum_{m=0}^{n-k-1} \nu_m 2^m$), and:

$$D_\nu(l) = \prod_{l=0}^{n-1} (\rho_l)^{t_\nu(l)}$$

$t_\nu(l)$ being the l-th bit of the ν-th vector of the dual of the code, that is, therefore:

$$t_\nu(l) = \left(\sum_{m=0}^{n-k-1} \nu_m h_{ml}\right) \bmod 2 = \langle \nu, h_l \rangle \bmod 2$$

The calculation of $D_\nu(l)$ normally requires n multiplications. But using the dual code and applying a fast Hadamard transform, it is possible to lower this cost to one term of the order of $n-k$ multiplications. If the coding rate is high, then n is much higher than $n-k$ and the gain in terms of computation complexity is high.

To do this, we re-write the general term of $D_\nu(l)$ in the form:

$$(\rho_l)^{t_\nu(l)} = \exp\{t_\nu(l)\ln|\rho_l|\}\exp\{j\pi q_l t_\nu(l)\}$$

where q_l is such that $\rho_l = (-1)^{q_l}|\rho_l|$.
We then have:

$$D_\nu(l) = \exp\left\{\sum_{l=0}^{n-1}(t_\nu(l)\ln|\rho_l|) + (j\pi q_l t_\nu(l))\right\}$$

$$= \exp\left\{\sum_{l=0}^{n-1}(t_\nu(l)\ln|\rho_l|)\right\}\exp\left\{\sum_{l=0}^{n-1}(j\pi q_l t_\nu(l))\right\}$$

Put:

$$F_\rho(w) = \sum_{l=0}^{n-1}\ln(|\rho_l|)\exp\left\{j\pi\sum_{m=0}^{n-k-1}w_m h_{ml}\right\}$$

for any integer $0 \le w \le 2^{n-k}-1$. We have, therefore:

$$F_\rho(w) = \sum_{l=0}^{n-1}\ln(|\rho_l|)\exp\{j\pi t_w(l)\}$$

with, in particular, $F_\rho(0) = \sum_{l=0}^{n-1}\ln(|\rho_l|)$.

On the other hand, if $t = 0$ or 1, then $\frac{1-\exp\{j\pi t\}}{2} = t$ and

$$\frac{F_\rho(0) - F_\rho(\nu)}{2} = \sum_{l=0}^{n-1}(t_\nu(l)\ln|\rho_l|)\ .$$

Likewise, if we put $F_q(w) = \sum_{l=0}^{n-1} q_l \exp\left\{j\pi\sum_{m=0}^{n-k-1}w_m h_{ml}\right\}$, we have:

$$\frac{F_q(0) - F_q(\nu)}{2} = \sum_{l=0}^{n-1}(q_l\ln|\rho_l|)$$

and therefore:

$$D_\nu(l) = \exp\left(\frac{1}{2}(F_\rho(0) - F_\rho(\nu))\right)\exp\left(\frac{1}{2}j\pi(F_q(0) - F_q(\nu))\right)$$

The two terms $F_\rho(\nu)$ and $F_q(\nu)$ have a common expression of the form:

$$F(w) = \sum_{l=0}^{n-1} f_l \exp\left\{j\pi\sum_{m=0}^{n-k-1}w_m h_{ml}\right\}$$

where f_l are real numbres that depend only on l.

We define function g on the set $\{0, \cdots, 2^{n-k}-1\}$ with by:

$$g(p) = \begin{cases} f_l & \text{if } \exists l, \quad p = \sum\limits_{m=0}^{n-k-1} h_{ml} 2^m \\ 0 & \text{otherwise} \end{cases}$$

Function g is well defined since the columns of $\mathbf{H}$ are linearly independent and therefore *a fortiori* two-by-two distinct. The Hadamard transform of $\mathbf{G}$ is then a function with real values defined on the interval $[0..2^{n-k}-1]$ by:

$$G(w) = \sum_{p=0}^{2^{n-k}-1} g(p)(-1)^{<p,w>}$$

Now, function g is null except for points $p_l = \sum\limits_{l=0}^{n-k-1} h_{ml} 2^m$ for $l \in [0, \cdots, n-1]$. Thus, we have:

$$G(w) = \sum_{l=0}^{n-1} f_l (-1)^{<\sum\limits_{m=0}^{n-k-1} h_{ml} 2^m, w>} = \sum_{l=0}^{n-1} f_l \exp\left(j\pi \sum_{m=0}^{n-k-1} w_m h_{ml}\right) = F(w)$$

The two terms $F_\rho(\nu)$and $F_q(\nu)$ are therefore expressed as Hadamard transforms and can be calculated by means of the fast Hadamard transform.

Fast Hadamard transform

Let $\mathbf{R} = [R_0, R_1, ..., R_{2^n-1}]$ be a vector with real components. The vector obtained from $\mathbf{R}$ by the Hadamard transform is vector $\hat{\mathbf{R}} = [\hat{R}_0, \hat{R}_1, \cdots, \hat{R}_{2^n-1}]$ such that

$$\hat{R}_j = \sum_{i=0}^{2^n-1} R_i (-1)^{<i,j>} \tag{8.3}$$

The scalar product $< i, j >$ is, as above, the bit-by-bit scalar product of the binary developments of i and j. We also write in vector form, $\hat{\mathbf{R}} = \mathbf{R}\mathbf{H}_{2^n}$ where $\mathbf{H}_{2^n}$ is the Hadamard matrix of order 2^n whose coefficient $(H_{2^n})_{i,j} = (-1)^{<i,j>}$.

Let $\mathbf{A}$ be a matrix size $a_1 \times a_2$ and $\mathbf{B}$ a matrix size $b_1 \times b_2$ with real coefficients. Then the Kronecker product of $\mathbf{A}$ by $\mathbf{B}$, denoted $\mathbf{A} \otimes \mathbf{B}$, is a matrix size $(a_1 b_1) \times (a_2 b_2)$ such that $(\mathbf{A} \otimes \mathbf{B})_{i,j} = A_{q_1 q_2} B_{r_1 r_2}$ where $i = b_1 q_1 + r_1$ and $j = b_2 q_2 + r_2$ with $0 \leq r_1 < b_1$ and $0 \leq r_2 < b_2$.

If $N = 2^n$, we show that ([8.11]):

$$\mathbf{H}_N = (\mathbf{H}_2 \otimes \mathbf{I}_{N/2})(\mathbf{I}_2 \otimes \mathbf{H}_{N/2})$$

where $\mathbf{I}_k$ is the unit matrix of order k.
Developing recursively, we obtain:

$$\mathbf{H}_N = \prod_{i=1}^{n} \left(\mathbf{I}_{2^{i-1}} \otimes \mathbf{H}_2 \otimes \mathbf{I}_{N/2^i}\right)$$

The fast Hadamard transform is in fact the use of this factorization to calculate $\hat{\mathbf{R}}$.

Example 8.6

Let us calculate the Hadamard transform of vector

$$\mathbf{R} = [0.2; 0.5; -0.7; 1.3; 0.1; -1.1; 0.8; -0.3]$$

We have $\hat{\mathbf{R}} = \mathbf{R}\mathbf{H}_8$ with

$$\mathbf{H}_8 = \begin{bmatrix} 1 & 1 & 1 & 1 & 1 & 1 & 1 & 1 \\ 1 & -1 & 1 & -1 & 1 & -1 & 1 & -1 \\ 1 & 1 & -1 & -1 & 1 & 1 & -1 & -1 \\ 1 & -1 & -1 & 1 & 1 & -1 & -1 & 1 \\ 1 & 1 & 1 & 1 & -1 & -1 & -1 & -1 \\ 1 & -1 & 1 & -1 & -1 & 1 & -1 & 1 \\ 1 & 1 & -1 & -1 & -1 & -1 & 1 & 1 \\ 1 & -1 & -1 & 1 & -1 & 1 & 1 & -1 \end{bmatrix}$$

The direct calculation gives:

$$\mathbf{R} = [0.8; 0.0; -1.4; 1.8; 1.8; -4.6; 1.6; 1.6]$$

Now, according to the above, $\mathbf{H}_8 = \mathbf{G}_1\mathbf{G}_2\mathbf{G}_3$ where $\mathbf{G}_i = \mathbf{I}_i \otimes \mathbf{H}_2 \otimes \mathbf{I}_{8/2^i}$, we have:

$$\mathbf{G}_1 = [1] \otimes \begin{bmatrix} 1 & 1 \\ 1 & -1 \end{bmatrix} \otimes \begin{bmatrix} 1 & 0 & 0 & 0 \\ 0 & 1 & 0 & 0 \\ 0 & 0 & 1 & 0 \\ 0 & 0 & 0 & 1 \end{bmatrix} = \begin{bmatrix} 1 & 0 & 0 & 0 & 1 & 0 & 0 & 0 \\ 0 & 1 & 0 & 0 & 0 & 1 & 0 & 0 \\ 0 & 0 & 1 & 0 & 0 & 0 & 1 & 0 \\ 0 & 0 & 0 & 1 & 0 & 0 & 0 & 1 \\ 1 & 0 & 0 & 0 & -1 & 0 & 0 & 0 \\ 0 & 1 & 0 & 0 & 0 & -1 & 0 & 0 \\ 0 & 0 & 1 & 0 & 0 & 0 & -1 & 0 \\ 0 & 0 & 0 & 1 & 0 & 0 & 0 & -1 \end{bmatrix}$$

$$\mathbf{G}_2 = \begin{bmatrix} 1 & 0 \\ 0 & 1 \end{bmatrix} \otimes \begin{bmatrix} 1 & 1 \\ 1 & -1 \end{bmatrix} \otimes \begin{bmatrix} 1 & 0 \\ 0 & 1 \end{bmatrix} = \begin{bmatrix} 1 & 0 \\ 0 & 1 \end{bmatrix} \otimes \begin{bmatrix} 1 & 0 & 1 & 0 \\ 0 & 1 & 0 & 1 \\ 1 & 0 & -1 & 0 \\ 0 & 1 & 0 & -1 \end{bmatrix}$$

$$= \begin{bmatrix} 1 & 0 & 1 & 0 & 0 & 0 & 0 & 0 \\ 0 & 1 & 0 & 1 & 0 & 0 & 0 & 0 \\ 1 & 0 & -1 & 0 & 0 & 0 & 0 & 0 \\ 0 & 1 & 0 & -1 & 0 & 0 & 0 & 0 \\ 0 & 0 & 0 & 0 & 1 & 0 & 1 & 0 \\ 0 & 0 & 0 & 0 & 0 & 1 & 0 & 1 \\ 0 & 0 & 0 & 0 & 1 & 0 & -1 & 0 \\ 0 & 0 & 0 & 0 & 0 & 1 & 0 & -1 \end{bmatrix}$$

$$\mathbf{G}_3 = \begin{bmatrix} 1 & 0 & 0 & 0 \\ 0 & 1 & 0 & 0 \\ 0 & 0 & 1 & 0 \\ 0 & 0 & 0 & 1 \end{bmatrix} \otimes \begin{bmatrix} 1 & 1 \\ 1 & -1 \end{bmatrix} = \begin{bmatrix} 1 & 1 & 0 & 0 & 0 & 0 & 0 & 0 \\ 1 & -1 & 0 & 0 & 0 & 0 & 0 & 0 \\ 0 & 0 & 1 & 1 & 0 & 0 & 0 & 0 \\ 0 & 0 & 1 & -1 & 0 & 0 & 0 & 0 \\ 0 & 0 & 0 & 0 & 1 & 1 & 0 & 0 \\ 0 & 0 & 0 & 0 & 1 & -1 & 0 & 0 \\ 0 & 0 & 0 & 0 & 0 & 0 & 1 & 1 \\ 0 & 0 & 0 & 0 & 0 & 0 & 1 & -1 \end{bmatrix}$$

We can then calculate $\hat{\mathbf{R}}$ by three matrix multiplications:

$$\begin{aligned} \hat{\mathbf{R}} &= [0.8; 0.0; -1.4; 1.8; -1.0; -1.8; 0.0; 3.2] \cdot \mathbf{G}_1 \cdot \mathbf{G}_2 \cdot \mathbf{G}_3 \\ &= [0.8; 0.0; -1.4; 1.8; 1.8; -4.6; 1.6; 1.6] \end{aligned}$$

The matrices $\mathbf{G}_i$ are sparse matrices having only two non-null elements per column. Moreover, factorization of the Hadamard matrix $\mathbf{H}_N$ has a length proportional to $\log(N)$. The total computation cost is therefore in $N \log(N)$ by the fast transform, instead of N^2 by the direct method. Figure 8.6 presents the graph of the calculations for the fast Hadamard transform in the case $N = 8$.

In terms of error correcting performance, article [8.6] shows that we obtain the same performance as the Chase-Pyndiah algorithm for two times fewer iterations.

8.4.5 Other soft input decoding algorithms

There are still other algorithms for decoding block codes with weighted input and output. One of the oldest is Farrell *et al.*'s algorithm [8.5] for decoding by local

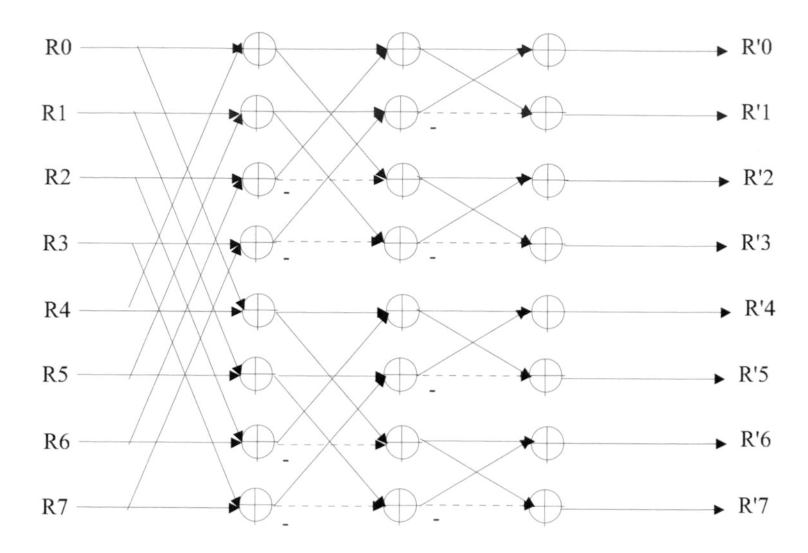

Figure 8.6 – Graph of the computation flow for the fast Hadamard transform $N = 8$.

optimization. This technique involves posing the problem of searching for the most probable word as a problem of minimizing a global cost function, having as its variables the information bits of the word. This optimization must be done theoretically in integers, which makes the problem very difficult. The technique used is to replace these integer variables by real variables. The problem is then no longer strictly equivalent to the initial problem but it becomes possible to use classical non-linear optimization techniques, like the gradient method, to search for the minimum of the cost function.

Another approach introduced by Kschischang and Sorokine [8.17] involves using the fact that most block codes used in turbo product codes are in fact trellis codes. It is then possible to use classical(MAP, Max-Log-MAP, ...) algorithms for decoding trellis codes to obtain the soft decisions of the decoder. This type of algorithm is an application in the case of the turbo product codes of the maximum likelihood decoding algorithms of Wolf [8.19].

Another more recently invented algorithm is Kötter and Vardy's [8.10]. This algorithm is only applicable to very specific codes, mainly including Reed-Solomon codes and algebraic codes. It is based on the Sudan algorithm [8.18] which is capable of determining the codewords in a neighbourhood close to the received word. It is then possible to use conventional weighting techniques. Although the initial version of this algorithm is relatively computation-costly, theoretical improvements have been made and this algorithm is even more promising since Reed-Solomon codes are widely used in the domain of telecommunications. There are also decoding algorithms based on sub-codes [8.12]. Although slightly complex at implementation level, these algorithms provide excellent performance. Finally, recent studies have shown that decoding algorithms based on belief propagation can be applied to linear codes in general [8.9].

8.5 Implantation of the Chase-Pyndiah algorithm

The Fang-Battail algorithm requires the systematization of the parity matrix which is a very costly operation. The Hartmann-Nazarov algorithm has been implemented on *DSP* [8.6], but the precision necessary for the computation of the metrics is too great for it to be envisaged to implement the decoder on a reasonably sized *ASIC*. This is the reason why the Chase-Pyndiah algorithm is the most commonly used decoding algorithm with weighted output for implementing dedicated circuits [8.1], since by a judicious use of memories, it allows turbo product codes architectures to be elaborated that are adapted to high-rate transmissions [8.3].

Turbo decoding using the Chase-Pyndiah algorithm alternates weighted decoding of the rows and columns. This iterative process leads to the architecture of Figure 8.7. Between each half-iteration is inserted a phase for reconstructing the matrix in order to obtain the decoded word.

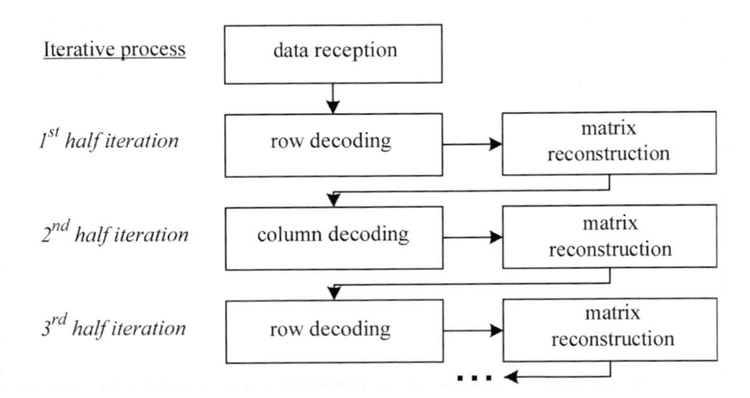

Figure 8.7 – Global architecture of a turbo product decoder.

Each half-iteration processor takes at its input the channel data as well as the extrinsic values produced by the previous half-iteration. At the output, the processor transfers the channel data (to ensure a *pipeline* operation of the decoding chain), the hard decisions of the most probable codeword and the extrinsic values calculated by the Chase-Pyndiah algorithm. The architecture of the processor is illustrated in Figure 8.8. The FIFOs (*First-In/First-Out*) are used to synchronize the input and output data of the SISO decoder which has a certain latency. Small sized FIFOs are generally implemented by rows of D flip-flops. When size increases, for reasons of hardware space and consumption, it becomes fairly quickly worthwhile using a RAM memory and two pointers (one for writing and one for reading), incremented at each new datum, with the addressing managed circularly. We can also make a remark about the multiplication of the extrinsic values by α. In a conventional implementation, values W_k are generally integers of reduced size (5 or 6 bits) but the hardware cost of a real multiplier is prohibitive, and we generally prefer to substitute it by a simple table.

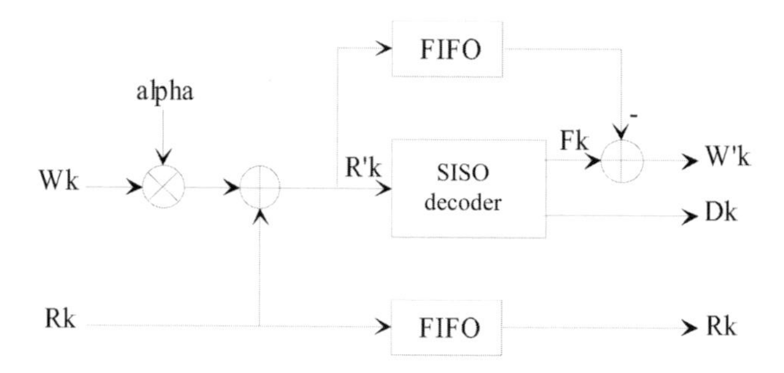

Figure 8.8 – Internal architecture of the half-iteration processor.

The SISO decoder, described by Figure 8.9, performs the steps of the Chase-Pyndiah algorithm. The decoder is made up of five parts:

- *The module for sequential processing of the data* calculates in parallel the input codeword syndrome and the least reliable positions in the frame.
- *The algebraic decoding module* performs the algebraic decoding of the words built from input $R_{k'}^{'}$ data and knowledge of the least reliable places.
- *The selection module* determines the most probable codeword as well as the closest concurrent word or words.
- *The module for calculating the weightings* determines the reliability of the decoded bits.
- *The memory module* stores the total input weightings that are used to calculate the weightings.

The module for processing the data receives the sample bits, one after the other. If the code is cyclic (BCH, for example) calculating the syndrome is then very simply done by using the factorization of the generator polynomial following the Hörner scheme. Determining the least reliable positions is often done by sequentially managing the list of not so reliable positions in a small local RAM. There are also other solutions that are more economical in size, like Leiserson's systolic array, but the gain obtained is small.

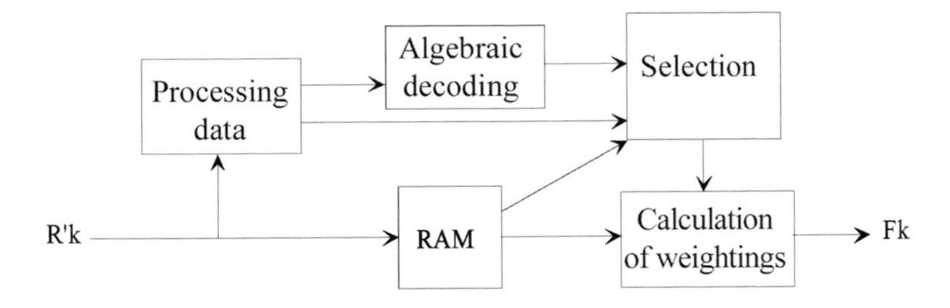

Figure 8.9 – Internal architecture of the SISO decoder.

The algebraic decoding module uses the value of the syndrome to determine the erroneous places in the concurrent vectors. In the case of BCH codes, it is possible to use the Berlekamp-Massey algorithm or the extended Euclid algorithm to make the correction. It should, however, be noted that this solution is really only economical for block codes with high correcting power. For codes with low error correction capability, it is less costly to store the bits to be corrected in a local ROM, for each possible value of the syndrome.

The selection module must sort among the words generated by the algebraic decoding module to determine the most probable ones. It must therefore calculate the metric of each of these words (which it does sequentially by additions) and determine, by computation of the minimum value, the most probable among them (their number is generally limited, for the sake of space).

Finally, the module for calculating the weightings uses the list of concurrent words chosen above to generate the weightings from the equation of step 6 of the Chase-Pyndiah algorithm. This module has low complexity since the calculations to be done are relatively simple and, for each bit, it must keep only two values sequentially (the smallest metric among the concurrent words having 0 as the corresponding value for this bit in its binary development, and the smallest metric for the candidate words having 1 as their binary value). This module also contains value β. In the case of an FPGA (*Field Programmable Gate Array*) implantation, all the iterations are generally executed on the same hardware which is re-used from half-iteration to half-iteration. We must therefore anticipate a procedure for loading value β.

Bibliography

[8.1] P. Adde, R. Pyndiah, and O. Raoul. Performance and complexity of block turbo decoder circuits. In *Proceedings of Third International Conference on Electronics, Circuits and Systems, (ICECS'96)*, pages 172–175, Rhodes, Greece, 1996.

[8.2] C. Berrou, A. Glavieux, and P. Thitimajshima. Near shannon limit error-correcting coding and decoding: turbo-codes. In *Proceedings of IEEE International Conference on Communications (ICC'93)*, pages 1064–1070, GENEVA, May 1993.

[8.3] J. Cuevas, P. Adde, and S. Kerouedan. Information theory turbo decoding of product codes for gigabit per second applications and beyond. *European Transactions on Telecommunications*, 17(1):345–55, Jan. 2006.

[8.4] P. Elias. Error-free coding. *IEEE Transactions on Information Theory*, 4(4):29–39, Sept. 1954.

[8.5] K. Farrell, L. Rudolph, C. Hartmann, and L. Nielsen. Decoding by local optimization. *IEEE Transactions on Information Theory*, 29(5):740–743, Sept. 1983.

[8.6] A. Goalic, K. Cavalec-Amis, and V. Kerbaol. Real-time turbo decoding of block turbo codes using the hartmann-nazarov algorithm on the dsp texas tms320c6201. In *Proceedings of International Conference on Communications (ICC'2002)*, New-York, 2002.

[8.7] J. Hagenauer, E. Offer, and L. Papke. Iterative decoding of binary block and convolutional codes. *IEEE Transactions on Information Theory*, IT-42:429–445, 1996.

[8.8] C.R.P. Hartmann and L.D. Rudolph. An optimum symbol-by-symbol decoding rule for linear codes. *IEEE Transactions on Inforamtion Theory*, IT-22:514–517, 1974.

[8.9] A. Kothiyal and O.Y. Takeshita. A comparison of adaptative belief propagation and the best graph algorithm for the decoding of linear block codes. In *Proceedings of IEEE International Symposium on Information Theory*, pages 724–728, 2005.

[8.10] R. Kotter and A. Vardy. Algebraic soft-decision decoding of reed-solomon codes. In *Proceedings og IEEE International Symposium on Information Theory*, page 61, 2000.

[8.11] M. Lee and M. Kaveh. Fast hadamard transform based on a simple matrix factorization. *IEEE Transactions on Acoustics, Speech, and Signal Processing*, 34:1666–1667, 1986.

[8.12] C.Y. Liu and S. Lin. Turbo encoding and decoding of reed-solomon codes through binary decomposition and self-concatenation. *IEEE Transactions on Communications*, 52(9):1484–1493, Sept. 2004.

[8.13] L.E. Nazarov and V.M. Smolyaninov. Use of fast walsh-hadamard transformation for optimal symbol-by-symbol binary block-code decoding. *Electronics Letters*, 34:261–262, 1998.

[8.14] R. Pyndiah, A. Glavieux, A. Picart, and S. Jacq. Near optimum decoding of product codes. In *Proceedings of IEEE Global Communications Conference (GLOBECOM'94)*, San Francisco, 1994.

[8.15] S. M. Reddy and J. P. Robinson. Random error and burst corrections by iterated codes. *IEEE Transactions on Information Theory*, 18(1):182–185, 1972.

[8.16] N. Sendrier. *Codes correcteurs à haut pouvoir de correction.* PhD thesis, Université Paris VI, 1991.

[8.17] V. Sorokine, F.R. Kschischang, and V. Durand. Trellis based decoding of binary linear block codes. *Lecture Notes in Computer Science*, 793:270–286, 1994.

[8.18] M. Sudan. Decoding of reed-solomon codes beyond the error correction bound. *Journal of Complexity*, 12:180–193, 1997.

[8.19] J. Wolf. Efficient maximum likelihood decoding of linear block codes using a trellis. *IEEE Transactions on Information Theory*, 24:76–80, 1978.

Chapter 9

LDPC codes

Low Density Parity Check (LDPC) codes make up a class of block codes that are characterized by a sparse parity check matrix. They were first described in Gallager's thesis at the beginning of the 60s [9.21]. Apart from the hard input decoding of LDPC codes, this thesis proposed iterative decoding based on belief propagation (BP). This work was forgotten for 30 years. Only a few rare studies referred to it during this dormant period, in particular, Tanner's which proposed a generalization of the Gallager codes and a bipartite graph [9.53] representation.

After the invention of turbo codes, LDPC codes were rediscovered in the middle of the 90s by MacKay *et al.* [9.39], Wilberg [9.64] and Sipser *et al.* [9.52]. Since then, considerable progress concerning the rules for building good LDPC codes, and coding and decoding techniques, have enabled LDPC codes to be used, like turbo codes, in practical applications.

This chapter gives an overview of the encoding and decoding of LDPC codes, and some considerations about hardware implementations.

9.1 Principle of LDPC codes

LDPC codes are codes built from the simplest elementary code: the single parity check code. We therefore begin this chapter by describing the single parity check code and its soft in soft out decoding before dealing with the construction of LDPC codes.

9.1.1 Parity check code

Definition

A parity equation, represented graphically by Figure 9.1, is an equation linking n binary data to each other by the *exclusive or*, denoted $\oplus$ operator. It is satisfied if the total number of 1s in the equation is even or null.

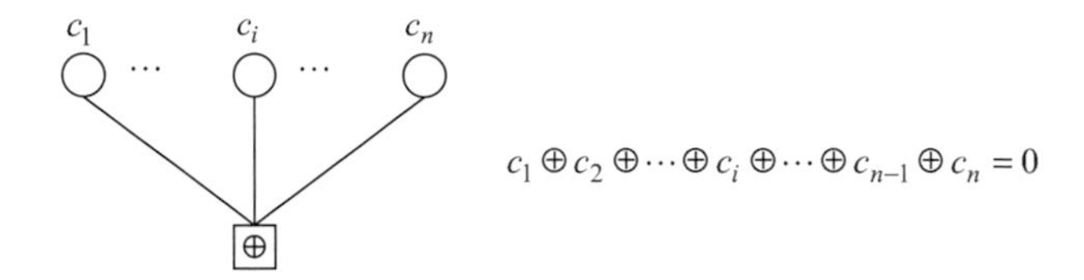

Figure 9.1 – Graphic representation of a parity equation.

The circles represent the binary data c_i, also called *variables*. The rectangle containing the *exclusive or* operator represents the parity equation (also called the parity constraint, or parity). The links between the variables and the operator indicate the variables involved in the parity equation.

Parity code with three bits

We consider that the binary variables c_1, c_2 and c_3 are linked by the parity constraint $c_1 \oplus c_2 \oplus c_3 = 0$, and that they make up the codeword (c_1, c_2, c_3). We assume that we know the log likelihood ratio (LLR) $L(c_1)$ and $L(c_2)$ of variables c_1 and c_2: what can we then say about the LLR $L(c_3)$ of variable c_3?
We recall that $L(c_j)$ is defined by the equation:

$$L\left(c_j\right) = \ln\left(\frac{\Pr\left(c_j = 1\right)}{\Pr\left(c_j = 0\right)}\right) \tag{9.1}$$

There are two codewords in which bit c_3 is equal to 0: codewords (0,0,0) and (1,1,0). Similarly, there are two codewords in which bit c_3 is equal to 1: codewords (1,0,1) and (0,1,1). We deduce from this the following two equations in the probability domain:

$$\left\{\begin{array}{l} \Pr(c_3 = 1) = \Pr(c_1 = 1) \times \Pr(c_2 = 0) + \Pr(c_1 = 0) \times \Pr(c_2 = 1) \\ \Pr(c_3 = 0) = \Pr(c_1 = 0) \times \Pr(c_2 = 0) + \Pr(c_1 = 1) \times \Pr(c_2 = 1) \end{array}\right. \tag{9.2}$$

Using the expression of each probability according to the likelihood ratio function, deduced from Equation (9.1):

$$\left\{\begin{array}{l} \Pr(c_j = 1) = \frac{\exp(L(c_j))}{1+\exp(L(c_j))} \\ \\ \Pr(c_j = 0) = 1 - \Pr(c_j = 1) = \frac{1}{1+\exp(L(c_j))} \end{array}\right.$$

we have:

$$L(c_3) = \ln\left[\frac{1+\exp(L(c_2)+L(c_1))}{\exp(L(c_2))+\exp(L(c_1))}\right] \triangleq L(c_1) \oplus L(c_2) \tag{9.3}$$

Equation (9.3) enables us to define the switching operator $\oplus$ between the two LLRs of the variables c_1 and c_2.

Applying function $\tanh(x/2) = \frac{\exp(x)-1}{\exp(x)+1}$ to Equation (9.3), the latter becomes:

$$\begin{aligned} \tanh\left(\tfrac{L(c_3)}{2}\right) &= \tfrac{\exp(L(c_0))-1}{\exp(L(c_0))+1} \times \tfrac{\exp(L(c_1))-1}{\exp(L(c_1))+1} \\ &= \prod_{j=0}^{1} \tanh\left(\tfrac{L(c_j)}{2}\right) \end{aligned} \tag{9.4}$$

It is practical (and frequent) to separate the processing of the sign and the magnitude in Equation (9.4) which can then be replaced by the following two equations:

$$\operatorname{sgn}(L(c_3)) = \prod_{j=0}^{1} \operatorname{sgn}(L(c_j)) \tag{9.5}$$

$$\tanh\left(\frac{|L(c_3)|}{2}\right) = \prod_{j=0}^{1} \tanh\left(\frac{|L(c_j)|}{2}\right) \tag{9.6}$$

where the sign function $\operatorname{sgn}(x)$ is such that $\operatorname{sgn}(x) = +1$ if $x \geq 0$ and $\operatorname{sgn}(x) = -1$ otherwise.

Processing the magnitude given by Equation (9.6) can be simplified by taking the inverse of the logarithm of each of the terms of the equation, which gives:

$$|L(c_3)| = f^{-1}\left(\sum_{j=1,2} f\left(|L(c_j)|\right)\right) \tag{9.7}$$

where function f, satisfying $f^{-1}(x) = f(x)$, is defined by:

$$f(x) = \ln\left(\tanh(x/2)\right) \tag{9.8}$$

These different aspects of the computation of function $\oplus$ will be developed in the architecture part of this chapter.

Expression (9.6) in fact corresponds to the computation of (9.3) in the Fourier domain. Finally, there is also a third writing of the LLR of variable c_2 [9.65, 9.23]:

$$\begin{aligned} L(c_3) = \operatorname{sign}(L(c_1)) \operatorname{sign}(L(c_2)) \min\left(|L(c_1)|, |L(c_2)|\right) \\ - \ln\left(1+\exp\left(-|L(c_1)-L(c_2)|\right)\right) \\ + \ln\left(1+\exp\left(-|L(c_1)+L(c_2)|\right)\right) \end{aligned} \tag{9.9}$$

This other expression of operator $\oplus$ can easily be processed by using a look-up table for function g defined by:

$$g(x) = \ln\left(1 + \exp\left(-\left|x\right|\right)\right) \tag{9.10}$$

Practical example
Let us assume that $\Pr(c_1 = 1) = 0.8$ and $\Pr(c_2 = 1) = 0.1$. We then have,

$$\Pr(c_1 = 0) = 0.2 \quad \text{et} \quad \Pr(c_2 = 0) = 0.9$$

It is therefore more probable that $c_1 = 1$ and $c_2 = 0$. A direct application of Equation (9.2) then gives

$$\Pr(c_3 = 0) = 0.26 \quad \text{and} \quad \Pr(c_3 = 1) = 0.74$$

c_3 is therefore more probably equal to 1, which intuitively is justified since the number of 1s belonging to the parity check equation must be even. Using Equation (9.3) gives

$$L(c_3) = L(c_1) \oplus L(c_2) = (-1.386) \oplus (2.197) = -1.045$$

that is, $\Pr(c_3 = 0) = 0.26$ again. We find the same result again using (9.7) and (9.9).

Parity check code with n bits

We can now proceed to the parity check equation with n bits. We consider that the binary variables $c_1, \cdots, c_n$ are linked by the parity constraint $c_1 \oplus \cdots \oplus c_n = 0$ and that they make up the codeword $(c_1, \cdots, c_n)$. The LLR of the variables $\{c_j\}_{j=1..n, j\neq i}$ is assumed to be known and we search for the LLR of variable c_i. It is then simple to generalize the equations obtained for the parity code with 3 bits. Thus, taking the operator defined by (9.2) again:

$$L(c_i) = L(c_1) \oplus L(c_2) \oplus \cdots \oplus L(c_{j\neq i}) \oplus \cdots \oplus L(c_n) = \bigoplus_{j\neq i} L(c_j) \tag{9.11}$$

Similarly, the hyperbolic tangent rule is expressed by:

$$\tanh\left(\frac{L(c_i)}{2}\right) = \prod_{j\neq i} \tanh\left(\frac{L(c_j)}{2}\right) \tag{9.12}$$

or, separating the sign and the magnitude:

$$\operatorname{sgn}\left(L(c_i)\right) = \prod_{j\neq i} \operatorname{sgn}\left(L(c_j)\right) \tag{9.13}$$

$$\left|L(c_i)\right| = f^{-1}\left(\sum_{j\neq i} f\left(\left|L(c_j)\right|\right)\right) \tag{9.14}$$

where f is defined by Equation (9.8).

9.1.2 Definition of an LDPC code

Linear block codes

Linear block codes (see Chapter 4) can be defined by a parity check matrix **H** of size $m \times n$, where $m = n - k$. This matrix can be seen as a linear system of m parity check equations. The words **c** of the code defined by **H** are the binary words whose n bits simultaneously satisfy the m parity check equations. This system of linear equations is represented graphically in Figure 9.2 for the case of the Hamming binary block code of length $n = 7$.

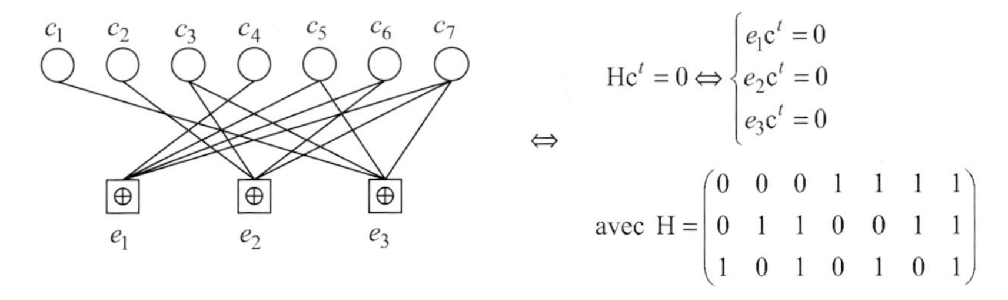

Figure 9.2 – Graphic representation of a block code: example of a Hamming code of length 7.

Such a representation is called the *bipartite graph* of the code. In this graph, branches link two different classes of nodes to each other:

- The first class of nodes called *variable nodes*, correspond to the bits of the codewords (c_j, $j \in \{1, \cdots, n\}$), and therefore to the columns of **H**.
- The second class of nodes, called *parity check nodes*, correspond to the parity check equations (e_p, $p \in \{1, \cdots, m\}$), and therefore to the rows of **H**.

Thus, to each branch linking a variable node c_j to a parity check node e_p corresponds the 1 that is situated at the intersection of the j-th column and the p-th row of the parity check matrix.

By convention, we denote $P(j)$ (respectively $J(p)$) all the indices of the parity nodes (respectively variable nodes) connected to the variable with index j (respectively to the parity with index p). We denote by $P(j)\backslash p$ (respectively $J(p)\backslash j$) all the $P(j)$ not having index p (respectively, all the $J(p)$ not having index j). Thus, in the example of Figure 9.2, we have

$$P(5) = \{1, 3\} \quad \textit{and} \quad J(1)\backslash 5 = \{4, 6, 7\}$$

A *cycle* on a bipartite graph is a path on the graph which makes it possible to leave a node and to return to this same node without passing twice through the same branch. The size of a cycle is given by the number of branches contained in the cycle. The graph being bipartite, the size of the cycles is even. The size of the shortest cycle in a graph is called the *girth*. The presence of cycles in the graph may degrade the decoding performance by a phenomenon of self-confirmation during the propagation of the messages. Figure 9.3 illustrates two cycles of sizes 4 and 6.

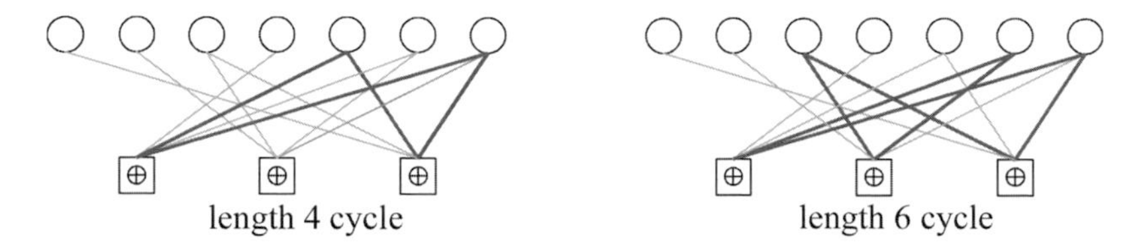

Figure 9.3 – Cycles in a bipartite graph.

Low density parity check codes

LDPC codes are linear block codes, the term low density coming from the fact that parity check matrix **H** contains a low number of non null values: it is a *sparse* matrix. In the particular case of binary LDPC codes studied here, the parity check matrix contains a small number of 1s. In other words, the associated bipartite graph contains a small number of branches. The adjective "low" mathematically means that when the length n of a codeword increases, the number of 1s in the matrix increases in $O(n)$ (compared to an increase in $O(n^2)$ of the number of elements of the matrix if the rate remains fixed).

The class of LDPC codes generates a very large number of codes. It is convenient to divide them into two sub-classes:

- regular LDPC codes
- irregular LDPC codes

An LDPC code is said to be regular in the particular case where parity check matrix **H** contains a constant number d_c of 1s in each row, and a constant number d_v of 1s in each column. We then say that the variables are of degree d_v and that the parities are of degree d_c. The code is denoted a (d_v,d_c) regular LDPC code. For example, the parity check matrix of a regular LDPC code (3,6) contains only 3 non zero values in each column, and 6 non zero values in each row. Figure 9.4 presents an example of a regular (3,6) code of size $n = 256$ obtained by drawing randomly the non zero positions. Out of the 256x128 inputs of the matrix, only 3x256 are non zero, that is, around 2.3%. This percentage tends towards 0 if the size of the code, for a fixed rate, tends towards infinity.

The irregularity profile of the variables of an irregular LDPC code is defined by the polynomial $\lambda(x) = \sum \lambda_j x^{j-1}$ where coefficient λ_j is equal to the ratio between the accumulated number of 1s in the columns (or variable) of degree j

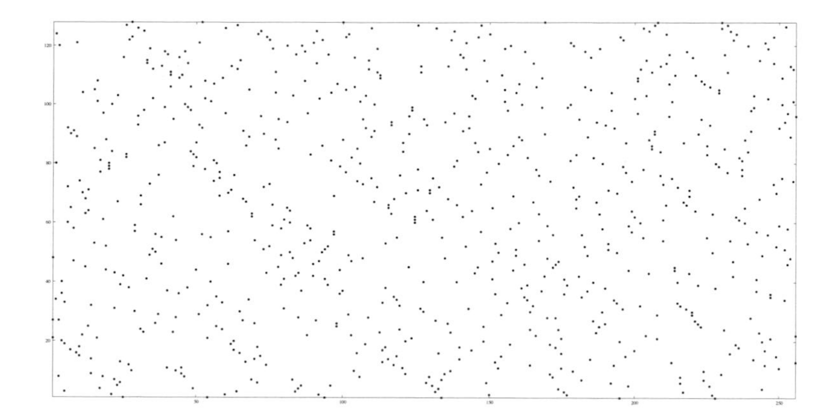

Figure 9.4 – Parity check matrix of a regular (3,6) LDPC code of size $n = 256$ and rate $R = 0,5$.

and the total number E of 1s in matrix $\mathbf{H}$. For example, $\lambda(x) = 0,2x^4 + 0,8x^3$ indicates a code where 20% of the 1s are associated with variables of degree 5 and 80% with variables of degree 4. Note that, by definition, $\lambda(1) = \sum \lambda_j = 1$. Moreover, the proportion of variables of degree j in the matrix is given by

$$\bar{\lambda}_j = \frac{\lambda_j / j}{\sum\limits_k \lambda_k / k}$$

Symmetrically, the irregularity profile of the parities is represented by the polynomial $\rho(x) = \sum \rho_p x^{p-1}$, coefficient ρ_p being equal to the ratio between the accumulated number of 1s in the rows (or parity) of degree p and the total number of 1s denoted E. Similarly, we obtain $\rho(1) = \sum \rho_j = 1$. The proportion $\bar{\rho}_p$ of columns of degree p in matrix $\mathbf{H}$ is given by

$$\bar{\rho}_p = \frac{\rho_p / p}{\sum\limits_k \rho_k / k}$$

Irregular codes have more degrees of freedom than regular codes and it is thus possible to optimize them more efficiently: their asymptotic performance is better than that of regular codes.

Coding rate

Consider a parity equation of degree d_c. It is possible to arbitrarily fix the values of the $d_c - 1$ first bits; only the last bit is constrained and corresponds to the redundancy. Thus, in a parity matrix $\mathbf{H}$ of size (m,n), each of the m rows corresponds to 1 redundancy bit. If the m rows of $\mathbf{H}$ are independent, the code then has m redundancy bits. The total number of bits of the code being

n, the number of information bits is then $k = n - m$ and the coding rate is $R = (n - m)/n = 1 - m/n$. Note that in the case where the m rows are not independent (for example, two identical rows), the number of constrained bits is lower than m. We then have $R > 1 - m/n$.

In the case of a regular LDPC code (d_v,d_c), each of the m rows has d_c non zero values, that is, a total of $E = md_c$ non zero values in matrix $\mathbf{H}$. Symetrically, each of the n columns contains d_v non zero values. We deduce from this that E satisfies $E = nd_v = md_c$, that is, $m/n = d_v/d_c$. The rate of such a code then satisfies $R \geqslant (1 - d_v/d_c)$.

In the case of an irregular code, the expression of the rate is generalized, taking into account each degree weighted by:

$$R \geqslant 1 - \frac{\sum_p \rho_p/p}{\sum_j \lambda_j/j} \tag{9.15}$$

Equality is reached if matrix $\mathbf{H}$ is of rank m.

9.1.3 Encoding

Encoding an LDPC code can turn out to be relatively complex if matrix $\mathbf{H}$ does not have a particular structure. There exist generic encoding solutions, including an algorithm with complexity in $O(n)$, requiring complex preprocessing on the matrix $\mathbf{H}$. Another solution involves directly building matrix $\mathbf{H}$ so as to obtain a systematic code very simple to encode. It is this solution, in particular, that was adopted for the standard DVB-S2 code for digital television transmission by satellite.

Generic encoding

Encoding with a generator matrix

LDPC codes being linear block codes, the coding can be done via the generator matrix $\mathbf{G}$ of size $k \times n$ of the code, such as defined in Chapter 4. As we have seen, LDPC codes are defined from their parity check matrix $\mathbf{H}$, which is generally not systematic. A transformation of $\mathbf{H}$ into a systematic matrix $\mathbf{H}_{sys}$ is possible, for example with the Gaussian elimination algorithm. This relatively simple technique, however, has a major drawback: the generator matrix $\mathbf{G}_{sys}$ of the systematic code is generally not sparse. The coding complexity increases rapidly in $\mathrm{O}\left(n^2\right)$, which makes this operation too complex for usual length codes.

Coding with linear complexity

Richardson *et al.* [9.49] proposed a solution enabling quasi-linear complexity encoding, as well as *greedy* algorithms making it possible to preprocess parity check matrix $\mathbf{H}$. The aim of the preprocessing is to put $\mathbf{H}$ as close as possible to a lower triangular form, as illustrated in Figure 9.5, using only permutations

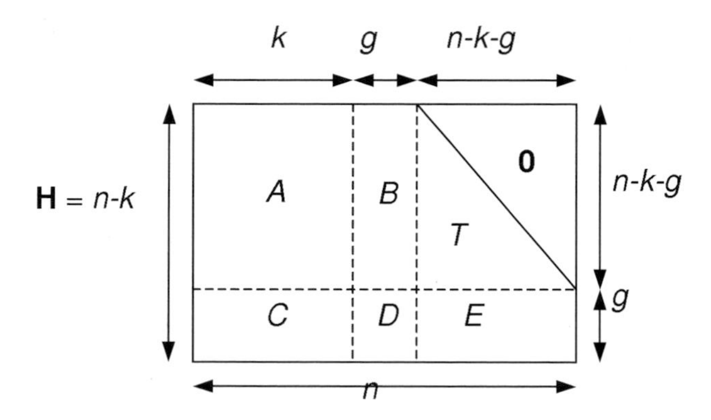

Figure 9.5 – Representation in the lower pseudo-triangular form of the parity check matrix H.

of rows or columns. This matrix is made up of 6 sparse sub-matrices, denoted A,B,C,D,E and T, the latter being a lower triangular sub-matrix. Once the preprocessing of **H** is finished, the encoding principle is based on solving the system represented by the following matrix equation:

$$\mathbf{c}\mathbf{H}^T = 0 \tag{9.16}$$

The codeword searched for is decomposed into three parts: $\mathbf{c} = (\mathbf{d}, \mathbf{r}_1, \mathbf{r}_2)$, where $\mathbf{d}$ is the systematic part that is known and where the redundancy bits searched for are split into two vectors $\mathbf{r}_1$ and $\mathbf{r}_2$, of respective size g and $n - k - g$. After multiplication on the right-hand side by matrix $\begin{pmatrix} I & 0 \\ -ET^{-1} & I \end{pmatrix}$, Equation (9.16) becomes:

$$A\mathbf{d}^t + B\mathbf{r}_\mathbf{1}^t + T\mathbf{r}_\mathbf{2}^t = 0 \tag{9.17}$$

$$\left(-ET^{-1}A + C\right)\mathbf{d}^t + \left(-ET^{-1}B + D\right)\mathbf{r}_\mathbf{1}^t = 0 \tag{9.18}$$

Equation (9.18) enables $\mathbf{r}_1$ to be found by inverting $\Phi = -ET^{-1}B + D$. Then Equation (9.17) enables $\mathbf{r}_2$ to be found by inverting T. Many time-consuming operations can be done once and for all during preprocessing. All the operations repeated during the encoding have a complexity in $\mathrm{O}(n)$ except the multiplication of $\left(-ET^{-1}A + C\right)\mathbf{d}^t$ by square matrix $\left(-\Phi^{-1}\right)$ of size $g \times g$ which after inversion is no longer sparse, hence a complexity in $\mathrm{O}\left(g^2\right)$. It is shown in [9.49] that we can obtain a value of g equal to a small fraction of n: $g = \alpha n$ where α is a sufficiently low coefficient for $\mathrm{O}\left(g^2\right) << \mathrm{O}(n)$ for values of n up to 10^5.

Specific constructions

Coding with a sparse generator matrix

One idea proposed by Oenning *et al.* [9.45] involves directly building a sparse systematic generator matrix, so the coding is performed by simple multiplication and the parity check matrix remains sparse. These codes are called Low-Density Generator-Matrix (LDGM) codes. Their performance is however poor [9.36], even if it is possible to optimize their construction [9.22] and lower the error floor.

Encoding by solving the system $cH^T = 0$ *obtained by substitution*

Mackay *et al.* [9.40] propose to constrain the parity matrix so that it is composed of the three sub-matrices A, B and C arranged as in Figure 9.6.

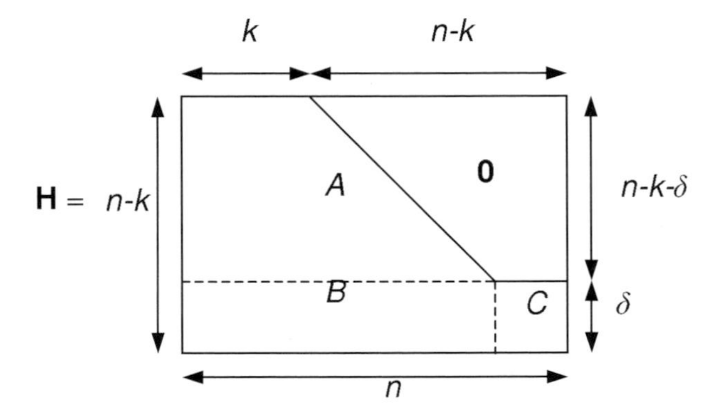

Figure 9.6 – Specific construction of parity check matrix H facilitating the encoding.

Systematic encoding is performed by solving Equation (9.16), which means solving $(n - k - \delta)$ equations by substitution. Each row of the parity matrix containing a low number of 1s, this operation is linear with n. The remaining δ equations are solved by inversion of matrix C defined in Figure 9.4. That leads to multiplication by a non-sparse matrix and therefore a complexity in $\mathrm{O}\left(\delta^2\right)$. Bond *et al.* [9.7] and Hu *et al.* [9.29] have proposed building parity check matrices with $\delta = 0$. In [9.25] Haley *et al.* define a class of codes enabling equation 9.12 to be solved by an iterative algorithm similar to the one used for the decoding.

Cyclic coding

The classes of LDPC codes defined by finite geometry or by projective geometry [9.29, 9.46, 9.34, 9.1, 9.58] enable cyclic or pseudo-cyclic codes to be obtained (see Section **??** for further details about this type of construction).The codes thus obtained can be encoded efficiently by using shift registers. In addition, they offer good properties in terms of the distribution of cycle length ($\geqslant 6$). The main drawback is that the cardinal of these classes of code is relatively small. These classes therefore offer only a very limited number of possible *size – rate – irregularity profile* combinations.

Summary

Table 9.1 summarizes the different possible types of coding encountered in the literature. In practice, the conventional encoding of block codes by a generator matrix is not used for LDPC codes due to the large the length of the codewords. The codes obtained by projective or finite geometry cannot be optimized (optimal design of the irregularity profiles). There therefore remain only codes built to facilitate the encoding by solving the equation $cH^T = 0$ by substitution, such as the one chosen for the DVB-S2 standard.

Type of encoding		Complexity	Remarks
Generic	Generator matrix	$\sim O(n^2)$	Not used in practice
	Pseudo-linear encoding	$\sim O(n)$	A lot of preprocessing
Ad hoc construction	Solving $\mathbf{cH}^T = 0$ by substitution	$\sim O(n)$ (si $\delta = 0$)	Possible loss of performance
	Cyclic or pseudo-cyclic	$\sim O(n)$	Limited number of possible combinations of the different parameters

Table 9.1 – Summary of the different possible encodings.

Analogy between an LDPC code and a turbo code

Figure 9.7 presents a turbo code in the form of a bipartite graph proposed by Tanner [9.53], thus showing the very close relation which links the family of turbo codes and that of LDPC codes.

In the case of a turbo code, the constraints are greater than in the case of an LDPC code since elementary codes are convolutional codes. But in the same way as for LDPC codes, a word is a codeword if and only if the two constraints of the graph are respected. Note here that the degree of the bits of a turbo code is two for the information bits and one for the redundancy bits.

Similarly, a product code can also be represented by a bipartite graph. The number of constraint nodes is then $2\sqrt{n}$ (compared with 2 for the turbo code and $n/2$ for an LDPC code with rate 0,5) and the latter have an intermediate complexity between that of the turbo code and that of the LDPC code. Turbo codes and LDPC codes are thus the two extremities of the spectrum of "composite" codes. The former contain only two very complex constraint nodes, the latter, very many constraint nodes, each node being made up of the simplest linear code possible (parity code). Note that from this representation of the bipartite graph, an infinite number of more or less exotic codes can be built.

It is interesting to note that encoding LDPC codes is tending to be performed more and more like the encoding of turbo codes (in a serial concatenation). The precursors were, without doubt, the *Repeat Accumulate* codes proposed in [9.16] whose encoding is composed of a repetition code, an interleaver and an accumulator. These codes are then decoded by an algorithm of the LDPC type with an adapted schedule. In the literature, we can now find many variants of this type of encoding, which involves combining elementary encoders and interleavers.

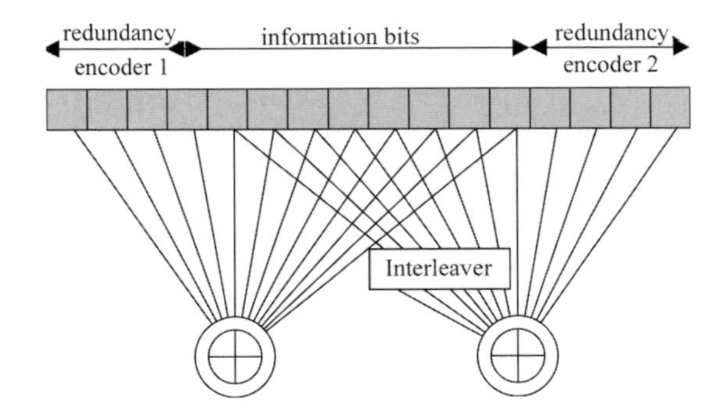

Figure 9.7 – Representation of a turbo code in the form of a bipartite graph.

The similarity between turbo codes and LDPC codes is even greater than can be assumed from the representations in the form of bipartite graphs. Indeed, it is shown in [9.36], [9.44] and in Section 6.2 that it is possible to represent a turbo code in the form of an LDPC matrix. The resemblance stops there. Parity check matrix **H** of a turbo code contains many rectangular patterns (four 1s making a rectangle in matrix **H**), that is, many cycles of length 4, which make the algorithms for decoding LDPC codes, to be described below, inefficient.

9.1.4 Decoding LDPC codes

Decoding an LDPC code is done using the same principle as decoding a turbo code by an iterative algorithm called a belief propagation algorithm. Each variable node sends to the parity nodes with which it is associated a message about the estimated value of the variable (*a priori* information). The set of *a priori* messages received enables the parity constraint to compute then return the extrinsic information. The successive processing of the variable then parity nodes make up one iteration. At each iteration, there is therefore a bilateral exchange of messages between the parity nodes and variable nodes, on the arcs of the bipartite graph representing the LDPC code. At the level of the receiver, the method for quantifying the sequence received, $\mathbf{r}$, determines the choice of decoding algorithm.

Hard input algorithm

Quantization on one bit involves processing only the sign of the samples received. Hard input decoding algorithms are based on the one proposed by Gallager under the name of algorithm A [9.21]. These decoders of course offer lower performance than those of soft input decoders. They are only implemented for very particular applications like optical communications, for example [9.17]. These algorithms will not be considered in the remainder of this chapter.

Belief propagation algorithm

When quantization is done on more than one bit, the decoding may use soft inputs: the *a priori* probability of the received symbols. In the case of binary codes and in the logarithm domain, we use the *a priori* log likelihood ratio (LLR) of samples r_j :

$$L\left(r_j \left| c_j \right.\right) = \ln\left(\frac{p\left(r_j \left| c_j = 0\right.\right)}{p\left(r_j \left| c_j = 1\right.\right)}\right) \tag{9.19}$$

where c_j is the j-th bit of the codeword and $r_j = 2c_{j-1} + b_j$. In the case of the additive white Gaussian noise channel, the noise samples b_j follow a centred Gaussian law with variance σ^2, that is:

$$p\left(r_j \left| c_j \right.\right) = \frac{1}{\sqrt{2\pi\sigma^2}} \exp\left[\frac{\left(r_j - (2c_j - 1)\right)^2}{2\sigma^2}\right] \tag{9.20}$$

Combining (9.19) and (9.20), intrinsic information I_j can be defined :

$$I_j \triangleq L\left(r_j \left| c_j \right.\right) = -\frac{2r_j}{\sigma^2} \tag{9.21}$$

Each iteration of the BP algorithm is decomposed into two steps:

1. Processing the parities:

$$Z_{j,p} = 2\tanh^{-1}\left[\prod_{j' \in J(p)/j} \tanh\frac{L_{j',p'}}{2}\right] \tag{9.22}$$

2. Processing the variables:

$$L_{j,p} = I_j + \sum_{p' \in P(j)/p} Z_{j',p'} \tag{9.23}$$

The iterations are repeated until the maximum number of iterations N_{it} is reached. It is possible to stop the iterations before N_{it} when all the parity equations are satisfied. This enables either a gain in mean throughput, or a limit in consumption.

We call L_j the total information or the LLR of bit j. This is the sum of the intrinsic information I_i and the total extrinsic information Z_j which is by definition the sum of the extrinsic information of branches $Z_{j,p}$:

$$Z_j \triangleq \sum_{p \in P(j)} Z_{j,p} \tag{9.24}$$

We therefore have $L_j = I_j + Z_j$ and Equation (9.23) can then be written:

$$L_{j,p} = L_j - Z_{j,p} = I_j + Z_j - Z_{j,p} \tag{9.25}$$

The BP algorithm is optimal in the case where the graph of the code does not contain any cycle: all the schedules[1] give the same result. As LDPC codes involve cycles, their decoding by the BP algorithm can lead to phenomena of self-confirmation of the messages which degrade the convergence and make the BP algorithm distinctly sub-optimal. However, these phenomena can be limited if the cycles are large enough.

The first schedule proposed is called the "flooding schedule" [9.35]. It involves successively processing all the parities then all the variables.

Flooding schedule algorithm

Initialization :

1- $n_{it} = 0,\ Z_{j,p}^{(0)} = 0 \quad \forall p \quad \forall j \in J(p),\ I_j = 2y_j/\sigma^2 \quad \forall j$

Repeat until $n_{it} = N_{it}$ or until the system has converged towards a codeword :

2- $n_{it} = n_{it} + 1$

3- $\forall j \in \{1, \cdots, n\}$ do: {Computation of the variable towards parity messages}

4- $Z_j^{(n_{it})} = \sum_{p \in P(j)} Z_{j,p}^{(n_{it}-1)}$ and $L_j^{(n_{it})} = I_j + Z_j^{(n_{it})}$

5- $\forall p \in P(j)$:

$$L_{j,p}^{(n_{it})} = I_j + \sum_{p' \in P(j)/p} Z_{j,p'}^{(n_{it}-1)} = I_j + Z_j^{(n_{it})} - Z_{j,p}^{(n_{it}-1)}$$

[1] By schedules, we mean the order in which each parity and each variable is processed.

6- $\forall p \in \{1, \cdots, m\}$ faire : {Computation of the parity towards variable message}

7- $\forall j \in J(p)$: $Z_{j,p}^{(n_{it})} = \bigoplus_{j' \in J(p)/j} L_{j,p'}^{(n_{it})}$

The bits decoded are then estimated by $\text{sgn}\left(-L_j^{(n_{it})}\right)$.
It is interesting to note that it is possible to modify the algorithm by "ordering" the flooding schedule depending on the parity nodes. The latter are then processed serially, and the algorithm becomes:

3'- $\forall j \in \{1, \cdots, n\}$: $Z_j^{(n_{it}+1)} = 0$

4'- $\forall p \in \{1, \cdots, m\}$ do:

5'- Computation of the input messages

$$\forall j \in J(p) \quad L_{j,p}^{(n_{it})} = I_j + Z_j^{(n_{it})} - Z_{j,p}^{(n_{it}-1)}$$

6'- Computation of the extrinsic information

$$\forall j \in J(p) \quad Z_{j,p}^{(n_{it})} = \bigoplus_{j' \in J(p)/j} L_{j,p'}^{(n_{it})}$$

7'- Update for the following iteration

$$\forall j \in J(p) \quad Z_j^{(n_{it}+1)} = Z_j^{(n_{it}+1)} + Z_{j,p}^{(n_{it})}$$

A similar organization of computations for the variable nodes will be called "distributed computation" since the computations linked with a variable node will be distributed during one iteration. In Section 9.2, the different types of schedule will be detailed then generalized.

It must also be noted that the notion of iteration (the computation of all the messages of the graph once and only once) is not strict. Thus, Mao *et al.* [9.42] proposed a variant of the flooding schedule in order to limit the impact of the effect of the cycles on the convergence. This variant, called "probabilistic scheduling", involves not processing some variables at each iteration. The choice of these variables is random and depends on the size of the smallest cycle associated with this variable: the smaller the cycle, the lower the probability of processing the variables involved in this cycle. This method limits phenomena of self-confirmation introduced by short cycles. It enables convergence to be obtained more rapidly than with the flooding schedule. The architectures linked with this schedule will not be discussed in this chapter.

9.1.5 Random construction of LDPC codes

The construction of an LDPC code (or of a family of LDPC codes) must naturally be done so as to optimize the performance of the code while minimizing the hardware complexity of the associated decoder.

Building a code remains a delicate problem so we refer the reader wishing to explore the subject further to the references given in this chapter. The problem of building an LDPC code adapted to decoding hardware will be dealt with in Section 9.2.

Optimizing an LDPC code is carried out in three steps:

- *a priori* optimization of the irregularity profiles of the parity and variable nodes;
- construction of matrices **H** of an adequate size respecting the irregularity profile and maximizing the length of the cycles;
- if necessary, selection or rejection of the codes using the minimum distance criterion or the performance computed by simulation.

Optimization of irregularity profiles

We make the hypothesis of codes with infinite size and an infinite number of iterations. Indeed, this enables optimization of their asymptotic characteristics (irregularity profile, rate) as a function of the channel targeted. Two techniques exist: the density evolution algorithm and its Gaussian approximation, and the extrinsic information transfer chart.

The *density evolution* algorithm was proposed by Richardson [9.48]. This algorithm calculates the probability density of messages $L_{j,p}$ and $Z_{j,p}$ after each new iteration. The algorithm is initialized with the probability density of the input samples, which depends on the level of noise σ^2 of the channel. Using this algorithm enables us to know the maximum value of σ^2 below which the algorithm converges, that is, such that the error probability is lower than a given threshold. It is also possible to determine by linear programming an irregularity profile that gives the lowest possible threshold.

A simplification of the density evolution algorithm proposed by Chung *et al.* [9.14, 9.13], is obtained by approximating the real probability densities by Gaussian densities. The interest of Gaussian density approximation is that it suffices to calculate the evolution of a single parameter by making the hypothesis that these Gaussian densities are consistent, that is, the variance is equal to two times the mean. Indeed, assuming that the "all zero"word was sent, at initialization we have ($n_{it} = 0$):

$$L_{j,p}^{(0)} = -\frac{2r_j}{\sigma^2} \quad \text{with} \quad r_j \sim N\left(-1, \sigma^2\right) \tag{9.26}$$

$$\text{therefore } L_{j,p}^{(0)} \sim N\left(\frac{2}{\sigma^2}, \frac{4}{\sigma^2}\right)$$

We denote:

$m_j^{(0)} = \frac{2}{\sigma^2}$ the average of the consistent Gaussian probability density of variable c_j of degree d_v sent to parities e_p of degree d_c which are connected to it,

$\mu_p^{(n_{it})}$ the mean of messages $Z_{j,p}^{(n_{it})}$.

To follow the evolution of the average $m_j^{(n_{it})}$ during the iterations n_{it} , it then suffices to take the mathematical expectation of Equations (9.22) and (9.23) relative to the variable and parity processing , which gives:

$$\Psi\left(\mu_p^{(n_{it})}\right) = \Psi\left(m_j^{(n_{it})}\right)^{d_c-1} \text{with } \Psi(x) = E\left[\tanh(x/2)\right],\ x \sim N(m, 2m) \quad (9.27)$$

$$m_j^{(n_{it}+1)} = \frac{2}{\sigma^2} + (d_v - 1)\,\mu_p^{(n_{it})} \quad (9.28)$$

Thus for a regular (d_v, d_c) LDPC code and for a given noise with variance σ^2, Equations (9.27) and (9.28) enable us, by an iterative computation, to know if the mean of the messages tends towards infinity or not. If such is the case, it is possible to decode without errors with a codeword of infinite size and an infinite number of iterations. In the case of an irregular code, it suffices to make the weighted mean on the different degrees of Equations (9.27) and (9.28).

The maximum value of σ for which the mean tends towards infinity, and therefore for which the error probability tends towards 0, is the threshold of the code. For example, the threshold of a regular code (3,6), obtained with the density evolution algorithm, is $\sigma_{\max} = 0.8809$ [9.13], which corresponds to a minimum signal to noise ratio of $\frac{E_b}{N_0}_{\min} = 1.1\text{dB}$.

Another technique derived from extrinsic information transfer (EXIT) charts[2] proposed by ten Brink [9.55, 9.56] enables the irregularity profiles to be optimized. Whereas the density evolution algorithm is interested in the evolution of the probability densities of the messages during the iterations, these charts are interested in the transfer of mutual information between the input and the output of the decoders of the constituent codes [9.56]. The principle of these charts has also been used with parameters other than mutual information, like the signal to noise ratio or error probability [9.3, 9.2]. It has also been applied to other types of channels [9.15].

[2] The principle of building EXIT charts is described in Section 7.6.3

Optimization of cycle size

Optimization of the irregularity profiles being asymptotic, we now have to build a parity check matrix of finite size. This phase can be performed randomly: we draw the non zero inputs of the parity check matrix at random, respecting as far as possible the irregularity profile of the nodes. It is also possible to build codes by randomly drawing permutations of an elementary matrix which are then concatenated. Another way to build LDPC codes is the deterministic construction of a matrix (finite and projective geometry).

In all cases, we must pay attention to the cycles present in the graph of the code. The belief propagation decoding algorithm assumes that the cycles that would deteriorate the independence of the messages entering a node do not exist. In practice, the presence of cycles in the graph is inevitable, but if they are large enough, the independence of the messages remains a good approximation. Building good LDPC codes must therefore ensure the absence of smaller cycles, those of size 4. Very many solutions are proposed in the literature to build LDPC codes. For example, Campello *et al.* [9.9] propose optimizing the size of the minimum cycle for a given rate. Hu *et al.* [9.65] suggest building the graph branch by branch in order to avoid at maximum the lowest cycle sizes (*Progressive Edge Geometry* or PEG). Zhang *et al.* [9.67] build LDPC codes whose smallest cycles are size 12, 16 or 20, but the variables are only degree 2. Tian *et al.* [9.57] use the fact that not all the small size cycles have the same influence and omit only those that are the most penalizing.

Selecting the code by the impulse method

The decoding performance using the belief propagation algorithm is improved by avoiding small sized cycles. But it is also important to have "good" error correcting codes, that is to say, those that have a large minimum distance. The impulse method was first proposed by Berrou *et al.* [9.4, 9.5] to evaluate the minimum Hamming distance of a turbo code. It was then adapted to the case of LDPC codes by Hu *et al.* [9.26]. It thus enables us to simply verify that the minimum distance of the code designed is sufficient to reach the error rate targeted for the application required.

Selecting the code by simulation

Two codes of the same size and the same rate, built with the same irregularity profiles, not having a short cycle and having the same minimum distance can nevertheless have a fairly different performance. These differences can be explained by two phenomena: the existence of "parasitic" fixed points introduced by the sub-optimality of the iterative decoding algorithm which increase the binary error rate in relation to the theoretical value [9.33]. The number of codewords with minimum distance also influences the performance of the code.

Figure 9.8 shows the performance of an LDPC code for different sizes and different rates in the case of a DVB-S2 decoder implemented on an Altera Stratix80 FPGA.

LDPC codes therefore have an excellent theoretical performance. This must however be translated by simplicity in their hardware implementation to enable these codes to be used in practice. That is why particular attention must be paid to LDPC decoder architectures and implementations.

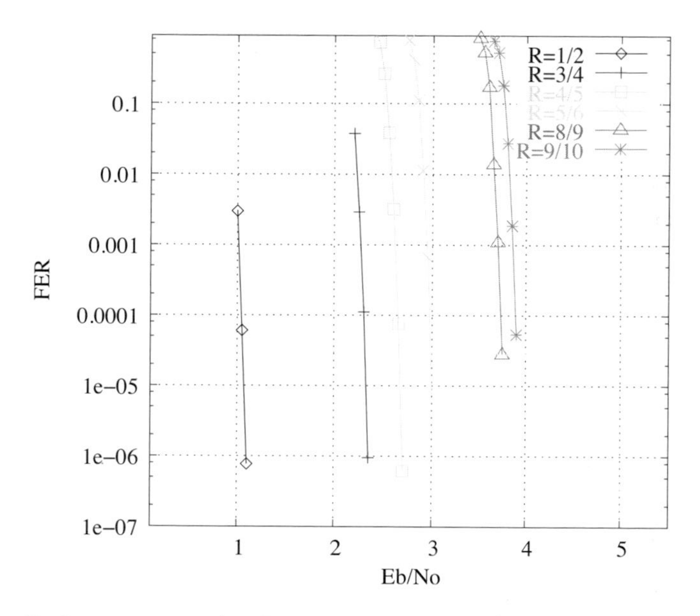

Figure 9.8 – Packet error rate (or Frame error rate, FER) obtained for codeword sizes of 64 kbits and different rates of the DVB-S2 standard (50 iterations, fixed point). With the permission of TurboConcept S.A.S, France.

9.1.6 Some geometrical constructions of LDPC codes

To complete the random constructions presented above, we list below some deterministic constructions leaving much less room, if any, for random ones.

Cayley / Ramanujan constructions

Margulis [9.43] was the first to propose algebraic LDPC codes. Then Rosenthal and Votonbel [9.50, 9.51] extended these results to obtain high expansion factor graphs with high girths, using Ramanujan graphs instead of Cayley graphs. Some drawbacks were raised by MacKay and Postol [9.38] about these codes (error floor and low minimum distance for some sets of parameters).

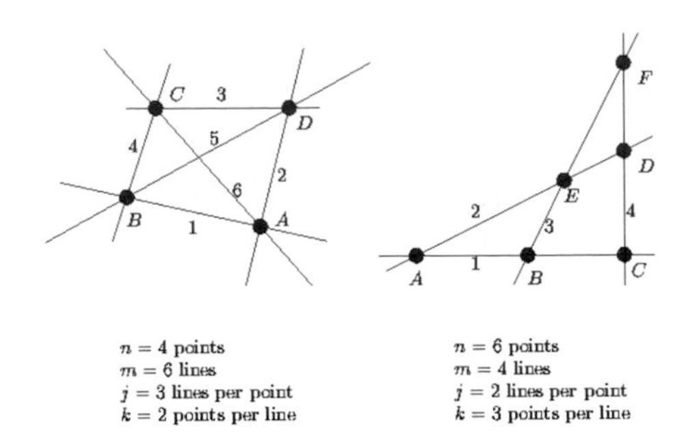

Figure 9.9 – Simple illustration of finite geometry

Kou, Lin and Fossorier's Euclidian / Projective Geometry LDPC

These LDPC codes are built on finite geometry [9.34]. Finite geometry is based on n points and m rows, such that each row contains k points and each point is on j lines. Two lines are either parallel or they have only one common point. A parity check matrix $\mathbf{H} = (h_{ij})$ can be built by assuming that $h_{ij} = 1$ if and only if the i-th row contains the j-th point. Of course, j and k have to be small compared to m and n in order to build an LDPC code that is called *type I*. An example of simple finite geometry is illustrated in Figure 9.9.

These codes are cyclic. When considering the transposed version of $\mathbf{H}$, we can obtain quasi-cyclic LDPC codes which are referred to as *type II* codes. Two kinds of finite geometry are used: Euclidean geometry (EG) and projective geometry (PG).

Constructions based on permutation matrices

Tanner *et al.* have proposed an algebraic construction of LDPC parity-check matrices based on an idea of Tanner [9.54]. A regular (j, k) code of length $n = kp$ and $m = jp$ parity checks can be obtained, assuming that p is a prime number and j, k are chosen among the prime factors of $p - 1$. The structure of the parity check matrix is:

$$H = \begin{pmatrix} I_1 & I_a & I_{a^2} & \cdots & I_{a^{k-1}} \\ I_b & I_{ab} & I_{a^2b} & \cdots & I_{a^{k-1}b} \\ \vdots & \vdots & \vdots & \ddots & \vdots \\ I_{b^{j-1}} & I_{ab^{j-1}} & I_{a^2b^{j-1}} & \cdots & I_{a^{k-1}b^{j-1}} \end{pmatrix}$$

where I_x is the identity matrix whose columns have been right-shifted x times, and where a and b are two non-zero elements in $\mathbf{F}_p$ of order j and k respectively.

Matrices based on Pseudo random generators

An important drawback of random constructions is that the parity check matrix has to be saved in memory, which takes up a lot of room for long codes. Prabhakar and Narayanan [9.47] found an interesting solution to circumvent this issue by using linear congruential sequences to design the parity-check matrix. Hence, after a non-complex computation, the generator outputs the address of the non-zero entries of the matrix. This solution has been implemented by Verdier *et al.* [9.62]. Girths of at least 6 can be obtained by correctly choosing the parameters of the generator.

Array-based LDPC

Array codes are two-dimensional codes that have been proposed for detecting and correcting burst errors [9.6]. When viewed as binary codes, the parity check matrix of array codes exhibit sparseness, which can be exploited for decoding them as LDPC codes using the BP algorithm [9.19]. Therefore, array codes provide the framework for defining a family of LDPC codes that lend themselves to deterministic constructions [9.18]. The parity check matrix of an array-based LDPC code is:

$$H = \begin{pmatrix} I & I & I & \cdots & I \\ I & \alpha & \alpha^2 & \cdots & \alpha^{k-1} \\ \vdots & \vdots & \vdots & \ddots & \vdots \\ I & \alpha^{j-1} & \alpha^{2(j-1)} & \cdots & \alpha^{(j-1)(k-1)} \end{pmatrix}$$

where I is the $p \times p$ identity matrix, p being an odd prime number, α is the I matrix whose rows have been shifted once, and $j, k \geq p$.

BIBDs, Latin rectangles

A block design is an incidence system [9.63] (v, k, λ, r, b) in which a set X of v points is partitioned into a family A of b subsets (blocks) in such a way that any two points determine λ blocks with k points in each block, and each point is contained in r different blocks. It is also generally required that $k < v$, which leads to a balanced incomplete block design (BIBD) of the LDPC code. The five parameters are not independent, but satisfy the two relations

$$\begin{aligned} vr &= bk \\ \lambda(v-1) &= r(k-1) \end{aligned}$$

A BIBD is therefore commonly simply written as (v, k, λ), since b and r are given in terms of v, k, and λ by

$$b = \frac{v(v-1)\lambda}{k(k-1)} \qquad r = \frac{\lambda(v-1)}{k-1}$$

A BIBD is said to be symmetric if $b = v$ (or, equivalently, $r = k$).

These constructions have been widely studied in the literature. For example, MacKay and Davey [9.37] first proposed the use of Steiner triple systems to design short LDPC codes. Johnson and Weller [9.30, 9.31], and also B. Vasic [9.60], presented a family of LDPC codes based on Kirkman triple systems. A design based on anti-Pasch Steiner systems is also presented in [9.59], and mutually orthogonal Latin rectangles (MOLR) are used in [9.61].

9.2 Architecture for decoding LDPC codes for the Gaussian channel

When the belief propagation algorithm is implemented, the general architecture of decoders for LDPC codes can be performed with the help of generic node processors (GNP) modelling either the parity processing, or the variable processing. This section describes the different possible implementations of these processors after analysing the decoding complexity of LDPC codes. The different possibilities for controlling this GNP-based architecture enables us to define three classes of schedule for the belief propagation algorithm: a two-pass schedule, a "vertical" schedule and a "horizontal" schedule. This original, unified presentation of the architectures of decoders for LDPC codes enables us to cover many existing architectures published so far, and to synthesize innovatory architectures.

9.2.1 Analysis of the complexity

The decoding complexity of LDPC codes is directly linked with the number of branches in the bipartite graph of the code, or with the number of 1s in the parity check matrix. The iterative decoding belief propagation algorithm has two steps. At each step, we have to calculate information $L_{j,p}$ or $Z_{j,p}$ which is associated with the branch linking variable j to parity p. Let us denote B the number of branch in the bipartite graph of the LDPC code. For example, in the case of a regular code (d_v, d_c) of size n, the number of branches B is given by:

$$B = d_v n = d_c m \tag{9.29}$$

The computing power P_c necessary to decode LDPC codes is then defined as the number of branches to process per clock cycle. This parameter depends on:

- the number k of information bits to transmit per codeword,
- the number of branches B,
- the data rate D of information desired,
- the maximum number of iterations N_{it},

- the clock frequency f_{clk}.

In one second, the number of codewords to process in order to obtain an information data rate D is equal to D/k (words/second). In the worst case, decoding a codeword requires computing $B \times N_{it}$ branches. To guarantee a data rate D, an architecture must provide the power to compute $D \times B \times N_{it}/k$ branches per second. The minimum computing power P_c to provide per clock cycle is therefore:

$$P_c = \frac{BN_{it}D}{kf_{clk}} \text{ (branches/cycle)} \tag{9.30}$$

Note that, for a fully parallel architecture in which each node of the graph is associated with a processor, all the branches of the graph are processed in one clock cycle. The computing power is then $(P_c)_{\max} = B$. There is no practical interest in trying to go beyond this power since the critical path then becomes longer.

9.2.2 Architecture of a generic node processor (GNP)

The computations performed in a variable node processor (VNP) and in a parity node processor (PNP) have an identical dependency between the inputs and the outputs. Indeed, for both the PNP and VNP, the d outputs are calculated from the d inputs, with the i-th output depending on all the inputs less the i-th input. It is thus possible to represent the different processor architectures abstractly by a generic node processor. The latter will then be specialized according to the decoding algorithm used. The GNP therefore receives at its input d messages $(e_i)_{i=1..d}$ and produces at its output d messages $(s_j)_{j=1..d}$ defined by:

$$s_j = \bigotimes_{i \neq j} e_i \tag{9.31}$$

Operator $\otimes$ is a generic associative-commutative operator for computations, whose implementation will be specified below. The condensed expression 9.31 means that the operator is applied to all the variables e_i for $i \neq j$.
Figure 9.10 illustrates the three main versions of GNP parallel architectures:

- *Direct architecture*: the computations of the d output messages are performed independently (Figure 9.10(a)). The computations of the different outputs can also be factorized. The number of $\otimes$ components traversed is of the order of $\log_2(d)$.

- *Trellis architecture (Forward-Backward type)*: This architecture corresponds to a particular factorized form of parallel architecture which has great regularity, but whose number of $\otimes$ operators is linear with d.

- *Total sum architecture*: this architecture is possible only if the generic operator $\otimes$ allows an inverse denoted $inv_{\otimes}$. In this case, the generic operator

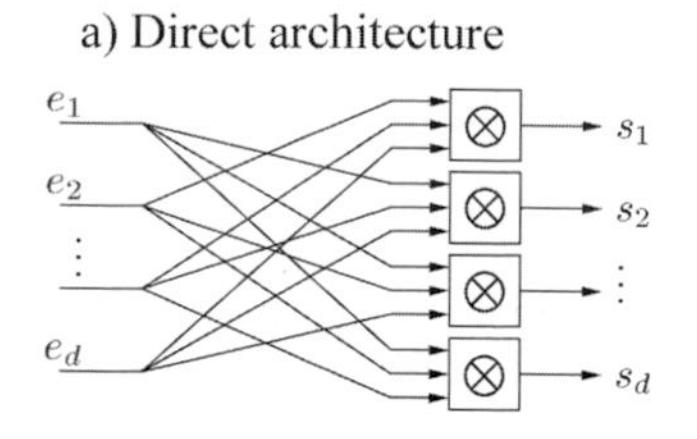

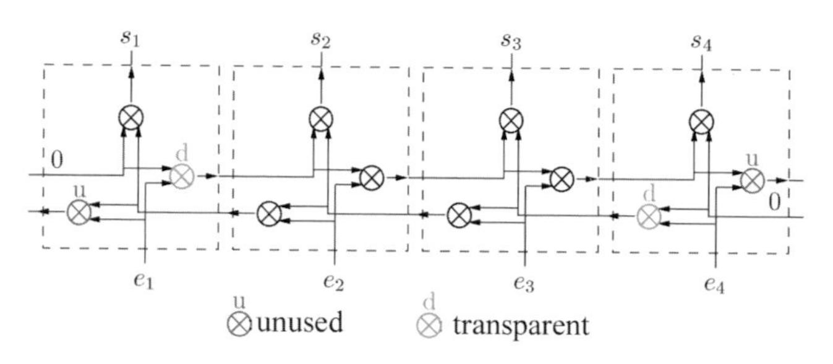

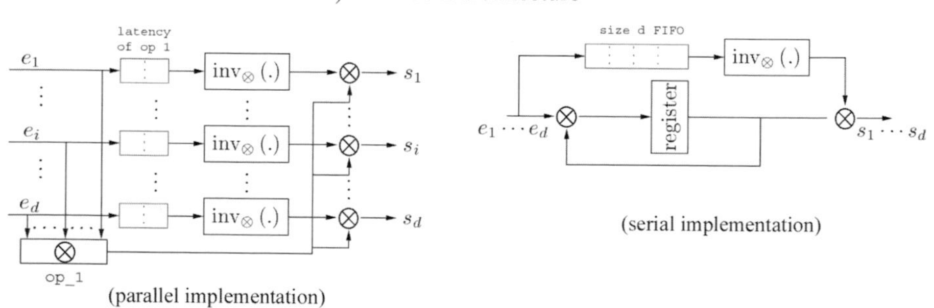

Figure 9.10 – The different "compact mode" architectures for implementing the generic operator $\otimes$.

is applied at all the inputs (total sum) then each output is calculated by eliminating the contribution of the corresponding input, with the help of the inverse operator.

It is possible to modify these architectures in order to introduce intermediate *pipeline* registers enabling the critical path to be reduced. There are also architectures of the serial type (Figure 9.10(c)).

In what follows, the degree of parallelism of a GNP will be denoted α_g. This is the number of cycles necessary to process a node (without considering latency due to the *pipeline* processing). Thus, for a parallel architecture capable of processing one node at each clock cycle, $\alpha_g = 1$, whereas for a serial architecture, $\alpha_g = d$.

Note that in all the GNP architectures presented, we implicitly made the hypothesis that all the inputs were available and that all the outputs had to be generated either simultaneously (parallel architecture), or grouped in time (serial architecture). This kind of GNP control mode is called the "compact mode".

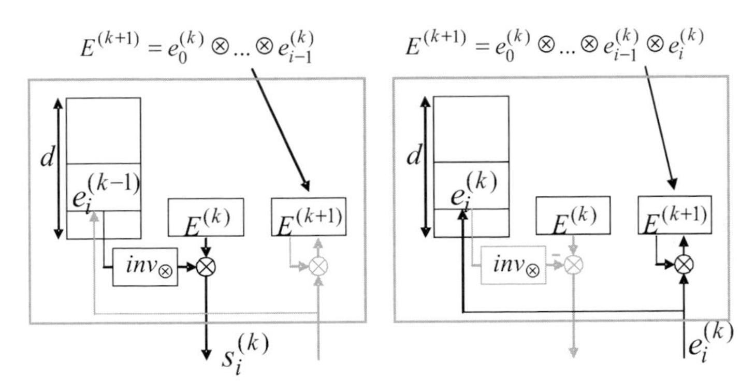

Figure 9.11 – Principle of the distributed mode (delayed update).

It is possible to imagine different execution modes, like the "distributed mode", in which the GNP inputs and outputs are distributed throughout the decoding iteration.

Figure 9.11 shows the distributed mode operation of a processor :

- During the current iteration n_{it}, we consider that the input variables e_i come from the previous iteration $n_{it} - 1$ whereas the output variables belong to the current iteration.

- At the end of an iteration, we assume that the d input variables $(e_i^{(n_{it}-1)})_{i=1..d}$ are memorized in a memory (internal or external to the GNP) as well as the value of $E^{(n_{it})} = \bigotimes_{i=1..d} e_i^{(n_{it}-1)}$.

- The GNP can therefore, at the request of the system, calculate the i-th output $s_i^{(n_{it})} = E^{(n_{it})} \otimes inv_\otimes(e_i^{(n_{it}-1)})$.

- This output is sent, via the interleaver, to the opposite node which, once the computation is over, returns $e_i^{(n_{it})}$.

- This new value then replaces $e_i^{(n_{it}-1)}$ in the memory and is also accumulated to obtain the value of $E^{(n_{it}+1)}$ at the end of the iteration. Two accumulation modes are possible:

1. The first mode – delayed update (Figure 9.11) – involves using an accumulation register initialized to zero at each new iteration. This register enables $E^{(n_{it}+1)} = \bigotimes_{i=1..d} e_i^{(n_{it})}$ to be calculated directly. This architecture therefore has $d+2$ memory words, d for the inputs $(e_i^{(n_{it}-1)})_{i=1..d}$, a word for $E^{(n_{it})}$ and a word for the accumulation of $E^{(n_{it}+1)}$.
2. The second mode – immediate update – involves replacing the contribution of $e_i^{(n_{it}-1)}$ in $E^{(n_{it})}$ by that of $e_i^{(n_{it})}$ as soon as a new input $e_i^{(n_{it})}$ arrives, that is:

$$E^{(n_{it})} = E^{(n_{it})} \otimes e_i^{(n_{it})} \otimes inv_{\otimes}\left(e_i^{(n_{it}-1)}\right) \tag{9.32}$$

At the end of the iteration, we thus have $E^{(n_{it}+1)} = E^{(n_{it})}$. This solution offers two advantages in relation to delayed updating:

- one less memory word;
- an acceleration in the convergence of the algorithm as the new values of the inputs are taken into account sooner.

Choice of a generic operator

Figure 9.12 gives a "cross-section" view of the belief propagation algorithm on the bipartite graph of the LDPC code. We assume that each branch is split into two to differentiate the variable towards parity messages from the parity towards variable messages. This view shows the great resemblance between processing variables and processing the parities and enables us to imagine other positions of the interconnection networks in the computation cycle. Each position of the interconnection graph is thus translated by a different processing of the parity nodes and the variable nodes. Table 9.2 gives the different computations to be carried out according to the position of the interconnection network.

When the interconnection network is in position 1 (Table 9.2), we again have the classical separation between a variable processor and a parity processor. The latter can then either be performed in the frequency domain (as indicated in Figure 9.11), or directly in the domain of the LLRs via the $\oplus$ operator defined in equation (9.3).

9.2.3 Generic architecture for message propagation

Presentation of the model

PNPs and GNPs are characterized by their architecture and their generic operator, depending on the position of the interconnection network. The architecture presented in Figure 9.13 enables the exchange of messages between these different processors.

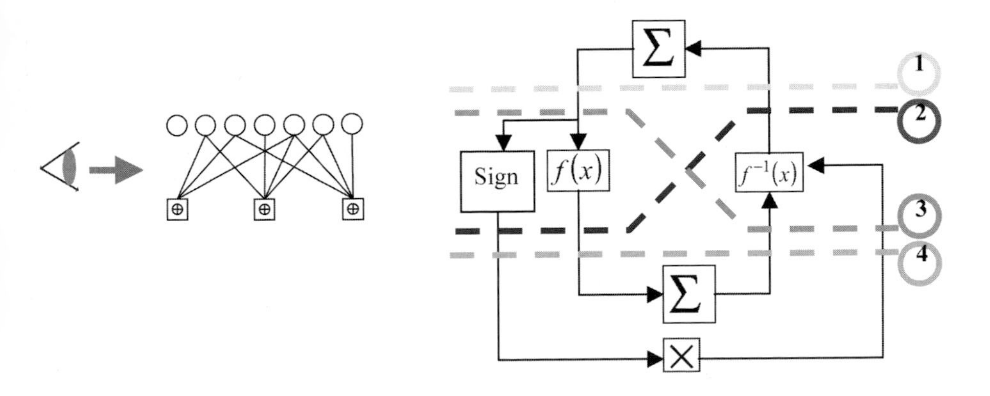

Figure 9.12 – Different positions of the interconnection network obtained by processing the parity nodes in the Fourier domain. These positions separate those parts of the iteration to be performed in the VNP from those performed in the PNP.

Position of the network		1	2	3	4
VNP		Σ	$f \circ \Sigma$	$\Sigma \circ f^{-1}$	$f \circ \Sigma \circ f^{-1}$
PNP module	Fourier	$f^{-1} \circ \Sigma \circ f$	$f^{-1} \circ \Sigma$	$\Sigma \circ f$	Σ
	Direct	$\oplus$			
PNP sign		Product of the signs			

Table 9.2 – Value of the generic operator associated with the variable processors (VNPs) and parity processors (PNPs) as a function of the position of the interconnection network.

This architecture is composed of P PNPs which each generate $d = d_p$ messages in α_p clock cycles. These processors therefore have $d'_c = d_c/\alpha_p$ inputs and outputs. They are connected to an interleaving network that is direct and inverse. On the other side of this interleaving network are placed $\left(Pd'_p/d'_v\right)$ VNPs. These processors similarly generate $d = d_v$ messages in α_v clock cycles, and therefore have $d'_v = d_v/\alpha_v$ inputs and outputs.

The degree of parallelism of the architecture is defined by the three parameters P, α_p and α_v. It is possible to obtain all the degrees of parallelism possible, ranging from the completely parallel architecture where $P = m$, $\alpha_p = 1$ and $\alpha_v = 1$ to the completely serial architecture where $P = 1$, $\alpha_c = d_c$ and $\alpha_v = d_v$. Note that such an architecture has a computing power (equation (9.30)) $P_c = P \times d'_c$.

Direct inverse interleaving networks enable the messages associated with the different VNPs to be routed towards different PNPs and vice-versa. This kind of network generally makes it possible to perform several permutations, called space permutations. Another type of permutation, called a time permutation,

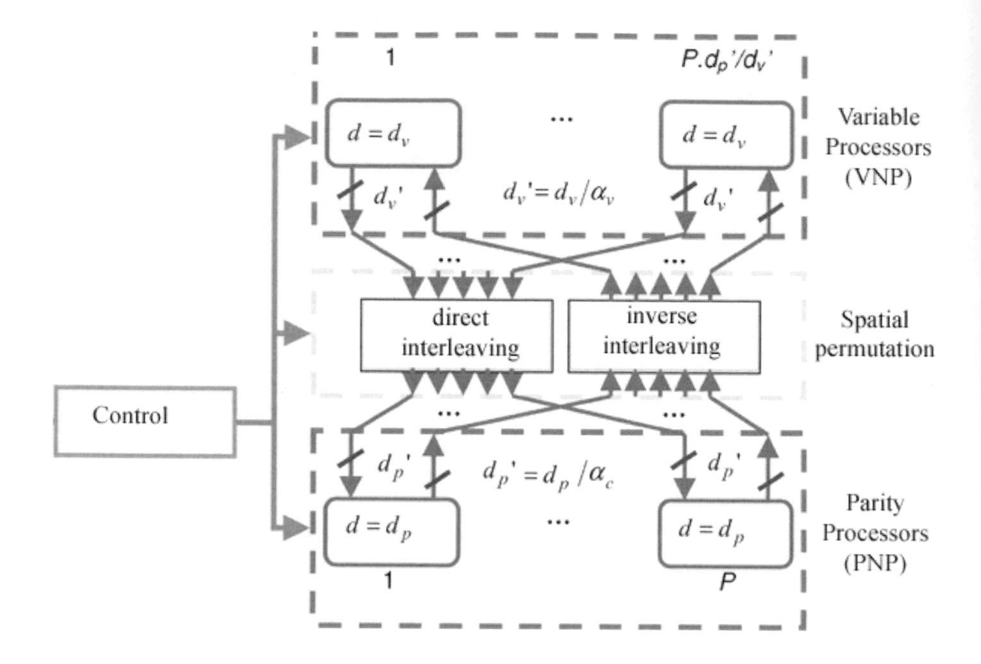

Figure 9.13 – Generic serial-parallel architecture.

makes it possible to randomly access the nodes associated with a same processor of nodes, by memory addressing, for example. The combination of these two types of permutations enables a random interconnection such as that existing between the variable nodes and the parity nodes of the LDPC code.

Example of an implementation

To help clarify ideas about the way to organize the computations and the propagation of the messages in the decoder, and to truly understand the link between the organization of the propagation of the messages and the structure of the LDPC code, Figure 9.14 shows a simple example of decoding an LDPC code of length $n = 12$ and rate $R = 0,5$ (therefore $m = 6$), with $P = 2$, $d_c = 3$, $\alpha = 1$ and $\beta = d_v$. There are therefore $P = 2$ parity node processors and $n/P = 6$ variable node processors. One iteration is performed in $m/P = 3$ steps:

- At the first cycle, reading the information relative to the bits is done in each of the VNPs, each of them containing $n/P = 2$ bits of the codeword (in practice, n/P can be much higher). These bits are shaded in grey in each VNP: this is space permutation.

- This information is then sent to the PNPs via the permutation network, whose address was generated from the cycle number (read into a memory for example): this is time permutation.
- Combining the two describes the random interleaving between the variables and the first two parities of the bipartite graph shown on the right-hand part of the figure.

In a single cycle, the first two parities will therefore be able to be processed. The following two will be processed at the second cycle and so on and so forth, until all the parities of the code have been processed. Note that this technique where the PNP information arrives simultaneously prevents two bits contained in the same VNP being involved in the same parity. Thus, for example, bits 1 and 2 cannot be involved in the same parity otherwise that would lead to a memory conflict. This solution therefore imposes constraints on matrix $\mathbf{H}$, if we want it to be decodable by this structure. One solution to relax the constraints involves, for example, entering the data serially into the parities.

9.2.4 Combining parameters of the architecture

A certain number of parameters characterizing LDPC decoder architectures have been defined above:

- Node processors:
  - 3 possible architectures (direct, trellis, total sum)
  - 4 possible positions of the interconnection network (see Figure 9.12)
  - 3 input-output control modes (compact, distributed with delayed or immediate update)
- Message propagation architecture:
  - 3 parameters characterizing the level of parallelism (P, α, β)

All the combinations of these different parameters are possible to describe or create an LDPC decoder architecture. Of course, some of these combinations are more or less of interest, depending on the specifications required. For example, the combinations of the control modes between VNPs and PNPs, showing the different possible decoding schedules, are given in Table 9.3.

In the case where the controls on the two processors are of the compact flow of inputs-outputs type, the schedule performed is of the flooding type: all the PNPs are processed then all the VNPs. This schedule can easily be used with completely parallel ($P = m$) architectures. For mixed ($P < m$) architectures, the parities cannot all be processed completely before processing the variables. The control of the VNPs in distributed mode with delayed update allows the processing to be done since it guarantees that the new outputs will

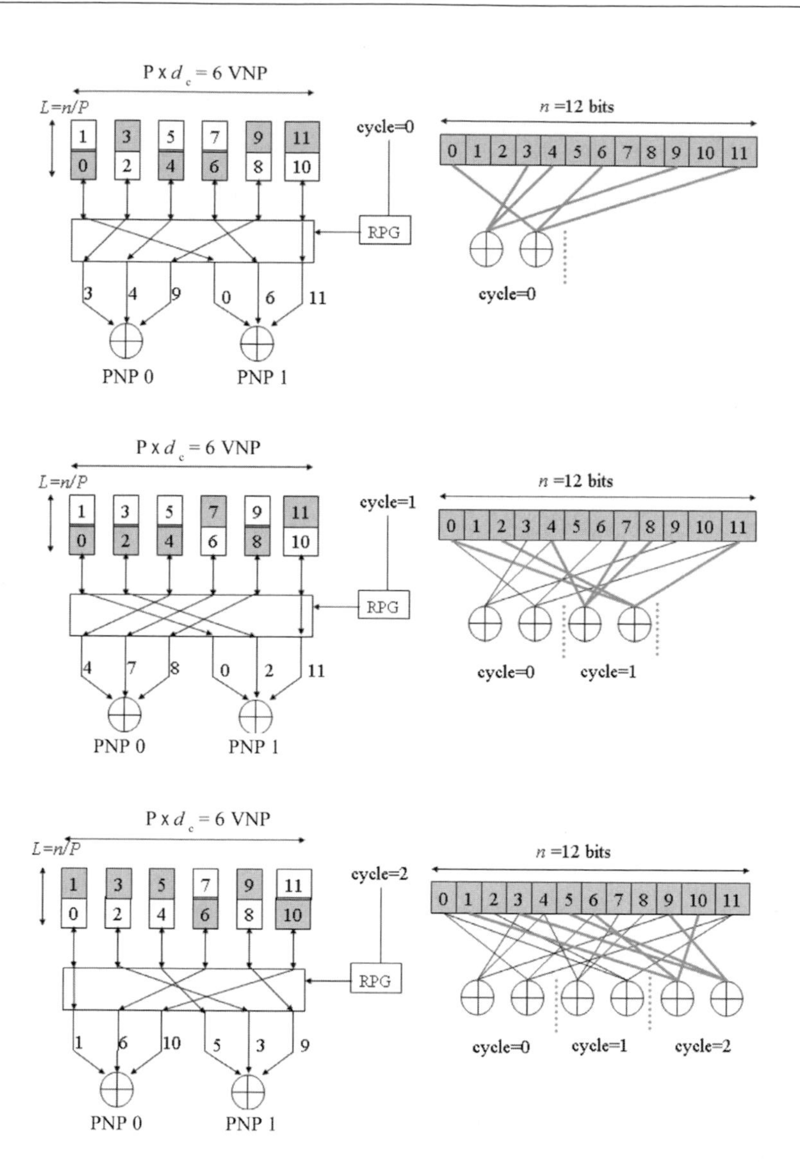

Figure 9.14 – Example of a message propagation architecture: link between the decoder's addressing and code structure.

only be computed when all the parities have been processed. This control mode implements a flooding schedule according to the parities. Symmetrically, we make a flooding schedule appear according to the variables, when the PNPs are in distributed mode and the VNPs are in compact mode. These three types of

<table>
<tr><td colspan="3" rowspan="3"></td><td colspan="3">VNP</td></tr>
<tr><td rowspan="2">Compact</td><td colspan="2">distributed</td></tr>
<tr><td>Delayed update</td><td>Immediate update</td></tr>
<tr><td rowspan="3">PNP</td><td colspan="2">Compact</td><td>Flooding</td><td>Flooding (parity)</td><td>Interleaving (horizontal)</td></tr>
<tr><td rowspan="2">distributed</td><td>Delayed update</td><td>Flooding (variable)</td><td colspan="2" rowspan="2">Branches</td></tr>
<tr><td>Immediate update</td><td>interleaving (vertical)</td></tr>
</table>

Table 9.3 – Schedules associated with the different combinations of the node processor controls.

schedules converge towards the same values: They do not change the information propagation operation.

When one of the two types of processor is controlled in compact mode and the other in distributed mode with immediate update, we implement a schedule of the horizontal or vertical interleaving (*shuffle*) type. The order in which the processors are activated is similar to the flooding schedule according to the variables or the parities. Only the update of information changes since it is performed as soon as a new input has arrived, thus accelerating the convergence of the code.

The case where the two VNP and PNP processors are controlled in distributed mode is not of great interest. It would in fact correspond to controlling the decoding, branch by branch.

The memory required to implement these different combinations is given in Table 9.4.

<table>
<tr><td colspan="3" rowspan="3"></td><td colspan="3">VNP</td></tr>
<tr><td rowspan="2">Compact</td><td colspan="2">Distributed</td></tr>
<tr><td>Delayed update</td><td>Immediate update</td></tr>
<tr><td rowspan="3">PNP</td><td colspan="2">Compact</td><td>$B+n$</td><td>$3n+g(B,d_c)$</td><td>$2n+g(B,d_c)$</td></tr>
<tr><td rowspan="2">Distributed</td><td>Delayed update</td><td>$B+n+2m$</td><td>$3n+2m+B$</td><td>$2n+2m+B$</td></tr>
<tr><td>Immediate update</td><td>$B+n+m$</td><td>$3n+m+B$</td><td>$2n+m+B$</td></tr>
</table>

Table 9.4 – Quantity of memory necessary as a function of the combinations of the different node processor controls.

Parameter B designates, like above, the number of branches in the graph. Each extrinsic branch information must be memorized, whatever the schedule used. In all cases the intrinsic variable information must also be memorized, that is, n values. When the parity check mode is the compact one, the accumulation of the messages of the n variables in each VNP must be memorized, that is, n memories if we update them immediately, and $2n$ in the opposite case. The reasoning is the same if the VNPs are in compact mode and the PNPs are in distributed mode, but in this case, it is the accumulations of messages in the m parities that must be memorized.

It is sometimes possible, as we shall see later, to memorize the $Z_{j,p}$ messages in a compressed way. The number of messages to memorize then passes from B to $g(B, d_c)$, with g representing a compression function ($g(B, d_c) < B$).

9.2.5 Example of synthesis of an LDPC decoder architecture

The two examples described in this part allow us to show two LDPC decoder architectures using two different schedules. For each of these examples, we will give the values of the parameters characterizing these architectures.

Parameters		Values
Message propagation architecture		($\alpha_p = 1$, $\alpha_v = d_v = 3$, $P = 3$)
Position of the interconnection network		1
VNP	Control	Distributed, delayed update
	Data path	Total sum, serial
PNP	Control	Compact
	Data path	Trellis, parallel

Table 9.5 – Parameters characterizing the flooding schedule architecture.

Flooding schedule (according to parities)

The architecture described here to illustrate the flooding schedule is based on the one proposed initially by Boutillon *et al.* [9.8]. It is schematized in Figure 9.15. In this example, $P = 3$ PNP operate simultaneously in compact mode. As $\alpha_p = 1$ and $\alpha_v = 3$, there are 12 VNPs that operate simultaneously but in distributed mode (only one VNP and one PNP are shown in the figure). Note that the computing power of such an architecture is 12 branches per cycle.

The architecture of the PNPs is of the trellis type, with parallel implementation. The chronogram at the bottom of Figure 9.15 indicates that at time T_1, $d_v = 4$ messages $L_{j,p}(T_1)$ are produced at each PNP. After a latency of $T_2 - T_1$ clock cycles, the $Z_{j,p}(T_2)$ messages leaving are sent to the VNPs. This operation is reproduced m/P times to carry out one complete iteration. The data path of the VNPs is of the total sum type, with a serial implementation. The delayed update is shown by using two memory blocks, one for the extrinsic information during accumulation (Lacc), and another for the total extrinsic information of

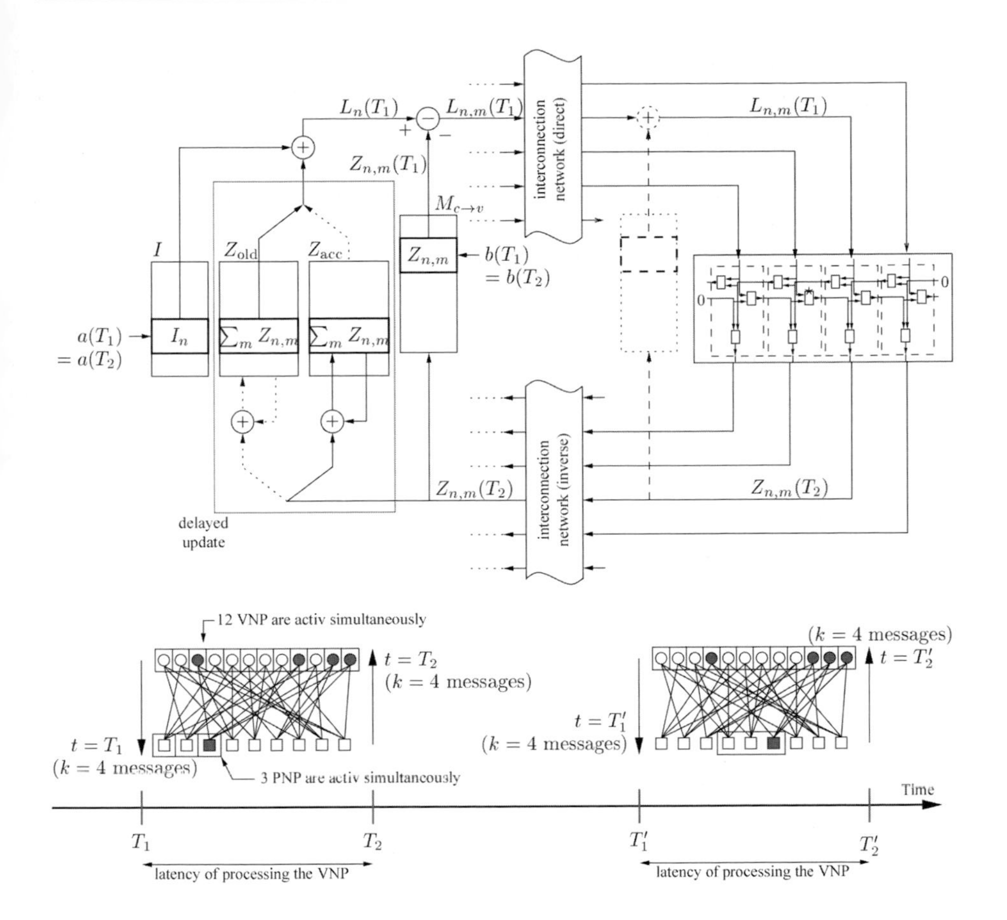

Figure 9.15 – Example of architecture for a flooding schedule (according to parities).

the previous iteration (Lold). At the end of each iteration, the role of these two memories is exchanged. In this architecture, the extrinsic branch information $Z_{j,p}$ can be saved either on the VNP side (solid line in the figure) or on the PNP side (dotted line in the figure), like in the Chen *et al.* [9.12] and Guilloud *et al.* [9.24] architectures.

Horizontal interleaving schedule

This second type of architecture illustrates the horizontal interleaving schedule, proposed by Mansour *et al.* [9.41] in a particular case of the turbo decoding of LDPC codes. In this example, illustrated in Figure 9.16, there are $P = 3$ PNPs that operate simultaneously in compact mode. As $\alpha_p = 4$ and $\alpha_v = 3$, there are 3 VNPs that operate simultaneously in distributed mode, which gives a computing power of 3 branches per cycle.

The data paths of the VNPs and PNPs are both of the total sum type, with a serial implementation. From time T_1, $d_c = 4$ messages $L_{j,p}$ enter serially into the PNP. After a computation latency of $T_2 - T_1$, the messages $Z_{j,p}$ calculated are sent back, again serially, to the VNPs which are controlled in distributed mode. But in this case, the update of the information is immediate. This is translated by using a single block of *Lacc* memory. Thus, the sum of the extrinsic information of the j bits is updated as soon as a new input $Z_{j,p}$ arrives.

Parameters		Values
Message propagation architecture		$(\alpha_p = d_c = 4,\ \alpha_v = d_v = 3,\ P = 3)$
Position of the interconnection network		4
VNP	Control	Compact
	Data path	Total sum, serial
PNP	Control	Distributed, immediate update
	Data path	Total sum, serial

Table 9.6 – Values of the parameters characterizing vertical interleaving architecture.

9.2.6 Sub-optimal decoding algorithm

In order to reduce the complexity of the LDPC decoder, many "sub-optimal" decoding algorithms have been proposed. These algorithms are based on the same principle: reduction in complexity and in memory of the (parity or variable) node processors, by replacing the individual computation of the d output messages (with d the degree of the node) by the computation of Δ ($\Delta < d$) distinct values. Of course, using a sub-optimal algorithm generally degrades the performance of the code. A compromise thus has to be found between performance and complexity.

Single message decoding algorithm ($\Delta = 1$)

This is the simplest algorithm since all the outputs of the (variable or parity) node processor are assigned a same single value at each step in the iterative process.

VNP with $\Delta = 1$

In this technique, the VNP simply returns L_j to the parity constraints to which it is connected. Thus, it is no longer necessary to memorize the $Z_{j,p}$ messages since the latter are no longer used by the VNP. There results a significant economy in memory. This algorithm, APP algorithm, was first proposed by Fossorier *et al.* in [9.20] , and taken up again by E. Yeo *et al.* in [9.66].

Note that the hypothesis of independence between the messages leaving and entering a parity node is absolutely not verified. That is why the iterative

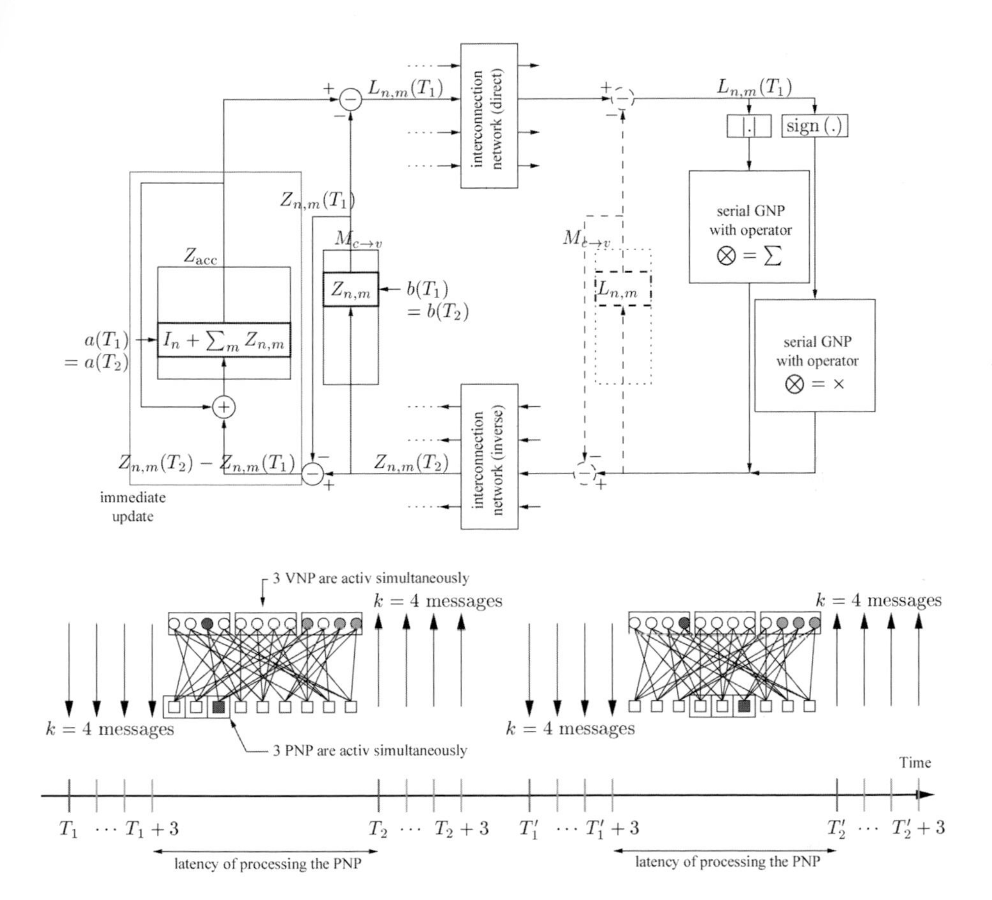

Figure 9.16 – Example of architecture for a vertical interleaving schedule. The serial implementation of the PNP is not detailed in this figure.

decoding algorithm diverges very rapidly as it is subject to the self-confirmation phenomenon: the propagation of the information occurs as if cycles with length 2 existed in the graph.

PNP with $\Delta = 1$

This is the algorithm symmetric to the previous one: the PNP returns a unique value. This technique, which is very efficient in terms of complexity, enables the algorithm to reach its correction capacity in very few iterations, typically 5. Although its correction capacity is very low in relation to the BP algorithm, it is interesting to note that for 5 iterations, such an algorithm is more efficient than the BP algorithm after this same number of iterations. Thus, such procedures can be successfully applied for high data-rate applications where only a reduced number of iterations can be performed.

Sub-optimal PNP algorithms ($\Delta > 1$)

In the state of the art there are three algorithms for Δ >1 concerning the PNP.

Min-sum or BP-Based algorithm ($\Delta = 2$)

This algorithm proposed by Fossorier *et al.* [9.20] requires no computations in the PNP. Indeed, the authors suggest approximating parity processing algorithm (9.22) by:

$$\left\{\begin{array}{l} |Z_{j,p}| = \underset{j' \in J(p)/j}{\text{Min}} \left(|L_{j',p}|\right) \\ \text{sign}\left(Z_{j,p}\right) = \prod\limits_{j' \in J(p)/j} \text{sign}\left(L_{j',p}\right) \end{array}\right. \tag{9.33}$$

Only the computation of the magnitude changes: it is approximated by excess by the minimum of the magnitudes of the messages entering the PNP. Processing in the PNP therefore involves only computing the sign and sorting the two lowest magnitudes of the input messages. Note that this approximation makes the iterative decoding processing independent of the knowledge of the level of noise σ^2 of the channel. The loss in performance is of the order of around 1 dB compared to the *BP* algorithm.

This approximation by excess of the Min-Sum algorithm can however be compensated by simple methods. It is thus possible to reduce the value of $|Z_{j,p}|$ by assigning it a multiplicative factor A strictly lower than 1. It is also possible to subtract from it an offset B (B>0), taking the precaution, however, of saturating the result to zero if the result of $|Z_{j,p}| - B$ is negative. The value of $|Z_{j,p}|$ corrected $|Z_{j,p}|^c$ is therefore:

$$\left\{\begin{array}{l} |Z_{j,p}|^c = A \times \max(|Z_{j,p}| - B, 0) \\ \text{sign}\left(Z_{j,p}\right) = \prod\limits_{j' \in J(p)/j} \text{sign}\left(L_{j',p}\right) \end{array}\right. \tag{9.34}$$

These two variants of the Min-sum algorithm are called *Offset BP-based* and *Normalized BP-Based* [9.10] respectively. The optimization of coefficients A and B enables decoders to be differentiated. They can be constant or variable according to the signal to noise ratio, the degree of the parity constraint, or the processed iteration number, etc.

$\lambda - \min algorithm(\Delta = \lambda + 1)$

This algorithm was presented initially by Hu *et al.* [9.27, 9.28] then reformulated independently by Guilloud *et al.* [9.24]. Function f, defined by Equation (9.35), is such that $f(x)$ is large for low x, and low when x is large. Thus, the sum in (9.35) can be approximated by its λ highest values, that is to say, by the λ lowest values of $|L_{j,p}|$. Once the set denoted $J_\lambda(p)$ of minima λ is obtained, the PNP will calculate $\Delta = \lambda + 1$ distinct magnitudes:

$$|Z_{j,p}| = f\left(\sum_{j' \in J_\lambda(p)/j} f\left|L_{j',p}\right|\right) \quad \text{with} \quad f(x) = \ln\,\tanh\left(\frac{x}{2}\right) \tag{9.35}$$

Indeed, if j' is the index of a bit having sent one of the values of the set of minima, the magnitude is calculated on all the $\lambda - 1$ other minima (λ computations on $\lambda - 1$ values). However, for all the bits, the same magnitude is returned (a computation on λ values). It must be noted that the performance of the $\lambda -$ min algorithm can be improved by adding a correction factor A and B as defined in equation (9.34).

$A - \min *algorithm(\Delta = 2)$

The last sub-optimal algorithm published so far is called the "$A - \min *$" algorithm and was proposed by Jones *et al.* [9.32]. Here also, the first step is to find index j_0 of the bit having the message with the lowest module: $j_0 = Arg\,\mathrm{Min}_{j \in J(p)}\left(|L_{j,p}|\right)$. Then, two distinct messages are calculated:

$$\text{If} \quad j = j_0 : |Z_{j_0,p}| = f\left(\sum_{j \in J(p)/j_0} f\,|L_{j,p}|\right) \tag{9.36}$$

$$\text{If not} \quad j \neq j_0 : |Z_{j,p}| = f\left(\sum_{j \in J(p)} f\,|L_{j,p}|\right) \tag{9.37}$$

A comparison of performance in terms of binary error rate and packet error rate between the sub-optimal algorithms and the BP algorithm is presented in Figure 9.17. The code simulated is an irregular code with size $n = 1008$ and rate $R = 0,5$, built by Hu (*IBM Zurich Research Labs*) with the help of the PEG (*Progressive Edge Growth*) technique [9.29]. The degrees of the parities are thus almost all equal to 8. The degrees of the variables vary from 2 to 15. We see that the sub-optimal algorithm is not necessarily less efficient than the BP algorithm (typically for the $A - \min *$ algorithm). No compensation of extrinsic information was used for these simulations. It is possible to greatly improve the performance of these algorithms by adding this compensation to them, like for the *offset* and *normalized BP-based* algorithms cited above.

9.2.7 Influence of quantization

Quantization is likely to create a degradation in performance by essentially affecting two points: the computation of intrinsic information I_i defined by (9.21) and the computation of branch (9.22) and total (9.24) extrinsic information.

We distinguish two quantization parameters: the number of quantization bits n_q and the quantization dynamic δ. The representable values thus lie between $-\delta$ and $+\delta$. This means that the position of the decimal point is not necessarily defined in the n_q bits or, to put it another way, that the coding dynamic is not necessarily a power of two. Thus, the bit with the lowest weight will be equal to:

$$q_{\mathrm{LSB}} = \frac{\delta}{2^{n_q - 1}} \tag{9.38}$$

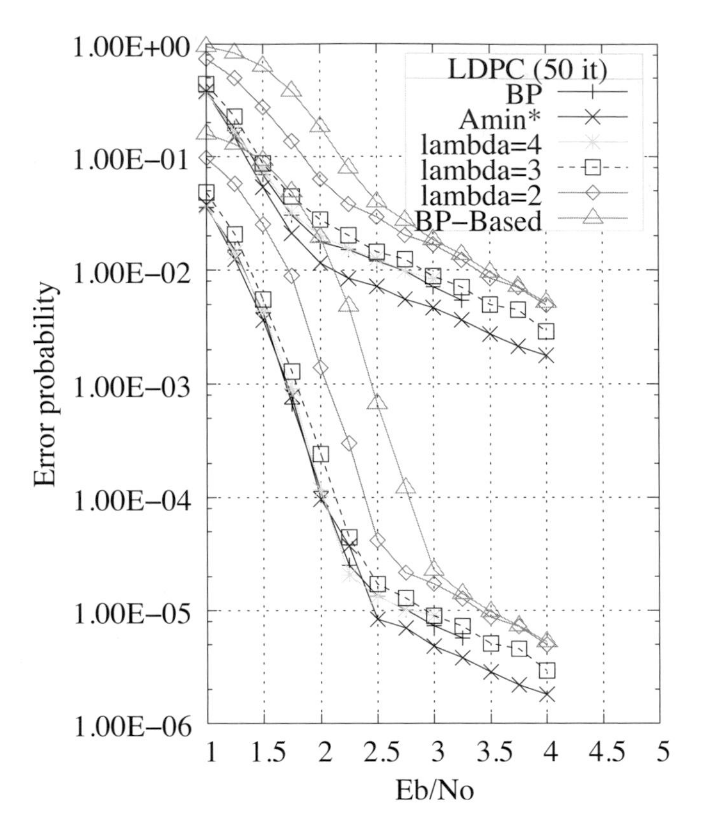

Figure 9.17 – Comparison of performance between the $3-\min$ and $A-\min*$ algorithms in the case of decoding an irregular code C_3.

and we pass from a quantified scalar a_q to a non-quantified scalar a by the relations:

$$\begin{cases} a_q = \text{trunc}\left(a\frac{2^{n_q-1}}{\delta} + 0.5\right), \\ a = a_q\frac{\delta}{2^{n_q-1}} \end{cases} \tag{9.39}$$

where *trunc* designates the truncature operation.
These two parameters can influence the decoding performance. Too low a dynamic allows an error floor to appear on the error rate curves. This floor appears much earlier than that associated with the minimum distance of the code.

However it is important to note that increasing the dynamic without increasing the number of quantization bits increases the value of the bit with the lowest weight, and consequently decreases the precision of the computations done in the PNP. Increasing the dynamic without increasing the number of bits degrades the decoding performance in the convergence zone.

The decoding performance obtained in practice is very close to that obtained with floating decimal points for quantizations on 4 to 6 bits (see the table in Figure 9.18). The influence of the parameters can be studied by the density evolution algorithm [9.15, 9.11].

9.2.8 State of the art of published LDPC decoder architectures

The table in Figure 9.18 groups the main characteristics of LDPC decoder circuits published in the literature so far. The inputs of the table are as follows:

- Circuit: description of the type of circuit used (ASIC or FPGA).
- Authors: reference to authors and articles concerning the platform.
- Architecture:
  - Decoder: indication of the type of decoder (serial, parallel or mixed) with parameters (P, α_p, α_v) associated with the message propagation architecture.
  - Data path: indication for each node processor (variable and constraint) of type of architecture used (direct, trellis or total sum).
  - Control: indication for each node processor (variable and constraint) of type of control used, compact or distributed and, if applicable, of type of update.
  - Position of the interconnection network (between 1 and 4).
- Characteristics of the LDPC code: size, rate and regularity of the LDPC code.
- quantization format: number of bits used to represent the data in the decoder.
- Clock frequency of the chip in MHz.
- Data rate (in bits per second). The information binary rate is obtained by multiplying by the coding rate.
- Maximum number of iterations.

So far, no architecture has been published that describes in detail a decoder using a sub-optimal algorithm. However, we can mention that of Jones *et al.* [9.32], but which contains too little information to be classified here.

Circuit	Authors	Architecture								Characteristics of the code				Coding format (bits)	Clock frequency (MHz)	Rate (Mb/s)	Max number of iterations
		Decoder			Data path		Control		position of inter-connection network	n	m	R	degrees				
		P	α	B	PNP	PNV	PNP	PNV									
Synthesis	Levine et. al. 2000 [9.53]	Serial 1	1	1	Parallel (no LLR)					1200	900	0,25	dv=3; dc=4	8	50	4	10
DSP - TMS320C6201	Bhatt et. al. 2000 [9.54]	Serial (?)			Serial (?)				1	200	100		dv=3; dc=6	16	200	0,133	
ASIC 0.16um (7.5x7 mm²)	Blanksby et. al. 2002 [9.55]	Parallel M	1	1	Total-Sum		Compact			1024	512		dv=3,6,7,8 dc=6,7	4	64	1000	64
FPGA Xilinx Virtex-E 2600	Zhang et. al.2002 [9.56]	Mixed 3.dc	1	1	Trellis	direct	Compact	Compact	3	9216	4608	0,5	dv=3; dc=6	5	56	54	18
ASIC 0.18um (3.1x4.2 mm²)	Mansour et. al. sept 2003 [9.57]	Mixed 64	1	1	Trellis	Total-Sum	Compact	Distributed, immediate update		2048	1024			4	125	1600	1
ASIC 0.18um (9.41 mm²)	Mansour et. al. dec. 2003 [9.58]	Mixed 64	1	1					1	2304	1536	2/3	irregular	5	200	192	10
FPGA Xilinx Virtex-II 8000	Chen. Y. et. al. 2003 [9.41]	Mixed 24	1	J	Total-Sum		compact	Distributed, delayed update		8088	4044	0,5	irregular	6	44	80	25
ASIC 0.11um (2.61 mm²)															212	376	
ASIC 0.13nm (49.6mm²) 0.09nm (15.8mm²)	P. Urard et. al. 2005 [9.59]	360								DVB-S2 + BCH				6	200 300	90 135	
ASIC 0.13nm (22.74mm²)	F. Kienle et. al. 2005 [9.60]	360								DVB-S2				6	270	255	30

Figure 9.18 – State of the art of the different platforms published.

Bibliography

[9.1] PB. Ammar, B. Honary, Y. Kou, and S. Lin. Construction of low density parity check codes: a combinatoric design approach. In *Proceedings of IEEE International Symposium on Information Theory (ISIT'02)*, July 2002.

[9.2] M. Ardakani, T.H. Chan, and F.R. Kschischang. Properties of the exit chart for one-dimensional ldpc decoding schemes. In *Proceedings of Canadian Workshop on Information Theory*, May 2003.

[9.3] M. Ardakani and F.R. Kschischang. Designing irregular lpdc codes using exit charts based on message error rate. In *Proceedings of International Symposium on Information Theory (ISIT'02)*, July 2002.

[9.4] C. Berrou and S. Vaton. Computing the minimum distance of linear codes by the error impulse method. In *Proceedings of IEEE International Symposium on Information Theory*, July 2002.

[9.5] C. Berrou, S. Vaton, M. Jézéquel, and C. Douillard. Computing the minimum distance of linear codes by the error impulse method. In *Proceedings of IEEE Global Communication Conference (Globecom'2002)*, pages 1017–1020, Taipei, Taiwan, Nov. 2002.

[9.6] M. Blaum, P. Farrel, and H. Van Tilborg. Chapter 22: Array codes. In *Handbook of Coding Theory*. Elsevier, 1998.

[9.7] J.W. Bond, S. Hui, and H. Schmidt. Constructing low-density parity-check codes. *EUROCOMM 2000. Information Systems for Enhanced Public Safety and Security, IEEE AFCEA*, May 2000.

[9.8] E. Boutillon, J. Castura, and F.R. Kschischang. Decoder-first code design. In *Proceedings of 2nd International Symposium on Turbo Codes & Related Topics*, pages 459–462, Brest, France, 2000.

[9.9] J. Campello and D.S. Modha. Extended bit-filling and ldpc code design. *Proceedings of IEEE Global Telecommunications Conference (GLOBECOM'01)*, pages 985–989, Nov. 2001.

[9.10] J. Chen and M. Fossorier. Near optimum universal belief propagation based decoding of low-density parity check codes. *IEEE Transactions on Communications*, 50:406–414, March 2002.

[9.11] J. Chen and M.P.C. Fossorier. Density evolution for two improved bp-based decoding algorithms of ldpc codes. *IEEE Communications Letters*, 6:208–210, May 2002.

[9.12] Y. Chen and D. Hocevar. A fpga and asic implementation of rate 1/2, 8088-b irregular low density parity check decoder. In *Proceedings of IEEE Global Telecommunications Conference (GLOBECOM'03)*, 1-5 Dec. 2003.

[9.13] S.-Y. Chung. *On the Construction of some Capacity-Approaching Coding Schemes*. PhD thesis, MIT, Cambridge, MA, 2000.

[9.14] S.-Y. Chung, T.J. Richardson, and R.L. Urbanke. Analysis of sum-product decoding of low-density parity-check codes using a gaussian approximation. *IEEE Transactions on Information Theory*, 47, Feb. 2001.

[9.15] D. Declercq. Optimisation et performances des codes ldpc pour des canaux non standards. Master's thesis, Université de Cergy Pontoise, Dec. 2003.

[9.16] D. Divsalar, H. Jin, and R. J. McEliece. Coding theorems for turbo-like codes. In *Proceedings of 36th Allerton Conference on Communication, Control, and Computing*, pages 201–210, Sept. 1998.

[9.17] I.B. Djordjevic, S. Sankaranarayanan, and B.V. Vasic. Projective-plane iteratively decodable block codes for wdm high-speed long-haul transmission systems. *Journal of Lightwave Technology*, 22, March 2004.

[9.18] E. Eleftheriou and S. Olcer. Low-density parity-check codes for digital subscriber lines. In *Proceedings of International Conference on Communications*, pages 1752–1757, 28 Apr.-2 May 2002.

[9.19] J. L. Fan. Array codes as low-density parity-check codes. In *Proceedings of 2nd Symposium on Turbo Codes*, pages 543–546, Brest, France, Sept. 2000.

[9.20] M.P.C. Fossorier, M. Mihaljevic, and I. Imai. Reduced complexity iterative decoding of low-density parity-check codes based on belief propagation. *IEEE Transactions on Commununications*, 47:673–680, May 1999.

[9.21] R. G. Gallager. *Low-Density Parity-Check Codes*. MIT Press, Cambridge, MA, 1963.

[9.22] J. Garcia-Frias and Wei Zhong. Approaching shannon performance by iterative decoding of linear codes with low-density generator matrix. *IEEE Communications Letters*, 7:266–268, June 2003.

[9.23] F. Guilloud. *Generic Architecture for LDPC Codes Decoding*. PhD thesis, ENST Paris, July 2004.

[9.24] F. Guilloud, E. Boutillon, and J.-L. Danger. Lambda-min decoding algorithm of regular and irregular ldpc codes. In *Proceedings of 3rd International Symposium on Turbo Codes & Related Topics*, 1-5 Sept. 2003.

[9.25] D. Haley, A. Grant, and J. Buetefuer. Iterative encoding of low-density parity-check codes. In *Proceedings of IEEE Global Telecommunications Conference (GLOBECOM'02)*, Nov. 2002.

[9.26] X.-Y. Hu, M.P.C. Fossorier, and E. Eleftheriou. On the computation of the minimum distance of low-density parity-check codes. In *Proceedings of IEEE International Conference on Communications (ICC'04)*, 2004.

[9.27] X.-Y. Hu and R. Mittelholzer. An ordered-statistics-based approximation of the sum-product algorithm. In *Proceedings of IEEE International Telecommunications Symposium*, Natal, Brazil, 8-12 Sept. 2002.

[9.28] X.-Y. Hu and R. Mittelholzer. A sorting-based approximation of the sum-product algorithm. *Journal of the Brazilian Telecommunications Society*, 18:54–60, June 2003.

[9.29] X.Y. Hu, E. Eleftheriou, and D.-M. Arnold. Progressive edge-growth tanner graphs. In *Proceedings of IEEE Global Telecommunications Conference (GLOBECOM'01)*, Nov. 2001.

[9.30] S.J. Johnson and S.R. Weller. Construction of low-density parity-check codes from kirkman triple systems. In *Proceedings of IEEE Global Telecommunication Conference (GLOBECOM'01)*, volume 2, pages 970–974, 25-29 Nov. 2001.

[9.31] S.J. Johnson and S.R. Weller. Regular low-density parity-check codes from combinatorial designs. In *Proceedings of Information Theory Workshop*, pages 90–92, Sept. 2001.

[9.32] C. Jones, E. Vallés, M. Smith, and J. Villasenor. Approximate min* constraint node updating for ldpc code decoding. In *Proceedings of IEEE Military Communications Conference (MILCOM'03)*, 13-16 Oct. 2003.

[9.33] R. Koetter and P.O. Vontobel. Graph-covers and the iterative decoding of finite length codes. In *Proceedings of 3rd International Symposium on turboCodes & Related Topics*, Sept. 2003.

[9.34] Y. Kou, S. Lin, and M.P.C. Fossorier. Low-density parity-check codes based on finite geometries: A rediscovery and new results. *IEEE Transactions on Information Theory*, 47:2711–2736, Nov. 2001.

[9.35] F.R. Kschischang and B.J. Frey. Iterative decoding of compound codes by probability propagation in graphical models. *IEEE Journal on Selected Areas in Commununications*, 16:219–230, 1998.

[9.36] D. J. C. MacKay. Good error-correcting codes based on very sparse matrices. *IEEE Transactions on Information Theory*, 45(2):399–431, March 1999.

[9.37] D. J. C. MacKay and M. C. Davey. Evaluation of gallager codes for short block length and high rate applications. In B. Marcus and J. Rosenthal, editors, *Codes, Systems and Graphical Models*, volume 123 of *IMA Volumes in Mathematics and its Applications,*, pages 113–130. Springer, New York, 2000.

[9.38] David J.C. MacKay and Michael S. Postol. Weaknesses of margulis and ramanujan-margulis low-density parity-check codes. *Electronic Notes in Theoretical Computer Science*, 74, 2003.

[9.39] D.J.C MacKay and R.M. Neal. Good codes based on very sparse matrices. In *Proceedings of 5th IMA Conference on CryproGraphy and Coding*, Berlin, Germany, 1995.

[9.40] D.J.C MacKay, S.T. Wilson, and M.C. Davey. Comparison of constructions of irregular gallager codes. *IEEE Transactions on Communications*, 47:1449–1454, Oct. 1999.

[9.41] M.M. Mansour and N.R. Shanbhag. Turbo decoder architectures for low-density parity-check codes. In *Proceedings of IEEE Global Telecommunications Conference (GLOBECOM'02)*, 17-21 Nov. 2002.

[9.42] Y. Mao and A.H. Banihashemi. Decoding low-density parity-check codes with probabilistic scheduling. *IEEE Communications Letters*, 5:414–416, Oct. 2001.

[9.43] G. A. Margulis. Explicit construction of graphs without short cycles and low density codes. *Combinatorica*, 2(1):71–78, 1982.

[9.44] R. J. McEliece, D. J. C. MacKay, and J.-F. Cheng. Turbo decoding as an instance of pearle's belief propagation algorithm. *IEEE Journal on Selected Areas in Commununications*, 16:140–152, Feb. 1998.

[9.45] T.R. Oenning and Jaekyun Moon. A low-density generator matrix interpretation of parallel concatenated single bit parity codes. *IEEE Transactions on Magnetics*, 37:737– 741, 2001.

[9.46] T. Okamura. Designing ldpc codes using cyclic shifts. In *Proceedings of IEEE International Symposium on Information Theory (ISIT'03)*, July 2003.

[9.47] A. Prabhakar and K. Narayanan. Pseudorandom construction of low density parity-check codes using linear congruential sequences. *IEEE Transactions on Communications*, 50:1389–1396, Sept. 2002.

[9.48] T.J. Richardson, M.A. Shokrollahi, and R.L. Urbanke. Design of capacity-approaching irregular low-density parity-check codes. *IEEE Transactions on Information Theory*, 47:619–637, Feb. 2001.

[9.49] T.J. Richardson and R.L Urbanke. Efficient encoding of low-density parity-check codes. *IEEE Transactions on Information Theory*, 47:638–656, Feb. 2001.

[9.50] J. Rosenthal and P. Vontobel. Constructions of ldpc codes using ramanujan graphs and ideas from margulis. In *Proceedings of the 38-th Annual Allerton Conference on Communication, Control, and Computing*, pages 248–257, 2000.

[9.51] J. Rosenthal and P. Vontobel. Constructions of regular and irregular ldpc codes using ramanujan graphs and ideas from margulis. In *Proceedings of IEEE International Symposium on Information Theory (ISIT'01)*, page 4, June 2001.

[9.52] M. Sipser and D.A. Spielman. Expander codes. *IEEE Transactions on Information Theory*, 42:1710–1722, Nov. 1996.

[9.53] R. M. Tanner. A recursive approach to low complexity codes. *IEEE Transactions on Information Theory*, IT-271:533–547, Sept. 1981.

[9.54] R.M. Tanner. A [155, 64, 20] sparse graph (ldpc) code. In *Proceedings of IEEE International Symposium on Information Theory*, Sorrento, Italy, June 2000.

[9.55] S. ten Brink. Convergence of iterative decoding. *IEE Electronics Letters*, 35:806–808, May 1999.

[9.56] S. ten Brink. Iterative decoding trajectories of parallel concatenated codes. In *Proceedings of 3rd IEEE ITG Conference on Source and Channel Coding*, pages 75–80, Munich, Germany, Jan. 2000.

[9.57] T. Tian, C. Jones, J.D. Villasenor, and R.D. Wesel. Construction of irregular ldpc codes with low error floors. In *Proceedings of IEEE International Conference on Communications (ICC'03)*, 2003.

[9.58] B. Vasic. Combinatorial constructions of low-density parity check codes for iterative decoding. In *Proceedings of IEEE International Symposium on Information Theory (ISIT'02)*, July 2002.

[9.59] B. Vasic. High-rate low-density parity check codes based on anti-pasch affine geometries. In *Proceedings of IEEE International Conference on Communications (ICC'02)*, volume 3, pages 1332–1336, 2002.

[9.60] B. Vasic, E.M. Kurtas, and A.V. Kuznetsov. Kirkman systems and their application in perpendicular magnetic recording. *IEEE Transactions on Magnetics*, 38:1705–1710, July 2002.

[9.61] B. Vasic, E.M. Kurtas, and A.V. Kuznetsov. Ldpc codes based on mutually orthogonal latin rectangles and their application in perpendicular magnetic recording. *IEEE Transactions on Magnetics*, 38:2346–2348, Sept. 2002.

[9.62] F. Verdier and D. Declercq. A ldpc parity check matrix construction for parallel hardware decoding. In *Proceedings of 3rd International Symposium on Turbo Codes & related topics*, 1-5 Sept. 2003.

[9.63] Eric W. Weisstein. Weisstein. from Mathworld, http://mathworld.wolfram.com/BlockDesign.html.

[9.64] N. Wiberg. *Codes and Decoding on General Graphs.* PhD thesis, Linköping University, 1996.

[9.65] X.-Y.Hu, E. Eleftheriou, D.-M. Arnold, and A. Dholakia. Efficient implementations of the sum-product algorithm for decoding ldpc codes. In *Proceedings of IEEE Global Telecommunications Conference (GLOBECOM'01)*, pages 1036–1036, Nov. 2001.

[9.66] E. Yeo, B. Nikolic, and V. Anantharam. High throughput low-density parity-check decoder architectures. In *Proceedings of IEEE Global Telecommunications Conference (GLOBECOM'01)*, San Antonio, 25-29 Nov. 2001.

[9.67] H. Zhang and J.M.F. Moura. The design of structured regular ldpc codes with large girth. In *Proceedings of IEEE Global Telecommunications Conference (GLOBECOM'03)*, Dec. 2003.

Chapter 10

Turbo codes and large spectral efficiency transmissions

Transporting information in telecommunication systems is carried out at higher and higher data rates and in narrower and narrower frequency bands. Consequently, we wish to maximize the ratio of the useful data rate to bandwidth, that is to say, the spectral efficiency of the transmissions. To do this, it seems natural to couple digital modulations having large constellations with powerful high-rate error correcting codes like turbo codes.

The studies undertaken in this domain are essentially based on two approaches: *turbo trellis coded modulation* and *pragmatic turbo coded modulation.*

10.1 Turbo trellis coded modulation (TTCM)

Turbo trellis coded modulation or TTCM was introduced by Robertson and Wörz in 1995 [10.5, 10.6]. It uses the notion of parallel concatenation, which is at the origin of turbo coding, applied to two trellis coded modulations, or TCMs.

TCM, introduced by Ungerboeck at the beginning of the 80s [10.7] is based on the joint optimization of error correction coding and modulation. The coding is performed directly in the signal space so that the error correcting code and the bit to signal mapping of the modulation can be represented jointly using a single trellis. The criterion for optimizing a TCM thus involves maximizing the minimum Euclidean distance between two coded sequences. To do this, Ungerboeck proposed a two-step approach: partitioning the constellation of the modulation into sub-constellations presenting increasing minimum Euclidean distances, then assigning to each branch of the trellis a signal belonging to the constellation, respecting a set of rules such as those described in [10.7].

The TTCM scheme presented by Robertson and Wörz is shown in Figure 10.1. Each TCM encoder is made up of a recursive systematic convolutional encoder, or RSC encoder, with rate $q/(q+1)$, and a modulation without memory of order $Q = 2^{q+1}$. The binary symbols coming from the source are grouped into symbols of q bits. These symbols are encoded by the first TCM in the order in which are produced by the source and by the second TCM after interleaving.

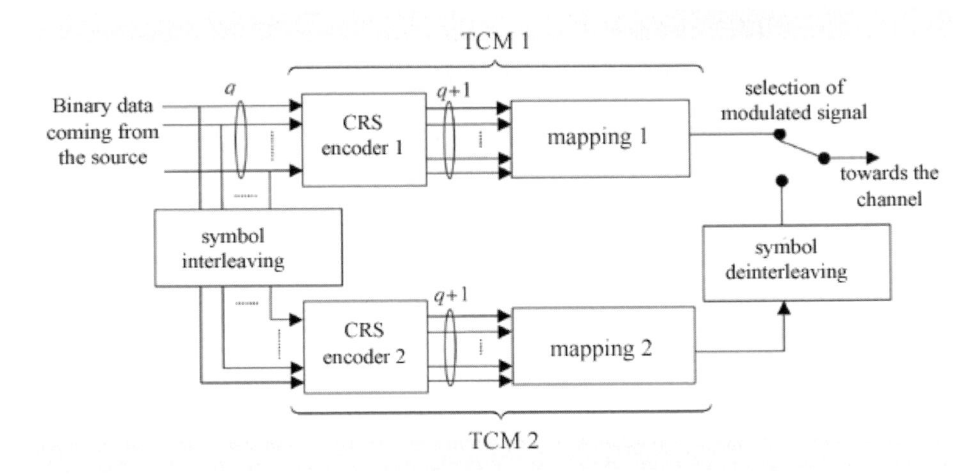

Figure 10.1 – Diagram of the principle of turbo trellis coded modulation (TTCM), according to Robertson and Wörz [10.5, 10.6]. Spectral efficiency $\eta = q$ bit/s/Hz.

Each q-tuple coming from the source being encoded two times, a selection operator alternatively transmits the output of one of the two TCM encoders, in order to avoid the double transmission of information, which would lead to a spectral efficiency of the system of $q/2$ bit/s/Hz. This in fact amounts to puncturing half of the redundancy sequence for each convolutional code.

At reception, the TTCM decoder is similar to a turbo decoder, except that the former directly processes the $(q+1)$-ary symbols coming from the demodulator. Thus, the calculation of the transition probabilities at each step of the MAP algorithm (see Section 7.4) uses the Euclidean distance between the received symbol and the symbol carried by each branch of the trellis. If the decoding algorithm operates in the logarithmic domain (Log-MAP, Max-Log-MAP), it is the branch metrics that are taken equal to the Euclidean distances. Computing an estimate of the bits carried by each demodulated symbol, before decoding, would indeed be a sub-optimal implementation of the receiver.

Similarly, for efficient implementation of turbo decoding, the extrinsic information exchanged by the elementary decoders must directly concern the q-tuples of information transmitted and not the binary elements that they are made up of. At each decoding instant, the elementary decoders thus exchange 2^q values of extrinsic information.

Figure 10.2 provides two examples of elementary RSC codes used in [10.5, 10.6] to build an 8-PSK TTCM with 8 states of spectral efficiency $\eta = 2$ bit/s/Hz and a 16-QAM TTCM with spectral efficiency $\eta = 3$ bit/s/Hz.

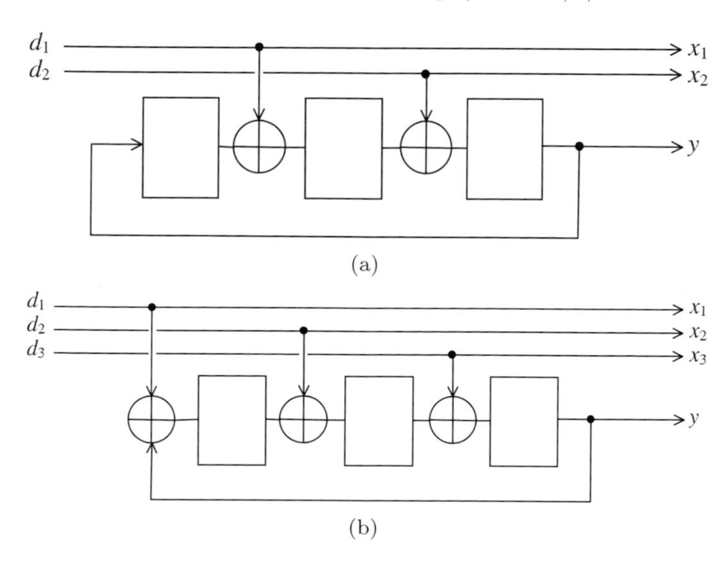

Figure 10.2 – Examples of elementary RSC codes used in [10.5, 10.6] for the construction of a 8-PSK turbo trellis (a) and a 16-QAM turbo trellis (b) coded modulations.

Figures 10.3 and 10.4 show the performance of these two TTCMs in terms of binary error rates (BER) as a function of the signal to noise ratio for transmission over a Gaussian channel. At high and average error rates, these schemes show correction performance close to capacity: a BER of 10^{-4} is reached at around 0.65 dB from Shannon's theoretical limit for the transmission of packets of 5,000 coded modulated symbols. On the other hand, as the interleaving function of the TTCM has not been the object of any particular optimization in [10.5, 10.6], the error rates curves presented reveal early changes in slope (BER$\sim 10^{-5}$) that are very pronounced.

A variant of this technique, proposed by Benedetto *et al.* [10.1] made it possible to improve its asymptotic performance. An alternative method to build a TTCM with spectral efficiency q bit/s/Hz involves using two RSC codes with rate $q/(q+1)$ and for each of them to puncture $q/2$ information bits (q is assumed to be even). For each elementary code we thus transmit only half the information bits and all the redundancy bits. The bits at the output of each encoder are associated with a modulation with $2^{(q/2)+1}$ points. The same operation is performed for the two RSC codes, taking care that each systematic bit is transmitted once and only once, so that the resulting turbo code is system-

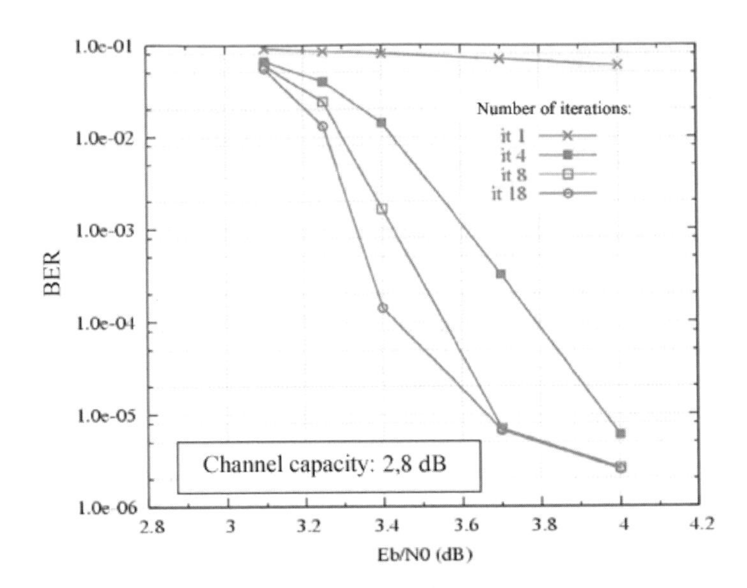

Figure 10.3 – Binary error rate (BER) as a function of the signal to noise ratio E_b/N_0 of the 8-PSK TTCM with 8 states using the RSC code of Figure 10.2(a). Transmission over a Gaussian channel. Spectral efficiency $\eta = 2$ bit/s/Hz. Blocks of 10,000 information bits, 5,000 modulated symbols. MAP decoding algorithm. Curves taken from [10.6].

atic. On the other hand, this technique uses interleaving at bit level, and not at symbol level like in the previous approach.

The criterion for optimizing the TTCM proposed in [10.1] is based on maximizing the *effective Euclidean distance*, defined as the minimum Euclidean distance between two encoded sequences whose information sequences have a Hamming weight equal to 2. Figures 10.5 and 10.6 show two examples of TTCMs built on this principle.

The correction performance of these two TTCMs over a Gaussian channel are presented in Figures 10.7 and 10.8. At high and average error rates, they are close to those given by the scheme of Robertson and Wörz; on the other hand, using interleavers operating on the bits rather than on the symbols has made it possible to significantly improve the behaviour at low error rates.

TTCMs lead to excellent correction performance over a Gaussian channel, since they are an *ad hoc* approach to turbo coded modulation. However, they have the main drawback of very limited flexibility: a new code must be defined for each coding rate and each modulation considered. This drawback is cumbersome in any practical system requiring a certain degree of adaptability. On the other hand, although they are a quasi-optimal solution to the problem of coded modulations for the Gaussian channel, their behaviour over fading channels like Rayleigh channels leads to mediocre performance [10.9].

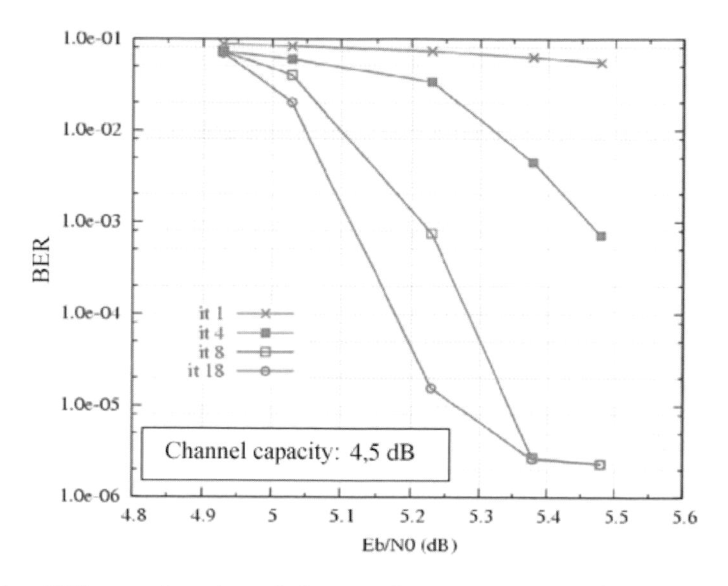

Figure 10.4 – BER as a function of the signal to noise ratio E_b/N_0 of the 16-QAM TTCM with 8 states using the RSC code of Figure 10.2(b). Transmission over a Gaussian channel. Spectral efficiency $\eta = 3$ bit/s/Hz. Blocks of 15,000 information bits, 5,000 modulated symbols. MAP decoding algorithm. Curves taken from [10.6].

10.2 Pragmatic turbo coded modulation

The so-called *pragmatic* approach was chronologically the first implementation. It was introduced by Le Goff *et al.* [10.4] in 1994. This technique takes its name from its similarities with the technique of associating a convolutional code and modulation proposed by Viterbi [10.2] as an alternative solution to Ungerboeck's TCMs. The coding and modulation functions are processed independently, without joint optimization. It uses a "good" turbo code, a bit to signal mapping which minimizes the probability of binary error at the output of the corresponding demapper (Gray coding) and associates the two functions via puncturing and multiplexing to adapt the whole scheme to the spectral efficiency targeted. Figures 10.9 and 10.10 present the general diagram for the principle of the transmitter and the receiver for the pragmatic association of a turbo code and modulation with $Q = 2^q$states.

With this pragmatic approach to turbo coded modulation, the encoder and the decoder used are standard turbo encoders and decoders, identical for all coding rates and modulations considered. If the size of the blocks of data transmitted is variable, simple parametering of the code's permutation function must allow it to adapt to different sizes.

When the targeted coding rate is higher than the natural rate of the turbo code, the puncturing operation enables it to erase, that is to say, not transmit,

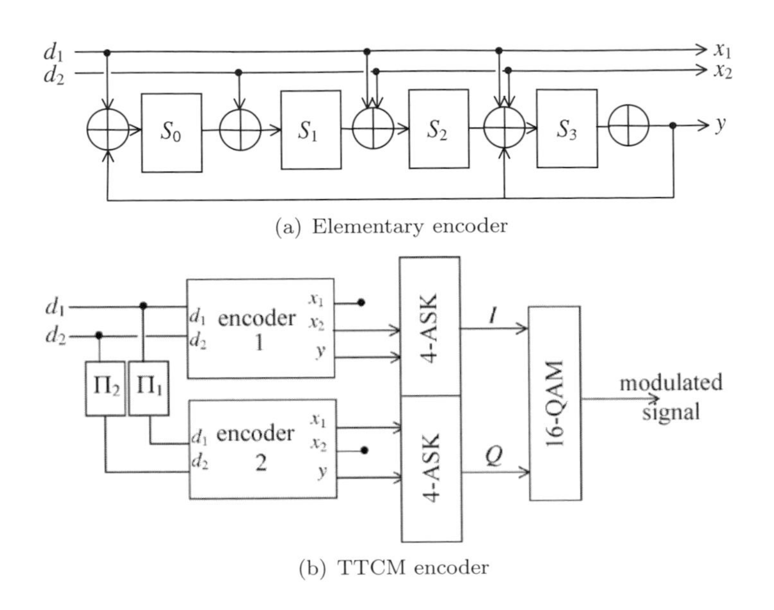

Figure 10.5 – Construction of a 16-QAM TTCM according to the method described in [10.1]. Spectral efficiency $\eta = 2$ bit/s/Hz.

certain coded bits. In practice, for practical reasons of hardware implementation, the puncturing pattern is periodic or quasi-periodic. If possible, only the parity bits are punctured. Indeed, puncturing the systematic bitsleads to a rapid degradation in the decoding convergence threshold, as these bits take part in the process of decoding the two codes, unlike the redundancy bits. When the coding rate is high, a slight puncturing of the data can nevertheless improve the asymptotic behaviour of the system.

The presence of interleaving functions $\Pi^{'}$ and $\Pi^{'-1}$ is justified by the need to decorrelate the data at the input of the turbo decoder. In fact, it is shown in [10.4] that inserting this interleaving has no significant effect on the error rates at the output of the decoder in the case of a transmission on a Gaussian channel. However, in the case of fading channels, the interleaving is necessary as we must prevent bits coming from the same coding instant from belonging to a same symbol transmitted over the channel, so that they will not be affected simultaneously by fading. The studies carried out in this domain [10.9, 10.4, 10.8, 10.2] have shown that in the case of fading channels, the best performance is obtained by using independent interleavers at bit level. This technique is called Bit-Interleaved Coded Modulation (BICM). When the interleaver is placed at modulation symbol level, the order of diversity of the coded modulation is equal to the minimum number of different symbols between two coded modulated sequences. With independent interleavers placed at each output of the encoder, it

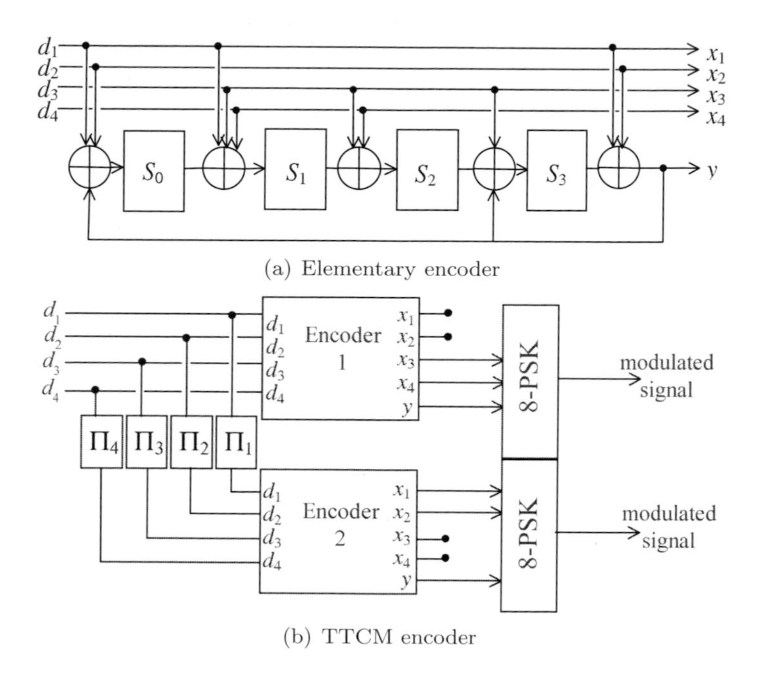

Figure 10.6 – Construction of an 8-PSK TTCM according to the method described in [10.1]. Spectral efficiency $\eta = 2$ bit/s/Hz

can ideally reach the Hamming distanceof the code. Consequently, transmission schemes using the BICM principle in practice have better performance on fading channels than TTCMs have.

The code and the modulation not being jointly optimized, unlike a TTCM scheme, we choose binary mapping of the constellation points which minimizes the mean binary error rates at the input of the decoder. When it can be envisaged, Gray encoding satisfies this condition. For simplicity in implementing the modulator and demodulator, in the case of square QAM (q even), the in-phase and in-quadrature axes, I and Q, are mapped independently.

In Figure 10.9, the role of the "Multiplexing / symbol composition" block is to distribute the encoded bits, after interleaving for fading channels, into modulation symbols. This block, the meeting point between the code and the modulation, enables a certain level of adjustment of the coded modulation according to the performance targeted. This adjustment is possible since the code and the modulation do not play the same role in relation to all the bits transmitted.

On the one hand, we can distinguish two distinct families of encoded bits at the output of the encoder: systematic bits and redundancy bits. These two families of bits play a different role in the decoding process: the systematic bits, coming directly from the source, are used by the two elementary decoders

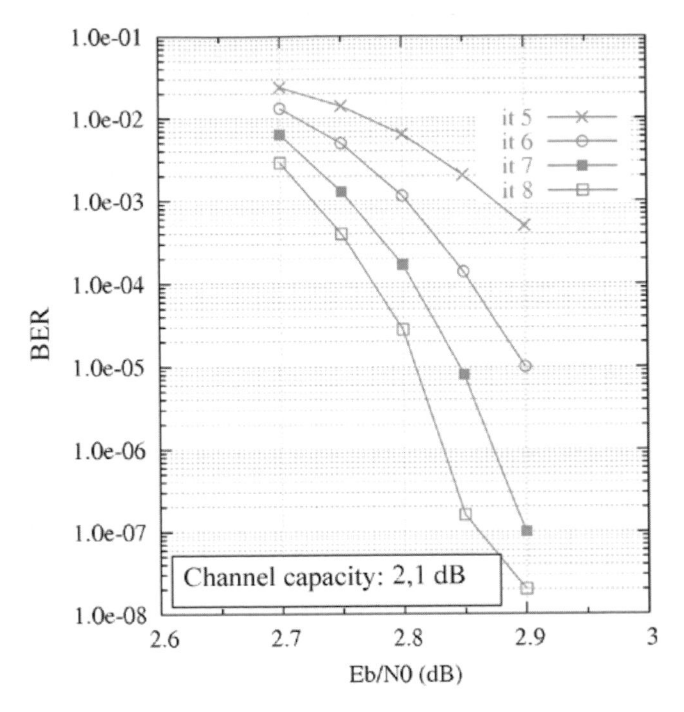

Figure 10.7 – BER as a function of the signal to noise ratio E_b/N_0 of the 16-QAM TTCM with 16 states using the RSC code of Figure 10.5. Transmission over a Gaussian channel. Spectral efficiency $\eta = 2$ bit/s/Hz. 2×16,384 information bits. MAP decoding algorithm. Curves taken from [10.1].

at reception whereas the redundancy bits, coming from the two elementary encoders, are used by only one of the two decoders.

On the other hand, the binary elements contained in a modulated symbol are not, in general, all protected identically by the modulation. For example, in the case of PSK or QAM modulation with Gray encoding, only modulations with two or four points offer the same level of protection to all the bits of a same symbol. For higher order modulations, certain bits are better protected than others.

As an illustration, consider a 16-QAM modulation, mapped independently and in an analogue manner on the in-phase and in-quadrature axes by Gray encoding. The projection of this modulation on each of the paths is amplitude shift keying (ASK) with 4 symbols (see Figure 10.11).

We can show that, for a transmission over a Gaussian channel, the error probabilities on the binary positions s_1 and s_2, given the transmitted symbol (± 3 or ± 1), are expressed in the form:

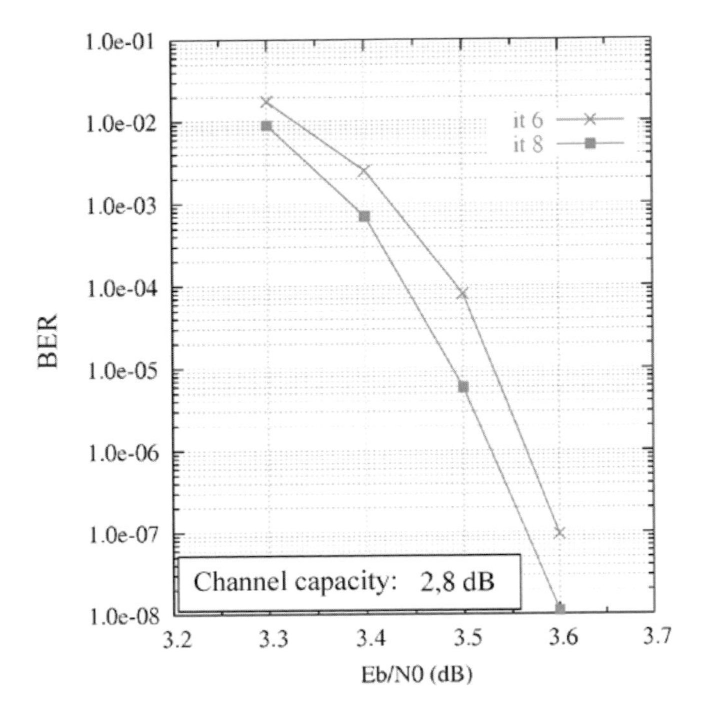

Figure 10.8 – BER as a function of the signal to noise ratio E_b/N_0 of the 8-PSK TTCM with 16 states using the RSC code of Figure 10.6. Transmission over a Gaussian channel. Spectral efficiency $\eta = 2$ bit/s/Hz. 4×4,096 information bits. MAP decoding algorithm. Curves taken from [10.3].

$$\begin{aligned}
P_{eb}(s_2 \mid \pm 3) &= \frac{1}{2}\text{erfc}\left(\frac{3}{\sigma\sqrt{2}}\right)\\
P_{eb}(s_2 \mid \pm 1) &= \frac{1}{2}\text{erfc}\left(\frac{1}{\sigma\sqrt{2}}\right)\\
P_{eb}(s_1 \mid \pm 3) &= \frac{1}{2}\text{erfc}\left(\frac{1}{\sigma\sqrt{2}}\right) - \frac{1}{2}\text{erfc}\left(\frac{5}{\sigma\sqrt{2}}\right) \approx \frac{1}{2}\text{erfc}\left(\frac{1}{\sigma\sqrt{2}}\right)\\
P_{eb}(s_1 \mid \pm 1) &= \frac{1}{2}\text{erfc}\left(\frac{3}{\sigma\sqrt{2}}\right) + \frac{1}{2}\text{erfc}\left(\frac{1}{\sigma\sqrt{2}}\right) \approx \frac{1}{2}\text{erfc}\left(\frac{1}{\sigma\sqrt{2}}\right)
\end{aligned} \tag{10.1}$$

where erfc represents the complementary error function and σ^2 designates the noise variance on the channel. We observe that binary position s_2 is on average better protected by the modulation than position s_1.

Consequently, it is possible to define several strategies for building modulation symbols by associating as a matter of priority the systematic bits or redun-

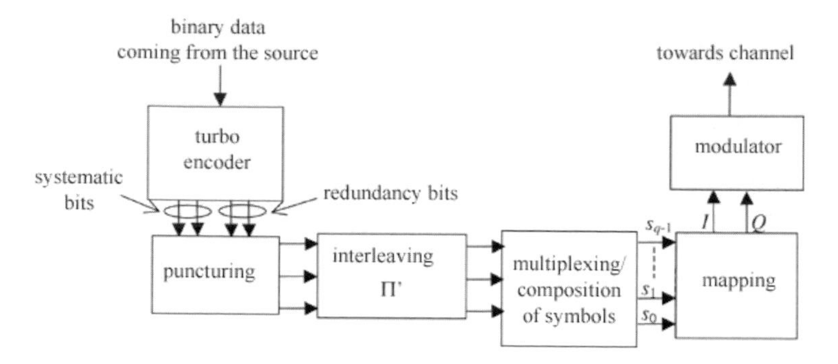

Figure 10.9 – Diagram of the principle of the transmitter in the case of the pragmatic association of a turbo code and modulation with $Q = 2^q$ states.

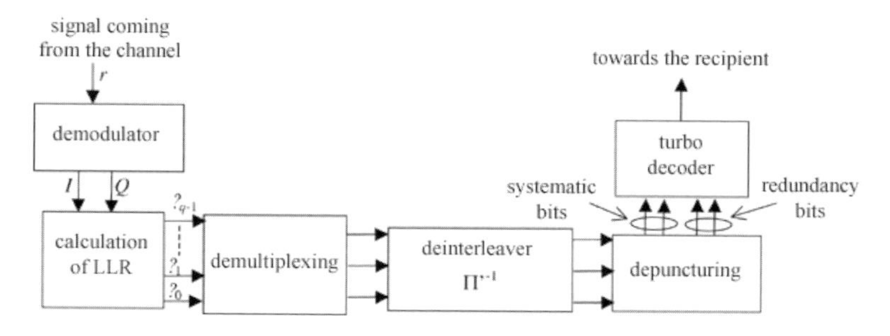

Figure 10.10 – Diagram of the principle of the receiver for the pragmatic turbo coded modulation scheme of Figure 10.9.

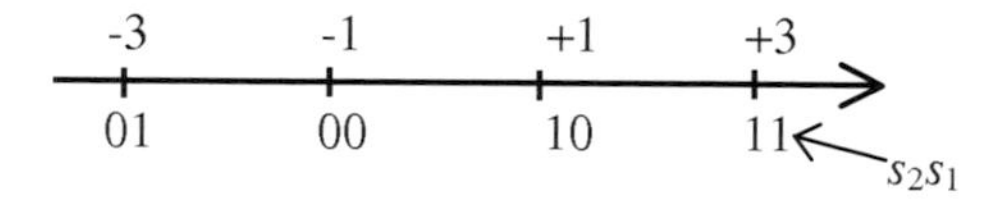

Figure 10.11 – Diagram of the signals of 4-ASK modulation with Gray encoding.

dant bits with the positions that are the best protected by the modulation. Two extreme strategies can thus be defined in all cases:

- so-called "systematic" scheme: the bits best protected by the modulation are associated as a matter of priority with the systematic bits;
- so-called "redundant" scheme: the bits best protected by the modulation are associated as a matter of priority with the redundancy bits.

Modulations of orders higher than 16-QAM offer more than two levels of protection for the different binary positions. 64-QAM modulation, for example, gives three different levels of protection, if the in-phase and in-quadrature axes are mapped independently and in an analogue manner by using a Gray code.

In this case, other schemes, falling in between "systematic" and "redundant" schemes, can be defined.

The reception scheme corresponding to the transmitter of Figure 10.9 is described in Figure 10.10. A standard turbo decoder is used, which requires calculating a weighted estimation of each of the bits contained in the symbols at the output of the demodulator before carrying out the decoding.

The weighted estimation of each bit s_i is obtained by calculating the log likelihood ratio (LLR) defined by:

$$\hat{s}_i = \Lambda(s_i) = \frac{\sigma^2}{2} \ln\left(\frac{\Pr(s_i = 1 \mid r)}{\Pr(s_i = 0 \mid r)}\right) = \frac{\sigma^2}{2} \ln\left(\frac{\Pr(s_i = 1 \mid I, Q)}{\Pr(s_i = 0 \mid I, Q)}\right) \tag{10.2}$$

The sign of the LLR provides the binary decision on s_i (0 if $\Lambda(s_i) \leq 0$, 1 otherwise) and its absolute value represents the weight, that is to say, the reliability, associated with the decision. In the case of a transmission over a Gaussian channel, $\hat{s}_i$ can be written:

$$\hat{s}_i = \frac{\sigma^2}{2} \ln\left(\frac{\sum\limits_{s \in S_{i,1}} \exp\left(-\frac{d_{r,s}^2}{2\sigma^2}\right)}{\sum\limits_{s \in S_{i,0}} \exp\left(-\frac{d_{r,s}^2}{2\sigma^2}\right)}\right) \tag{10.3}$$

where $S_{i,1}$ and $S_{i,0}$ represent the sets of points s of the constellation such that the ith bit s_i is equal to 1 or 0, and $d_{r,s}$ is the Euclidean distance between the received symbol r and the constellation point considered s.

In practice, the Max-Log approximation is commonly used to simplify the calculation of the LLRs:

$$\ln(\exp(a) + \exp(b)) \approx \max(a, b) \tag{10.4}$$

and the weighted estimations are calculated as:

$$\hat{s}_i = \frac{1}{4}\left(\min_{s \in S_{i,0}}(d_{r,s}^2) - \min_{s \in S_{i,1}}(d_{r,s}^2)\right). \tag{10.5}$$

We note that it is not necessary to know the noise variance on the channel, when this simplification is used.

The other blocks of the receiver perform the inverse operations of the blocks of Figure 10.9. The depuncturing operation corresponds to the insertion of an LLR equal to 0, this is to say, a zero reliability decision, at the input of the decoder for all the non-transmitted bits.

The pragmatic turbo coded modulation approach enables performance very close to that of TTCMs to be obtained. Figures 10.12 and 10.13 show the performance of two pragmatic turbo coded modulations using the double-binary turbo code with 16 states presented in Section 7.5 for transmission conditions similar to those leading to the curves obtained in Figures 10.3 and 10.4. We observe that after 8 decoding iterations, the performance of the two turbo coded modulation families are equivalent down to BERs from 10^{-4} to 10^{-5}. The better behaviour of the pragmatic solution at lower error rates is due, on the one hand, to the use of 16-state elementary codes and, on the other hand, to the careful design of the turbo code interleaver.

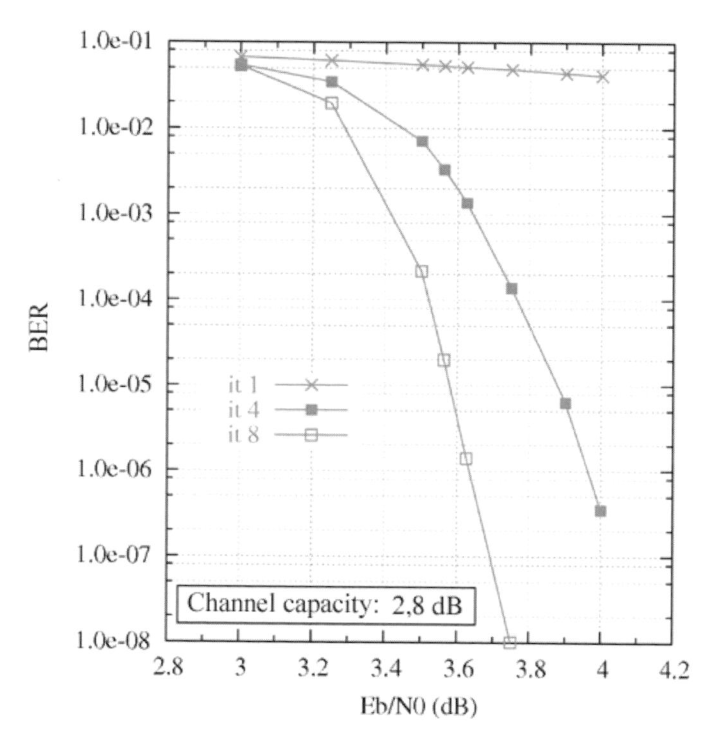

Figure 10.12 – BER as a function of the signal to noise ratio E_b/N_0 of pragmatic turbo-coded 8-PSK using a 16-state double-binary code. Transmission over a Gaussian channel. Spectral efficiency $\eta = 2$ bit/s/Hz. Blocks of 10,000 information bits, 5,000 modulated symbols. MAP decoding algorithm. "Systematic" scheme.

The curves of Figure 10.14 show the influence of the strategy of constructing symbols on the performance of turbo coded modulation. They show the behaviour of the association of a 16-state double-binary turbo code and a 16-QAM mapped independently on the in-phase and in-quadrature axes using the Gray code. The two extreme strategies for building the symbols described above were simulated. The size of the blocks, 54 bytes, and the simulated rates 1/2 and 3/4, are representative of concrete applications in the wireless technology sector (IEEE 802.16 standard, *Wireless Metropolitan Area Network*). Figure 10.14 also shows the theoretical limits of the transmission studied. These limits take into

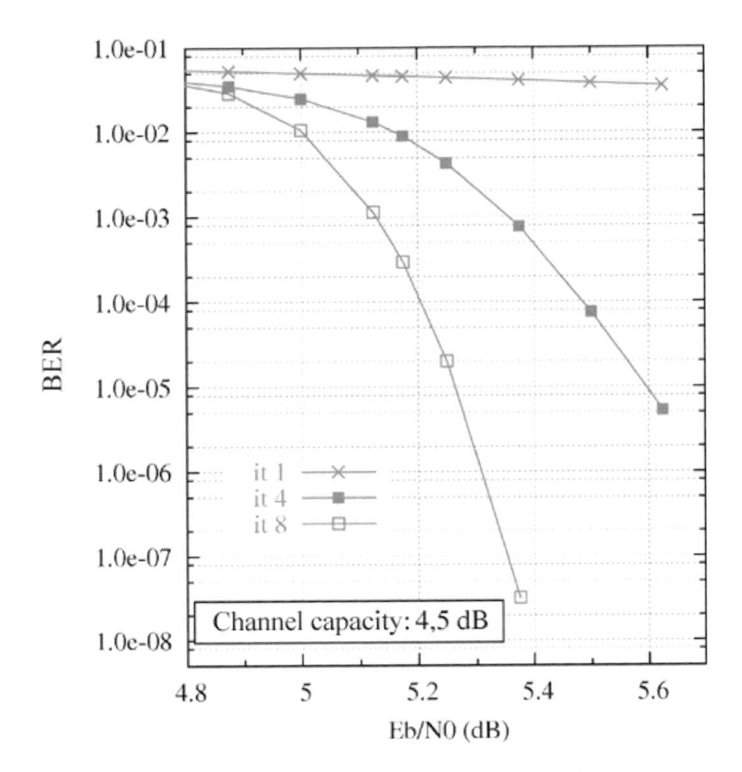

Figure 10.13 – BER as a function of the signal to noise ratio E_b/N_0 of pragmatic turbo-coded 16-QAM using a 16-state double-binary code. Transmission over a Gaussian channel. Spectral efficiency $\eta = 3$ bit/s/Hz. Blocks of 15,000 information bits, 5,000 modulated symbols. MAP decoding algorithm. "systematic" scheme.

account the size of the blocks transmitted as well as the packet error rates (PER) targeted. They are obtained from the value of the capacity of the channel, to which we add a correcting term obtained with the help of the so-called "sphere packing" bound, (see Section 3.3).

We observe that at high or average error rates, the convergence of the iterative decoding process is favoured by a better protection of the systematic bits. This result can be explained by the fact that, in the decoding process, each systematic data is used at the input of the two decoders. Consequently, an error on a systematic bit at the output of the channel causes an error at the input of the two elementary decoders, whereas erroneous redundancy only affects the input of one of the two elementary decoders. Consequently, reinforcing the protection of the systematic bits benefits the two elementary decoders simultaneously.

The higher the proportion of redundancy bits transmitted, that is to say, the lower the coding rate, the greater the gap in performance between the two schemes. As an example, at a binary error rate of 10^{-4}, we observe a gap of 0.9 dB for a coding rate $R = 1/2$, and 0.2 dB for $R = 3/4$.

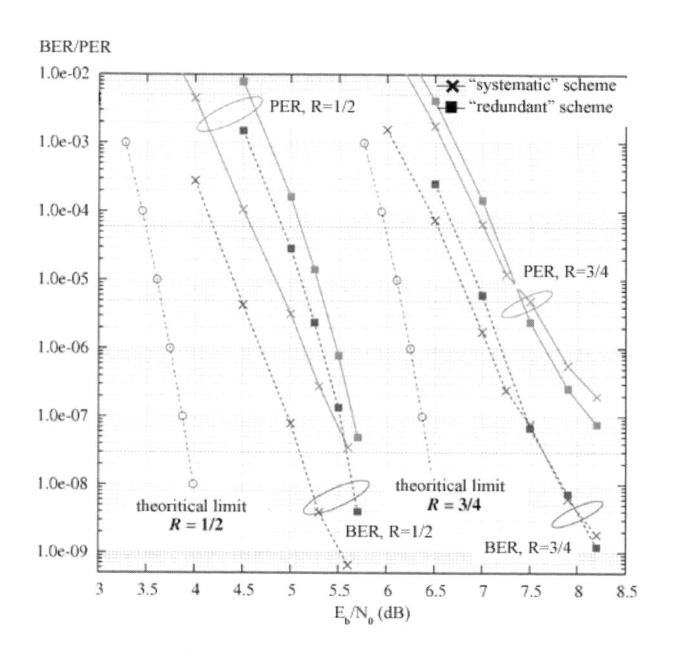

Figure 10.14 – Performance in binary error rate (BER) and packet error rate (PER) of the pragmatic association of a 16-QAM and a 16-state double-binary turbo code, for the transmission of blocks of 54 bytes over a Gaussian channel. Coding rates 1/2 and 3/4. Max-Log-MAP decoding, inputs of the decoder quantized on 6 bits, 8 decoding iterations.

For low and very low error rates, the scheme favouring the protection of the redundancy gives the best performance. This behaviour is difficult to prove by simulation for the lowest rates, as the assumed crossing point of the curves is situated at an error rate that is difficult to obtain by simulation ($PER \approx 10^{-8}$ for $R = 1/2$). The interpretation of this result requires analysis of the erroneous paths in trellises with a high signal to noise ratio. We have observed that, in the majority of cases, the erroneous sequences contain a fairly low number of erroneous systematic bits and a rather high number of erroneous redundancy bits. In other words, the erroneous sequences generally have a low input weight. In particular, the erroneous paths in question mainly correspond to rectangular patterns of errors (see Section 7.3.2). The result, from the point of view of the asymptotic behaviour of turbo coded modulation, is that it is preferable to ensure better protection of the parity bits.

The curves shown in Figure 10.14 were obtained with the help of the simplified Max-Log-MAP decoding algorithm, using data quantized on 6 bits at the input of the decoder. These conditions correspond to a hardware implementation of the decoder. In spite of these constraints, the performance obtained is fairly close to the theoretical limits: only 1 dB at a PER of 10^{-4} and 1.5 dB at a PER of 10^{-6}.

Bibliography

[10.1] S. Benedetto, D. Divsalar, G. Montorsi, and F. Pollara. Parallel concatenated trellis codes modulation. In *Proceeedings of International Conference on Communications (ICC'96)*, pages 974–978, Dallas, USA, 1996.

[10.2] G. Caire, G. Taricco, and E. Biglieri. Bit-interleaved coded modulation. *IEEE Transactions on Information Theory*, 44(3):927–946, May 1998.

[10.3] C. Fragouli and R. Wesel. Bit vs. symbol interleaving for parallel concatenated trellis coded modulation. In *Proceedings of Global Telecommunications Conference (Globecom'01)*, pages 931–935, San Antonio, USA, Nov. 2001.

[10.4] S. Le Goff, A. Glavieux, and C. Berrou. Turbo codes and high spectral efficiency modulation. In *Proceedings of International Conference on Communications (ICC'94)*, pages 645–649, New Orleans, USA, May 1994.

[10.5] P. Robertson and T. Wörz. Coded modulation scheme employing turbo codes. *Electronics Letters*, 31(2):1546–1547, Aug. 1995.

[10.6] P. Robertson and T. Wörz. Bandwidth-efficient turbo trellis-coded modulation using punctured component codes. *IEEE Journal on Selected Areas in Communications*, 16(2):206–218, Feb. 1998.

[10.7] G. Ungerboeck. Channel coding with mutilevel/phase signals. *IEEE Trans. Info. Theory.*, IT-28(1):55–67, Jan. 1982.

[10.8] A. J. Viterbi, J. K. Wolf, E. Zehavi, and R. Padovani. A pragmatic approach to trellis-coded modulation. *IEEE Communications Magazine*, 27(7):11–19, July 1989.

[10.9] E. Zehavi. 8-psk trellis codes for a rayleigh channel. *IEEE Transactions on Communications*, 40(5):873–884, May 1992.

Chapter 11

The turbo principle applied to equalization and detection

The invention of turbo codes at the beginning of the 90s totally revolutionized the field of error correcting coding. Codes relatively simple to build and decode, making it possible to approach Shannon's theoretical limit very closely, were at last available. However, the impact of this discovery was not limited to one single coding domain. More generally, it gave birth to a new paradigm for designing digital transmission systems, today commonly known as the "turbo principle". To solve certain very complex *a priori* signal processing problems, we can envisage dividing these problems into a cascade of elementary processing operations, simpler to implement. However, today we know that the one-directional succession of these processing operations leads to a loss of information. To overcome this sub-optimality, the turbo principle advocates establishing an exchange of probabilistic information, "in the two directions", between these different processing operations. All of the information available is thus taken into account in solving the global problem and a consensus can be found between all the elementary processing operations in order to elaborate the final decision.

The application of the turbo principle to a certain number of classical problems in digital transmission has provided impressive gains in performance in comparison to traditional systems. Therefore its use rapidly became popular within the scientific community. This chapter presents the first two systems having historically benefited from the application of the turbo principle to a context other than error correction coding. The first system, called turbo equalization, iterates between the equalization function and a decoding function to improve the processing of the intersymbol interference for data transmission over multipath channels. The second, commonly called turbo CDMA, exploits the turbo principle to improve the discrimination between users in the case of a radio-mobile communication between several users based on the *Code Division Multiple Access* technique.

11.1 Turbo equalization

Multipath channels have the particularity of transforming a transmitted signal into a linear superposition of several different copies (or echoes) of this signal. Turbo equalization is a digital reception technique that makes it possible to detect data deteriorated by multipath transmission channels. It combines the work of an equalizer and a channel decoder using the turbo principle. Schematically, this digital reception system involves a repetition of the equalization-interleaving-decoding processing chain. First, the equalization performs an initial estimation of the transmitted data. Second, the estimation is transmitted to the decoding module which updates this information. Then the information updated by the decoder is sent to the equalization module. Thus, over the iterations, the equalization and decoding processing operations exchange information in order to reach the performance of a transmission on a channel with a single path.

The purpose of this section is to present the turbo equalization principle and its implementation in two versions: turbo equalization according to the Maximum *A Posteriori* (MAP) criterion and turbo equalization according to the Minimum Mean Square Error (MMSE) criterion. We will describe the algorithms associated with these two techniques, as well as their respective complexity. This will lead us to present the possible architectures and give examples of implementation. Finally, potential and existing applications for these techniques will be shown.

11.1.1 Multipath channels and intersymbol interference

This section is dedicated to transmissions on multipath channels whose particularity is to generate one or several echoes of the signal transmitted. Physically these echoes can, for example, correspond to reflections off a building. The echoes thus produced come and superpose themselves on the signal initially transmitted and thus degrade the reception. The equivalent discrete channel model allows a mathematically simple representation of these physical phenomena in the form of a linear filtering of the transmitted discrete-time symbol sequence. Let x_i be the symbol transmitted at discrete instant i, and y_i be the received symbol at this same instant. The channel output is then given by

$$y_i = \sum_{k=0}^{L-1} h_k(i) x_{i-k} + w_i \tag{11.1}$$

where $h_k(i)$ represents the action of the channel (echo) at instant i on a symbol transmitted at instant $i - k$. The impulse response of the channel at instant i

is then written in the following way in the form of a z-transform:

$$h(z) = \sum_{k=0}^{L-1} h_k(i) z^{-k} \tag{11.2}$$

The impulse response of the channel is assumed to have finite duration (L coefficients), which is a realistic hypothesis in practice in most scenarios.

Equation (11.1) shows that generally, received symbol y_i is a function of the symbols transmitted before, or after (if the channel introduces a propagation delay) information symbol x_i considered at instant i. In accordance with what was introduced in Chapter 2, we then say that the received signal is spoiled by intersymbol interference (ISI). If we now assume that the transmission channel does not vary (or very little) on the duration of a transmitted block of information, model(11.1) can be simplified as follows:

$$y_i = \sum_{k=0}^{L-1} h_k x_{i-k} + w_i \tag{11.3}$$

where we have suppressed the time dependency from the coefficients of the equivalent discrete channel. The representation of the equivalent discrete channel in the form of a digital filter with finite impulse response presented in Figure 11.1 comes directly from (11.3). The coefficients of the filter are precisely those of the impulse response of the channel.

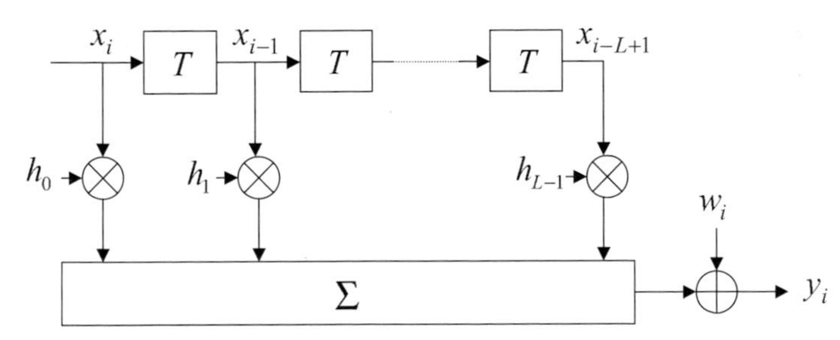

Figure 11.1 – Representation of the equivalent discrete channel in the form of a digital filter.

ISI can be a major obstacle for establishing a good quality digital transmission, even in the presence of very low noise. As an illustration, we have shown in Figure 11.2 the constellation of the symbols received at the output of a channel highly perturbed by ISI, for a signal to noise ratio of 20 dB[1], given that we have transmitted a sequence of discrete symbols with four phase states (QPSK modulation). We thus observe that when the ISI is not processed by an

[1] We recall that a signal to noise ratio of 20 dB corresponds to a power of the transmitted signal 100 times higher than the power of the additive noise on the link.

adequate device, it can lead to great degradation in the error rate at reception, and therefore in the general quality of the transmission.

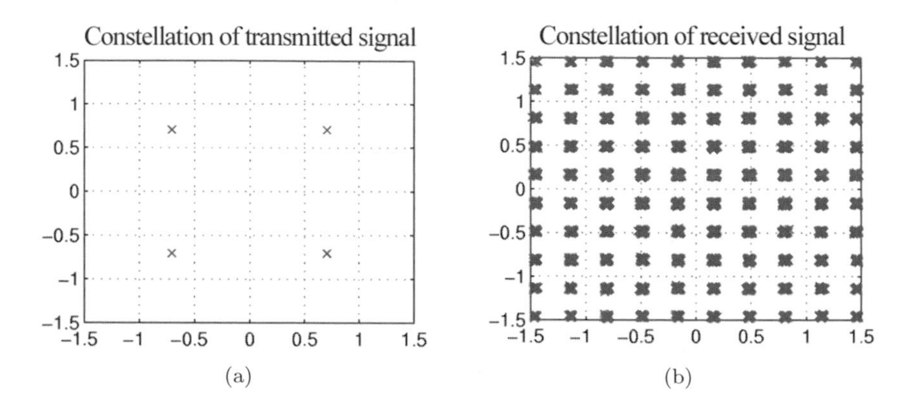

Figure 11.2 – Illustration of the phenomenon of ISI in the case of a 5-path highly frequency-selective channel, for a signal to noise ratio of 20 dB.

We now study the characteristics of a multipath channel in the frequency domain. We show in Figure 11.3 the frequency response of the channel generating the constellation presented in Figure 11.2. The latter is highly perturbed by ISI. We note that the frequencies of the signal will not be attenuated and delayed in the same way over the whole frequency band. Thus, a signal having a band W between 0 and 3 kHz will be distorted by the channel. We then speak of a frequency selective channel in opposition to a flat non-frequency selective channel, for which all the frequencies undergo the same distortion. To resume, when a multipath channel generates intersymbol interference in the time domain, it is then frequency selective in the frequency domain.

We mainly have three different techniques to combat the frequency selectivity of transmission channels: multi-carrier transmissions, spread spectrum and equalization. In this chapter, we deal only with the third solution, applied here to transmissions on a single carrier frequency ("single-carrier" transmissions). Note, however, that some of the concepts tackled here can be transposed relatively easily to systems of the multi-carrier type (*Orthogonal Frequency Division Multiplex*, or OFDM systems).

11.1.2 The equalization function

In its most general form, the purpose of the equalization function is to give an estimation of the transmitted sequence of symbols from the sequence observed at the output of the channel, the latter being perturbed both by intersymbol interference and additive noise, assumed to be Gaussian. We distinguish different equalization strategies. Here we limit ourselves to a succinct overview of the

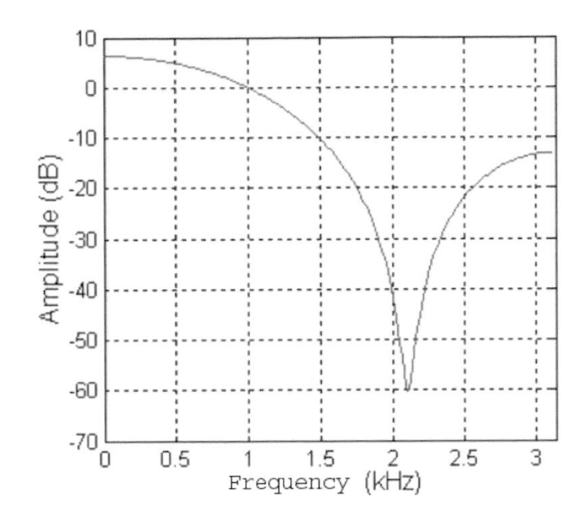

Figure 11.3 – Frequency response of the 5-path discrete-time channel.

main techniques usually implemented in systems. The interested reader can find additional information in Chapters 10 and 11 of [11.44], in articles [11.45] and [11.54] or in book [11.11], for example.

A first solution, called *Maximum Likelihood Sequence Detection*, or MLSD, involves searching for the most probable sequence transmitted relatively to the observation received at the output of the channel. We can show that this criterion amounts to choosing the candidate sequence at the minimum Euclidean distance from the observation, and that it thus minimizes the error probability per sequence, that is to say, the probability of choosing a candidate sequence other than the sequence transmitted. A naive implementation of this criterion involves listing the set of admissible sequences in such a way as to calculate the distance between each sequence and the observation received, then to select the sequence closest to this observation. However, the complexity of this approach increases exponentially with the size of the message transmitted, which turns out to be unacceptable for a practical implementation.

In a famous article dating from 1972 [11.27], Forney noted that a frequency selective channel presents a memory effect whose content characterizes its state at a given instant. More precisely, state s of the channel at instant i is perfectly defined by the knowledge of the $L-1$ previous symbols, which we denote $s = (x_i, \ldots, x_{i-L+2})$. This fact is based on the representation of the channel in the form of a shift register (see Figure 11.1). The evolution of the state of the channel over time can then be represented by trellis diagram having M^{L-1} states, where M denotes the number of discrete symbols in the modulation alphabet. As an illustration, we have represented in Figure 11.4 the trellis associated with a channel having $L = 3$ coefficients $\mathbf{h} = (h_0, h_1, h_2)$ in the case of a binary phase-shift-keying (BPSK) modulation.

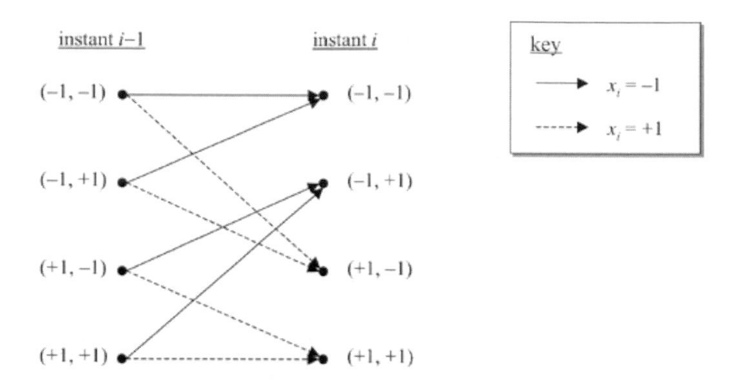

Figure 11.4 – Trellis representation for BPSK transmission on a frequency-selective discrete-time channel with $L = 3$ paths.

Each candidate sequence takes a single path in the trellis. Searching for the sequence with the minimum Euclidean distance from the observation can then be performed recursively, with a linear computation cost depending on the size of the message, by applying the Viterbi algorithm on the trellis of the channel.

The MLSD equalizer offers very good performance. However, the complexity of its implementation increases proportionally with the number of states in the trellis, and therefore exponentially with duration L of the impulse response of the channel and size M of the modulation alphabet. Its practical utilization is consequently limited to transmissions using modulations with a small number of states (2, 4, or 8) on channels with few echoes. On the other hand, it should be noted that this equalizer requires prior estimation of the impulse response of the channel in order to build the trellis. The MLSD solution has been adopted by many manufacturers to perform the equalization operation in mobile telephones for the worldwide GSM (*Global System Mobile*) standard.

In the presence of modulations with a large number of states or on channels whose impulse response length is large, the MLSD equalizer has an unacceptable computation time for real-time applications. An alternative strategy then involves combating the ISI with the help of equalizers having less complexity, implementing digital filters.

In this perspective, the simplest solution involves applying a linear transverse filter at the output of the channel. This filter is optimized so as to compensate ("equalize") the irregularities of the frequency response of the channel, with the aim of converting the frequency selective channel into an equivalent ideally ISI-free (or frequency-flat) channel, perturbed only by additive noise. The transmitted message is then estimated thanks to a simple operation of symbol by symbol decision (threshold detector) at the output of the equalizer, optimal on an additive white Gaussian noise (AWGN) channel. This equalizer, shown in Figure 11.5, is called a *linear equalizer* or LE.

Figure 11.5 – Linear equalizer.

We distinguish several optimization criteria to define the coefficients of the transverse filter. The optimal criterion involves minimizing the symbol error probability at the output of the filter, but its application leads to a system of equations difficult to solve. In practice, we prefer criteria sub-optimal in terms of performance, but leading to solutions easily implementable, like the *Minimum Mean Square Error* or MMSE criterion [11.44]. The linear MMSE equalizer is an attractive solution due to its simplicity. However, this equalizer suffers from the problem of amplification of the noise level on highly selective channels having strong attenuations at certain points in the frequency spectrum.

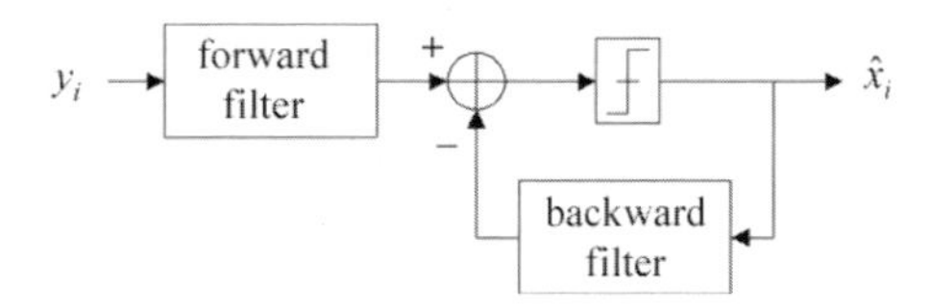

Figure 11.6 – Decision-feedback equalizer.

Examining the diagram of the principle of the linear equalizer, we can note that when we take a decision on symbol x_i at instant i, we have an estimation on the previous symbols $\hat{x}_{i-1}$, $\hat{x}_{i-2}$, ... We can therefore envisage rebuilding the (causal) interference caused by these data and therefore cancel it, in order to improve the decision. The equalizer which results from this reasoning is called a *Decision-Feedback Equalizer* or DFE. The diagram of the principle of the device is illustrated in Figure 11.6. It is made up of a forward filter, in charge of converting the impulse response of the channel into a purely causal response, followed by a decision device and a feedback filter, in charge of estimating the residual interference at the output of the feedback filter in order to cancel it via a feedback loop.

As a general rule, the DFE provides performance higher than that of the linear equalizer, particularly on channels that are highly frequency selective. However, this equalizer is non-linear in nature, due to the presence of the decision device in the feedback loop, which can give rise to an error propagation phenomenon (particularly at low signal to noise ratio) when some of the estimated data are incorrect. In practice, the filter coefficients are generally optimized following the MMSE criterion, by assuming that the estimated data are equal to the transmitted data, in order to simplify the computations (see Chapter 10 in [11.44], for example).

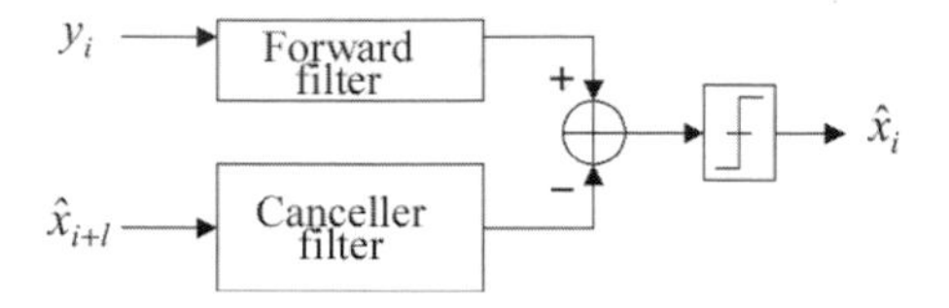

Figure 11.7 – Interference canceller.

If we assume now that we have a estimation $\hat{x}_{i+l}$ on the transmitted data both before ($l < 0$) and after ($l > 0$) the symbol considered at instant i, we can then envisage removing the whole of the ISI at the output of the channel. The equalization structure obtained is called an *interference canceller*, or IC [11-6,11-7]. It is detailed in Figure 11.7. This structure is made up of two digital transverse filters, with finite impulse response: a forward filter (matched to the channel) whose aim is to maximize the signal to noise ratio before the decision, and a canceller filter, in charge of rebuilding the ISI present at the output of the matched filter. Note that by construction, the central coefficient of the canceller filter is necessarily null in order to avoid subtracting the useful signal. With the reserve that the estimated data $\hat{x}_{i+l}$ be equal to the transmitted data, we can show that this equalizer eliminates all the ISI, without any increase in noise level. We thus reach the matched-filter bound, which represents what we can best do with an equalizer on a frequency selective channel. Of course, we never know *a priori* the transmitted data in practice. The difficulty then lies in building an estimation of the data that is sufficiently reliable to keep performance close to optimal.

None of the equalizer structures presented so far take into account the presence of a possible error correcting code on transmission. We shall now see how we can best combine the equalization and decoding functions to improve the global performance of the receiver.

11.1.3 Combining equalization and decoding

Most single-carrier digital transmission systems operating on frequency selective channels incorporate an error correction coding function before the actual modulation step at transmission. The error correcting code is traditionally inserted to combat the errors caused by the additive noise on the link. However, coupled with a carefully built interleaving function, the encoder also offers an additional degree of protection faced with power fading caused by the channel, when the characteristics of the latter vary over time. We saw in the previous section that independently of the nature of the equalizer used, the ISI systematically leads to a loss in performance compared with a non-selective AWGN channel. The presence of the encoder can then be exploited to reduce this gap in performance, by benefiting from the coding gain at reception.

In the following part of this section, we are going to examine a transmission system shown in Figure 11.8. The modulation and transmission operations on the channel are here represented in equivalent baseband, in order not to have to consider a carrier frequency.

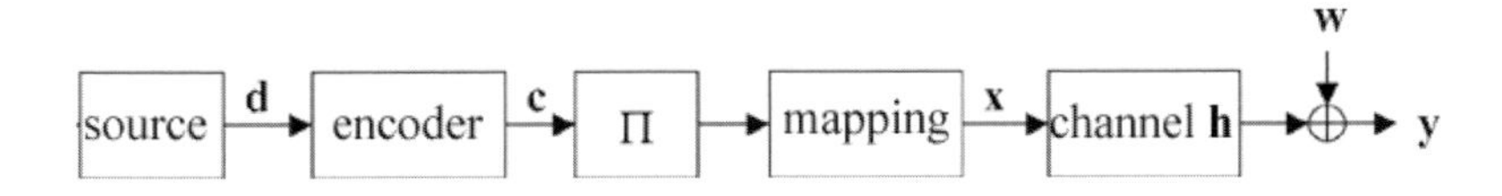

Figure 11.8 – Baseband model of the transmission system considered.

The source sends a sequence of Q information bits, $\mathbf{d} = (d_0, \ldots, d_{Q-1})$. This message is protected by a convolutional code of rate R, to provide a sequence $\mathbf{c} = (c_0, \ldots, c_{P-1})$ of $P = Q/R$ coded bits. The coded bits are then interleaved following a permutation Π, then finally converted in groups of m successive bits into discrete complex symbols chosen in an alphabet with $M = 2^m$ elements, that we will denote $\{X_1, \ldots, X_M\}$. This is the mapping operation. We thus obtain a sequence of $N = P/m$ complex symbols: $\mathbf{x} = (x_0, \ldots, x_{N-1})$. Later, the vector of m coded and interleaved binary elements associated with symbol x_i at instant i will be denoted $(x_{i,1}, \ldots, x_{i,j}, \ldots, x_{i,m})$. The transmitted symbols are discrete random variables with zero mean and variance $\sigma_x^2 = E\{|x_i|^2\}$.

This transmission scheme is called *Bit Interleaved Coded Modulation* or BICM. We still find it today in many systems: the mobile telephony standard GSM, for example. For voice transmission at 13 kbits/s (TCH/FS channel), the specifications of the radio interface indicate that at the output of the speech encoder, the most sensitive bits (class 1A and 1B bits) are protected by a convolutional code with rate $R = 1/2$ and generator polynomials (33,23) in octal [11.1]. The coded message is then interleaved on 8 consecutive packets (or *bursts*), to benefit from time-diversity in the presence of fading, then finally modulated following a waveform of the *Gaussian Minimum Shift Keying* (GMSK) type.

If we now return to the scenario of Figure 11.8, sequence $\mathbf{x}$ is transmitted within a frequency selective channel with L coefficients, with discrete impulse response $\mathbf{h} = (h_0, \ldots, h_{L-1})$. The resulting signal is then perturbed by a vector $\mathbf{w}$ with complex centred AWGN samples and variance $\sigma_w^2 = E\{|w_i|^2\}$. The noisy sequence observed at the output of the channel is finally denoted $\mathbf{y} = (y_0, \ldots, y_{N-1})$, the expression of sample y_i at instant i being given by Equation (11.3).

In this context, the problem that faces the designer of the receiver is the following: how can we best combine equalization and decoding, so that each processing benefits from the result of the other processing?

In reply to this question, estimation theory tells us that to minimize the error probability in this case, the equalization and decoding operations must be performed jointly, following the maximum likelihood criterion. Conceptually, implementing the optimal receiver then amounts to applying the Viterbi algorithm, for example, on a "super-trellis" simultaneously taking into account the constraints imposed by the code, the channel and the interleaver. However, the "super-trellis" has a number of states that increases exponentially with the size of the interleaver, which excludes a practical implementation of the optimal receiver. It is therefore legitimate to question the feasibility of such a receiver in the absence of an interleaver. Historically, this question has been asked in particular in the context of data transmission over twisted-pair telephone cables (voice-band modems). These systems implement error correction coding in Euclidean space (trellis coded modulations), without interleaving, and the telephone channel is a typical example of a frequency-selective, time-invariant channel. Assuming an encoder with S states, a constellation of M points and a discrete channel with L coefficients, the studies undertaken in this context have shown that the corresponding "super-trellis" has exactly $S(M/2)^{L-1}$ states [11.13]. It is then easy to verify that in spite of the absence of an interleaver, the complexity of the optimal receiver again rapidly becomes prohibitive, whenever we wish to transmit a high rate of information (with modulations having a large number of states) or when we are confronted with a channel having large delays.

To counter the unaffordable complexity of the optimal receiver, the solution commonly adopted in practice involves performing the equalization and decoding operations disjointly, sequentially in time. If we again take the example of GSM, the received data are thus first processed by an MLSD equalizer. The estimated sequence provided by the equalizer is then transmitted, after deinterleaving, to a Viterbi decoder. The permutation function then plays a twofold role in this context: not only combating slow fading on the channel, but also dispersing error packets at the input of the decoder. This strategy presents the advantage of simplicity of implementation, since the total complexity is then given by the sum of the individual complexities of the equalizer and the decoder. However, it necessarily leads to loss in performance compared with the optimal receiver since the equalization operation does not exploit all the available information. To be more precise, the estimation sent by the equalizer will not necessarily correspond to a valid coded sequence since the equalizer does not take into account the constraints imposed by the code. The performance of the disjoint solution can be improved when we introduce the passing of weighted (probabilistic) information instead of an exchange of binary data between the equalizer and the decoder. By propagating a reliability measure on the decisions of the equalizer, the decoder thus benefits from additional information to produce its own estimation of the message, and we benefit from a correction gain generally of the order of several dB (see for example [11.28, 11.23] or Chapter 3 in [11.15]). Despite this, the drawback of this solution remains: only one-way communication between the equalizer and the decoder.

Therefore, does a strategy exist that can somehow produce the best of both worlds, capable of reconciling both good performance of the optimal joint receiver and simplicity in implementation of the sub-optimal disjoint receiver? Today, it is possible to reply in the affirmative, thanks to what we have called "turbo equalization".

11.1.4 Principle of turbo equalization

The concept of turbo equalization first saw the light of day in the laboratories of ENST Bretagne at the beginning of the 90s, under the impulsion of the spectacular results obtained with turbo codes. It was the outcome of a very simple realization: the transmission scheme in Figure 11.8 can be seen as the serial concatenation of two codes (Chapter 6), separated by an interleaver, the second code being formed by cascading the mapping operation with the channel[2]. Seen from this angle, it would then seem natural to apply a decoding strategy of the "turbo" type at reception, that is, a reciprocal, iterative exchange of probabilistic information (extrinsic information) between the equalizer and the decoder. The first turbo equalization scheme was proposed in 1995 by Douillard *et al.* [11.12]. This scheme implements a weighted input and output (*Soft Input Soft Output*, or SISO) Viterbi equalizer according to the *Soft Output Viterbi Algorithm* (SOVA). The principle was then used in 1997 by Bauch *et al.*, substituting the SOVA equalizer by a SISO equalizer that was optimal in the sense of the MAP criterion, using the algorithm developed by Bahl *et al.* (BCJR algorithm [11.7]) .

The simulation results quickly showed that the turbo equalizer was capable of totally removing ISI, under certain conditions. Retrospectively, this excellent performance can be explained by the fact that this transmission scheme brings together two key ingredients which are the force of the turbo principle:

1. The implementation of iterative decoding at reception, introducing an exchange of probabilistic information between the processing operations, about which we today know that, when the signal to noise ratio exceeds a certain "convergence threshold", it converges towards the performance of the optimal joint receiver after a certain number of iterations.

2. The presence of an interleaver at transmission, whose role here mainly involves breaking up the error packets at the output of the equalizer (to avoid the phenomenon of error propagation), and decorrelating as far as

[2] Note that, strictly speaking, transmission on a selective channel does not represent a coding operation in itself, despite its convolutional character, as it does not provide any gain. Indeed, it only degrades performance. Nevertheless, this analogy makes sense from the iterative decoding point of view.

possible the probabilistic data exchanged between the equalizer and the channel decoder. The turbo equalizer is then capable of totally compensating the degradation caused by the ISI, with the reserve that the interleaver be large enough and carefully constructed.

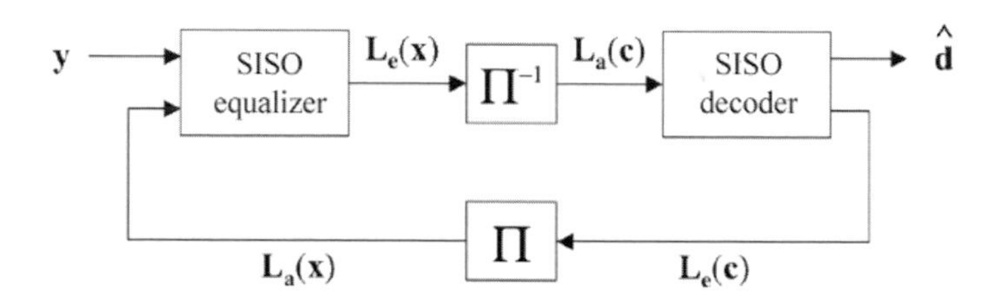

Figure 11.9 – Turbo equalizer for BICM transmission systems.

Generally, the turbo equalizer corresponding to the transmission scenario in Figure 11.8 takes the form shown in Figure 11.9. It is first made up of a SISO equalizer, which at the input takes both vector $\mathbf{y}$ of the data observed at the output of the channel, and *a priori* probabilistic information on the set of coded, interleaved bits $x_{i,j}$, here formally denoted $\mathbf{L}_a(\mathbf{x}) = \{L_a(x_{i,j})\}$. The probabilistic information is propagated in the form of log likelihood ratios (LLRs), the definition of which we recall here for a binary random variable d with values in $\{0, 1\}$:

$$L(d) = \ln\left(\frac{\Pr(d=1)}{\Pr(d=0)}\right) \tag{11.4}$$

The notion of LLR provides twofold information since the sign of the quantity $L(d)$ gives the hard decision on d, while its absolute value $|L(d)|$ measures the reliability this decision can be given.

From the two pieces of information $\mathbf{y}$ and $\mathbf{L}_a(\mathbf{x})$, the SISO equalizer produces extrinsic information denoted $\mathbf{L}_e(\mathbf{x}) = \{L_e(x_{i,j})\}$ on the coded, interleaved binary message. This vector $\mathbf{L}_e(\mathbf{x})$ is then deinterleaved to give a new sequence $\mathbf{L}_a(\mathbf{c})$, which is the *a priori* information on the sequence coded for the SISO decoder. The latter then deduces two pieces of information from this: a hard decision on the information message transmitted, here denoted $\hat{\mathbf{d}}$, and some new extrinsic information on the coded message, denoted $\mathbf{L}_e(\mathbf{c})$. This information is then re-interleaved and sent back to the SISO equalizer where it is exploited as *a priori* information for a new equalization step at the following iteration.

The turbo equalization scheme that we have presented above corresponds to BICM transmitters. It is however important to note that the turbo equalization principle also applies in the case of a system implementing traditional coded modulation, that is to say, a system where the coding and modulation operations are jointly optimized, on condition however that the symbols to be transmitted are interleaved before being modulated and sent on the channel (Figure 11.10).

The main difference with the previous scheme thus lies in the implementation of the SISO equalizer and SISO decoder. Indeed, these latter no longer exchange probabilistic information at binary level but at symbol level, whether in LLR form or directly in probability form. The interested reader can find further details on this subject in [11.8], for example.

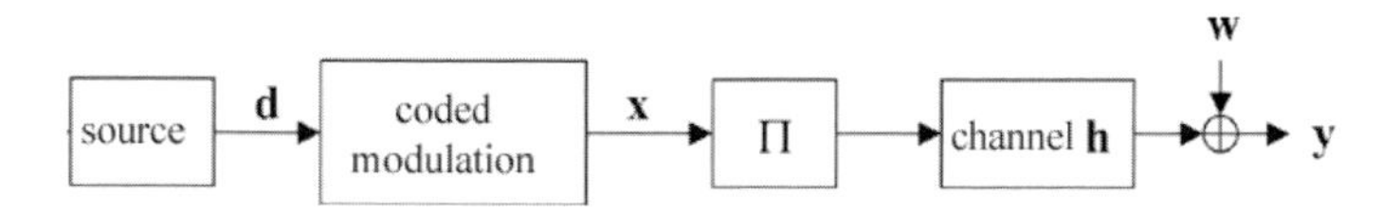

Figure 11.10 – Baseband model of traditional coded modulation systems.

As a general rule, the channel code is a convolutional code and the channel decoder uses a soft-input soft-output decoding algorithm of the MAP type (or its derivatives in the logarithmic domain: Log-MAP and Max-Log-MAP). Again, we will not consider the hardware implementation of the decoder since this subject is dealt with in Chapter 7. Note, however, that unlike classical turbo decoding schemes, the channel decoder here does not provide extrinsic information on the information bits, but instead on the coded bits.

On the other hand, we distinguish different optimization criteria to implement the SISO equalizer, leading to distinct families of turbo equalizers. The first, sometimes called "turbo detection" and what we call MAP turbo equalization here, uses an equalizer that is optimal in the Maximum *A Posteriori* sense. The SISO equalizer is then typically performed thanks to the BCJR-MAP algorithm. As we shall see in the following section, this approach leads to excellent performance, but like the classical MLSD equalizer, it has a very high computation cost which excludes any practical implementation in the case of modulations with a large number of states and for transmissions on channels having large time delays. We must then turn towards alternative solutions, with less complexity but that will necessarily be sub-optimal in nature. One strategy that can be envisaged in this context involves reducing the number of branches to examine at each instant in the trellis. This approach is commonly called "reduced complexity MAP turbo equalization". We know different methods to reach this result, which will be briefly presented in the following section. Another solution is inspired by classical equalization methods and implements an optimized SISO equalizer following the minimum mean square error (MMSE) criterion. We thus obtain an MMSE (filtering-based) turbo equalizer, a scheme described in Section 11.1.6 and that appears as a very promising solution today for high data rates transmissions on highly frequency-selective channels.

11.1.5 MAP turbo equalization

MAP turbo equalization corresponds to the turbo equalization scheme originally introduced by Douillard *et al.* [11.12]. In this section, we first present the equations for implementing the SISO equalizer. The good performance of the MAP turbo equalizer is illustrated by simulation. We also introduce solutions of less complexity derived from the MAP criterion. Finally, we examine the problems encountered during a circuit implementation of the turbo equalizer, as well as potential applications of this reception technique.

Implementation of the BCJR-MAP equalizer

The MAP equalizer shown in Figure 11.11 takes at its input vector $\mathbf{y}$ of the discrete symbols observed at the output of the channel, as well as *a priori* information denoted $\mathbf{L}_a(\mathbf{x})$ on the coded interleaved bits. This information comes from the channel decoder and is produced at the previous iteration. In the particular case of the first iteration, we do not generally have any *a priori* information other than the hypothesis of equiprobability on the bits transmitted, which leads us to put $L_a(x_{i,j}) = 0$.

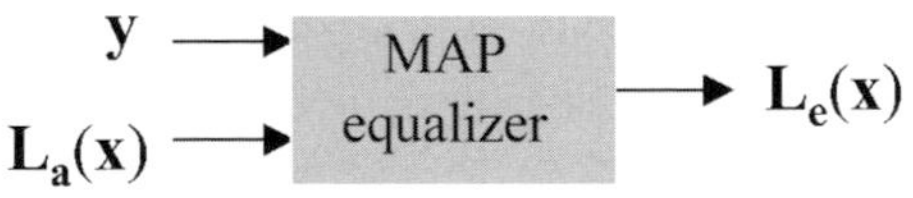

Figure 11.11 – Block diagram of the MAP equalizer.

The purpose of the MAP equalizer is to evaluate the *a posteriori* LLR $L(x_{i,j})$ on each coded interleaved bit $x_{i,j}$, defined as follows:

$$L(x_{i,j}) = \ln\left(\frac{\Pr(x_{i,j} = 1\,|\mathbf{y})}{\Pr(x_{i,j} = 0\,|\mathbf{y})}\right) \tag{11.5}$$

Using conventional results in detection theory, we can show that this equalizer is optimal in the sense of the minimization of the symbol error probability. To calculate the *a posteriori* LLR $L(x_{i,j})$, we will use the trellis representation associated with transmission on the frequency selective channel. Applying Bayes' relation, the previous relation can also be written:

$$L(x_{i,j}) = \ln\left(\frac{\Pr(x_{i,j} = 1, \mathbf{y})}{\Pr(x_{i,j} = 0, \mathbf{y})}\right) \tag{11.6}$$

Among the set of possible sequences transmitted, each candidate sequence traces a single path in the trellis. The joint probability $\Pr(x_{i,j} = 0 \text{ or } 1, \mathbf{y})$ can then be evaluated by summing the probability $\Pr(s', s, \mathbf{y})$ of passing through a particular transition in the trellis linking a state s' at instant $i-1$ to a state s

at instant i, on all of the transitions between instants $i-1$ and i for which the j-th bit making up the symbol associated with these transitions equals 0 or 1. Thus,

$$L(x_{i,j}) = \ln\left(\frac{\sum\limits_{(s',s)/x_{i,j}=1} \Pr(s',s,\mathbf{y})}{\sum\limits_{(s',s)/x_{i,j}=0} \Pr(s',s,\mathbf{y})}\right) \tag{11.7}$$

Adopting a similar approach now to the one presented in the original article by Bahl *et al.* [11.3], we can show that the joint probability $\Pr(s',s,\mathbf{y})$ associated with each transition considered can be decomposed into a product of 3 terms:

$$\Pr(s',s,\mathbf{y}) = \alpha_{i-1}(s')\gamma_{i-1}(s',s)\beta_i(s) \tag{11.8}$$

Figure 11.12 shows the conventions of notation used here.

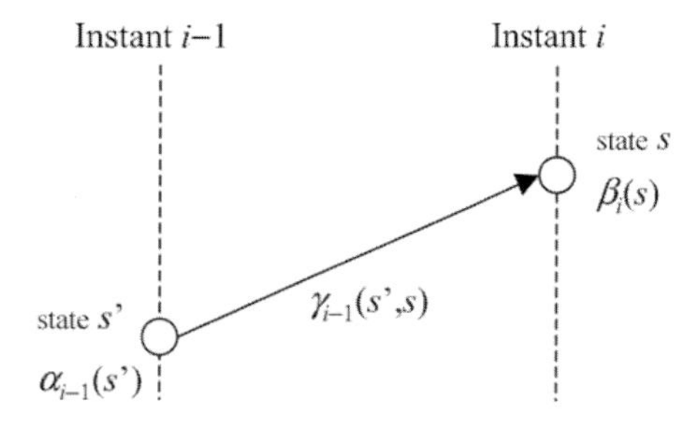

Figure 11.12 – Conventions of notation used to describe the MAP equalizer.

Forward and backward state probabilities $\alpha_{i-1}(s')$ and $\beta_i(s)$ can be calculated recursively for each state and at each instant in the trellis, by applying the following update equations:

$$\alpha_i(s) = \sum_{(s',s)} \alpha_{i-1}(s')\gamma_{i-1}(s',s) \tag{11.9}$$

$$\beta_i(s') = \sum_{(s',s)} \gamma_i(s',s)\beta_{i+1}(s) \tag{11.10}$$

These two steps are called *forward recursion* and *backward recursion*, respectively. Summations are performed over all the couples of states with indices (s', s) for which there is a valid transition between two consecutive instants in the trellis. Forward recursion uses the following initial condition:

$$\alpha_0(0) = 1, \quad \alpha_0(s) = 1 \text{ for } s \neq 0 \tag{11.11}$$

This condition translates the fact that the initial state in the trellis (with index 0, by convention) is perfectly known. Concerning the backward recursion, we usually assign the same weight to each state at the end of the trellis since the arrival state is generally not known *a priori*:

$$\beta_N(s) = \frac{1}{M^{L-1}} \quad \forall s \tag{11.12}$$

In practice, we see that the dynamic of values $\alpha_{i-1}(s')$ and $\beta_i(s)$ increases during the progression in the trellis. Consequently, these values must be normalized at regular intervals in order to avoid overflow problems in the computations. One natural solution involves dividing these metrics at each instant by constants K_α and K_β chosen in such a way as to satisfy the following normalization condition:

$$\frac{1}{K_\alpha}\sum_s \alpha_i(s) = 1 \quad \text{and} \quad \frac{1}{K_\beta}\sum_s \beta_i(s) = 1 \tag{11.13}$$

the sums here concerning all possible states s of the trellis at instant i.

To complete the description of the algorithm, it remains for us to develop the expression of the term $\gamma_{i-1}(s', s)$. This term can be assimilated to a branch metric. We can show that it is decomposed into a product with two terms:

$$\gamma_{i-1}(s', s) = \Pr(s|s')P(y_i|s', s) \tag{11.14}$$

$\Pr(s|s')$ represents the *a priori* probability of going through the transition between state s and state s', that is to say, the *a priori* probability $P_a(X_l) = \Pr(x_i = X_l)$ of having transmitted at time instant i the constellation symbol X_l labeling the branch considered in the trellis. Owing to the presence of the interleaver at transmission, bits $x_{i,j}$ composing symbol x_i can be assumed statistically independent. Consequently, probability $P_a(X_l)$ has the following decomposition:

$$P_a(X_l) = \Pr(x_i = X_l) = \prod_{j=1}^{m} P_a(X_{l,j}) \tag{11.15}$$

where we have written $P_a(X_{l,j}) = \Pr(x_{i,j} = X_{l,j})$, binary element $X_{l,j}$ taking the value 0 or 1 according to the symbol X_l considered and the mapping rule. Within the turbo equalization iterative process, the *a priori* probabilities $P_a(X_{l,j})$ are deduced from the *a priori* information available at the input of the equalizer. From the initial definition (11.4) of the LLR, we can in particular show that probability $P_a(X_{l,j})$ and corresponding *a priori* LLR $L_a(x_{i,j})$ are linked by the following expression:

$$P_a(X_{l,j}) = K \exp\left(X_{l,j} L_a(x_{i,j})\right) \quad \text{with} \quad X_{l,j} \in \{0, 1\} \tag{11.16}$$

The term K is a normalization constant that we can omit in the following computations without compromising the final result in any way.

Conditional probability $\Pr(s|s')$ is therefore finally given by:

$$\Pr(s\,|s'\,) = \exp\left(\sum_{j=1}^{m} X_{l,j} L_a(x_{i,j})\right) \tag{11.17}$$

As for the second term $P(y_i|s',s)$, it quite simply represents the likelihood $P(y_i|z_i)$ of observation y_i relative to branch label z_i associated with the transition considered. The latter corresponds to the symbol that we would have observed at the output of the channel in the absence of noise:

$$z_i = \sum_{k=0}^{L-1} h_k x_{i-k} \tag{11.18}$$

The sequence of symbols $(x_i, x_{i-1}, \ldots, x_{i-L+1})$ occurring in the computation of z_i is deduced from the knowledge of initial state s' and of information symbol X_l associated with transition $s' \to s$. In the presence of zero-mean circularly-symmetric complex additive white Gaussian noise with total variance σ_w^2, we obtain:

$$P(y_i\,|s', s\,) = P(y_i\,|z_i\,) = \frac{1}{\pi\sigma_w^2}\exp\left(-\frac{|y_i - z_i|^2}{\sigma_w^2}\right) \tag{11.19}$$

Factor $1/\pi\sigma_w^2$ is common to all the branch metrics and can therefore be omitted in the computations. On the other hand, we see here that calculating branch metrics $\gamma_{i-1}(s',s)$ requires both knowledge of the impulse response of the equivalent discrete channel and knowledge of the noise variance. In other words, in the context of a practical implementation of the system, the MAP equalizer will have to be preceded by a channel estimation procedure.

To summarize, after computing branch metrics $\gamma_{i-1}(s', s)$ then performing the forward and backward recursions, the *a posteriori* LLR $L(x_{i,j})$ is finally given by:

$$L(x_{i,j}) = \ln \frac{\sum\limits_{(s',s)/x_{i,j}=1} \alpha_{i-1}(s')\gamma_{i-1}(s', s)\beta_i(s)}{\sum\limits_{(s',s)/x_{i,j}=0} \alpha_{i-1}(s')\gamma_{i-1}(s', s)\beta_i(s)} \tag{11.20}$$

In reality and in accordance with the turbo principle, it is not this *a posteriori* information that is propagated to the SISO decoder, but rather the extrinsic information. Here, the latter measures the equalizer's own contribution in the global decision process, excluding the information relating to the bit considered coming from the decoder at the previous iteration, that is to say, the *a priori* LLR $L_a(x_{i,j})$. If we develop the expression of branch metric $\gamma_{i-1}(s', s)$ in the computation of $L(x_{i,j})$, we obtain:

$$L(x_{i,j}) = \ln \left[\frac{\sum\limits_{(s',s)/x_{i,j}=1} \alpha_{i-1}(s') \exp\left(-\frac{|y_i - z_i|^2}{\sigma_w^2} + \sum\limits_{k=1}^{m} X_{l,k} L_a(x_{i,k}) \right) \beta_i(s)}{\sum\limits_{(s',s)/x_{i,j}=0} \alpha_{i-1}(s') \exp\left(-\frac{|y_i - z_i|^2}{\sigma_w^2} + \sum\limits_{k=1}^{m} X_{l,k} L_a(x_{i,k}) \right) \beta_i(s)} \right] \quad (11.21)$$

We can then factorize the *a priori* information term in relation to the bit $x_{i,j}$ considered, both in numerator ($X_{l,j} = 1$) and in denominator ($X_{l,j} = 0$), which gives:

$$L(x_{i,j}) = L_a(x_{i,j}) + \underbrace{\ln \left[\frac{\sum\limits_{(s',s)/x_{i,j}=1} \alpha_{i-1}(s') \exp\left(-\frac{|y_i - z_i|^2}{\sigma_w^2} + \sum\limits_{k \neq j} X_{l,k} L_a(x_{i,k}) \right) \beta_i(s)}{\sum\limits_{(s',s)/x_{i,j}=0} \alpha_{i-1}(s') \exp\left(-\frac{|y_i - z_i|^2}{\sigma_w^2} + \sum\limits_{k \neq j} X_{l,k} L_a(x_{i,k}) \right) \beta_i(s)} \right]}_{L_e(x_{i,j})} \quad (11.22)$$

Finally, we see that the extrinsic information is obtained quite simply by subtracting the *a priori* information from the *a posteriori* LLR calculated by the equalizer:

$$L_e(x_{i,j}) = L(x_{i,j}) - L_a(x_{i,j}) \quad (11.23)$$

This remark concludes the description of the MAP equalizer. As we have presented it, this algorithm proves to be difficult to implement on a circuit due to the presence of numerous multiplication operations. In order to simplify the computations, we can then envisage transposing the whole algorithm into the logarithmic domain (Log-MAP algorithm), the advantage being that the multiplications are then converted into additions, which are simpler to do. If we wish to further reduce the processing complexity, we can also use a simplified (but sub-optimal) version, the Max-Log-MAP (or Sub-MAP) algorithm. These two variants were presented in the context of turbo codes in Chapter 7. The derivation is quite similar in the case of the MAP equalizer. Reference [11.5] presents a comparison in performance between these different algorithms in a MAP turbo equalization scenario. In particular, it turns out that the Max-Log-MAP equalizer offers the best performance/complexity compromise when the estimation of the channel is imperfect.

Example of performance

In order to illustrate the good performance offered by MAP turbo equalization, we chose to simulate the following transmission scenario: a binary source generates messages of 16382 bits of information, which are then protected by a rate

$R = 1/2$ non-recursive non-systematic convolutional code with 4 states, and with generator polynomials (5,7) in octal. Two null bits (tail-bits) are inserted at the end of the message in order to force the termination of the trellis in state 0. Thus we obtain a sequence of 32768 coded bits, which are then randomly interleaved and mapped into a sequence of BPSK symbols. These symbols are transmitted on a 5-path discrete-time channel with impulse response:

$$\mathbf{h} = (0.227, 0.460, 0.688, 0.460, 0.227)$$

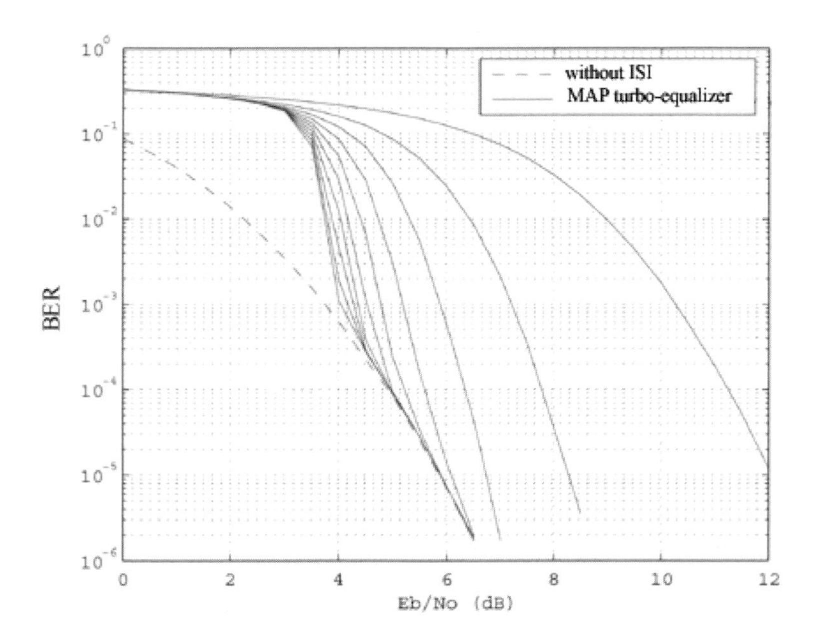

Figure 11.13 – Performance of the MAP turbo equalizer for BPSK transmission on the Proakis C channel, using a rate $R = 1/2$ 4-state non-recursive non-systematic convolutional code and a pseudo-random interleaver of size 32768 bits.

This channel model, called Proakis C, taken from Chapter 10 in [11.44], is relatively difficult to equalize. At reception, we implement 10 iterations of the MAP turbo equalizer described above. The SISO decoder is performed using the BCJR-MAP algorithm. Figure 11.13 presents the bit error rate after decoding, measured at each iteration, as a function of the normalized signal to noise ratio E_b/N_0 on the channel. For reference, we have also shown the performance obtained after decoding on a non-frequency selective AWGN channel. This curve shows the optimal performance of the system. We see that beyond a signal to noise ratio of 5 dB, the turbo equalizer suppresses all the ISI after 5 iterations, and we reach the ideal performance of the AWGN channel. Furthermore, for a target bit error rate of 10^{-5}, the iterative process provides a gain of the order of 6.2 dB compared with the performance of the conventional receiver performing the equalization and decoding disjointly, given by the curve at the 1st iteration. This performance is very similar to that presented in reference [11.7].

These results give rise to a certain number of remarks, since the example considered here presents the characteristic behaviour of turbo systems. In particular, we see that the gain provided by the iterative process only appears beyond a certain signal to noise ratio (convergence threshold, equal to 3 dB here). Beyond this threshold, we observe a rapid convergence of the turbo equalizer towards the asymptotic performance of the system, given by the error probability after decoding on a non-selective AWGN channel. To improve the global performance of the system, we can envisage using a more powerful error correcting code. Experience shows that we then come up against the necessity of finding a compromise in choosing the code, between rapid convergence of the iterative process and good asymptotic performance of the system (at high signal to noise ratios). The greater the correction capacity of the code, the higher the convergence threshold. On this topic, we point out that today there exist semi-analytical tools such as EXIT (*EXtrinsic Information Transfer*) charts [11.49], enabling the value of the convergence threshold to be predicted precisely, as well as the error rate after decoding for a given transmission scenario, under the hypothesis of ideal interleaving (infinite size). A second solution involves introducing a feedback effect in front of the equivalent discrete-time channel, by inserting an adequate precoding scheme at transmission. Cascading the pre-encoder with the channel produces a new channel model, recursive in nature, leading to a performance gain that is greater, the larger the dimension of the interleaver considered. This phenomenon is known as "interleaving gain" in the literature dedicated to serial turbo codes. Subject to carefully choosing the pre-encoder, we can then exceed the performance of classical non-recursive turbo equalization schemes as has been shown in [11.35] and [11.39].

Complexity of the MAP turbo equalizer and alternative solutions

The complexity of the MAP turbo equalizer is mainly dictated by the complexity of the MAP equalizer. Now, the latter increases proportionally with the number of branches to examine at each instant in the trellis. Considering a modulation with M states and a discrete channel with L coefficients, the total number of branches per section of the trellis rises to $M \times M^{L-1} = M^L$. We therefore see that the processing cost associated with the MAP equalizer increases exponentially with the number of states of the modulation and the length of the impulse response of the channel. As an illustration, EDGE (*Enhanced Data Rate for GSM Evolution*) introduces the use of 8-PSK modulation on channels with 6 coefficients maximum, that is, slightly more than 262000 branches to examine at each instant! MAP turbo equalization is therefore an attractive solution for modulations with a low number of states (typically BPSK and QPSK) on channels having ISI limited to a few symbol periods. Beyond that, we must turn to less complex, but less efficient, solutions.

There are several ways to deal with this problem. If we limit ourselves to using equalizers derived from the MAP criterion, one idea is to reduce the number of paths examined by the algorithm in the trellis. A first approach performs a truncation of the channel impulse response in order to keep only the $J < L$ first coefficients. The number of states in the trellis will then be decreased. The ISI terms ignored in the definition of the states are then taken into account when calculating the branch metrics, using past decisions obtained from the knowledge of the survivor path in each state. This strategy is called *Delayed Decision Feedback Sequence Estimation* (DDFSE). It offers good performance provided most of the channel's energy be concentrated in its first coefficients which, in practice, requires the implementation of a minimum-phase pre-filtering operation. Applying this technique to turbo equalization has, for example, been studied in [11.2]. A refinement of this algorithm involves grouping some states of the trellis together, in accordance with the set-partitioning rules defined by Ungerboeck [11.52] for designing trellis coded modulations. This improvement, called Reduced State Sequence Estimation (RSSE), includes DDFSE as a particular case [11.19]. In a similar way, we can also envisage retaining more than one survivor path in each state to improve the robustness of the equalizer and if necessary to omit the use of pre-filtering [11.42]. Rather than reduce the number of states of the trellis by truncation, it can also be envisaged to examine only a non-exhaustive list of the most likely paths at each instant. The resulting algorithm is called the "M algorithm", and its extension to SISO equalization was studied in [11.17]. Whatever the case, the search for efficient equalizers with reduced complexity regularly continues to give rise to new contributions.

All the strategies that we have mentioned above enter into the category of MAP *turbo equalizers* with reduced complexity. Generally, these solutions are interesting when the number of states of the modulation is not too high. On the other hand, in the case of high data rate transmissions on channels with long delay spreads, it is preferable to envisage filter-based turbo equalizers of the MMSE type.

Architectures and applications

When systems based on MAP turbo equalization require real time processing with relatively high data rates (of the order of several Mbits/s), a software implementation cannot be envisaged. In this case, we must resort to specific ASIC circuits. The circuit implementation of a MAP turbo equalizer poses problems similar to those encountered in the context of the hardware implementation of a turbo decoder. Two architectural solutions can be envisaged:

- The first uses an implementation of the turbo decoder in the form of a *pipeline* by cascading several elementary modules, each module implementing one detection and one decoding iteration.
- The second uses a single hardware module, implementing the successive iterations sequentially, by looping back on itself.

The first architecture presents a smaller latency, so is better adapted to applications requiring high data rates. On the other hand, the second solution enables an economy in the number of transistors and therefore in the silicon surface. In order to further reduce the surface used, some authors have proposed sophisticated architectures enabling part of the SISO algorithm to be shared between the equalizer and the decoder, despite the different structure of the trellises concerned [11.36]. This approach also enables a reduction in length of the critical path, and therefore in the global latency of the system. This last factor can be a major obstacle to the practical implementation of turbo equalization (and turbo systems more generally) since not all applications may tolerate an increase in the processing delay at reception. Resorting to analogue electronics will perhaps soon enable this obstacle to be overcome. An analogue implementation of a simplified MAP turbo equalizer has thus been reported in [11.24], with promising results.

From the algorithmic point of view, the application of MAP turbo equalization to the GSM system has been the subject of several studies [11.15, 11.43, 11.6, 11.18]. The traditional turbo equalization scheme must thus be revised in order to take into account the specificities of the standard (inter-frame interleaving, different levels of protection applied to the bits at the output of the speech encoder, GMSK modulation, ...). Simulation results show generally moderate gains in performance, in return for a large increase in the complexity of the receiver. This can be partly explained by the fact that the conventional GSM system faces only a limited level of ISI on the majority of the test channels defined in the standard. On the other hand, the introduction of 8-PSK modulation in the context of EDGE greatly increases the level of interference. This scenario therefore seems more appropriate for the use of turbo equalization, and has given rise to several contributions. In particular, the authors of [11.40][3] have studied the implementation of a complete turbo equalization system relying on a SISO equalizer of the DDFSE type with pre-filtering, coupled to a channel estimator. They have obtained gains of the order of several dB, depending on the modulation and coding scheme considered, compared with the performance of the classical receiver. Furthermore, they have also proved the fact that the equalization and iterative decoding principle could be carefully exploited in the context of the ARQ retransmission protocol defined in EDGE (the *Incremental Redundancy* scheme) to improve the global quality of service at reception.

[3] See also the references cited in this article.

11.1.6 MMSE turbo equalization

The increase in data rates, in response to current multimedia service requirements, combined with the infatuation with mobility and wireless infrastructures, present receivers with severe propagation conditions. Thus, if we take the example of the radio interface of the Wireless MAN (*Metropolitan Area Network*) 802.16a standard normalized by IEEE during 2003 and operating in the 2-11 GHz band, the ISI encountered is likely to recover up to 50 symbol durations, or even more. Underwater acoustic communications is another example. The application of turbo equalization to such scenarios involves using low complexity SISO equalizers. MMSE turbo equalization is an attractive solution in this context.

In contrast with the approaches described in the previous section, MMSE turbo equalization mainly involves substituting for the MAP equalizer an equalizer structure based on digital filters, optimized according to the minimum mean square error criterion[4]. This solution presents a certain number of advantages. First of all, simulations show that the MMSE turbo equalizer gives very good performance on average, sometimes very close to the performance offered by MAP turbo equalization. On the other hand, the complexity of the MMSE equalizer increases linearly (and not exponentially) with the length of the channel impulse response, independently of the order of the modulation. Finally, as we shall see in what follows, this approach naturally lends itself well to an implementation in adaptive form, appropriate for tracking the time variations of the channel.

Historically, the first MMSE turbo equalization scheme was proposed by Glavieux *et al.* in 1997 [11.20, 11.32, 11.34]. This original contribution laid down the bases of MMSE turbo equalization, particularly for the design of a filter-based soft-input soft-output equalizer. Indeed, classical equalizers based on digital filters do not naturally lend themselves to handling probabilistic information. This difficulty was overcome by inserting a binary to M-ary conversion operation at the input of the equalizer, in charge of rebuilding a soft estimation of the symbols transmitted using the *a priori* information sent by the decoder. In addition, a SISO demapping module placed at the output of the equalizer converts the equalized data (complex symbols) into extrinsic LLR on the coded bits, which are then sent to the decoder. This initial scheme relied on the implementation of an equalization structure of the interference canceller type, whose coefficients were updated adaptively thanks to the *Least Mean Square* (LMS) algorithm.*Least Mean Square* Remarkable progress was then achieved with the work of Wang and Poor [11.56], taken up by Reynolds and Wang [11.47] then by Tüchler *et al.* [11.51, 11.50]. These contributions have made it possible to

[4] Equalizers optimized according to the Zero Forcing criterion could also be envisaged. However these equalizers usually introduce significant noise enhancement on channels with deep nulls in their frequency response, and thus generally turn out to be less efficient than MMSE equalizers.

establish a theoretical expression for the coefficients of the equalizer, explicitly taking into account the presence of *a priori* information on the transmitted data. This progress has proved to be particularly interesting for packet mode transmissions, in which the coefficients of the equalizer are precalculated once from an estimation of the impulse response of the channel, and applied to the whole received block.

MMSE turbo equalization relies on a soft-input soft-output linear equalization scheme optimized according to the MMSE criterion. This type of equalizer is also sometimes known as a "linear MMSE equalizer with *a priori* information" in the literature. This section describes the principle of this equalizer, assuming that we know the parameters of the channel, which enables the filter coefficients to be calculated directly. Its implementation in adaptive form is also discussed. We next present some examples of MMSE turbo equalizer performance, and we describe the (*Digital Signal Processor*, or DSP) implementation of this solution. This part ends with a reflection on the potential applications of MMSE turbo equalization.

Principle of soft-input soft-ouput linear MMSE equalization

Generally, the linear soft-input soft-output MMSE equalizer can formally be decomposed into three main functions (Figure 11.14).

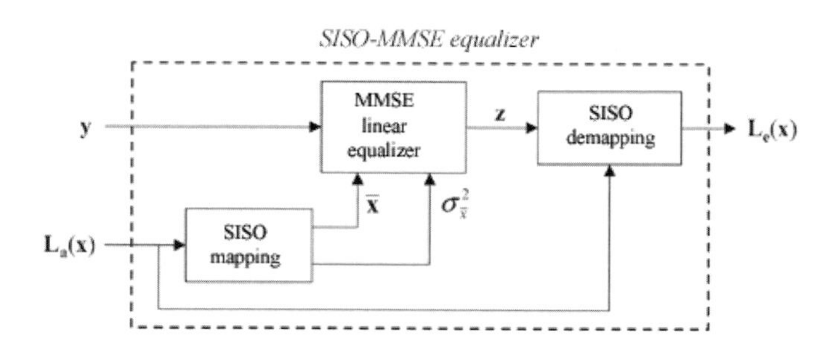

Figure 11.14 – Soft-input soft-output linear equalizer optimized according to the MMSE criterion.

1. The first operation, the SISO mapping, calculates a soft estimate for the transmitted symbols, denoted $\bar{\mathbf{x}} = (\bar{x}_0, \ldots, \bar{x}_{N-1})$, from the *a priori* information $\mathbf{L}_a(\mathbf{x})$ coming from the decoder at the previous iteration.

2. The linear equalizer then uses estimated data $\bar{x}_i$ to rebuild and cancel the ISI affecting the received signal. The resulting signal is filtered in order to eliminate residual interference. The filter coefficients are optimized so as to minimize the mean square error between the equalized data and

the corresponding transmitted data. However, unlike the classical linear MMSE equalizer, the reliability information coming from the decoder is here explicitly taken into account when calculating the coefficients.

3. The equalizer is finally followed by a soft-input soft-output demapping module whose role is to convert the equalized symbols into extrinsic LLRs on the (interleaved) coded bits.

We now examine in greater detail the implementation of each of these three functions.

- *SISO mapping*

This operation involves calculating the soft estimate $\bar{x}_i$, defined as the mathematical expectation of symbol x_i transmitted at instant i:

$$\bar{x}_i = E_a\{x_i\} = \sum_{l=1}^{M} X_l \times P_a(X_l) \tag{11.24}$$

The sum here concerns all of the discrete symbols in the constellation. The term $P_a(X_l)$ denotes the *a priori* probability $\Pr(x_i = X_l)$ of symbol X_l being transmitted at instant i. We have put index a at the level of the expectation term to highlight the fact that these probabilities are deduced from the *a priori* information at the input of the equalizer. Indeed, provided the m bits making up symbol x_i are statistically independent, it is possible to write:

$$P_a(X_l) = \prod_{j=1}^{m} P_a(X_{l,j}) \tag{11.25}$$

where binary element $X_{l,j}$ takes the value 0 or 1 according to the symbol X_l considered and the mapping rule. On the other hand, starting from the general definition (11.4) of the LLR, we can show that the *a priori* probability and the *a priori* LLR are linked by the following relation:

$$P_a(X_{l,j}) = \frac{1}{2}\left(1 + (2X_{l,j} - 1)\tanh\left(\frac{L_a(x_{i,j})}{2}\right)\right) \quad \text{with } X_{l,j} \in \{0, 1\} \tag{11.26}$$

In the particular case of a BPSK modulation, the above computations are greatly simplified. We then obtain the following expression for the soft estimate $\bar{x}_i$:

$$\bar{x}_i = \tanh\left(\frac{L_a(x_i)}{2}\right) \tag{11.27}$$

In the classical situation where we make the hypothesis of equiprobability on the transmitted symbols, we have $L_a(x_{i,j}) = 0$ and $\bar{x}_i = 0$. On the other hand, in the ideal case of perfect *a priori* information, $L_a(x_{i,j}) \rightarrow \pm\infty$ and the soft estimate $\bar{x}_i$ is then strictly equal to the transmitted symbol x_i (perfect estimate). To summarize, the value of the soft estimate $\bar{x}_i$ evolves as a function of the reliability of the *a priori* information provided by the decoder. This explains the name of "soft" (or probabilistic) estimate for $\bar{x}_i$. By construction, the estimated data $\bar{x}_i$ are random variables. In particular, we can show (see [11.33] for example) that they satisfy the following statistical properties:

$$E\left\{\bar{x}_i\right\} = 0 \tag{11.28}$$

$$E\left\{\bar{x}_i x_j^*\right\} = E\left\{\bar{x}_i \bar{x}_j^*\right\} = \sigma_{\bar{x}}^2 \delta_{i-j} \tag{11.29}$$

The parameter $\sigma_{\bar{x}}^2$ here denotes the variance of estimated data $\bar{x}_i$. In practice, this quantity can be estimated using the sample variance estimator on a frame of N symbols as follows:

$$\sigma_{\bar{x}}^2 = \frac{1}{N}\sum_{i=0}^{N-1} \left|\bar{x}_i\right|^2 \tag{11.30}$$

We easily verify that under the hypothesis of equiprobable *a priori* symbols, $\sigma_{\bar{x}}^2 = 0$. Conversely, we obtain $\sigma_{\bar{x}}^2 = \sigma_x^2$ in the case of perfect *a priori* information on the transmitted symbols. To summarize, the variance of the estimated data offers a measure of the reliability of the estimated data. This parameter plays a major role in the behaviour of the equalizer.

- *Calculating the linear equalizer coefficients*

As explained above, the equalization step can be seen as the cascading of an interference cancellation operation followed by a filtering operation. The filter coefficients are optimized so as to minimize the mean square error $E\{|z_i - x_{i-\Delta}|^2\}$ between the equalized symbol z_i at instant i and symbol $x_{i-\Delta}$ transmitted at instant $i - \Delta$. The introduction of a delay Δ enables the anti-causality of the solution to be taken into account. Here we will use a matrix formalism to derive the optimal form of the equalizer coefficients. Indeed, digital filters always have a finite number of coefficients in practice. The matrix formalism takes this aspect into account and thus enables us to establish the optimal coefficients under the constraint of a finite-length implementation.

Here we consider a filter with F coefficients: $\mathbf{f} = (\mathbf{f}_0, \ldots, \mathbf{f}_{F-1})$. The channel impulse response and the noise variance are assumed to be known, which requires prior estimation of these parameters in practice. Starting from the expression (11.3) and grouping the F last samples observed at the output of the channel up until instant i in the form of a vector column $\mathbf{y}_i$, we can write:

$$\mathbf{y}_i = \mathbf{H}\mathbf{x}_i + \mathbf{w}_i \tag{11.31}$$

with $\mathbf{y}_i = (y_i, \ldots, y_{i-F+1})^{\mathbf{T}}$, $\mathbf{x}_i = (x_i, \ldots, x_{i-F-L+2})^{\mathbf{T}}$ and $\mathbf{w}_i = (w_i, \ldots, w_{i-F+1})^{\mathbf{T}}$. Matrix $\mathbf{H}$, of dimensions $F \times (F + L - 1)$, is a Toeplitz matrix[5]

[5] The coefficients of the matrix are constant along each of the diagonals.

describing the convolution by the channel:

$$\mathbf{H} = \begin{pmatrix} h_0 & \cdots & h_{L-1} & 0 & \cdots & 0 \\ 0 & h_0 & & h_{L-1} & & \vdots \\ \vdots & & \ddots & & \ddots & 0 \\ 0 & \cdots & 0 & h_0 & \cdots & h_{L-1} \end{pmatrix} \tag{11.32}$$

With these notations, the interference cancellation step from the estimated signal $\bar{\mathbf{x}}$ can then be written formally:

$$\tilde{\mathbf{y}}_i = \mathbf{y}_i - \mathbf{H}\tilde{\mathbf{x}}_i \tag{11.33}$$

where the vector $\tilde{\mathbf{x}}_i = (\bar{x}_i, \ldots, \bar{x}_{i-\Delta+1}, 0, \bar{x}_{i-\Delta-1}, \ldots, \bar{x}_{i-F-L+2})^{\mathbf{T}}$ is of dimension $F+L-1$. The component related to symbol $x_{i-\Delta}$ is set to zero in order to cancel only the ISI and not the signal of interest. At the output of the forward filter, the expression of the equalized sample at instant i is given by:

$$z_i = \mathbf{f}^{\mathbf{T}}\tilde{\mathbf{y}}_i = \mathbf{f}^{\mathbf{T}}\left[\mathbf{y}_i - \mathbf{H}\tilde{\mathbf{x}}_i\right] \tag{11.34}$$

It remains to determine the theoretical expression of the coefficients of the filter $\mathbf{f}$ minimizing the mean square error $E\{|z_i - x_{i-\Delta}|^2\}$. In the most general case, these coefficients vary in time. The corresponding solution, developed in detail by Tüchler *et al.* [11.51, 11.50], leads to an equalizer whose coefficients must be recalculated for each received symbol. This equalizer represents what can best be done currently for MMSE equalization in the presence of *a priori* information. On the other hand, the computation load associated with updating the coefficients symbol by symbol increases quadratically with the number F of coefficients, which again turns out to be too complex for real time implementations. The equalizer that we present here can be seen as a simplified, and therefore sub-optimal, version of the solution cited above. The coefficients of filter $\mathbf{f}$ are calculated only once per frame (at each iteration) and then applied to the whole block, which considerably decreases the implementation cost. On the other hand, and despite this reduction in complexity, this equalizer retains performance close to the optimal one[6], which makes it an excellent candidate for practical realizations. This solution was derived independently by several authors, including [11.51] and [11.33].

With these hypotheses, the optimal form of the set of coefficients $\mathbf{f}$ is obtained using the projection theorem, which stipulates that the estimation error must be orthogonal to the observations[7] :

$$E\left\{(z_i - x_{i-\Delta})\tilde{\mathbf{y}}_i^{\mathbf{H}}\right\} = \mathbf{0} \tag{11.35}$$

[6] The degradation measured experimentally in comparison with Tüchler's original time-varying solution is at most 1 dB, depending on the channel model considered.

[7] We recall here that the notation $\mathbf{A}^{\mathbf{H}}$ denotes the *Hermitian* (conjugate) transpose of matrix $\mathbf{A}$.

We then obtain the following solution:

$$\mathbf{f}^* = E\{\tilde{\mathbf{y}}_i\tilde{\mathbf{y}}_i^{\mathbf{H}}\}^{-1}E\{x_{i-\Delta}^*\tilde{\mathbf{y}}_i\} \tag{11.36}$$

Using the statistical properties of the estimated data $\bar{x}_i$, we note that:

$$E\{x_{i-\Delta}^*\tilde{\mathbf{y}}_i\} = E\{x_{i-\Delta}^*\mathbf{H}(\mathbf{x}_i - \tilde{\mathbf{x}}_i)\} = \mathbf{H}\mathbf{e}_{\boldsymbol{\Delta}}\sigma_x^2 \tag{11.37}$$

where we have introduced the unit vector $\mathbf{e}_\Delta$ with dimension $F + L - 1$ that has a 1 in coordinate Δ and 0 elsewhere. Denoting by $\mathbf{h}_\Delta$ the Δ-th column Δ of matrix $\mathbf{H}$, the previous expression can also be written:

$$E\left\{x_{i-\Delta}^*\tilde{\mathbf{y}}_i\right\} = \mathbf{h}_{\boldsymbol{\Delta}}\sigma_x^2 \tag{11.38}$$

In addition,

$$\begin{aligned} E\left\{\tilde{\mathbf{y}}_i\tilde{\mathbf{y}}_i^{\mathbf{H}}\right\} &= \mathbf{H}E\left\{(\mathbf{x}_i - \tilde{\mathbf{x}}_i)(\mathbf{x}_i - \tilde{\mathbf{x}}_i)^{\mathbf{H}}\right\}\mathbf{H}^{\mathbf{H}} + \sigma_w^2\mathbf{I} \\ &= (\sigma_x^2 - \sigma_{\bar{x}}^2)\mathbf{H}\mathbf{H}^{\mathbf{H}} + \sigma_{\bar{x}}^2\mathbf{h}_{\boldsymbol{\Delta}}\mathbf{h}_{\boldsymbol{\Delta}}^{\mathbf{H}} + \sigma_w^2\mathbf{I} \end{aligned} \tag{11.39}$$

To summarize, the optimal form of the equalizer coefficients can finally be written:

$$\mathbf{f}^* = \left[(\sigma_x^2 - \sigma_{\bar{x}}^2)\mathbf{H}\mathbf{H}^{\mathbf{H}} + \sigma_{\bar{x}}^2\mathbf{h}_{\boldsymbol{\Delta}}\mathbf{h}_{\boldsymbol{\Delta}}^{\mathbf{H}} + \sigma_w^2\mathbf{I}\right]^{-1}\mathbf{h}_{\boldsymbol{\Delta}}\sigma_x^2 \tag{11.40}$$

By bringing into play a simplified form of the matrix inversion lemma[8], the previous solution can then be written:

$$\mathbf{f}^* = \frac{\sigma_x^2}{1 + \beta\sigma_{\bar{x}}^2}\tilde{\mathbf{f}}^* \tag{11.41}$$

where we have introduced vector $\tilde{\mathbf{f}}$ and scalar quantity β defined as follows:

$$\tilde{\mathbf{f}}^* = \left[(\sigma_x^2 - \sigma_{\bar{x}}^2)\mathbf{H}\mathbf{H}^{\mathbf{H}} + \sigma_w^2\mathbf{I}\right]^{-1}\mathbf{h}_{\boldsymbol{\Delta}} \quad \text{and} \quad \beta = \tilde{\mathbf{f}}^{\mathbf{T}}\mathbf{h}_{\boldsymbol{\Delta}} \tag{11.42}$$

By means of this new expression, we note that the computation of the coefficients of the equalizer is mainly based on the inversion of the matrix $(\sigma_x^2 - \sigma_{\bar{x}}^2)\mathbf{H}\mathbf{H}^{\mathbf{H}} + \sigma_w^2\mathbf{I}$, with dimensions $F \times F$. This matrix has a rich structure since it is a Toeplitz matrix with Hermitian symmetry. Consequently, matrix inversion can be performed efficiently with the help of dedicated algorithms, with a computation cost in $O(F^2)$ (see Chapter 4 in [11.21], for example). In order to reduce even further the complexity of determining the coefficients, the authors of [11.33] have proposed a sub-optimal, but nevertheless efficient, method using the *Fast Fourier Transform*, (or FFT), with a cost in $O(F\log_2(F))$. However, the number of coefficients F must be a power of 2.

[8] $\left[\mathbf{A} + \mathbf{u}\mathbf{u}^{\mathbf{H}}\right]^{-1} = \mathbf{A}^{-1} - \frac{\mathbf{A}^{-1}\mathbf{u}\mathbf{u}^{\mathbf{H}}\mathbf{A}^{-1}}{1+\mathbf{u}^{\mathbf{H}}\mathbf{A}^{-1}\mathbf{u}}$

It is particularly instructive to study the limiting form taken by the equalizer in the classical case where the transmitted symbols are assumed to be equiprobable (which corresponds to the 1st iteration of the turbo equalizer). In this case, $\sigma_{\bar{x}}^2 = 0$ and the equalizer coefficients can be written:

$$\mathbf{f}^* = \left[\sigma_x^2 \mathbf{H}\mathbf{H}^{\mathbf{H}} + \sigma_w^2 \mathbf{I}\right]^{-1} \mathbf{h}_{\boldsymbol{\Delta}} \sigma_x^2 \tag{11.43}$$

Here we can recognize the form of a classical linear MMSE equalizer with finite length. Inversely, under the hypothesis of perfect *a priori* information on the transmitted symbols, we have $\sigma_{\bar{x}}^2 = \sigma_x^2$. The equalizer then takes the following form:

$$\mathbf{f} = \frac{\sigma_x^2}{\sigma_x^2 \left\|\mathbf{h}\right\|^2 + \sigma_w^2} \mathbf{h}_{\boldsymbol{\Delta}}^* \quad \text{with} \quad \left\|\mathbf{h}\right\|^2 = \mathbf{h}_{\boldsymbol{\Delta}}^{\mathbf{H}} \mathbf{h}_{\boldsymbol{\Delta}} = \sum_{k=0}^{L-1} \left|h_k\right|^2 \tag{11.44}$$

and the equalized signal z_i can be written:

$$z_i = \frac{\sigma_x^2 \left\|\mathbf{h}\right\|^2}{\sigma_x^2 \left\|\mathbf{h}\right\|^2 + \sigma_w^2} \left(x_{i-\Delta} + \mathbf{h}_{\boldsymbol{\Delta}}^{\mathbf{H}} \mathbf{w}_i\right) \tag{11.45}$$

We recognize here the output of a classical MMSE interference canceller, fed by a perfect estimation of the transmitted data. The equalized signal can be decomposed as the sum of the useful signal $x_{i-\Delta}$, up to a scale factor that is characteristic of the MMSE criterion, and a coloured noise term. In other words, the equalizer suppresses all the ISI without raising the noise level and thus reaches the theoretical matched-filter bound corresponding to ISI-free transmission.

To summarize, we see that the SISO MMSE linear equalizer adapts the equalization strategy according to the reliability of the estimated data, measured here by parameter $\sigma_{\bar{x}}^2$.

To conclude this description of the equalizer, we point out that the interference cancellation operation defined formally by Equation (11.33) has no physical reality in the sense that it cannot be performed directly in this way using transverse linear filters. In practice, we prefer to use one of the two architectures presented in Figure 11.15, strictly equivalent from a theoretical point of view. The coefficient g_Δ appearing in implementation (1) is the central coefficient $g_\Delta = \mathbf{f}^T \mathbf{h}_\Delta$ of the global filter formed by the cascade of the channel with filter $\mathbf{f}$. In the case of implementation (2), we again find the classical structure of an interference canceller type equalizer, operating here on the estimated signal $\bar{\mathbf{x}}$. Filter $\mathbf{g} = \mathbf{f}^T \mathbf{H}$ is given by the convolution of filter $\mathbf{f}$ with the impulse response of the channel, the central coefficient g_Δ then being forced to zero.

• *SISO demapping*

The role of this module is to convert the equalized data z_i into extrinsic LLRs on the interleaved coded bits, which will be then transmitted to the channel

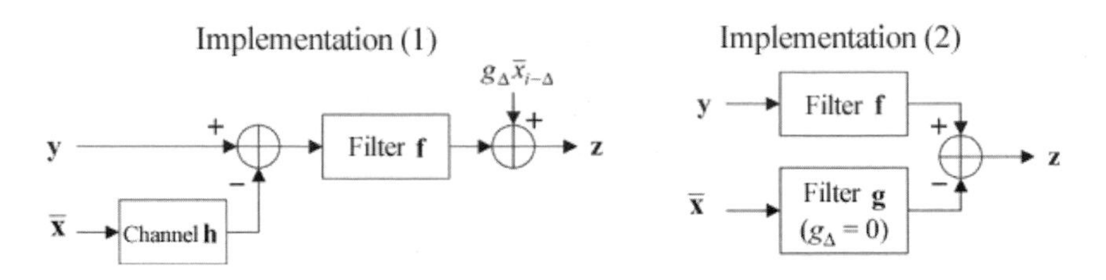

Figure 11.15 – Practical implementation of the equalizer using transverse filters.

decoder. Generally, we can always decompose the expression of z_i as the sum of two quantities:

$$z_i = g_\Delta x_{i-\Delta} + \nu_i \tag{11.46}$$

The term $g_\Delta x_{i-\Delta}$ represents the useful signal up to a constant factor g_Δ. We recall that this factor quite simply corresponds to the central coefficient of the cascading of the channel with the equalizer. The term ν_i accounts for both residual interference and noise at the output of the equalizer. In order to perform the demapping operation, we make the hypothesis[9] that interference term ν_i follows a complex Gaussian distribution, with zero mean and total variance σ_ν^2. Parameters g_Δ and σ_ν^2 are easy to deduce from the knowledge of the set of equalizer coefficients. We can thus show ([11.51, 11.33]) that we have:

$$g_\Delta = \mathbf{f}^{\mathbf{T}}\mathbf{h}_\Delta \text{ and } \sigma_\nu^2 = E\left\{\left|z_i - g_\Delta x_{i-\Delta}\right|^2\right\} = \sigma_x^2 g_\Delta (1 - g_\Delta) \tag{11.47}$$

Starting from these hypotheses, the demapping module calculates the *a posteriori* LLR on the coded interleaved bits, denoted $L(x_{i,j})$ and defined as follows:

$$L(x_{i,j}) = \ln\left(\frac{\Pr(x_{i,j} = 1\,|z_i)}{\Pr(x_{i,j} = 0\,|z_i)}\right) \tag{11.48}$$

The values present in the numerator and denominator can be evaluated by summing the *a posteriori* probability $\Pr(x_i = X_l|z_i)$ of having transmitted a particular symbol X_l of the constellation on all the symbols for which the j-th bit making up this symbol takes the value 0 or 1 respectively. Thus, we can write:

[9] This hypothesis rigorously only holds on condition that the equalizer suppresses all the ISI, which assumes perfect knowledge of the transmitted data. Nevertheless, it is a good approximation in practice, particularly in a turbo equalization context where the reliability of the decisions at the output of the decoder increases along the iterations which, in its turn, improves the equalization operation.

$$L(x_{i,j}) = \ln\left(\frac{\sum\limits_{X_l/X_{l,j}=1} \Pr(x_i = X_l | z_i)}{\sum\limits_{X_l/X_{l,j}=0} \Pr(x_i = X_l | z_i)}\right) = \ln\left(\frac{\sum\limits_{X_l/X_{l,j}=1} P(z_i | x_i = X_l) P_a(X_l)}{\sum\limits_{X_l/X_{l,j}=0} P(z_n | x_i = X_l) P_a(X_l)}\right) \tag{11.49}$$

The second equality results from applying Bayes' relation. It shows the *a priori* probability $P_a(X_l) = \Pr(x_i = X_l)$ of having transmitted a given symbol X_l of the modulation alphabet. This probability is calculated from the *a priori* information available at the input of the equalizer (relations (11.25) and (11.26)). By exploiting the above hypotheses, the likelihood of observation z_i conditionally to the hypothesis of having transmitted the symbol X_l at instant i can be written:

$$P(z_i | x_i = X_l) = \frac{1}{\pi\sigma_\nu^2} \exp\left(-\frac{|z_i - g_\Delta X_l|^2}{\sigma_\nu^2}\right) \tag{11.50}$$

After simplification, the *a posteriori* LLR calculated by the demapping operation becomes:

$$L(x_{i,j}) = \ln\left(\frac{\sum\limits_{X_l/X_{l,j}=1} \exp\left(-\frac{|z_i - g_\Delta X_l|^2}{\sigma_\nu^2} + \sum\limits_{k=1}^{m} X_{l,k} L_a(x_{i,k})\right)}{\sum\limits_{X_l/X_{l,j}=0} \exp\left(-\frac{|z_i - g_\Delta X_l|^2}{\sigma_\nu^2} + \sum\limits_{k=1}^{m} X_{l,k} L_a(x_{i,k})\right)}\right) \tag{11.51}$$

Like in the case of the BCJR-MAP equalizer, we can factorize in the numerator and denominator the *a priori* information term in relation to the considered bit, in order to obtain the extrinsic information that is then provided to the decoder:

$$L(x_{i,j}) = L_a(x_{i,j}) + \underbrace{\ln\left(\frac{\sum\limits_{X_l/X_{l,j}=1} \exp\left(-\frac{|z_i - g_\Delta X_l|^2}{\sigma_\nu^2} + \sum\limits_{k\neq j} X_{l,k} L_a(x_{i,k})\right)}{\sum\limits_{X_j/X_{j,i}=0} \exp\left(-\frac{|z_i - g_\Delta X_l|^2}{\sigma_\nu^2} + \sum\limits_{k\neq j} X_{l,k} L_a(x_{i,k})\right)}\right)}_{L_e(x_{i,j})} \tag{11.52}$$

Finally, the extrinsic information is obtained quite simply by subtracting the *a priori* information from the *a posteriori* LLR calculated by the equalizer:

$$L_e(x_{i,j}) = L(x_{i,j}) - L_a(x_{i,j}) \tag{11.53}$$

In the particular case of BPSK modulation, the SISO demapping equations are simplified to give the following expression of the extrinsic LLR:

$$L(x_i) = \frac{4}{1 - g_\Delta} Re\left\{z_i\right\} \tag{11.54}$$

When Gray mapping rules are used, experience shows that we can reduce the complexity of the demapping by ignoring the *a priori* information coming from the decoder in the equations above[10], without really affecting the performance of the device. On the other hand, this simplification no longer applies when we consider other mapping rules, like the *Set Partitioning* rule used in coded modulation schemes. This point has been particularly well highlighted in [11.14] and [11.30].

This completes the description of the soft-input soft-output linear MMSE equalizer. Finally, we can note that unlike the BCJR-MAP equalizer, the complexity of the SISO mapping and demapping operations increases linearly (and not exponentially) as a function of size M of the constellation and of the number L of taps in the impulse response of the discrete-time equivalent channel model.

Adaptive implementation of the equalizer

Historically, the first MMSE turbo equalizer was proposed in 1997, directly in adaptive form [11.20, 11.32]. The closed-form expression (11.40) enabling the computation of the equalizer coefficients from the knowledge of the channel impulse response was not known at that time. The chosen solution thus involved determining the filter coefficients in an adaptive manner, using stochastic gradient descent algorithms aiming at minimizing the mean square error between the transmitted data and the equalizer output. As we shall see in the following, when evaluating performance, the adaptive MMSE turbo equalizer remains a very interesting solution for time-invariant or slowly time-varying channels. The purpose of this section is to show that, for such channels, the adaptive MMSE equalizer and the MMSE equalizer proposed in (11.40) have similar performance and characteristics.

The structure of the considered equalizer is shown in Figure 11.15 (implementation (2)). An adaptive procedure is used to obtain the filters' coefficients. This adaptive algorithm is composed of two distinct phases: the training phase and the tracking phase. The training phase makes use of sequences known by the receiver (training sequences) to initialize the equalizer coefficients. Next, during the tracking period, the coefficients are continuously updated in a decision-directed manner, based on the receiver estimate of the transmitted sequence.

Adaptive algorithms involve determining, for each symbol entering the equalizer, output z_i from the following relation:

$$z_i = \mathbf{f}_i^{\mathbf{T}}\mathbf{y}_i - \mathbf{g}_i^{\mathbf{T}}\tilde{\mathbf{x}}_i \tag{11.55}$$

where $\mathbf{y}_i = (y_{i+F}, \ldots, y_{i-F})^T$ is the vector of channel output samples and $\tilde{\mathbf{x}}_i = (\bar{x}_{i+G}, \ldots, \bar{x}_{i-\Delta+1}, 0, \bar{x}_{i-\Delta-1}, \ldots, \bar{x}_{i-G})^T$ is the vector of estimated symbols, with respective lengths $2F+1$ and $2G+1$. Note that the coordinate relative

[10] This amounts to assuming the transmitted symbols to be equiprobable, i.e. to putting $P_a(X_l) = 1/M$ whatever the symbol and the iteration considered.

to the soft estimate $\bar{x}_{i-\Delta}$ in $\tilde{\mathbf{x}}_i$ is set to zero in order not to cancel the signal of interest. Vectors $\mathbf{f}_i = (f_{i,F}, \ldots, f_{i,-F})^T$ and $\mathbf{g}_i = (g_{i,G}, \ldots, g_{i,-G})^T$ represent the coefficients of filters $\mathbf{f}$ and $\mathbf{g}$, respectively. Both vectors are a function of time since they are updated at each new received symbol.

The relations used to update the vectors of the coefficients can be obtained from a least-mean square (LMS) gradient algorithm:

$$\begin{aligned} \mathbf{f}_{i+1} &= \mathbf{f}_i - \mu\left(z_i - x_{i-\Delta}\right)\mathbf{y}_i^* \\ \mathbf{g}_{i+1} &= \mathbf{g}_i - \mu\left(z_i - x_{i-\Delta}\right)\tilde{\mathbf{x}}_i^* \end{aligned} \tag{11.56}$$

where μ is a small, positive, step-size that controls the convergence properties of the algorithm.

During the first iteration of the turbo equalizer, $\tilde{\mathbf{x}}_i$ is a vector all the components of which are null; the result is that the coefficients vector $\mathbf{g}_i$ is also null. The MMSE equalizer then converges adaptively towards an MMSE transversal equalizer. When the estimated data are very reliable and close to the transmitted data, the MMSE equalizer converges towards an ideal (genie) interference canceller, then having the performance of a transmission without intersymbol interference. The limiting forms of the adaptive equalizer are therefore totally identical to those obtained in (11.43) and (11.44), on condition of course that the adaptive algorithm can converge towards a local minimum close to the optimal solution.

Note, however, that for intermediate iterations where the estimated information symbols $\bar{x}_i$ are neither null nor perfect, filter $\mathbf{g}_i$ must not be fed directly with the transmitted symbols otherwise the equalizer will converge towards the solution of the genie interference canceller, which is not the aim searched for. To enable the equalizer to converge towards the targeted solution, the idea here involves providing filter $\mathbf{g}_i$ with soft estimates built from the transmitted symbols, during the training periods:

$$\left(\bar{x}_i\right)_{its} = \sigma_{\bar{x}} x_i + \sqrt{1 - \sigma_{\bar{x}}^2}\eta_i \tag{11.57}$$

where $\sigma_{\bar{x}}^2$ corresponds to the variance of the soft estimates $\bar{x}_i$ obtained from (11.24) and η_i a zero-mean complex circularly-symmetric additive white Gaussian noise with unit variance.

In the tracking period and in order to enable the equalizer to follow the variations of the channel, it is possible to replace the transmitted symbols x_i in relations (11.56) by the decisions $\hat{x}_i$ at the output of the equalizer, or by the decisions on the estimated symbols $\bar{x}_i$.

When the SISO MMSE equalizer is realized in adaptive form, we do not explicitly have access to the channel impulse response, and the updating relation of $\mathbf{g}_i$ does not enable g_Δ to be obtained since component $g_{i,\Delta}$ is constrained to be zero. To perform the SISO demapping operation, we must however estimate both bias g_Δ on the data z_n provided by the equalizer and variance σ_ν^2 of the residual interference at the output of the equalizer. As we will see, these two parameters can be estimated from the output of the equalizer. From relation (11.46) again, we can show the general following result:

$$E\left(|z_i|^2\right) = g_\Delta \sigma_x^2 \tag{11.58}$$

Assuming that the variance of the transmitted data is normalized to unity, an estimate of g_Δ is given by:

$$\hat{g}_\Delta = \frac{1}{N} \sum_{i=0}^{N-1} |z_i|^2 \tag{11.59}$$

Once we have estimated g_Δ, we immediately deduce the value of σ_ν^2 thanks to relation (11.47):

$$\sigma_\nu^2 = \sigma_x^2 \hat{g}_\Delta (1 - \hat{g}_\Delta) \tag{11.60}$$

One particularity of adaptive MMSE turbo equalization concerns the determination of the estimated symbols. Indeed, in accordance with the remarks of [11.20] and [11.55], using *a posteriori* information instead of extrinsic information at the output of the channel decoder in (11.27) can yield significant performance improvement.

We have therefore defined an adaptive MMSE turbo equalizer whose coefficients are obtained from a low complexity stochastic gradient descent algorithm, making it possible to track the slow time variations of the transmission channel. A drawback of this technique lies in the necessity to transmit training sequences, which lower the spectral efficiency. The size of training sequences can be significantly reduced by considering self-learning or blind algorithms. In particular, the equalizer in the first iteration can be advantageously replaced by a self-learning equalizer called *Self Adaptive Decision-Feedback Equalizer* (SADFE) [11.32] that requires a very small transmission overhead. The work of Hélard *et al.* [11.26] has shown that such a turbo equalizer can reach performance virtually identical to that of the adaptive MMSE turbo equalizer with learning sequence, while operating at a higher spectral efficiency. On the other hand, a higher number of iterations is then required.

Examples of performance

For comparison purposes, the performance of the MMSE turbo equalizer has been simulated by considering the same transmission scenario as for the turbo MAP equalizer.

First, the parameters of the channel are assumed to be perfectly estimated. The coefficients are calculated once per frame by matrix inversion, by considering a digital filter with $F = 15$ coefficients and a designed delay $\Delta = 9$. The simulation results, obtained after 10 iterations, are presented in Figure 11.16.

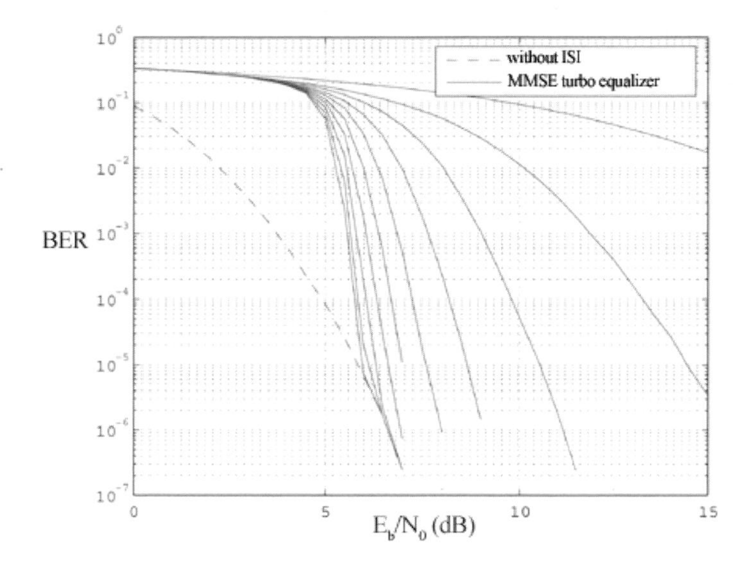

Figure 11.16 – Performance of the MMSE turbo equalizer for BPSK transmission on the Proakis C channel, with a 4-state rate $R = 1/2$ non-recursive non-systematic convolutional code and a 16384 bit pseudo-random interleaver.

Convergence of the iterative process occurs here at threshold signal to noise ratio of 4 dB, and the turbo equalizer suppresses all the ISI beyond a signal to noise ratio of 6 dB (after 10 iterations). Compared to the results obtained with the MAP turbo equalizer (Figure 11.13), we can therefore make the following remarks:

1. The convergence occurs later with MMSE turbo equalization (of the order of 1 dB here, compared to MAP turbo equalization).

2. The MMSE turbo equalizer requires more iterations than the MAP turbo equalizer to reach comparable error rates.

However, the MMSE turbo equalizer here shows its capacity to suppress all the ISI when the signal to noise ratio is high enough, even on a channel that is known to be difficult to equalize. It is therefore a serious alternative solution to the MAP turbo equalizer when the latter cannot be used for reasons of complexity.

Second, the hypothesis of perfect knowledge of the channel parameters has been removed and the turbo equalizer is simulated in the adaptive form, keeping the same transmission parameters. The communication begins with the transmission of an initial training sequence of 16384 symbols assumed to be perfectly known by the receiver. Then, frames composed of 1000 training symbols followed by 16384 information symbols are periodically sent into the channel.

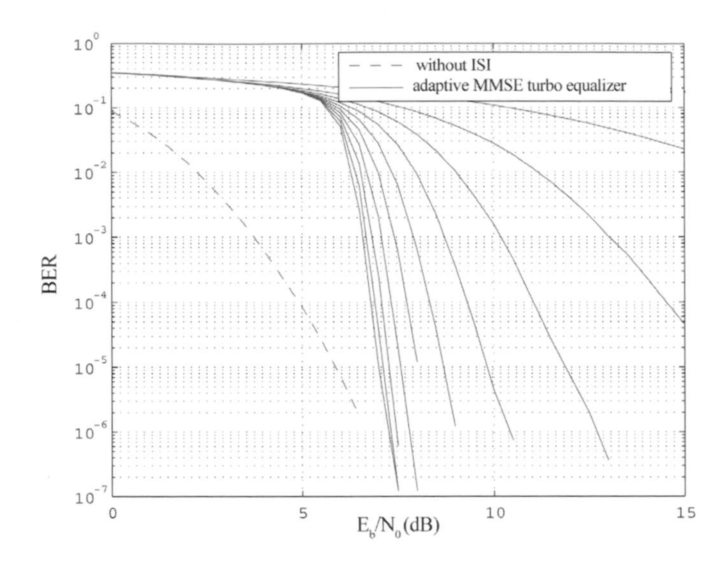

Figure 11.17 – Performance of the adaptive MMSE turbo equalizer for BPSK transmission on the Proakis C channel, with a 4-state rate $R = 1/2$ non-recursive non-systematic convolutional code and a 16384 bit pseudo-random interleaver.

During the processing of the 16384 information symbols, the turbo equalizer operates in a decision-directed manner. The equalizer filters each have 21 coefficients ($F = G = 10$). The coefficients are updated using the LMS algorithm. The step size is set to $\mu = 0,0005$ during the training period, and then to $\mu = 0,000005$ during the tracking period. Simulation results are given in Figure 11.17, considering 10 iterations at reception. We observe a degradation of the order of only 1 dB compared to the ideal situation where the channel is assumed to be perfectly known. We note that when the channel is estimated and used for the direct computation of the coefficients of the MMSE equalizer, losses in performance will also appear, which reduces the degradation in comparison to the ideal situation of Figure 11.16. Note also that, to track the performance of Figure 11.17, we have not taken into account the loss in the signal to noise ratio caused by the use of training sequences.

In the light of these results, we note that the major difference between adaptive MMSE turbo equalization and that which uses direct computation of the coefficients from the estimate of the channel lies in the way the filter coefficients are determined, since the structure and the optimization criterion of the equalizers are identical.

To finish, we point out that, in the same way as for the turbo MAP equalizer, we can use EXIT charts to predict the theoretical convergence threshold of the

MMSE turbo equalizer, under the hypothesis of ideal interleaving. The reader will find further information on this subject in [11.8] or [11.50], for example.

Example of implementation and applications

The implementation of an MMSE turbo equalizer on a signal processor was reported in [11.9]. The target was the TMS320VC5509 processor by Texas Instruments. This is a 16-bit fixed-point DSP with low power consumption, which makes it an ideal candidate for mobile receivers. The considered transmission scheme included a 4-state rate 1/2 convolutional encoder and a 1024 bit interleaver followed by a QPSK modulator. The whole turbo equalizer was implemented in C language on the DSP, with the exception of some processing optimized in assembly (filtering and FFT) provided by a specialized library. The equalizer included 32 coefficients. The decoding was performed using the Max-Log-MAP algorithm. The simulation results showed that, subject to carefully choosing the representation in fixed decimal points of the data handled (within the limit of 16 bits maximum granted by the DSP), data quantization did not cause any loss in performance in comparison with the corresponding floating-point receiver. The final data rate obtained was of the order of 42 kbits/s after 5 iterations, which shows the feasibility of such receivers using current technology. The challenge now involves defining appropriate circuit architectures, capable of operating at several Mbits/s, in order to respond to emerging demands for high data rate services.

MMSE turbo equalization is a relatively recent technology. Therefore, at the moment of writing this book, there have been few studies on the potential applications of this technique at reception. Generally, resorting to MMSE turbo equalization is an effective solution in the context of high data rate transmissions on highly frequency-selective channels. In particular, this system has shown excellent performance on the ionospheric channel typically used in the context of HF military communications. Indeed, the long echoes produced by this channel prevent the use of MAP equalizers. On the other hand, conventional linear equalization schemes do not make it possible to reach a transmission quality acceptable when high-order modulations (*e.g.* 8-PSK or 16-QAM) are considered. MMSE turbo equalization is thus an attractive solution to the problem of increasing the data rate of military transmissions. In the context of HF communications, the interest of MMSE turbo equalization for high spectral efficiency modulations has been validated by the work of Langlais [11.29] and Otnes [11.41], which shows that this technique can offer gains up to 5 dB compared to conventional receivers. To our knowledge, MMSE turbo equalization has not yet been implemented in standardized modems. However, it is important to note that this reception technique enables the transmission performance to be notably improved while keeping standardized transmitters.

11.2 Multi-user turbo detection and its application to CDMA systems

11.2.1 Introduction and some notations

In a Code Division Multiple Access (CDMA) system, such as the one shown in Figure 11.18, user k ($1 \leqslant k \leqslant K$) transmits a sequence of binary elements $\{d_k\} = \pm 1$ with an amplitude A_k. For each of the users, a channel encoder (CC_k) is used, followed by an external interleaver (π_k) before the spreading operation (multiplication by size N, normalized spreading code $\mathbf{s}_k$,) which provides binary symbols called *chips*. This code can vary at each symbol time.

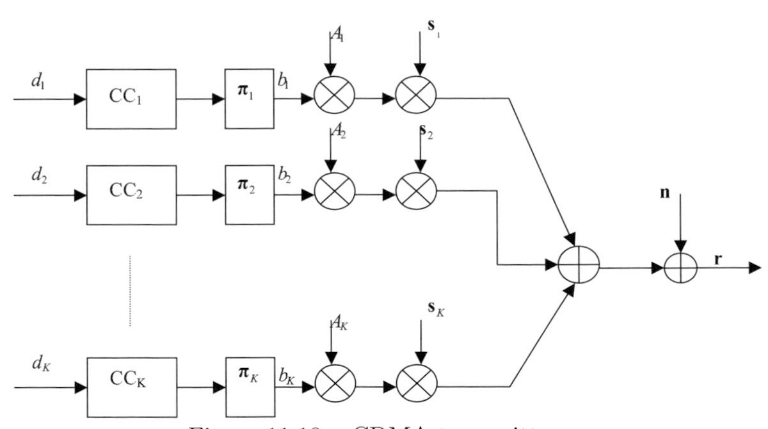

Figure 11.18 – CDMA transmitter.

The received signal, $\mathbf{r}$, can be written in matrix form by:

$$\mathbf{r} = \mathbf{SAb} + \mathbf{n} \tag{11.61}$$

where:

- $\mathbf{S}$ is the $N \times K$ matrix formed by the normalized codes of each user (the k-th column represents the k-th code $\mathbf{s}_k$ whose norm is equal to unity),
- $\mathbf{A}$ is a diagonal $K \times K$ matrix made up of the amplitudes A_k of each user.
- $\mathbf{b}$ is the vector of dimension K made up of the elements transmitted after coding channel by the K users.
- $\mathbf{n}$ is the N-dimensional centred Gaussian vector with covariance matrix $\sigma^2\mathbf{I}_N$.

The source data rates of the different users can be different. The size of the spreading code is such that the chip data rate (after spreading) is the same for all users. The received signal $\mathbf{r}$ is given by the contribution of all the K users plus a centred AWGN with variance σ^2. From observation $\mathbf{r}$, we wish to recover the information bits d_k of each user. Figure 11.19 gives the diagram of the receiver using a turbo CDMA type technique to jointly process the multi-detection and the channel decoding:

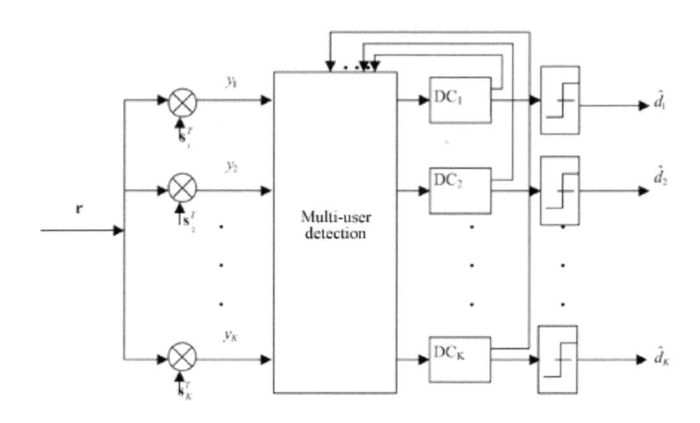

Figure 11.19 – Turbo CDMA receiver.

11.2.2 Multi-user detection

This section presents the main multi-user detection methods. In order to simplify the description of these methods, only the case of synchronous transmissions over Gaussian channels is considered.

Standard receiver

The simplest (conventional or standard) detector is the one which operates as if each user was alone on the channel. The receiver is quite simply made up of the filter adapted to the signature of the user concerned (this operation is also called despreading), see Figure 11.20.

At the output of the adapted filter bank the signal can be written in the form:

$$\mathbf{y} = \mathbf{S}^T\mathbf{r} = \mathbf{RAb} + \mathbf{S}^T\mathbf{n} \tag{11.62}$$

We note that the vector of the additive noise at the output of the adapted filter bank, is made up of correlated components. Its covariance matrix depends directly on the intercorrelation matrix of the spreading sequences used, $\mathbf{R} = \mathbf{S}^T\mathbf{S}$. We have $\mathbf{S}^T\mathbf{n} \sim N(0, \sigma^2\mathbf{R})$.

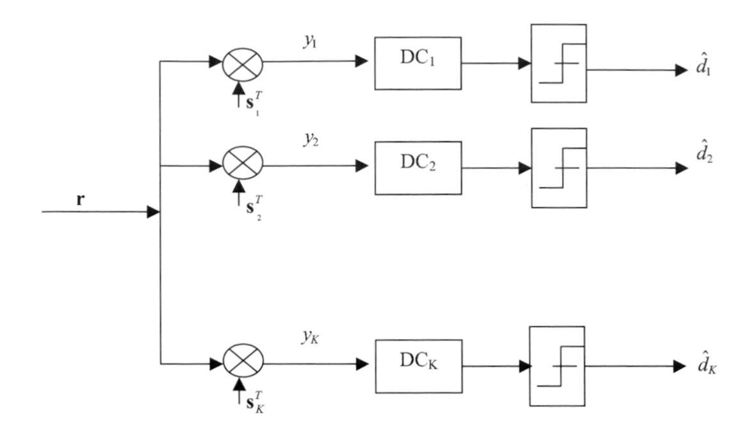

Figure 11.20 – Standard detector.

We can show that the error probability (before channel decoding) for the k-th user can be written in the form:

$$P_{e,k} = P\left(\hat{b}_k \neq b_k\right) = \frac{1}{2^{K-1}} \sum_{\mathbf{b}_{-k} \in \{-1,+1\}^{K-1}} Q\left(\frac{A_k}{\sigma} + \sum_{j \neq k} b_j \frac{A_j}{\sigma} \rho_{jk}\right) \quad (11.63)$$

where $\rho_{jk} = \mathbf{s}_j^T \mathbf{s}_k$ measures the intercorrelation between the codes of users j and k, with $\mathbf{b}_{-k} = (b_1, b_2, \cdots, b_{k-1}, b_{k+1}, \cdots, b_K)$.

Assuming that the spreading codes used are such that the intercorrelation coefficients are constant and equal to 0.2, Figure 11.21 gives the performance of the standard receiver, in terms of error probability of the first user as a function of the signal to noise ratio, for a number of users varying from 1 to 6. The messages of all the users are assumed to be received with the same power. We note of course that the higher the number of users, the worse the performance. This error probability can even tend towards $1/2$, while the signal to noise ratio increases if the following condition (*Near Far Effect*) is not satisfied:

$$A_k > \sum_{j \neq k} A_j \left|\rho_{jk}\right|$$

.

Optimal joint detection

Optimal joint detection involves maximizing the *a posteriori* probability (probability of vector $\mathbf{b}$ conditionally to observation $\mathbf{y}$). If we assume that the binary elements transmitted are equiprobable and given that $\mathbf{y} = \mathbf{S}^T\mathbf{r} = \mathbf{RAb} + \mathbf{S}^T\mathbf{n}$ with $\mathbf{S}^T\mathbf{n} \sim N(0, \sigma^2\mathbf{R})$, we can deduce that the optimal joint detection is given

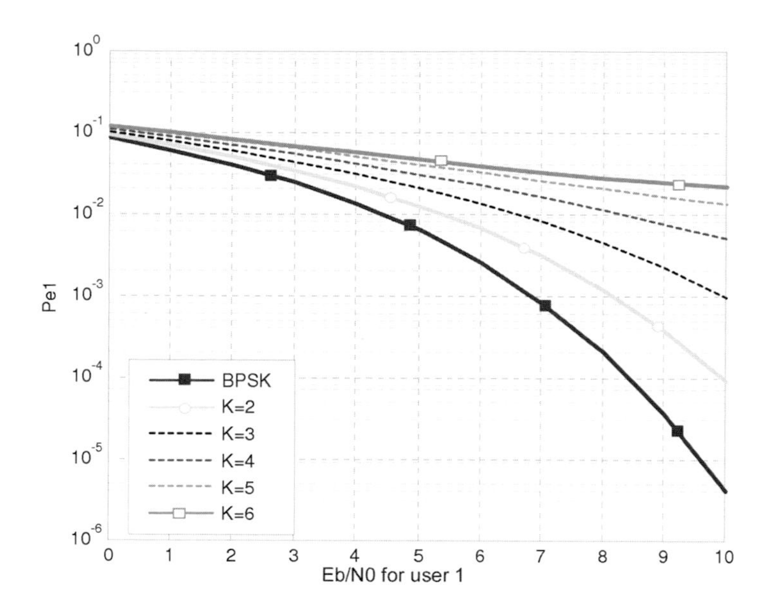

Figure 11.21 – Error probability of the first user as a function of the signal to noise ratio $Eb/N0$ for constant intercorrelation coefficients $\rho = 0.2$, for K=6 users sharing the resource.

by the equivalences:

$$\text{Max}_{\mathbf{b}}\left(f_{\mathbf{y}}^{\mathbf{b}}(\mathbf{y})\right) \Leftrightarrow \text{Min}_{\mathbf{b}}\left(\|\mathbf{y} - \mathbf{RAb}\|_{\mathbf{R}^{-1}}^{2}\right) \Leftrightarrow \text{Max}_{\mathbf{b}}\left(2\mathbf{b}^{T}\mathbf{Ay} - \mathbf{b}^{T}\mathbf{ARAb}\right) \tag{11.64}$$

An exhaustive search for the optimal solution is relatively complex.

Decorrelator detector

Decorrelator detectors involves multiplying observation $\mathbf{y}$ by the inverse of the intercorrelation matrix of the codes: $\mathbf{R}^{-1}\mathbf{y} = \mathbf{Ab} + \mathbf{R}^{-1}\mathbf{S}^{T}\mathbf{n}$. This equation shows that the decorrelator enables the multiple access interference to be cancelled completely, which makes it robust in relation to the *Near Far Effect*. On the other hand, the resulting additive Gaussian noise has greater variance. Indeed, we have: $\mathbf{R}^{-1}\mathbf{S}^{T}\mathbf{n} \sim N(0, \sigma^{2}\mathbf{R}^{-1})$. The error probability of the k-th user can then be written in the form:

$$P_{e,k} = P\left(\hat{b}_k \neq b_k\right) = Q\left(\frac{A_k}{\sigma\sqrt{(\mathbf{R}^{-1})_{kk}}}\right) \tag{11.65}$$

Linear MMSE detector

The MMSE detector involves finding the transformation $\mathbf{M}$ that minimizes the mean squared error: $\text{Min}_{\mathbf{M}\in R^{K\times K}} E\left(\|\mathbf{b} - \mathbf{My}\|^{2}\right)$. This transformation is no other than:

$$\mathbf{M} = \mathbf{A}^{-1}\left(\mathbf{R} + \sigma^2\mathbf{A}^{-2}\right)^{-1} \tag{11.66}$$

and consequently, the error probability of the k-th user can be written in the form:

$$P_{e,k} = \frac{1}{2^{K-1}} \sum_{\mathbf{b}_{-k}\in\{-1,+1\}^{K-1}} Q\left(\frac{A_k}{\sigma}\frac{(\mathbf{MR}_{k,k})}{\sqrt{(\mathbf{MRM})_{k,k}}}\left(1+\sum_{j\neq k}\frac{(\mathbf{MR}_{k,j})A_j b_j}{(\mathbf{MR}_{k,k})A_k}\right)\right) \tag{11.67}$$

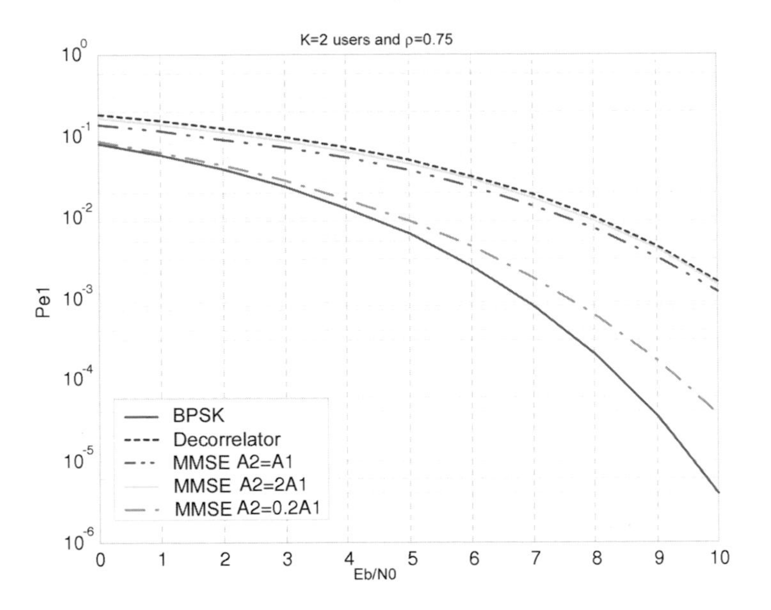

Figure 11.22 – Comparison between the decorrelator method and MMSE.

In order to compare the two techniques, the decorrelator detector and MMSE, the two receivers were simulated with 2 users whose spreading codes are highly correlated (intercorrelation coefficients equal to 0,75). Figure 11.22 shows the curves of the BER for the first user and parametred by the power of the 2nd user (or the user's amplitude). The performance of the MMSE receiver is always better than that of the decorrelator. For low power of the 2nd user, performance is close to that of the single-user. However, for a high power of the 2nd user, the MMSE performance will tend towards that of the decorrelator.

Iterative detector

Decorrelator receivers or MMSE receivers can be implemented with iterative matrix inversion (Jacobi, Gauss-Siedel, or relaxation) methods. The Jacobi method leads to the *Parallel Interference Cancellation*(PIC) method. The Gauss-Siedel method leads to the *Successive Interference Cancellation* (SIC) method. Figure 11.23 gives the diagram for implementing the *SIC* (with $K = 4$ users and $M = 3$ iterations) where $ICU_{m,k}$ is the *interference cancellation unit* (or ICU) for the k-th user at iteration m (see Figure 11.24). The binary elements are initialized to zero: at iteration $m = 0 \quad b_{0,k} = 0$ for $k = 1, \cdots, K$.

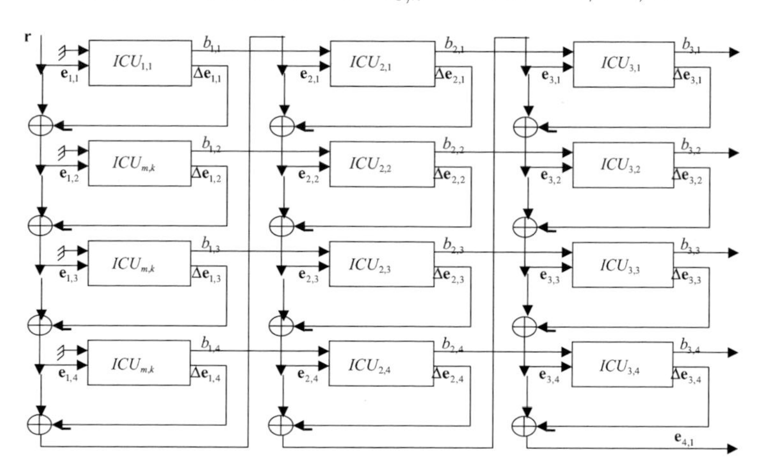

Figure 11.23 – Iterative SIC (*Successive Interference Cancellation*) detector, with $K = 4$ users and $M = 3$ iterations.

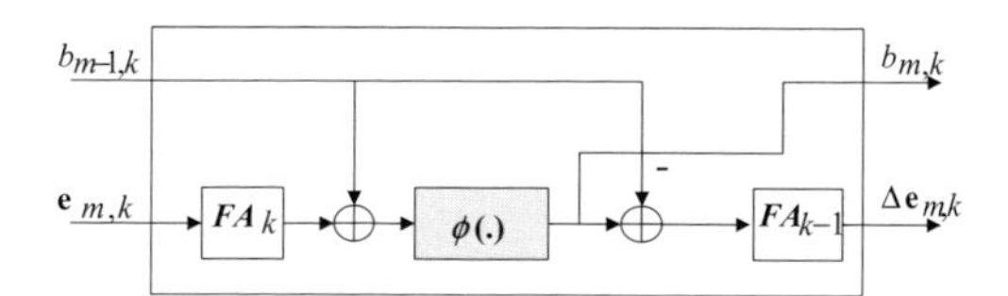

Figure 11.24 – Interference cancellation unit $ICU_{m,k}$ for user k, at iteration m.

Function FA_k (respectively FA_k^{-1}) is the despreading (respectively spreading) of the k-th user. Function $\phi(.)$ can be chosen as a non-linear function or quite simply as being equal to the identity function (see also the choice of $\phi(.)$ in the case of turbo CDMA). If we choose the identity function, the ICU unit of Figure 11.24 can of course be simplified and is easy to define. In this case, we can verify that, for user k and at iteration m, the output of the receiver can be written as the result of linear filtering:

$$b_{m,k} = \mathbf{s}_k^T \prod_{j=1}^{k-1} \left(\mathbf{I} - \mathbf{s}_j \mathbf{s}_j^T\right) \sum_{p=0}^{m-1} \mathbf{\Phi}_K^p \mathbf{r} = \mathbf{g}_{m,k}^T \mathbf{r} \tag{11.68}$$

with:

$$\mathbf{\Phi}_K = \prod_{j=1}^{K} \left(I - \mathbf{s}_j \mathbf{s}_j^T\right) \tag{11.69}$$

We can show that the error probability for the k-th user at iteration m can be written in the following form, where $\mathbf{S}$ is the matrix of the codes and $\mathbf{A}$ is the diagonal matrix of the amplitudes:

$$P_e(m,k) = \frac{1}{2^{K-1}} \sum_{\mathbf{b}/b_k=+1} Q\left(\frac{\mathbf{g}_{m,k}^T \mathbf{SAb}}{\sigma\sqrt{\mathbf{g}_{m,k}^T \mathbf{g}_{m,k}}}\right) \tag{11.70}$$

Figure 11.25 gives an example of simulations of the SIC method with 5 users, a spreading factor of 20 and an intercorrelation matrix given by:

$$\mathbf{R} = \begin{pmatrix} 1 & 0,3 & 0 & 0 & 0 \\ 0,3 & 1 & 0,3 & 0,3 & 0,1 \\ 0 & 0,3 & 1 & 0 & -0,2 \\ 0 & 0,3 & 0 & 1 & 0 \\ 0 & 0,1 & -0,2 & 0 & 1 \end{pmatrix}$$

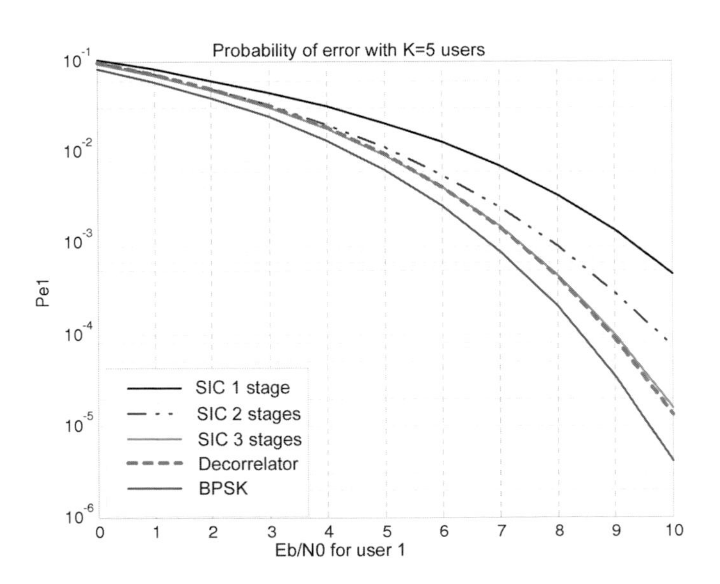

Figure 11.25 – Simulation of a SIC receiver.

We note that after 3 iterations, the SIC converges towards the result obtained with the decorrelator (we can prove mathematically that the SIC converges towards this result when the number of iterations M tends towards infinity).

11.2.3 Turbo CDMA

Several turbo CDMA type techniques have been proposed to jointly process multi-detection and channel decoding:

- Varanasi and Guess [11.53] have proposed (hard estimation) decoding and immediately recoding each user before subtracting this contribution from the received signal. The same operation is performed on the residual signal to decode the information of the second user, and so on, until the final user.

- Reed and Alexander [11.46] have proposed to use an adapted filter bank followed (in parallel) by different decoders before subtracting, for each user, the multiple access interference linked to the $K-1$ other users.

- Wang and Poor [11.56] have proposed a multi-user detector that involves implementing in parallel the MMSE filters associated with each user, followed by the corresponding channel decoders. These two elements exchange their extrinsic information iteratively.

- Tarable *et al.* [11.48] have proposed a simplification of the method presented in [11.56]. For the first iterations, an MMSE type multi-user detector is used, followed by channel decoders placed in parallel. For the final iterations, the MMSE filter is replaced by an adapted filter bank.

Turbo SIC detector

In this section, channel decoding is introduced into a new successive interference cancellation (SIC) structure. Figure 11.23 remains valid, only units $ICU_{m,k}$ change. Each interference cancellation unit $ICU_{m,k}$, relative to the k-th user and at iteration m, is given in Figure 11.26. The originality lies in the way in which this unit is designed: the residual error signal $\mathbf{e}_{m,k}$ is despread (by $\mathbf{s}_k$) then deinterleaved (π_k^{-1}) before adding the weighted estimation of the $b_{m-1,k}$ data of the same user calculated at the previous iteration. The signal thus obtained, $y_{m,k}$, passes through the channel decoder that provides the *a posteriori* log likelihood ratio, conditionally to the whole observation, of all the binary elements (both for the information bits and the parity bits):

$$\mathrm{LLR}(b_k/y_{m,k}) = \log\left(\frac{P\left[b_k = +1/y_{m,k}\right]}{P\left[b_k = -1/y_{m,k}\right]}\right) \tag{11.71}$$

This ratio is then transformed into a weighted estimation of the binary elements:

$$\tilde{b}_{m,k} = E\left[b_k/y_{m,k}\right] = \tanh\left(\frac{1}{2}\mathrm{LLR}\left(b_k/y_{m,k}\right)\right) \tag{11.72}$$

The soft estimation of user k at iteration m is given by $b_{m,k} = A_k\tilde{b}_{m,k}$. The difference $(b_{m,k} - b_{m-1,k})$ is interleaved by π_k before spreading by $\mathbf{s}_k$. The result

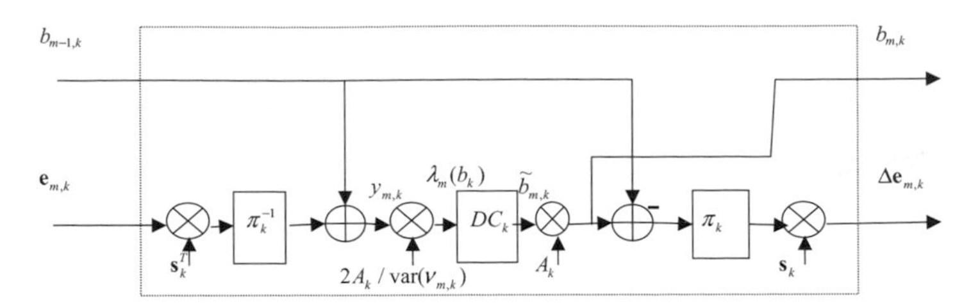

Figure 11.26 – Interference cancellation unit for the turbo SIC decoder in CDMA for the k-th user and at iteration m.

thus obtained $\Delta\mathbf{e}_{m,k}$ is subtracted from residual signal $\mathbf{e}_{m,k}$ to obtain the new residual signal $\mathbf{e}_{m,k+1}$ of the following user (if $k < K$) or to obtain the new residual signal $\mathbf{e}_{m+1,1}$ for the first user at the following iteration ($\mathbf{e}_{m,K+1} = \mathbf{e}_{m+1,1}$). Here, $y_{m,k}$ is written in the form $y_{m,k} = A_k b_k + \nu_{m,k}$ where $\nu_{m,k}$ (residual multiple access interference plus the additive noise) is approximated by a centred Gaussian random variable whose variance is given by:

$$\mathrm{var}(\nu_{m,k}) = \sum_{i<k} A_i^2 \rho_{i,k}^2 \left(1 - \tilde{b}_{m,i}^2\right) + \sum_{i>k} A_i^2 \rho_{i,k}^2 \left(1 - \tilde{b}_{m-1,i}^2\right) + \sigma^2 \qquad (11.73)$$

We show that the extrinsic information of user k at iteration m is given by:

$$\lambda_m(b_k) = \log\left(\frac{P\left[y_{m,k}/b_k = +1\right]}{P\left[y_{m,k}/b_k = -1\right]}\right) = \frac{2y_{m,k}A_k}{\mathrm{var}(\nu_{m,k})} \qquad (11.74)$$

This extrinsic information serves as the input at the decoder associated with the k-th user.

Some simulations

To give an idea of the performance of the turbo SIC decoder, Gold sequences of size 31 are generated. The channel turbo encoder (rate $R = 1/3$) normalized for UMTS [11.1] is used. We consider frames of 640 bits per user. The external interleavers of the different users are produced randomly. The BER and PERs are averaged over all the users. For the channel turbo encoder, the Max-Log-MAP algorithm is used, 8 being the number of iterations internal to the turbo decoder. Figure 11.27(a) gives the performance of the turbo SIC decoder for one, two and three iterations with $K = 31$ users (that is, 100% load rate) having the same power. The performance of the single-user detector and of the conventional detector are also indicated. Figure 11.27(b) shows performance in terms of PER.

Turbo SIC/RAKE detector

In the case where the propagation channel of the k-th user has an impulse response with multiple paths $c_k(t)$, it suffices to replace the despreading function

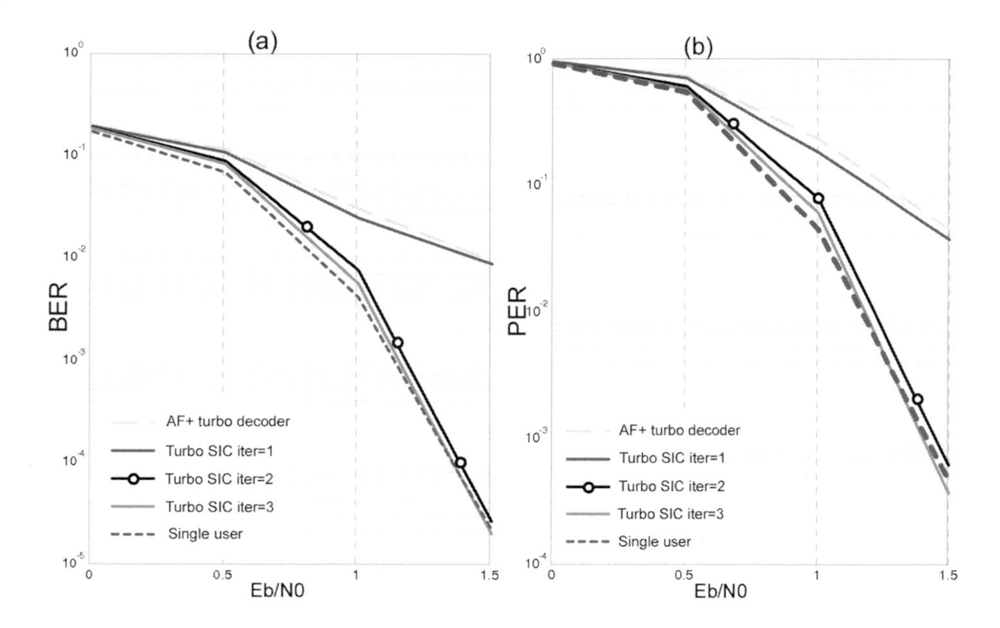

Figure 11.27 – Performance of the turbo SIC decoder: (a) mean Binary Error Rates (BER) (b) mean Packet Error Rates (PER). $K = 31$ users, spreading factor of 31, with frame size 640 bits.

by a RAKE filter (filter adapted to the spreading sequence convolved with the transfer function $c_k(t)$ in the $ICU_{m,k}$) unit, and to replace the spreading function by the spreading function convolved by $c_k(t)$. This new structure is called a turbo SIC/RAKE decoder.

The turbo *SIC/RAKE* decoder is used particularly in the context of the uplink in the UMTS-FDD system.

11.3 Conclusions

In this chapter, we have presented the first two systems to have benefited from applying the turbo principle to a context other than error correction coding. In the first part, we have described the principle of turbo equalization, which relies on an iterative exchange of probabilistic information between a SISO equalizer and a SISO decoder. The SISO equalizer can take different forms according to the chosen optimization criterion. We have presented two types of SISO equalizers: the BCJR-MAP equalizer, operating on the trellis representation of the ISI channel, and the MMSE equalizer, which uses linear filtering. The MAP turbo equalizer leads to excellent performance compared to the conventional receiver. However, this approach is often avoided in practice since it leads to a very high computation cost. We have discussed several solutions for reducing the complexity of the BCJR-MAP equalizer. As for the MMSE turbo equalizer, it offers a good compromise between performance and complexity. For many transmission configurations it leads to performance close to that offered by the BCJR-MAP turbo equalizer, with reasonable complexity. In addition, unlike the BCJR-MAP turbo equalizer, the MMSE turbo equalizer can be realized in adaptive form, thereby jointly performing equalization and tracking of the channel time variations.

In the second part, we have dealt with the application of the turbo principle to the domain of multi-user communications in code-division multiple access systems. We have presented a survey of conventional multi-user detection techniques. In particular, the PIC and SIC methods for cancellation of multi-user interference have been described. Their particular structures lead to a relatively simple exploitation of the turbo principle in a multi-user transmission context. Like for turbo equalization, different detectors can be implemented based on MMSE filters or matched-filter banks, for example.

In this chapter, we have deliberately limited ourselves to the presentation of two particular systems exploiting the turbo principle. However, more generally, any problem of detection or parameter estimation may benefit from the turbo principle. Thus, the range of solutions dealing with interference caused by a multi-antenna system at transmission and at reception (MIMO) has been enriched by iterative techniques such as the turbo BLAST (*Bell Labs layered space time*) [11.25]. The challenge involves proposing SISO detectors of reasonable complexity, without sacrificing data rates and/or the high performance of such systems.

We can also mention the efforts dedicated to receiver synchronization. Indeed, the gains in power provided by the turbo principle lead to moving the systems' operation point towards low signal to noise ratios. Now, conventional synchronization devices were not initially intended to operate in such difficult conditions [11.31]. One possible solution is to integrate the synchronization into the turbo process. A state of the art of turbo methods for timing synchronization was presented in [11.4]. More generally, when the choice of turbo processing at reception is performed, it seems interesting, or even necessary, to add a system to the receiver to iteratively estimate the transmission parameters, like channel turbo estimation or turbo synchronization.

Among other applications, the uplink of future radio-mobile communications systems will require higher and higher data rates, with an ever-increasing number of users. This is the one of the favourite applications of the turbo principle, the generalization of which will be essential in order to respond to the never-ending technological challenge posed by the evolution of telecommunications.

Understanding the turbo principle has led to the introduction of novel theoretical tools and concepts, like EXIT charts or factor graphs. While the former enable accurate prediction of the convergence threshold of iterative decoding schemes, the latter offer a graphical framework for representing complex detection/estimation problems and then deriving efficient turbo-like iterative algorithms for solving them. The interested reader will find good overviews of factor graphs and their applications in [11.37] and [11.38]. The use of factor graphs in a turbo equalization context has been considered in particular in [11.22] et[11.16] and an in-depth study of multi-user detection from a factor graph perspective has been presented in [11.10].

Bibliography

[11.1] Etsi digital cellular telecommunication system (phase 2+). GSM 05 Series, Rel. 1999.

[11.2] B. S. Ünal A. O. Berthet and R. Visoz. Iterative decoding of convolutionally encoded signals over multipath rayleigh fading channels. *IEEE Journal of Selected Areas in Communications*, 19(9):1729–1743, Sept. 2001.

[11.3] L. R. Bahl, J. Cocke, F. Jelinek, and J. Raviv. Optimal decoding of linear codes for minimizing symbol error rate. *IEEE Transactions on Information Theory*, IT-20:284–287, March 1974.

[11.4] J. R. Barry, A. Kavcic, S. W. McLaughlin, A. Nayak, and W. Zeng. Iterative timing recovery. *IEEE Signal Processing Magazine*, 21(1):89–102, Jan. 2004.

[11.5] G. Bauch and V. Franz. A comparison of soft-in/soft-out algorithms for turbo detection. *Proceedings of International Conference on Telecommunications (ICT'98)*, pages 259–263, June 1998.

[11.6] G. Bauch and V. Franz. Iterative equalization and decoding for the gsm system. In *Proceedings of IEEE Vehicular Technology Conference (VTC'98)*, pages 2262–2266, Ottawa, Canada, May 1998.

[11.7] G. Bauch, H. Khorram, and J. Hagenauer. Iterative equalization and decoding in mobile communication systems. In *Proceedings of 2nd European Personal Mobile Commununications Conference (EPMCC'97)*, pages 307–312, Bonn, Germany, Sept.-Oct. 1997.

[11.8] R. Le Bidan. *Turbo Equalization for Bandwidth-Efficient Digital Communication over Frequency-Selective Channels.* PhD thesis, INSA de Rennes, Nov. 2003.

[11.9] R. Le Bidan, C. Laot, and D. Leroux. Real-time mmse turbo equalization on the tms320c5509 fixed-point dsp. In *Proceedings of IEEE International Conference on Accoustics, Speech and Signal Processing ICCASP 2004*, volume 5, pages 325–328, Montreal, Canada, May 2004.

[11.10] J. Boutros and G. Caire. Iterative multiuser joint decoding: Unified framework and asymptotic analysis. *IEEE Transactions on Information Theory*, 48(7):1772–1793, July 2002.

[11.11] J.-M. Brossier. *Signal et Communication Numérique – Equalisation et Synchronisation*. Collection Traitement du Signal. Hermès, Paris, 1997.

[11.12] C. Berrou C. Douillard, M. Jézequel, A. Picart, P. Didier, and A. Glavieux. Iterative correction of intersymbol interference: Turbo equalization. *European Transactions on Telecommununication*, 6(5):507–511, Sept.-Oct. 1995.

[11.13] P. R. Chevillat and E. Eleftheriou. Decoding of trellis-encoded signals in the presence of intersymbol interference and noise. *IEEE Transactions on Commununications*, 37(7):669–676, July 1989.

[11.14] A. Dejonghe and L. Vandendorpe. Turbo equalization for multilevel modulation: An efficient low-complexity scheme. In *Proceedings of IEEE International Conference on Communications (ICC'2002)*, volume 3, pages 1863–1867, New-York City, NY, 28 Apr.-2 May 2002.

[11.15] P. Didier. *La Turbo-Egalisation et son Application aux Communications Radiomobiles*. PhD thesis, Université de Bretagne Occidentale, Déc. 1996.

[11.16] R. J. Drost and A. C. Singer. Factor-graph algorithms for equalization. *IEEE Transactions on Signal Processing*, 55(5):2052–2065, May 2007.

[11.17] C. Fragouli, N. Al-Dhahir, S. N. Diggavi, and W. Turin. Prefiltered space-time m-bcjr equalizer for frequency-selective channels. *IEEE Transactions on Commununications*, 50(5):742–753, May 2002.

[11.18] V. Franz. *Turbo Detection for GSM Systems – Channel Estimation, Equalization and Decoding*. PhD thesis, Lehrstuhl für Nachrichten Technik, Nov. 2000.

[11.19] G. Ferrari G. Colavolpe and R. Raheli. Reduced-state bcjr-type algorithms. *IEEE Journal of Selected Areas in Communications*, 19(5):849–859, May 2001.

[11.20] A. Glavieux, C. Laot, and J. Labat. Turbo equalization over a frequency selective channel. In *Proceedings of International Symposium on Turbo Codes & Related Topics*, pages 96–102, Brets, France, Sept. 1997.

[11.21] G. H. Golub and C. F. Van Loan. *Matrix Computations*. The Johns Hopkins University Press, Baltimore, 3rd edition, 1996.

[11.22] Q. Guo, L. Ping, and H.-A. Loeliger. Turbo equalization based on factor graphs. In *Proceedings of IEEE International Symposium on Information Theory (ISIT'05)*, pages 2021–2025, Sept. 2005.

[11.23] J. Hagenauer. Soft-in / soft-out – the benefits of using soft decisions in all stages of digital receivers. In *Proceedings of 3rd International Symposium on DSP Techniques applied to Space Communications*, Noordwijk, The Netherlands, Sept. 1992.

[11.24] J. Hagenauer, E. Offer, C. Measson, and M. Mörz. Decoding and equalization with analog non-linear networks. *European Transactions on Telecommunications*, pages 107–128, Oct. 1999.

[11.25] S. Haykin, M. Sellathurai, Y. de Jong, and T. Willink. Turbo mimo for wireless communications. *IEEE Communications Magazine*, pages 48–53, Oct. 2004.

[11.26] M. Hélard, P.J. Bouvet, C. Langlais, Y.M. Morgan, and I. Siaud. On the performance of a turbo equalizer including blind equalizer over time and frequency selective channel. comparison with an ofdm system. In *Proceedings of International Symposium on Turbo Codes & Related Topics*, pages 419–422, Brest, France, Sept. 2003.

[11.27] G. D. Forney Jr. Maximum-likelihood sequence estimation of digital sequences in the presence of intersymbol interference. *IEEE Transactions on Information Theory*, IT-18(3):363–378, May 1972.

[11.28] W. Koch and A. Baier. Optimum and sub-optimum detection of coded data disturbed by time-varying intersymbol interference. In *Proceedings of IEEE Global Telecommununication Conference (GLOBECOM'90)*, volume 3, pages 1679–1684, San Diego, CA, 2-5 Dec. 1990.

[11.29] C. Langlais. *Etude et amélioration d'une technique de réception numérique itérative: Turbo Egalisation.* PhD thesis, INSA de Rennes, Nov. 2002.

[11.30] C. Langlais and M. Hélard. Mapping optimization for turbo equalization improved by iterative demapping. *Electronics Letters*, 38(2), Oct. 2002.

[11.31] C. Langlais, M. Hélard, and M. Lanoiselée. Synchronization in the carrier recovery of a satellite link using turbo codes with the help of tentative decisions. In *Proceedings of IEE Colloqium on Turbo Codes in Digital Broadcasting*, Nov. 1999.

[11.32] C. Laot. *Egalisation Auto-didacte et Turbo-Egalisation – Application aux Canaux Sélectifs en Fréquence.* PhD thesis, Université de Rennes I, Rennes, France, July 1997.

[11.33] C. Laot, R. Le Bidan, and D. Leroux. Low-complexity mmse turbo equalization: A possible solution for edge. *IEEE Transactions on Wireless Communications*, 4(3), May 2005.

[11.34] C. Laot, A. Glavieux, and J. Labat. Turbo equalization: Adaptive equalization and channel decoding jointly optimized. *IEEE Journal of Selected Areas in communications*, 19(9):1744–1752, Sept. 2001.

[11.35] I. Lee. The effects of a precoder on serially concatenated coding systems with an isi channel. *IEEE Transactions on Communications*, 49(7):1168–1175, July 2001.

[11.36] S.-J. Lee, N. R. Shanbhag, and A. C. Singer. Area-efficient high-throughput vlsi architecture for map-based turbo equalizer. In *Proceedings of IEEE Workshop on Signal Processing Systems SIPS 2003*, pages 87–92, Seoul, Korea, Aug. 2003.

[11.37] H.-A. Loeliger. An introduction to factor graphs. *IEEE Signal Processing Magazine*, 21(1):28–41, Jan. 2004.

[11.38] H.-A. Loeliger, J. Dauwels, J. Hu, S. Korl, L. Ping, and F. R. Kschischang. The factor graph approach to model-based signal processing. *Proceedings of IEEE*, 95(6):1295–1322, June 2007.

[11.39] K. Narayanan. Effects of precoding on the convergence of turbo equalization for partial response channels. *IEEE Journal of Selected Areas in Communications*, 19(4):686–698, Apr. 2001.

[11.40] N. Nefedov, M. Pukkila, R. Visoz, and A. O. Berthet. Iterative receiver concept for tdma packet data systems. *European Transactions on Telecommunications*, 14(5):457–469, Sept.-Oct. 2003.

[11.41] R. Otnes. *Improved receivers for digital High Frequency communications: iterative channel estimation, equalization, and decoding (adaptive turbo equalization)*. PhD thesis, Norwegian University of Science and Technology, 2002.

[11.42] B. Penther, D. Castelain, and H. Kubo. A modified turbo detector for long delay spread channels. In *Proceedings of International Symposium on Turbo Codes & Related Topics*, pages 295–298, Brest, France, Sept. 2000.

[11.43] A. Picart, P. Didier, and A. Glavieux. Turbo detection: A new approach to combat channel frequency selectivity. In *Proceedings of IEEE International Conference on Communications (ICC'97)*, pages 1498–1502, Montreal, Canada, June 1997.

[11.44] J. G. Proakis. *Digital Communications.* McGraw-Hill, New-York, 4th edition, 2000.

[11.45] S. U. H. Qureshi. Adaptive equalization. *Proceedings of the IEEE,* 73(9):1349–1387, Sept. 1985.

[11.46] M.C. Reed and P.D. Alexander. Iterative multiuser detection using antenna arrays and fec on multipath channels. *IEEE Journal of Selected Areas in communications*, 17(12):2082–2089, Dec. 1999.

[11.47] D. Reynolds and X. Wang. Low-complexity turbo equalization for diversity channels. *Signal Processing*, 81(5):989–995, May 2001.

[11.48] A. Tarable, G. Montorsi, and S. Benedetto. A linear front end for iterative soft interference cancellation and decoding in coded cdma. In *Proceedings of IEEE International Conference on Communications (ICC'01)*, 11-14 June 2001.

[11.49] S. ten Brink. Designing iterative decoding schemes with the extrinsic information transfer chart. *AEÜ International Journal of Electronics Communications*, 54(6):389–398, Nov. 2000.

[11.50] M. Tüchler, R. Kötter, and A. C. Singer. Turbo equalization: Principles and new results. *IEEE Transactions on Communications*, 50(5):754–767, May 2002.

[11.51] M. Tüchler, A. C. Singer, and R. Kötte. Minimum mean-squared error equalization using a priori information. *IEEE Transactions on Signal Processing*, 50(3):673–683, March 2002.

[11.52] G. Ungerboeck. Channel coding with mutilevel/phase signals. *IEEE Trans. Info. Theory.*, IT-28(1):55–67, Jan. 1982.

[11.53] M.K. Varanasi and T. Guess. Optimum decision feedback multiuser equalization with successive decoding achieves the total capacity of the gaussian multiple-access channel. In *Conference Record of the Thirty-First Asilomar Conference on Signals, Systems & Computers*, volume 2, pages 1405–1409, 2-5 Nov. 1997.

[11.54] G. M. Vitetta, B. D. Hart, A. Mämmelä, and D. P. Taylor. Equalization techniques for single-carrier unspread digital modulations. In A. F. Molish, editor, *Wideband Wireless Digital Communications*, pages 155–308. Prentice-Hall, Upper Saddle River, NJ, 2001.

[11.55] F. Volgelbruch and S. Haar. Improved soft isi cancellation for turbo equalization using full soft output channel decoder's information. In *Proceedings of IEEE Global Telecommununication Conference (Globecom'2003)*, pages 1736–1740, San Francisco, USA, Dec. 2003.

[11.56] X. Wang and H. V. Poor. Iterative (turbo) soft interference cancellation and decoding for coded cdma. *IEEE Transactions on Communications*, 47(7):1046–1061, July 1999.

Index

Achevé d'imprimer sur les presses de la SEPEC

Dépôt légal : Juin 2010